Architecture and Engineering

Architecture and Engineering: The Challenges—Trends—Achievements

Editors

Oleg Kapliński
Wojciech Bonenberg

MDPI • Basel • Beijing • Wuhan • Barcelona • Belgrade • Manchester • Tokyo • Cluj • Tianjin

Editors
Oleg Kapliński
Poznań University of Technology
Poland

Wojciech Bonenberg
Poznań University of Technology
Poland

Editorial Office
MDPI
St. Alban-Anlage 66
4052 Basel, Switzerland

This is a reprint of articles from the Special Issue published online in the open access journal *Buildings* (ISSN 2075-5309) (available at: https://www.mdpi.com/journal/buildings/special_issues/ Architecture_Engineering).

For citation purposes, cite each article independently as indicated on the article page online and as indicated below:

LastName, A.A.; LastName, B.B.; LastName, C.C. Article Title. *Journal Name* **Year**, *Volume Number*, Page Range.

ISBN 978-3-03943-749-8 (Hbk)
ISBN 978-3-03943-750-4 (PDF)

Contents

About the Editors

Oleg Kapliński is currently Full Professor at the Faculty of Architecture, Poznań University of Technology, in Poznań, Poland. He received his Ph.D. and DSc. in Civil Engineering and has authored or co-authored 280 publications (articles, reports), including 12 books (academic scripts and monographs) on related topics. His academic achievements cover the theory of decision making, including multicriteria decision aiding; construction processes organization and modeling, including analysis of the phenomena of expectation, the phenomena of equilibrium, balancing of the construction processes in conditions of uncertainty, risk in management, network planning, and reliability of production systems; research on resentment and predilection to risk in the light of utility theory; work ethos; integrated design and management; and sustainable development. Currently, his research at the Faculty of Architecture includes interactions between architects and engineers. Kapliński is a member of the Civil Engineering Committee of the Polish Academy of Sciences and Doctor honoris causa of VGTU, Lithuania.

Wojciech Bonenberg is Full Professor at the Faculty of Architecture of the Poznań University of Technology, Poland. He is the promoter of 26 doctoral dissertations and 150 diploma theses. He is the author or co-author of over 100 scientific publications, including 9 books. His research interests are related to sustainable design, architectural revitalization and requirements engineering in architecture. He has been invited for seminars and a series of lectures by universities in Germany, Netherlands, Italy, Belarus, Ukraine, and China. He is the author of more than 200 architectural projects, many of them awarded. These include the projects of the largest trans-European road terminals: Terminal Swiecko–Frankfurt/O, Terminal Koroszczyn–Brest (route Moscow–Paris), Terminal Olszyna–Forst (route Kiev–Strasbourg). Bonenberg is a member of the Architecture and Urban Planning Committee of the Polish Academy of Sciences and of the Council of Scientific Excellence in Poland.

Preface to "Architecture and Engineering: The Challenges—Trends—Achievements"

There is always something interesting going on at the border of disciplines. This is also the case here, i.e., at the border of architecture and engineering. Design and research are areas connecting their activities. In this book, the reader is offered a collection of articles confirming that the border between architecture and civil engineering is multidimensional. A dynamically changing reality is the reason, supported by new design paradigms and new research techniques. The new design and research tools are an inspiration and a keystone bonding architects and engineers. We encourage the reader to learn about their achievements.

Oleg Kapliński , Wojciech Bonenberg
Editors

Editorial

Architecture and Engineering: The Challenges—Trends—Achievements

Oleg Kapliński * and Wojciech Bonenberg

Faculty of Architecture, Poznań University of Technology, ul. Jacka Rychlewskiego 2, 60-965 Poznań, Poland; wojciech.bonenberg@put.poznan.pl
* Correspondence: oleg.kaplinski@put.poznan.pl

Received: 11 October 2020; Accepted: 12 October 2020; Published: 13 October 2020

Abstract: The current Special Issue is addressed to architects and engineers. Design and research are areas connecting their activities. A review of 17 published articles confirms the fact that the interface between architecture and engineering is multidimensional. The ways of finding points of contact between the two industries are highlighted. This is favoured by the dynamically changing reality, supported by new design paradigms and new research techniques. The multi-threaded subject matter of the articles is reduced to six blocks: research scopes, methods, design aspects, context, nature of research, and economy and cost calculation. Each of the articles in these six blocks has its weight, and so, in the "Nature of research" block, the following areas have been underscored: laboratory tests, in situ research, field investigations, and street perception experiments. The "Design aspects" block includes design-oriented thinking, geometrical forms, location of buildings, cost prediction, attractor and distractor elements, and shaping spatial structures. The new design and research tools are an inspiration and a keystone bonding architects and engineers.

Keywords: architecture; engineering; design paradigms; research methods; circular building; sustainability; spatial structures; design-oriented thinking; BIM; MCDM; SVM

1. Introduction

The keystone binding the articles in the presented Special Issue is design understood as the activity that the architects and engineers engage in. Part of the dynamically changing reality is the emergence of new design paradigms. New architectural, functional, and technological solutions as well as research methods are constantly researched in order to (inter alia) ensure a good indoor climate while achieving energy and economic efficiency. This search coincides with the paradigms of sustainable development which, as the submitted articles show, have become a permanent fixture in our collective awareness.

We can see that the importance of knowledge, knowledge-based design, building physics, technical building equipment, and circular economy is constantly increasing. Modelling and digitization in architecture and civil engineering have become commonplace in research and design.

New, broadly understood technologies also involve changes in the forms of organizing a designer's work, organizational changes in all entities of the investment process, manifested in the Integrated Project Delivery (IPD), and integrated management. Concepts such as architect—engineer—contractor—user are becoming inseparable.

Individual industries participate in the advanced design process but they should not work in isolation, because it only favours linear design.

The intention of the current Special Issue is to indicate the possibilities of finding points of contact between these industries, especially between architects and engineers. The selected, constantly evolving design techniques and research methods presented in this issue are intended to foster this integration.

2. Contributions

The current Special Issue includes 17 articles. They are all original research papers; no review articles or technical reports have been published in this issue.

A total of 34 authors or co-authors from three countries, Belgium, Israel, and Poland, took part. Of this number, as many as 47% are architects or people associated with architectural institutions (universities, administration). In total, 10 of the 17 articles are those with the participation of architects. The fact that architects are actively involved in academic research, especially on the borderline with engineering, is encouraging. The reader is offered a decent dose of selected references. The total number is 827 quoted publications.

Despite the profiling of the subject matter of the articles, the interface between architecture and engineering is multi-threaded.

The scope of these articles, relations between the indicated threads, various aspects, and applications of research results can be grouped into several blocks. Their outline is presented in Figure 1. Six blocks have been distinguished: research scopes, methods, design aspects, context, nature of research, and economy and cost calculation. Each of the articles in these six blocks has its weight.

One can express satisfaction that almost all authors make a direct or indirect reference to sustainability. Of the few important articles (listed in the "Context" block), two are truly spectacular.

Majerska-Pałubicka and Latusek [1] have focused on the issue of development of a degraded riverside quay in an inner-city environment. The proposed research method includes in situ research, with consideration of legal and historical aspects, as well as the condition of the built-up and natural environment. The result of this inquisitive research is a multifunctional complex meeting the paradigm of sustainable development, while the historic city of Cracow (Podolski Boulevard) is the beneficiary of this research.

The field research presented in [2] answers the questions of how urban planning and architecture, i.e., how spatial and geophysical conditions specific to a given place, may affect the quality of the living environment. The presented in-depth research in an industrial piedmont village relates to the morphological structure of buildings, the degree of their modernization, and types of heating systems. The concept of urban ventilation, which is well explained here, is worth a closer look.

All the examples which show that the dilemmas (faced by architects in the area of sustainability) are still valid, as identified in [3]. However, they also clearly show that these dilemmas can be solved.

The block entitled "Methods" (Figure 1) contains many attractive methods and approaches, ranging from BIM (with aspects) to parametric design. The design thinking method is also presented. Let us discuss five of them here.

Tymkiewicz introduces us to the world of the image of sustainable smart cities. The author demands that the image be created primarily by architects. He claims that the roles and tasks faced by architects change, which he highlighted in his article [4]. He refers to the so-called six building blocks of a smart city. The author also recommends a kind of creative approach to design, i.e., the design thinking method. Pilot studies confirm this approach.

The issues of dynamically developing BIM appear in as many as three articles. Finally, we are dealing not with theory but with the implementation of BIM technology in the architectural and construction industry. In [5], Zima at al. perform a SWOT analysis (Strengths–Weaknesses–Opportunities–Threats) to assess how BIM application is used. The implementation of BIM (e.g., in Poland) currently has a favourable position on the market, resulting from the existence of its strengths. The authors indicate that the best strategic solution for the implementation of BIM technology is an aggressive development strategy, recommended in "maxi-maxi" situations. Today, such a topic does not need promotion because it promotes itself.

The leitmotif of the article (by Maskil-Leitan et al. [6]) is BIM and green buildings, but what is essential is BIM management benchmarking. The Green BIM Index has been introduced to assess the level of BIM and green building integration. The conclusions are drawn from the assessment of nine cases. The article bridges the gap between information modelling and the social system.

The issue of parametric design, so attractive in recent years, is the centrepiece of the next article [7]. Rhinoceros, Grasshopper, and Karamba 3D software is commented and the calculations are supported by the finite element method (FEM). However, genetic algorithms are the primary research instrument. These tools were used in shaping curvilinear steel bar structures that are hyperbolic paraboloid canopy roofs. The methods presented here are attractive to both designers and architects.

Evaluation of design solutions is the core of the design. Methods from the group multi-criteria decision-making (MCDM), among others, are used to this end. Ogrodnik [8] uses the application potential of these methods to locate single-family residential buildings with solar installations. The author points out that analytic hierarchy process (AHP) and its modification, i.e., the AHP fuzzy method, are the most suitable methods for these considerations. It is a significant example of the potential of MCDM methods in evaluating architectural designs. The author's approach is in line with the achievements of Professor E. K. Zavadskas [9]. The group of MCDA methods is indicated as particularly useful in solving problems at the interface of architecture and engineering (c.f. Zavadskas et al. [10], Saaty and De Paola [11]).

The group of papers identified in the "Economy and cost calculation" block looks interesting.

Construction costs are of interest to both engineers and architects, especially early estimates of the costs. Juszczyk [12] presents the results of his research on the development of a cost forecasting model, and the subject of his analysis design is bridge construction. The original cost forecasting model is based on machine learning, i.e., support vector machines (SVMs). Several SVM-based regression models were tested, and the proposed model was tested for the required accuracy.

However, Cambier et al. [13] clearly emphasise the design for circularity. Circulation and the function of the building are concepts from the realm of the circular economy. BIM is treated as a process here, with an indication for the integration of BIM with LCA. The required design aid tools, as well as three urgent research paths, have been identified. The article is addressed to architects, developers, and researchers. Circulation becomes a symbol of combining architecture and engineering, i.e., architectural design and engineering solutions.

Nowogońska and Korentz [14] present issues at the interface between technical solutions and costs. Buildings' age, or rather their technical condition, changes, measured by the degree of wear. Thus, we have two basic issues: costs of renovation and repair works of a building and methods of calculating the degree of technical wear. The article uses the prediction of reliability according to Rayleigh distribution (PRRD) model. Of course, the analysis of the technical condition and the application of this method was carried out in the area of housing construction. The method can be generalized and used in other types of construction.

From the collection of submitted articles, it is possible to sublimate a group in terms of the way the research was carried out, namely the "Nature of research" block. Here are the distinctive items (from perception experiments to laboratory research).

The first article in this group [15] focuses on perception—more precisely, on the visual perception of selected buildings. The research was based on the "eye-tracking research" method. These methods, known in psychology, have been successfully transferred to the area of architecture and urban planning. In this case, eye-tracking measurement is based on recording two types of information: fixations and saccades. The research was conducted on the perception of works of architects in Cologne and the following was established: (a) attractors (elements that attract one's attention) and (b) distractors (elements that distract one's attention). The so-called Generation Z participated in the research.

This group of articles also includes article [2], discussed earlier, on field research. Classic laboratory test results are included in [16,17].

Article [16] is a structural analysis of a brick building. New building materials and new technologies require specific solutions to work out the way to join walls. Jasiński and Galman discuss ways to join walls using autoclaved aerated concrete (AAC) masonry elements. The discussion was supported by laboratory tests and numerical analysis using FEM (finite element method). The analyses focus on the morphology and the mechanism of joint damage.

The next article (Kubissa et al. [17]) combines laboratory testing with ecological issues. Concrete with waste copper slag was tested, taking into account its natural radioactivity and types of cement. Concrete was assessed using the Ecological Index and Performance Index (EIPI) method, which takes into account the Ecological Index (EI) and Performance Index (PI). The Gross Ecological and Performance Indicator (GEPI) was used as a support indicator.

The Research scope block gathers all of the articles, some of which are covered under different, distinctive blocks. This block shows a variety of topics (c.f. Figure 1).

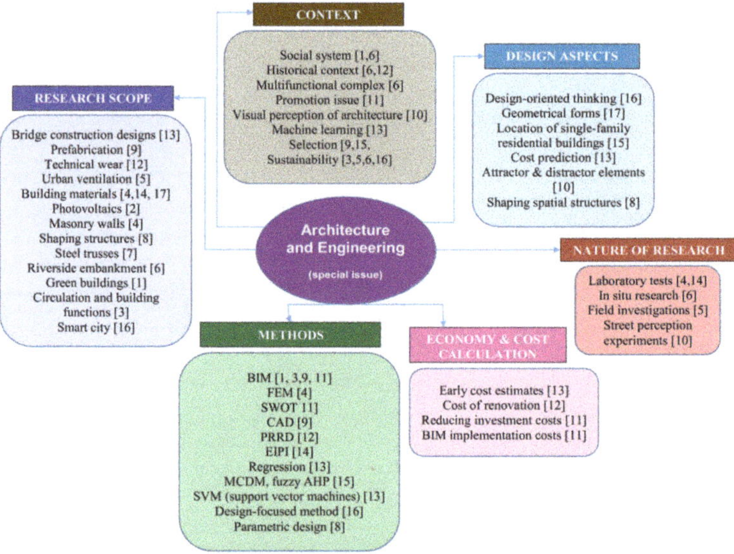

Figure 1. Research areas of case studies and their dominants.

The search for a symbiosis of balanced development and digitization is extremely valuable, and it is popular because it allows the use of intelligent systems—in this case, control at the stage of highly specialized assembly and maintenance of the structure [18]. The use of logical algorithms, Monte Carlo simulations, and CAD elements made it possible to develop an interactive method of computer-aided assembly planning. Moreover, the presented research is part of the trend of the increasing use of prefabrication in construction.

Architects and investors' distrust of photovoltaic solutions in construction is dispelled by Celadyn and Filipek [19]. The systematization of knowledge in the area of photovoltaic systems (modules) is addressed mainly to architects. The architect's attitude, the design process, savings, and the desired energy efficiency are underscored. The outlined vision is promising. Ultimately, this will allow designing structures with zero energy consumption.

In engineering and architecture, new technological developments which affect the design and implementation of high-rise buildings are commonplace. The architects Szołomicki and Golasz-Szolomicka [20] call attention not only to the geometric form but also to development trends; they bring up the issue of the growing importance of composite materials.

This block ends with article [21]. It is easy to notice that the hollow section structures embellish architecture. The article by Broniewicz and Broniewicz [21] is devoted to this issue, offering designers the most up-to-date information on welds for so-called lap joints. The article is addressed to designers of steel structures and is directly applicable in determining the quality of joints between bars of truss connections.

3. Discussion and Comments

Each of the published articles can be classified into the "Research scope" block—see Figure 1—and each of them includes at least several threads that could be included in several of the following five blocks, or even in additional blocks. All this proves that the presented issues are multi-threaded and multi-layered.

In this Special Issue, the design paradigms are less emphasized; research has clearly dominated this issue. There is a green light for digitization and applications such as IoT (Internet of Things), VR (Virtual Reality), AR (Augmented Reality), and the transition from 3D to 7D modelling. At the interface between architecture and engineering, we would be pleased to see solutions such as those originating from bionic engineering or kinetic architecture. The concept of architectural IQ should not be considered in terms of the future but in terms of the present.

The issues at this interface are broad and open. Requirements engineering, both in terms of product engineering and process engineering, might offer assistance here. The possibilities of using requirements engineering specifically in architectural design and its importance in introducing the principles of sustainable development into architectural practice are explained in [22].

Technical progress on this matter alone will not solve these problems. There are many contributing factors, even legal ones. For example, European Union directives bring an important dimension, including the directive on the energy performance of buildings (i.e., Directive 2010/31/EC), which introduced the concept and obligation to design and implement nearly zero energy buildings (nZEB).

An open society pays close attention to operating costs in building life cycle analysis. All of this creates new challenges for architects and engineers.

4. Conclusions

A review of the 17 published articles shows that the interface between architecture and engineering is multidimensional.

There are many paths at the interface between architecture and engineering and not all have been identified on this issue. There are many common and complementary problems. This is an area and an incentive for further research.

The implementation of Integrated Project Delivery (IPD) and BIM (especially BIM as a process) turned out to be an important element, conducive to the integration of architectural and engineering activities. This is why designing has evidently become a team game.

What the analysis of the research presented in this Special Issue has revealed is that apart from the sustainability paradigm, there are other elements which intensely affect the interface between architecture and engineering, namely (a) Circular Building (according to the Circular Economy principles), (b) multi-criteria decision support, e.g., MCDM (as a tool for evaluation, comparison, selection, and optimization), and (c) parametric design.

The presented research emphasizes the fact that new design tools do not divide but connect subjective industries. They constitute a strong current inspiring them and binding them together.

Author Contributions: Both authors contributed to every part of the research described in this paper. All authors have read and agree to the published version of the manuscript.

Funding: This research received no external funding.

Acknowledgments: The authors express their gratitude to the *Buildings* journal for offering them an academic platform for research where they can to contribute and exchange their recent findings in architecture and engineering.

Conflicts of Interest: The authors declare no conflict of interest.

References

1. Majerska-Pałubicka, B.; Latusek, E. A Concept of the Development of Riverside Embankment in the Context of Cracow (A Local Centre). *Buildings* **2020**, *10*, 56. [CrossRef]
2. Blazy, R. Living Environment Quality Determinants, Including $PM_{2.5}$ and PM_{10} Dust Pollution in the Context of Spatial Issues—The Case of Radzionków. *Buildings* **2020**, *10*, 58. [CrossRef]
3. Bonenberg, W.; Kapliński, O. The Architect and the Paradigms of Sustainable Development: A Review of Dilemmas. *Sustainability* **2018**, *10*, 100. [CrossRef]
4. Tymkiewicz, J. The Role of an Architect in Creating the Image of an Elderly-Friendly Sustainable Smart City. *Buildings* **2019**, *9*, 223. [CrossRef]
5. Zima, K.; Plebankiewicz, E.; Wieczorek, D. A SWOT Analysis of the Use of BIM Technology in the Polish Construction Industry. *Buildings* **2020**, *10*, 16. [CrossRef]
6. Maskil-Leitan, R.; Gurevich, U.; Reychav, I. BIM Management Measure for an Effective Green Building Project. *Buildings* **2020**, *10*, 147. [CrossRef]
7. Dzwierzynska, J. Multi-Objective Optimizing Curvilinear Steel Bar Structures of Hyperbolic Paraboloid Canopy Roofs. *Buildings* **2020**, *10*, 39. [CrossRef]
8. Ogrodnik, K. Multi-Criteria Analysis of Design Solutions in Architecture and Engineering: Review of Applications and a Case Study. *Buildings* **2019**, *9*, 244. [CrossRef]
9. Kaplinski, O.; Peldschus, F.; Nazarko, J.; Kaklauskas, A.; Baušys, R. MCDM, operational research and sustainable development in the trans-border Lithuanian-German-Polish co-operation. *Eng. Manag. Prod. Serv.* **2019**, *11*, 7–18. [CrossRef]
10. Zavadskas, E.K.; Antucheviciene, J.; Kaplinski, O. Multi-criteria decision making in civil engineering: Part I—A state-of-the-art survey. *Eng. Struct. Technol.* **2015**, *7*, 103–113. [CrossRef]
11. Saaty, T.L.; De Paola, P. Rethinking Design and Urban Planning for the Cities of the Future. *Buildings* **2017**, *7*, 76. [CrossRef]
12. Juszczyk, M. On the Search of Models for Early Cost Estimates of Bridges: An SVM-Based Approach. *Buildings* **2020**, *10*, 2. [CrossRef]
13. Cambier, C.; Galle, W.; De Temmerman, N. Research and Development Directions for Design Support Tools for Circular Building. *Buildings* **2020**, *10*, 142. [CrossRef]
14. Nowogońska, B.; Korentz, J. Value of Technical Wear and Costs of Restoring Performance Characteristics to Residential Buildings. *Buildings* **2020**, *10*, 9. [CrossRef]
15. Lisińska-Kuśnierz, M.; Krupa, M. Suitability of Eye Tracking in Assessing the Visual Perception of Architecture—A Case Study Concerning Selected Projects Located in Cologne. *Buildings* **2020**, *10*, 20. [CrossRef]
16. Jasiński, R.; Galman, I. Testing Joints between Walls Made of AAC Masonry Units. *Buildings* **2020**, *10*, 69. [CrossRef]
17. Kubissa, W.; Jaskulski, R.; Gil, D.; Wilińska, I. Holistic Analysis of Waste Copper Slag Based Concrete by Means of EIPI Method. *Buildings* **2020**, *10*, 1. [CrossRef]
18. Marcinkowski, R.; Banach, M. Computer Aided Assembly of Buildings. *Buildings* **2020**, *10*, 28. [CrossRef]
19. Celadyn, W.; Filipek, P. Investigation of the Effective Use of Photovoltaic Modules in Architecture. *Buildings* **2020**, *10*, 145. [CrossRef]
20. Szolomicki, J.; Golasz-Szolomicka, H. Technological Advances and Trends in Modern High-Rise Buildings. *Buildings* **2019**, *9*, 193. [CrossRef]
21. Broniewicz, M.; Broniewicz, F. Welds Assessment in K-Type Joints of Hollow Section Trusses with I or H Section Chords. *Buildings* **2020**, *10*, 43. [CrossRef]
22. Bonenberg, W. Requirements Engineering as a Tool for Sustainable Architectural Design. In *Advances in Human Factors, Sustainable Urban Planning and Infrastructure*; Charytonowicz, J., Ed.; Springer: Cham, Switzerland, 2018; Volume 600, pp. 218–227. ISBN 978-3-319-60449-7.

Article

A Concept of the Development of Riverside Embankment in the Context of Cracow (A Local Centre)

Beata Majerska-Pałubicka and Elżbieta Latusek *

Faculty of Architecture, Silesian University of Technology, Akademicka 7, 44—100 Gliwice, Poland; beata.majerska-palubicka@polsl.pl
* Correspondence: latusek.elzbieta@tlen.pl; Tel.: +48-608-035-396

Received: 27 December 2019; Accepted: 4 March 2020; Published: 13 March 2020

Abstract: The subject of this article is the presentation of site conditions and the authors' concept of the development of the degraded riverside area located in the city of Cracow-Kraków Zabłocie. The concept transforms the above-named area into a multifunctional complex including museum, coworking, business and hotel functions. The area subject to development borders three important districts of Cracow: Old Town (Stare Miasto), Grzegórzki and Podgórze on the bank of the Vistula (Wisła) river. In the land development and urban planning documents of the city of Cracow this area has been marked as the public space which is to become a local focal point or a local centre. The main objective of this work was to find answers to the posed research questions concerning the historic context, formal and legal state, significance for the community as well as economic and ecological implications of the area to be developed. The main purpose was to properly develop the degraded riverside embankment in the downtown environment. The research method was based on own mixed method which encompassed the studies of historical literature and the legal–formal status as well as in situ examinations, including the analyses of the condition of the built and natural environment, traffic and circulation as well as photographic documentation. The authors also familiarised themselves with the activities undertaken by the local community with a view to the area's regeneration. On the grounds of initial investigations, the SWOT analysis was performed and the evaluation of groups of prospective users was conducted. Comparative studies were conducted including selected examples of European riverside development projects. In its assumptions, the proposed concept of the riverside development in Kraków-Zabłocie is to meet the needs of the local community, enable further development of tourism, which is very important to Cracow, and satisfy the paradigm of sustainable development. The effect is a multi-functional complex that becomes an inherent part of the existing context.

Keywords: Kraków Zabłocie; Podolski Boulevard; development of riverside embankment; downtown riverside areas; urban local centre; community; historical context; multifunctional complex

---

1. Introduction

Cracow (Kraków), being an important point on the map of Polish historical heritage, is associated mainly with impressive buildings of historic significance well known in Poland and abroad. However, not all city districts have been developing as dynamically as the city centre. Very often the districts of crucial historic importance have been neglected. One of such places was Zabłocie, whose importance has been noticed only in recent years [1]. This area is located in a post-industrial part of the district of Podgórze (Figure 1), in the vicinity of the city's chief arterial roads. There are also two trestle bridges being built at the moment, which are to join the eastern and southern railway exit from Cracow (Kraków). In addition, there is a newly renovated interchange station Kraków Zabłocie adjacent to the subject-related area from the western direction. Former industrial areas of Zabłocie of some historic importance are being converted into residential areas. As a result, the number of district inhabitants and users has been steadily growing in recent years.

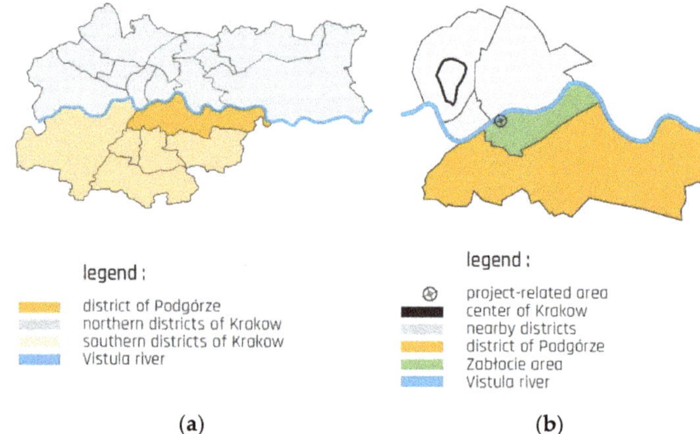

Figure 1. Diagrams showing the location of the project-related area: (**a**) Podgórze district area in the context of other Cracow districts; (**b**) Zabłocie area within the scope of Podgórze district and in proximity of the Old Town [2] (elaborated by E. Latusek).

The subject of this work is a fragment of the riverside embankment of the river Vistula (Wisła) in Cracow (Kraków). This riverfront is in many respects a very interesting area, and it has not been properly developed yet. This particular site was selected due to its purpose: in urban planning documents this place is intended for the function of a local centre (Figure 2). It is a socially significant place. This site requires a proper approach to designing with a focus on the creation of architecture. After the research and spatial analyses, it was decided that a multi-functional complex should be designed to meet the needs of Cracow's inhabitants and visiting tourists. This work and investigations concern the area which was designated as a local centre in the Study of Conditions and Directions of Spatial Development of the City of Cracow (Kraków).

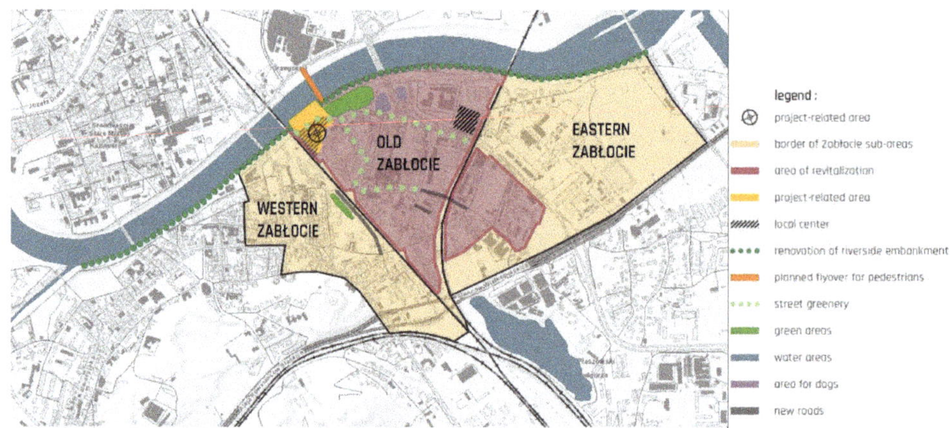

Figure 2. Fragment of the regeneration area, sub-area Stare Podgórze–Zabłocie, own elaboration by E. Latusek on the basis of the map extract from the Study.

In 2014, a section of Podolski Boulevard (Podolski Bulwar) was given a new name of Boulevard of the Allied Forces Pilots (Bulwar Lotników Alianckich) because of two important anniversaries: 1 August, the 70th anniversary of the outbreak of the Warsaw Uprising, and 1 September, the 85th anniversary of the outbreak of the Second World War [3]. To commemorate the Polish Air Forces a memorial is planned to be erected there. This fact does not significantly influence the way of development or utilisation of the embankment. Quoting M. Przybyła, "by analysing the course of events connected with the development of the urban space of Zabłocie after 1989, it can be stated that Zabłocie potentially constitutes one of the most important development-prone areas of the city of Cracow" [4].

Everything that is happening around the project-related area shows how interesting and valuable place this is. Two questions arise: Why is this area still neglected? Why has its potential not been used so far? It is worth emphasizing that "one of the crucial, however much delayed, elements of the programme of {Zabłocie regeneration} is the creation of a representative space having functions of a local centre [...] This newly defined showpiece of the area may not only be a magnet attracting tourists interested in the history of technology but also may increase the attractivity of the neighbouring areas" [5].

The main objective of this article is to present methods of the formulation of bespoke concepts of land development in the area of riverside embankment in the downtown zone with a multifunctional complex including the Museum of the Allied Forces Pilots, which will constitute a local centre in the context of the spatial development of a section of the Vistula riverside.

2. Materials and Methods

One of the fundamental assumptions of this work is a holistic multifaceted analysis of the study area. To make the most appropriate decisions concerning the selection of a research method, the authors analysed the suitability of a series of research methods usually applied to complex issues requiring an extensive analysis. The study was based on two, in the authors' opinion the most adequate, research methods in a given context:

- Method of logical argumentation—as a search for theoretical interpretation of developments (events) with the application of a logical description of reality, based on analysis and synthesis.
- Heuristic method—understood as ways and rules of proceedings serving the purpose of making the most appropriate decisions in complicated situations, requiring the analysis of available information and the prediction of future phenomena. The method is based on creative thinking and logical combinations [5].

Both the method of logical argumentation and heuristic methods help systemize the activities connected with investigations, beginning from the determination of the study subject and ending at the definition of expected results from the work conducted. In the above-mentioned methods, the subject and scope of the study defined aspects such as the search for key developmental factors, types of investigations (theoretical, analytical), humanistic and philosophical interpretation of architectural issues, etc. Next, the types of undertaken activities are determined (deduction, synthesis, analogies, drawing conclusions, abstract and logical analysis). The heuristic methodology specifies a study approach, for instance, if it is expected to share one's knowledge and opinions, as well as if the researcher should enter some social interactions. Both methods also define techniques applied in order to carry out investigations, such as logical interpretations, SWOT analyses, scenarios, Delphi technique, and foresight. The researcher also determines which tools will be used (architectural documentation, lists and comparisons, and tables). Finally, the expected results are predicted. In the heuristic method, determination of possible directions of development (empirical approach) and definition of methods of implementation of strategic goals (normative approach), or in the case of the method of logical argumentation, for example, if it is planned to publish the study results.

The combination of two methods into one mixed method (Table 1) aims at the systematization of the undertaken work which is supposed to yield specific effects. Although creating their own mixed method, the authors based it on the rules and methods used in deconstructivism, which challenges fixed patterns and ponders on a given issue anew, from scratch, following simultaneously technological changes. This approach was very helpful when it came to the questioning of the legitimacy of the so-far well-established, either by law or custom, decisions, and activities. It facilitated also the search for the balance between individual issues, which had been earlier deconstructed into separate elements. In Tischner's book we read that "Jackques Derrida proposed that the act of creation should be the goal in its own right, not a piece of creation itself, even in architecture. However, this seems to be an extreme view, suspending the centuries-long aim of building engineering" [6].

Table 1. Development of own mixed method on the basis of heuristic methods and the method of logical argumentation, (table by E. Latusek, based on [7]).

Type of Activity	Own Mixed Method
Subject	A concept of the development of the riverside called Podolski Boulevard (Bulwar Podolski) in Cracow-Kraków Zabłocie into a multifunctional complex.
Objective	Development of degraded riverside areas in the downtown zone.
Research issue	Proper development of degraded riverside areas in the downtown zone.
Thesis	A multifunctional complex including the Museum of Allied Forces Pilots may constitute a local centre in the context of the spatial development of a section of the Vistula riverside embankment.
Hypothesis	Will the change of the way of development of downtown riverside areas positively affect the surrounding social environment?
Scope of research issues	Theoretical and analytical research, humanistic interpretation of architectural issues, diagnosis of the present-day (social, environmental and economic) state and prediction of directions of development, search for key developmental factors.
Activities undertaken	Research on experts' opinions, use of logic, analysis, comparative analysis deduction, synthesis, analogies, drawing conclusions, logical abstract thinking, opinion surveys of local communities and analysis of undertaken local activities of revitalisation
Research approach	Sharing knowledge and opinions; search for theoretical interpretation of objective or abstract facts.
Techniques applied	Description, explanation, logical interpretation, comparative studies, scale of grades and SWOT analysis.
Tools used	Subject literature, architectural and urban planning documentation, computer and software programmes, comparative lists, tables and the Internet.
Effects expected	Description of the problem and its interpretation, development of a procedure. Presentation of logical conclusions (academic approach). Prediction of possible ways of development (empirical approach). Determination of ways of implementation of strategic goals (normative approach).

Similarly to two basic methods, the authors' own mixed method determined the types of activities that had to be undertaken to meet specific investigation needs. The initial phase of the work encompassed the definition of study subject and scope, types of activities, research approach to analyses and design, tools and techniques to be applied as well as expected results. To facilitate the general reception, Table 1 was supplemented with the aspects, such as: subject, work objective(s), research problem, thesis and hypothesis. Own mixed method aimed at the following.

- Initial determination of the study area resulting in

  - historical literature research,
  - examination of the formal and legal status,
  - analysis of the built and natural environment,
  - analysis of traffic and circulation,
  - photographic documentation and
  - familiarisation with activities undertaken by the local community with the purpose of the area revitalisation.

- Development of the following items on the basis of the initial research:

  - SWOT analysis,
  - assessment of the user groups and
  - performance of comparative studies with selected European examples of riverside development.

- Collection and systemization of analysed data according to certain order.
- Ordering and facilitation of the researchers' activities in the scope of undertaken studies.
- Facilitation of the researchers' insight into basic issues connected with the conducted analyses.

The research on the formal and legal state of the area of Zabłocie and the project-related land encompassed many urban planning documents, which had an impact on designing decisions in the scope of land development, position of the object on the building plot and its forms. The analysed formal and legal documentation first included documents such as the Building Law, types of flood risk and other legal acts. The legal–formal documentation analysed included first of all documents such as the Act on the Building Law and types of flooding risk. The analysed legal acts concerning the project-related area encompassed the following: A Study of Conditions and Directions of Spatial Development of the City of Cracow (Kraków), Prognosis of Environmental Impact, A Local Programme of Zabłocie Regeneration, Update of Municipal Regeneration Programme of the City of Cracow (Kraków), A Local Plan of the Vistula Riverside "Wisła Boulevards" and A Local Plan of Zabłocie.

2.1. Research Questions

- Why has the riverside area of Podolski Boulevard (Bulwar Podolski) not been properly developed yet?
- Will the change in the development of downtown riverside areas positively influence the surrounding community and contribute to the creation of a local centre?

2.2. Historical Research

Extensive historical research was conducted beginning with the district of Podgórze, through the area of Zabłocie, Podolski Boulevard (Bulwar Podolski) and Boulevard of the Allied Forces Pilots (Bulwar Lotników Alianckich). This article contains only a fragment of the above-mentioned study due to its length and broad scope. The name "Zabłocie" meant the land is located behind a muddy area, in relation to royal forests "circa Zabloczyc" [4]. The second half of the 19th century was a period of the greatest development of this area. On the eastern side of the town of Podgórze the main railway line of Galicia was constructed. That sparked the development of railway workshops, industrial plants and warehouses as well as a river port. The urban development was steadily increasing and was barred from the Vistula (Wisła) river with a flood embankment. In 1991, Zabłocie was incorporated into District 13-Podgórze. In the early 20th century, many new factories were built there, for instance, Minor Poland's Factory of Enamelware and Metal Products 'Record' (Małopolska Fabryka Naczyń

Emaliowanych i Wyrobów Blaszanych "Rekord"), later Schindler's Factory, today's museum of Oscar Schindler's Enamel Factory. After 1989, the neighbourhood went into decline as a result of liquidation of many state-owned plants and companies. In 1991, an administrative reform merged the areas of Podgórze, Płaszów and historic Zabłocie into the 13th District of the city of Cracow (Kraków) [8].

The list of objects and groups of objects under conservation and legal protection defined by the Municipal Plan of Spatial Development of 2006 clearly pointed out elements that required protection and inclusion in the project. In accordance with the decision on the enlisting of some objects in the project-related area in the heritage register, there are following objects and facilities in "ZONE A–Stare Podgórze (Old Podgórze)" which have been recorded in the heritage register [9]: restaurant "Zabłocie 13" and a cultural and community centre 'Workshop' ('Warsztat') (Figure 3).

Figure 3. Objects recorded in the heritage register (photo by E. Latusek).

2.3. Examination of the Contemporary State

Since the late 1990s, activities have been undertaken to regenerate this area, which has been one of the most dynamically developing districts of Cracow. In 2001, Kotlarski Bridge (Most Kotlarski) was built [10], resulting in better transport, traffic and circulation conditions of this part of the city with the city centre. At the beginning of the 21st century, a new educational centre was built in Zabłocie, namely, the Andrzej Frycz Modrzewski Higher School (Krakowska Szkoła Wyższa im. Andrzeja Frycza Modrzewskiego). Also, Cracow Artistic Schools (Krakowskie Szkoły Artystyczne) have been there since 1992. In 2006, Zabłocie was considered to be a strategic area in the development of the city of Cracow (Kraków) and a programme of activisation and regeneration was born [11]. For centuries, the area of Zabłocie played a function of industrial background, initially for the district of Kazimierz, later for Podgórze. Nowadays, there is tendency either to remove or adapt old post-industrial building development into residential objects. Zabłocie is acquiring new features and values thanks to subsequent cultural objects having interesting architectural forms and yet preserving the context of the site. Recent years have witnessed an increase in the number of the area inhabitants. Cultural actions organised by various associations and by the district dwellers are transforming this place into a thriving part of the city. In spite of these changes, the area has not lost its industrial character and is home to a considerable number of small and medium enterprises, printing shops, carpenter's, locksmith's or toolmaker's shops. On the other hand, newly built exclusive apartment buildings (Garden Residence) and modern office buildings (Diamante Plaza) are contributing greatly to the improved and favourable image of Zabłocie (Figure 4).

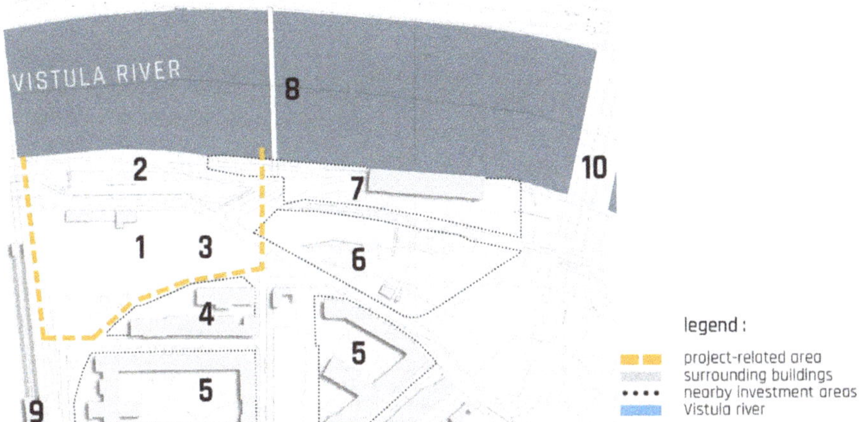

Figure 4. Potentially important places in the vicinity of the project-related area: (1) Project-related area; (2) Boulevard of the Allied Forces Pilots (Bulwar Lotników Alianckich); (3) Planned memorial of the Allied Forces Pilots; (4) The project-related area borders with a lighting company; (5) New residential quarters ATAL Residence and Garden Residence; (6) Park "Vistula Station" ("Park Stacja Wisła") which gained an award in the contest for the best developed space in Poland; (7) Planned Cracow Marina ("Marina Kraków"); (8) Planned flyover for pedestrians and bicycles; (9) Railway station Kraków Zabłocie; (10) Kotlarski Bridge (Most Kotlarski) (elaborated by E. Latusek).

The riverside embankment called Podolski Boulevard (Bulwar Podolski), which is the subject of this work, stretches along the right bank of the Vistula (Wisła) river, between the mouth of the Wilga river and the railway bridge in Zabłocie. The embankment is linked with the other bank by means of Józef Piłsudski Bridge (Most Józefa Piłsudskiego) and Silesian Insurgents Bridge (Most Powstańców Śląskich) as well as railway bridges and the flyover for bicycles and pedestrians. The topography of the riverside area is varied. From the direction of the mouth of the Wilga River there stretches a retaining wall built in the late 19th century as flood defence. From the boundaries of the project-related area, the lie of the land changes into a floodbank descending gradually in the direction of the river. The further east from there you go, the more neglected and overgrown with wild plants the landscape becomes. The Local Plan of Spatial Development (Miejscowy Plan Zagospodarowania Przestrzennego) [12] provides for a place where the flyover for bicycles and pedestrians is to be built in order to link two banks of the Vistula (Wisła) river. There is an inland sailing trail on the Vistula River in Cracow called "Waterway of the Upper Vistula River" ("Droga Wodna Górnej Wisły"), which will implement sailing facilities as a further element of the programme of the development of inland waterways in Poland. Prompted by the inhabitants, the authorities of the city of Cracow (Kraków) are planning to implement a project of the Cracow Marina ('Marina Kraków'). There is also an international cycling trail on the route of Cracow–Moravia–Vienna, which is an eco-tourism corridor revealing the cultural, natural and historic heritage of the Central Europe, including the longest alley of trees in Europe.

2.4. Studies of User Groups

The research question of whether the change in the ways of development of downtown riverside areas will positively influence the surrounding social environment aims to draw attention to the significance of transformations occurring in the public space. For this reason, the studies of user groups were conducted (Table 2). The Polish society is going to face changes which will reshuffle the labour market; the baby boom generation will retire, and will be replaced by only half the number of new employees. The majority of them will probably be from 'generation Y'. The notion of 'generation Y' appeared for the first time in 1993 in the magazine 'Advertising Age' denoting the last generation to be born in the 20th century; in Poland it means people born between 1984 and 2000. The expectations of generation Y differ from the expectations of previous generations: "older generations lived to work, whereas millennials work to live" [13]. People at the age of twenty and thirty will dominate the future labour market and because of this the functioning of the working world will depend on their lifestyle, needs and expectations. The characteristic features of the representatives of "generation Y" are as follows:

- they are familiar with technological novelties,
- thanks to the Internet access they live in a "global village",
- they have a less materialistic approach to life than previous generations,
- they are characterised by high self-esteem and high professional competences,
- they are well educated and ready for further development,
- they live longer with their parents delaying the entry into the adulthood and
- they were brought up in the realities of free market.

Table 2. User groups in the project-related area and their needs (table by E. Latusek).

New Residential Objects (Flats and Apartments)	Vistula Boulevards (Bulwary Wisły)	Cracow Academy and Film School	City Centre and the District of Kazimierz	Museums and Cultural Objects	Businesses and Industrial Objects
New inhabitants (Generation Y)	Tourists and visitors	Students	Tourists and visitors	Tourists and visitors	Employees
Children	Cracow inhabitants	Employees of schools and universities	Cracow inhabitants	Visiting school groups	Customers
New inhabitants of Cracow after completion of their studies	Customers at riverside restaurants and floating bar barges				
	Cycling tourism				

In combination with negative demographic trends and depopulation of Polish cities, it means a tremendous challenge for local authorities, which will be more and more interested in attracting young specialists in order to maintain their competitiveness [14]. Therefore, the question is not if, but how to compete for young talents? The understanding of the present-day situation of young Poles may provide some suggestions in relation to the project-related area.

Inhabitants living in newly-built residential buildings in Zabłocie will also use and contribute to the local centre. Despite the fact that a local centre was designated within the area of the nearby Cracow Academy College (Akademia Krakowska–Studium), it has no functions either within a wide range of possibilities of community integration or a big choice of entertainment and recreation.

2.5. Research on the Local Community's Initiatives

In 2014, with relation to the 70th anniversary of the outbreak of Warsaw Uprising and the 75th anniversary of the outbreak of the Second World War, the City Council passed a resolution to give a section of Podolski Boulevard (Bulwar Podolski) a new name: Boulevard of the Allied Forces Pilots (Bulwar Lotników Alianckich) [15,16]. It was done to commemorate the catastrophe of an Allied Forces' plane "Liberator" [17], which happened in significant historic circumstances, as well as to indicate the connection between this section of the Vistula Valley (Dolina Wisły) with the history of aerial operations over the Minor Poland (Małopolska) region as a result of global political and military developments.

In 1986, thanks to the efforts made by the Cracow Club of Aviation Seniors (Krakowski Klub Seniorów Lotnictwa), the catastrophe was commemorated by placing a memorial plaque in the wall of the Schindler's Factory at 4 Lipowa Street (ul. Lipowa 4). The plaque pays tribute to the crew of the shot-down aircraft [17].

In 1999, the Institute of Landscaping of the Cracow University of Technology (Instytut Architektury Krajobrazu Politechniki Krakowskiej) drew up a project of the development of the Vistula riverside from the side of Zabłocie. The above-mentioned concept was triggered by the planned re-building of the Vistula river floodbanks. The students' designs referred to, among other things, the creation of a memorial of the catastrophe site and the aircraft crew who met a tragic end. Since 2006, there have been anniversary walks "Liberator above Zabłocie" organised every year on the day of the catastrophe by the Association Podgorze.PL. These walks following the "traces" of the shot-down aircraft Liberator KG-933 attract several dozens of people interested in the past of the city and the district. The walks are advertised in the local and regional media, and the history often appears in the press [18].

Moreover, the local community's initiatives included the creation of a mural showing a Liberator aircraft on the wall of the building located at 14 Dąbrowskiego street (ul. Dąbrowskiego 14) in Cracow (Kraków) as well as the raising of a memorial obelisk at the exit of Przemysłowa street (ul. Przemysłowa) to commemorate the site of the catastrophe.

The professors and students of the Faculty of Landscaping of the Cracow University of Technology undertook the task of preparing the land development project and the concept of the memorial as continuation of the efforts of the aviation and pilot community aiming to commemorate the aerial combat over Cracow (Kraków) and Minor Poland (Małopolska) region [19].

2.6. Legal and Formal State

The Local Plan of Spatial Development 'Zabłocie' [12] prepared in 2006 is no longer up to date due to recent transformations the area of Zabłocie is undergoing. The discussed Podolski Boulevard (Bulwar Podolski) stretches between the mouth of the Wilga River, which flows into the Vistula and a railway bridge (Figure 5). The Local Plan of Spatial Development "Vistula Boulevards" ("Bulwary Wisły") (2013) [20] defines the rules and regulations of space formation, but only does so for the western part of the riverside area. In the document Study of Conditions and Directions of Spatial Development (2014), an analysis of the justification of the preparation of a new plan Zabłocie–East (Zabłocie–Wschód) was made, which demonstrates the Cracow City Council's interest in further development of this post-industrial part of Podgórze.

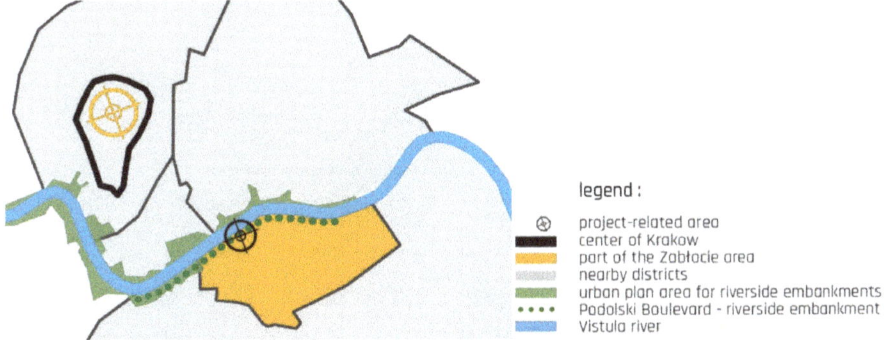

Figure 5. Boulevard of the Allied Forces Pilots (Bulwar Lotników Alianckich) within Podolski Boulevard (Bulwar Podolski) only partially covered by the Local Plan of Spatial Development of the area called 'Vistula Boulevards' ('Bulwary Wisły') [20], (elaborated by E. Latusek).

2.7. Research on Urban Development Composition

Both national and European roads run through Cracow (Kraków) (Figure 6). Typical traffic intensity during rush hours does not exceed critical limits. High traffic intensity occurs along the second ring road of Cracow, in Gustawa Herlinga-Grudzińskiego Street and in the vicinity of the Kotlarski Bridge and the Marshal Józef Piłsudski Bridge. However, traffic jams appear on the regional road no. 776 in Powstańców Wielkopolskich Street.

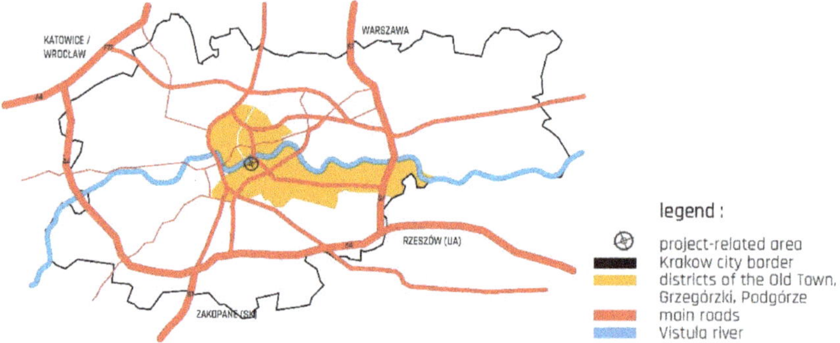

Figure 6. Main arterial roads of Cracow (Kraków), (elaborated by E. Latusek).

Cracow (Kraków) is one of the largest railway interchange stations in Poland (Figure 7). It is linked to the majority of cities in Poland, including express Pendolino links with Warsaw (Warszawa) and Gdańsk. In addition, it has international connections with Vienna, Prague, Budapest and Lviv. The Main Railway Station in Cracow along with the Małopolska Region Coach Station, municipal public transport (buses, underground fast tram) and the link to the Cracow-Balice International Airport make up a complex called the Cracow Public Transport Centre. By the end of 2020, four new rail tracks will have been built on two newly constructed railway trestle bridges on the crosstown line. The Polish Railways PKP Polskie Linie Kolejowe S.A. (Joint Stock Company) link the central railway station with the station Kraków-Płaszów facilitating thus the traffic of agglomeration and long-distance trains. The station Kraków-Zabłocie is currently under modernization, which is connected with the above-mentioned investment.

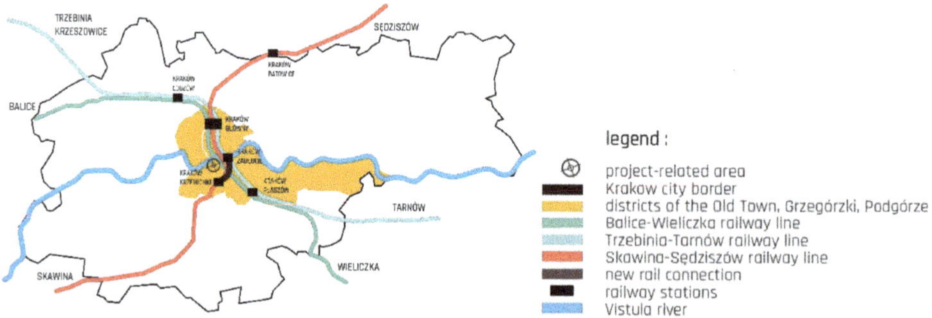

Figure 7. Railways in Cracow (Kraków), (elaborated by E. Latusek).

Poland is to ultimately house five green bicycle trails referred to as greenways. Local bicycle loops have been opened on the Amber Greenway (Szlak Bursztynowy: Budapest-Bańska Szczawnica-Cracow-Gdańsk). There is an international trail between Cracow–Moravia–Vienna, being an eco-touristic corridor exhibiting the cultural, natural and historic heritage of Central Europe. In the future, the aforesaid corridor should become the longest "alley of trees" in Europe. In the direct vicinity of the study area there are many local bicycle lanes and a public bike rental system.

Along the river Vistula (Wisła) in Kraków there is an inland shipping route known as "Waterway of the Upper Vistula River" (Droga Wodna Górnej Wisły). In 2018, the Ministry of Marine Economy and Inland Navigation signed an agreement for the development of a transport analysis, being the first study of this type in relation to inland water transport. The analysis should concern inland navigation on the river Odra and Wisła, as another element of the programme aimed to develop inland waterways in Poland. The city of Cracow, encouraged by its residents, is planning to implement a project named "Marina Kraków".

2.8. Urban Development Dominants

Near the study area there are three new high-standard residential complexes. South of the area there are spaces with strong historical connotations: Cricoteka, Ghetto Heroes Square, a concept to create the Planet Lem object, Oscar Schindler's Enamel Factory and Museum of Contemporary Art in Cracow-MOCAK (Figure 8). Nearby large educational establishments include the Andrzej Frycz Modrzewski Higher School in Cracow, the Institute of Ceramics and Building Materials, Glass and Building Materials Division in Kraków-Podgórze, AMA Film Academy, students' dormitory of the Academy of Music in Cracow and the Adam Mickiewicz Secondary School of General Education no. 4. Nearby hotels include 4-star standard Qubus Hotel, Hotel Galaxy, PURO Hotel Kraków Kazimierz and INX Design Hotel as well as many other hotels located in the district of Kazimierz. Nearby recreational facilities include shopping mall Galeria Kazimierz, Saturn Fitness, Gym Park fitness centre, FitNOW fitness centre and dietician's, Laserpark laser entertainment centre and, located by the river, Wisła: a water tram stop and a kayak rental point. On the opposite bank of the river Wisła there is Galeria Kazimierz shopping mall, which, in the future, will be accessible via a footbridge (for pedestrians and cyclists).

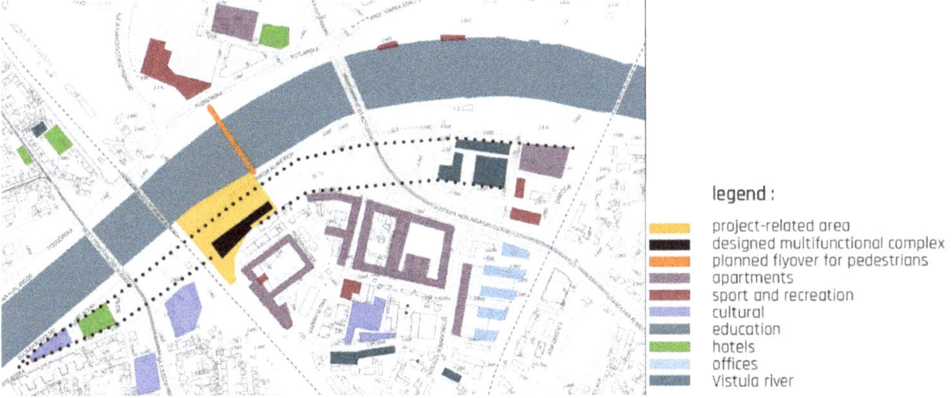

Figure 8. Significant characteristic building development surrounding the study area (elaborated by E. Latusek).

2.9. Major Vistas

The area of Podolski Boulevard (Bulwar Podolski) is located between the mouth of the Wilga river and the railway bridge in Zabłocie in the district of Podgórze. The Local Development Plan for the Area of the Vistula Riverside, the so-called "Wisła Boulevards", contains regulations related to land development, yet only in relation to the western part of Podolski Boulevard, without its eastern part located in Zabłocie (Figure 9). This part of riverside including areas located east of the railway bridge has not been regulated in terms of land development, supplementation of landscape architecture and lighting, adjustment of greenery. The general plan provides for related supplementation as well as the maintaining of main passageways and viewpoints.

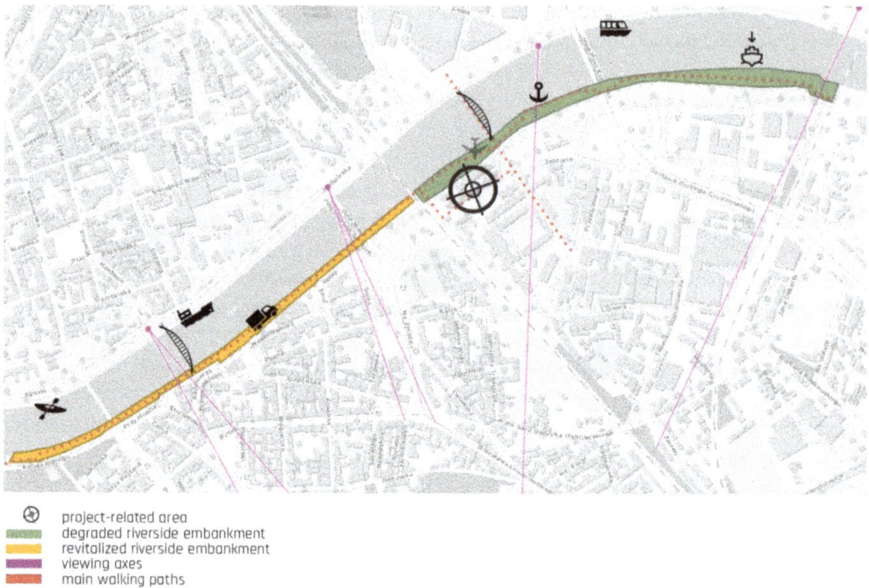

Figure 9. Main vistas in the context of Podolski Boulevard (Bulwar Podolski) (elaborated by E. Latusek).

2.10. Research on Greenery Elements

The preliminary analyses related to the study area provoked a number of questions, one of which is concerned with the lack of the appropriate development plan for Podolski Boulevard. The detailed assessment of the existing condition revealed that an intended space of historical commemoration was to be the Boulevard of Allied Forces Pilots (Bulwar Lotników Alianckich). Today, this area is still wasteland (Figure 10) (overgrown with grass and high greenery) despite the fact that nearby there are new apartment buildings and the Andrzej Frycz Modrzewski Higher School in Cracow. Only a plaque with the boulevard name stresses the significance of the area. Because of the fact that Zabłocie is characterised by the high-density housing development of the city centre, and yet does not have a payable parking zone, today the boulevard is often used as a "wild" car park.

Figure 10. Plots constituting the main part of the development of the design of a multi-functional complex with objects and sites under the conservation (photo by E. Latusek).

2.11. Architectural Research

Historic housing development of the district of Podgórze is undergoing gradual regeneration (Figure 11). In the area of post-industrial Zabłocie, there is one of the most important historic objects, namely, Oskar Schindler's Enamel Factory (Figure 12). The owner of the factory rescued Jews from the Holocaust during the Second World War, which was shown in the film "Schindler's List" made in 1993. The Municipal District Authorities incorporated this building into the Historical Museum of the City of Cracow (Muzeum Historyczne Miasta Krakowa) in 2005. The permanent exhibition held in this place "Cracow–Under Occupation 1939–1945" received an award for the best historical exhibition in Poland in the contest 'Sybilla 2010'. Monthly, as many as 15–20 thousand tourists from all over the world visit this exhibition [21].

Figure 11. Cultural dominants of Podolski Boulevard (Bulwar Podolski): (1) Project-related area; (2) Oskar Schindler's Enamel Factory; (3) Museum of Contemporary Art in Kraków-MOCAK; (4) Centre for the Documentation of the Art of Tadeusz Kantor–Cricoteka; (5) Centre of Literature and Language–'Planet Lem' ('Planeta Lem'); (6) Zabłocie Business Park; (7) Student Hall of Residence Livinn Cracow; (8) Park 'Vistula Station' ('Stacja Wisła'); (9) ATAL Residence; (10) Cracow Marina, (elaborated by E. Latusek).

Figure 12. Museum-Oskar Schindler's Enamel Factory in Cracow (Item 10, Figure 1) (photo by E. Latusek).

In contrast to the heavily historically-oriented Oskar Schindler's Enamel Factory, there is the Museum of Contemporary Art in Kraków (MOCAK) (Muzeum Sztuki Współczesnej w Krakowie) (Figure 13) focusing on ethical and cognitive values, and showing the connection art has with everyday life. The exhibitions encompass the latest international art, education as well as research and publishing projects.

Figure 13. Museum of Contemporary Art in Kraków-MOCAK, designed by Claudio Nardi Architette (Item 10, Figure 1) (photo by E. Latusek).

Another characteristic building facing the Vistula (Wisła) river consists of the facilities of the former Podgórze Power Plant and plays a function of the Centre for the Documentation of the Art of Tadeusz Kantor-Cricoteka (Centrum Dokumentacji Sztuki Tadeusza Kantora–Cricoteka) (Figure 14) [22]. The form of regeneration of the former power plant presented by designers (Architectural Office Vision-Biuro Architektoniczne Wizja and nsMoonStudio) represents an interesting way of activisation of the Vistula (Wisła) riverside.

Figure 14. Centre of the Documentation of the Art of Tadeusz Kantor-Cricoteka, designed by Biuro Architektoniczne Wizja and nsMoonStudio (Item 10, Figure 1), (photo by E. Latusek).

In March 2019, an architectural and urban planning competition was adjudicated. It aimed to select the best concept of a multi-functional centre of literature and language. The winning concept plans to regenerate the 18th century Salt Warehouse (Skład Solny) located at 8 Na Zjeździe street (ul. Na Zjeździe 8) in Cracow and the creation of the Centre of Literature and Language–Planet Lem (Centrum Literatury i Języka–Planeta Lem). The object is supposed to become an operational centre for the programme Cracow City of Literature (Kraków Miasto Literatury) UNESCO [23].

Zabłocie being one of the oldest Cracow districts lies in the vicinity of the city centre, a fashionable district of Kazimierz and numerous colleges, which undoubtedly is a great asset from a perspective of the localisation of office buildings. On the site of the Cracow Electronic Plants Telpod (Krakowskie Zakłady Elektroniczne Telpod) including shop floors of approximately 5 000 m^2 and 10 000 m^2, a new office complex Zabłocie Business Park is to be created [24]. The first 7-storey building of class A, having a BREEAM certificate and offering 11 300 m^2 of surface, was commissioned in 2017. Another, out of four objects, will be completed in mid-2020 [25]. In the same area, a student residence hall, Akademik Livinn Kraków, was built (Figure 15). The building includes 290 flats for over 700 students.

Figure 15. Student Residence Hall Livinn Kraków, renovation and modernisation by Unibep (Item 10, Figure 1) (photo by E. Latusek).

In June 2017, in the area of the former railway station Podgórze-Wisła (later Kraków-Wisła) a park was created called Park Vistula Station (Park Stacja Wisła) (Figure 16). The competition was won by the design devised by Michał Grzybowski, a post-graduate student of the Landscaping at the Cracow University of Technology (Architektura Krajobrazu, Politechnika Krakowska). The park obtained an award in the competition for the best developed space in Poland [26]. The nearby premises of the former factory "Miraculum" were earmarked to be the site of construction of high-class residential buildings ATAL Residence (Figure 16) [27]. The investment was implemented nearby the Vistula (Wisła) river in Zabłocie street. A character of this building development fuses modern and post-industrial features, matching a new face of Zabłocie to its historic landscape. This policy contributes to the fact that Zabłocie is one of the best developing parts of Cracow (Kraków).

Figure 16. Park Vistula Station (Park Stacja Wisła), designed by Michał Grzybowski. In the background, residential building ATAL Residence Kraków, designed by Biuro Rozwoju Krakowa S.A. (Item 10, Figure 1), (photo by E. Latusek).

Not far from the project-related area, in the vicinity of Kotlarski Bridge (Most Kotlarski), a new investment is being planned, namely, Cracow Marina (Marina Kraków). Due to its location in the city centre, the marina will attract tourists. It will also become a leisure and recreation centre for the Cracow inhabitants who want to spend their free time by the river. The marina facilities are planned to stretch over a distance of approximately 1 km [28].

2.12. SWOT Analysis

The SWOT analysis (Table 3) performed in this work summarises the conducted investigations and classifies all information obtained to identify assets and advantages of the space analysed as well as weaknesses showing design barriers. The analysis also shows opportunities and aspects bringing benefits for the area analysed. On the other hand, it reveals threats and risks connected with unfavourable changes.

Table 3. SWOT analysis concerning the conditions of the existing project-related area (table by E. Latusek).

Strengths	Weaknesses
1. Historic development of the district of Podgórze is undergoing gradual regeneration.	1. Post-industrial areas are neglected, littered and serve as wild parking spaces.
2. Local community's activities connected with the commemoration of the Liberator aircraft catastrophe.	2. Plan to combine the investment of the flyover for bicycles and pedestrians with the construction of the memorial delays their implementation.
3. Diverse topography of the riverside embankment (a vertical wall of flood defences vs. floodbank gradually going down towards the river) as attractive elements of the development.	3. Diverse topography of the riverside (a vertical wall of flood defences vs. floodbank gradually going down towards the river) as a design- related impediment and a cost-increasing factor.
4. Main traffic and transport arteries in the vicinity.	4. Increased traffic.
5. Podolski Boulevard was one of the main topics in the political campaign in the election of local self-governments in 2018.	5. Declarations made during the election campaign were not put into practice and implemented.
6. Attempts made to regenerate this part of Cracow proved very successful (cultural events, new cultural objects); at the same time, industrial character of the area was preserved.	6. Many historic industrial objects were demolished (for instance, Miraculum factory, Telpod industrial plants).
7. Tendency to remove or adapt old buildings for the purpose of multi-family residential buildings resulting in the increase in the district population.	7. Difficulties resulting from joining different plots of land.
Opportunities	**Risks**
1. Area borders three important districts: Kazimierz, Podgórze and Grzegórzki.	1. Traffic jams during rush hours.
2. Area features objects under conservation.	2. Impediments in the transformation process of land development.
3. Railway station Kraków Zabłocie was renovated, which may decrease traffic.	3. Noise made by trains will become difficult for the users of offices and residential buildings.
4. Two railway trestle bridges will be built.	4. Degradation of the environment of the ecological corridor of the Vistula (Wisła) river.
5. Coordination of the circulation routes with the planned cycling and pedestrian flyover and cycling paths along the Vistula (Wisła) riverside.	5. Lack of proper separation of the circulation of pedestrians, bicycles and cars may lead to traffic and transport difficulties.

Table 3. *Cont.*

6. Possibility of using the EU funds for the recultivation of a part of the riverside situated in a good location.	6. Necessity of meeting restrictive requirements of the EU projects.
7. A new Local Plan of Land Development is going to be devised.	7. Introduction of big changes in the district development without an updated Local Plan.
8. Enrichment of the embankment with public facilities having an interesting architectural form may positively influence the image of Cracow from the riverside.	8. Enrichment of the embankment with public facilities may unfavourably influence the natural environment of the Vistula riverside.
9. Tendency to remove or adapt old buildings for the purpose of multi-family residential buildings resulting in the increase in the district population.	9. Necessity of the removal of derelict or temporary buildings; flood risk.
10. A local centre for the integration of the local community was provided for in the Study of Conditions and Directions of Spatial Development of the City.	10. Conflict of interests between local community, investors, city authorities and environmental requirements.
11. Further land development should match the award winning public space of the Park Vistula Station (Park Stacja Wisła) and the concept of Cracow Marina (Marina Kraków).	11. Spatial and functional conflicts.

3. Conclusions Based on Investigations

The strategic analysis showed that the most of the issues connected with the strengths of the existing project-related area, as well as opportunities of its development, may result in the transformation of this area into well-functioning space with regard to spatial, economic, social and cultural aspects. On the other hand, there are weaknesses which result from long-term negligence. However, nowadays efforts are made to eliminate them. The risks can be eliminated already in the initial phase of designing new spatial development.

Necessity arises of the development of Boulevard of the Allied Forces Pilots (Bulwar Lotników Alianckich) and creation of an attractive multi-functional space, which would serve as a local centre attracting local residents, tourists and visitors. A museum space connected with the history of the Allied Forces Pilots should constitute a memorial commemorating bygone events and at the same time be an important element of this area. Other functions, such as offices, hotels and recreation facilities should be gradually completed. The designed development should constitute the continuation of attractive public facilities facing the Vistula (Wisła) river.

Guidelines

In the Study of Conditions and Directions of Spatial Development of the City of Cracow (Kraków), the analysed area is located within the boundaries of two structural urban planning units. The delineated roads are to correspond to the guidelines provided in the local development plans. The above-mentioned guidelines for the local development plans define also the category of the analysed area as the services areas (Figure 17). Their primary purpose is to become the building development intended for the following functions: offices, culture and other services along with some necessary ancillary buildings and accompanying greenery. The optional function includes arranged or unarranged green spaces, such as parks, squares, greenery spots, river parks as well as green belts around buildings and vegetation screens (the so-called "vegetation barriers").

The Resolution of the Council of the City of Cracow (Kraków) of 28 June 2006 on Passing the Local Plan of Spatial Development of the Zabłocie Area general regulations constitute that the existing building development should be preserved or rebuilt, new building development should be implemented and new investments made, the changes in the use and development of the land should be introduced. The existing valuable buildings and areas can be utilised in the previous way, until these areas are newly developed according to plan. The whole area covered by the plan is located on the terrain where there is a risk of landslide. No admissible noise levels in the environment were defined for this area in the development plan. The space of the site needs putting in order, that is, the integration of land plots and removal of some derelict buildings. The objects under conservation, namely, a restaurant (Zabłocie 13) and a culture and community centre 'Workshop' ('Warsztat') (Zabłocie 25) may be included in the development context as objects of historical significance.

Figure 17. Plots constituting the main part of the development of the concept of a multi-functional complex with objects and sites under the conservation (elaborated by E. Latusek).

4. Discussion/Proposed Solutions

The planned development of the project-related area located in Zabłocie Street is to provide answers to the above-posed research question: Whether the change in the development of downtown riverside areas will positively influence the surrounding community and contribute to the creation of a local centre? A newly designed multifunctional complex will fill in the space in the existing buildings providing services and having representative functions (Cricoteka, Qubus Hotel, Cracow Academy-Akademia Krakowska) (Figure 18). The conditions for the localisation of the new object are that it must have a supplementary character matching the existing building development, it must comply with the requirements of the Local Plan and must not cross the boundaries of the building development marked in the drawing of the local site plan.

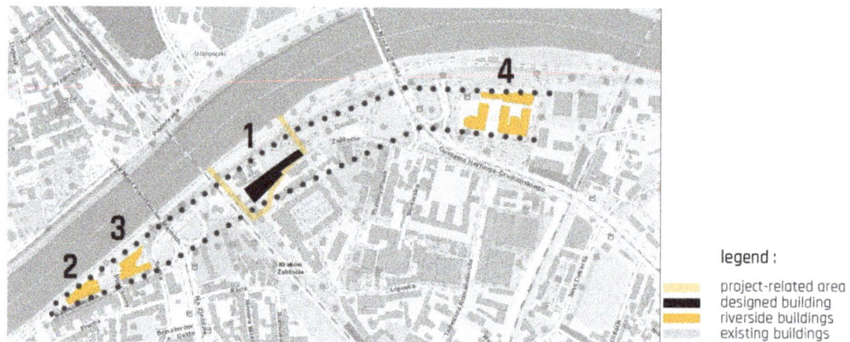

Figure 18. Design concept in the context of surrounding building development: (1) Project-related area; (2) Crikoteka; (2) Qubus Hotel; (3) Cracow Academy (elaborated by E. Latusek).

4.1. Greenery

The development of the document "Trends of the Development and the Management of Green Areas in Kraków in the Years 2017–2030" involved the development of the electronic space-related "Concept of the Public Green Areas System" and "Register of Green Areas", helpful in the continuous management and maintenance of the areas of greenery by the Department for Municipal Greenery Management. As can be seen in the diagram below (Figure 19), a large concentration of greenery is located in the Park 'The Vistula Station' (Park Stacja Wisła). Along the Podolski Boulevard there are single groups of trees. Only the area designated for the "Marina Kraków" investment is characterised by the greater density of riverside greenery and high trees. The study area is intended to be planted with many trees and shrubs. The study area will be maintained as the area of ecological and landscape greenery. Ecological aspects of the introduced changes are as follows.

- In accordance with the directions of development and the management of green areas in Cracow (Kraków), the green character of the study area has been maintained in the form of a green area of significant landscape and ecological values.
- It is planned to plant a large number of park trees and bushes in the study area as well as to implement small architecture and lighting in the scope required for the safety of use.

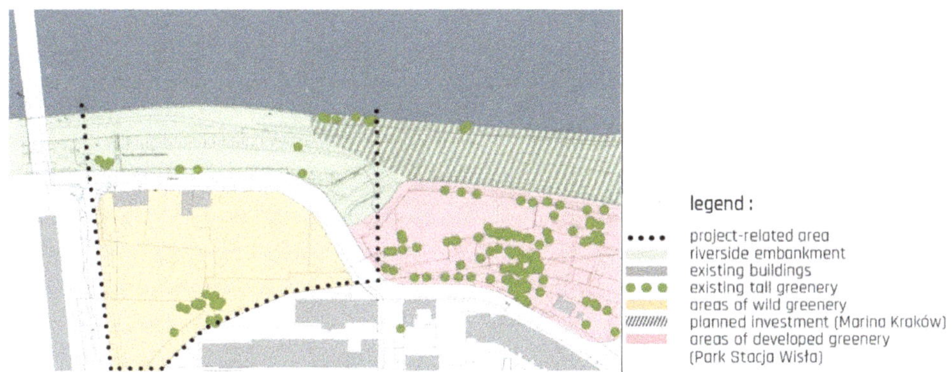

Figure 19. Analysis of greenery, (elaborated by E. Latusek).

4.2. Cubature Buildings

The designed multifunctional complex is located between historic buildings (from the northern direction) and buildings belonging to the lighting company (from the southern direction). The main limitations to the surface of the building development are as follows (Figure 20):

1. Areas of a potential flood risk (from the north). The outline of the building was designed in such a way so as not to cross the potential line of flooding in case the Vistula bursts its banks. For this reason, the surface of the planned building development amounts to ~25% of the whole project-related plot. A vast area in front of the designed building may practically play a function of riverside boulevards and constitute the local centre, or be a continuation of an attractive space of the Park Vistula Station "Park Stacja Wisła".
2. The proximity of the building complex, as close as possible, to the boundary of the plot and neighbouring objects belongs to the lighting company (from the south).

The design preserves the existing historic buildings located in the project-related area (Workshop "Warsztat" and "Zabłocie 13") and provides for a small cubature building, which is to serve as sports equipment rental, in their immediate vicinity.

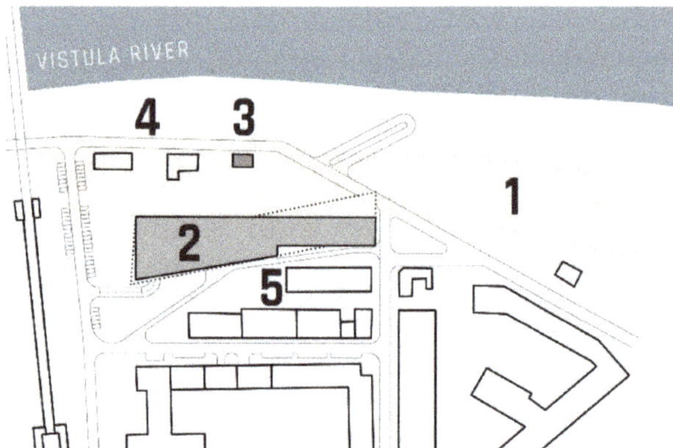

Figure 20. Embedding the object in the context: (1) Flood risk area, (2) Designed multifunctional complex, (3) Designed sports equipment rental, (4) Existing historic buildings, (5) Existing objects of the lighting company (elaborated by E. Latusek).

4.3. Transport, Traffic and Circulation

The design provides for the coordination of circulation routes with the planned cycling and pedestrian flyover as well as cycling paths along the Vistula (Wisła) riverside (Figure 21). The pedestrians' path (east–west) will facilitate the circulation of pedestrians to and from the existing railway station Kraków Zabłocie. A fragment of the main street ul. Zabłocie (east–west) was lowered, which enabled positioning of a smaller pedestrian flyover over the road. It is most probable that the traffic in Zabłocie street will be significantly increasing, therefore thanks to the flyover pedestrians and cyclists will be able to safely circulate between the designed multifunctional complex and the existing green areas located on the Vistula embankments. The designed underground car park with 70 places and newly-marked parking spaces located along the access road to the planned building development will prevent the creation of wild parking spaces in the riverside area. Ecological aspects of the introduced changes are as follows.

- Reduction of the CO_2 emission due to the restriction of motor traffic in favour of cycling.
- Expansion of biologically active areas thanks to the design of an underground car park and parking spaces along the access roads.
- Concept of a new footbridge for pedestrians and cyclists (a cycling link between the districts of Grzegórzki-Podgórze) to improve, in an ecological way, the accessibility of the area for new users coming from outside the district.

A social aspect is as follows:

- Increase in the safety of users by separation of motor traffic from pedestrians and cyclists thanks to a pedestrian and cycling footbridge over the street of Zabłocie, linking the study area with the adjacent areas of the Boulevard of the Allied Forces Pilots.

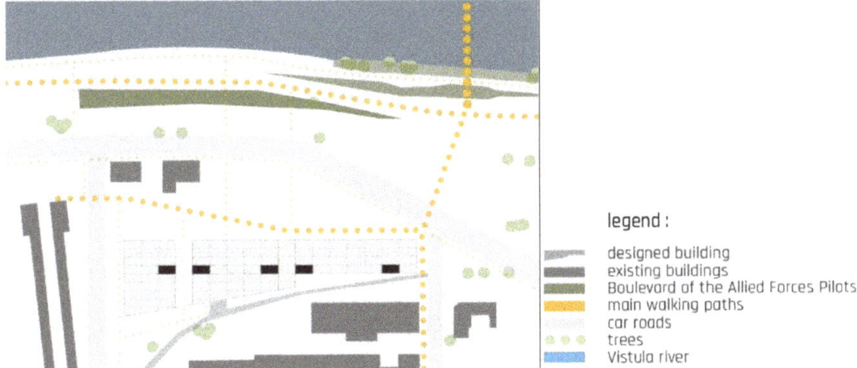

Figure 21. Diagrams showing the location of the study area–transport and circulation, (elaborated by E. Latusek).

4.4. Programme

The research conducted in this work contributed to the determination of needs and requirements within the scope of the following functions; museum, office, hotel and a local centre (Figure 22).

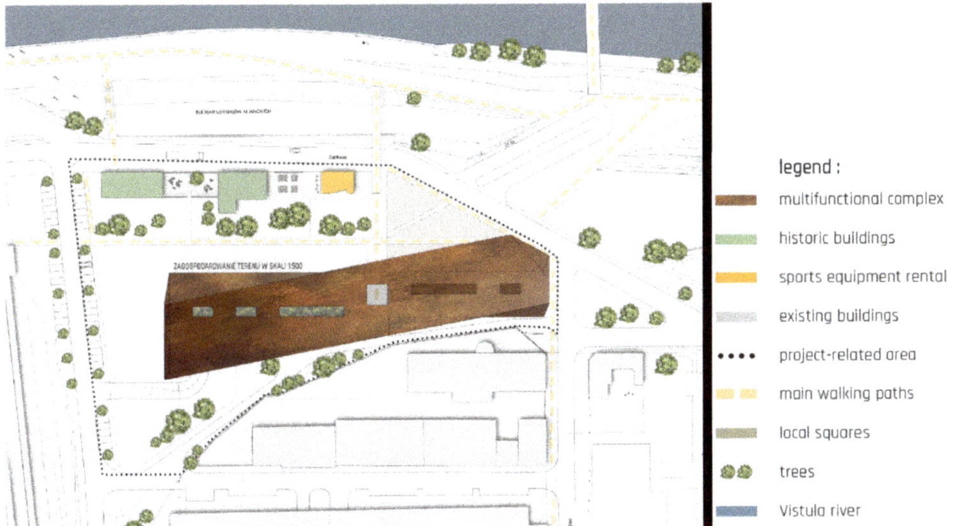

Figure 22. Designed land development, including the multifunctional complex and nearby objects, own elaboration (elaborated by E. Latusek).

Museum: Having conducted historical analysis, a decision was made to create a museum connected with the catastrophe of a Liberator aircraft and the history of the Allied Forces Pilots. This idea was based and preceded by activities that had been carried out since 1999, including the change of the name of the embankment's section into Boulevard of the Allied Forces Pilots (Bulwar Lotników Alianckich). The museum building may become a further unit of the Historical Museum of Cracow (Kraków) cooperating with other museums and cultural objects in the neighbourhood, such as MOCAK, Oskar Schindler's Enamel Factory, Cricoteka and Planeta Lem.

Offices: The conducted surveys of user groups indicated the necessity of the assumption that the newly designed offices are to feature functions of the space satisfying the needs of freelancers, as the space of a coworking type, workplaces used for a certain amount of time and the space enhancing start-ups. Coworking makes it possible to work in a rented room or space. It is the space used most often by freelancers because it gives a bigger comfort of work than at home. Coworking centres can be already found in almost all large cities, also Cracow witnesses the appearance of a bigger and bigger number of such spaces. For instance, in the vicinity of the project-related area there are two such facilities: Studio Zabłocie 2 and Biznes Lab. The designed office part of the multifunctional complex provides for similar spaces, however, not competing with the existing ones.

Hotel: The formation of a local centre and planned development of the area in the scope of culture and promotion of local history will cause the influx of tourists, visitors, customers of the industrial plants and people using the office facilities. Due to this fact, the design provides for the creation of a four-star hotel in this area. Although the project Cracow Marina Marina Kraków provides for the hotel connected with sports and recreational functions, the function of the newly-designed hotel will be connected with office and conference purposes. Economic aspects of the introduced changes are as follows.

- Multifunctional complex features functions satisfying the contemporary needs of users of this rapidly transforming district; this aims to attract young labour force and increase the attractiveness of this area in relation to other areas located in proximity.
- Introduction of workplaces into the district which has a developing residential function will limit the traffic and circulation of inhabitants and thus positively influence the quality of life in the district.
- Objects under conservation policy: restaurant (Zabłocie 13) as well as culture and community centre 'Warsztat' ('Workshop' Zabłocie 25) have been included in the development context as objects of historic value and constitute an added value to the developed space by attracting clients.

Social aspects of the introduced changes are as follows.

- Representative character of the public space was highlighted as a local centre for the district inhabitants and visiting tourists.
- Concept of the multifunctional complex enriches and revitalizes functions connected with culture, recreation and historical education. Being the "salon" of the local centre, this place and its well-adopted functions are expected to attract users and visitors with different needs and interests.
- Multifunctional complex faces the Vistula (Wisła) river in order to create an attractive panorama of Zabłocie, which can be seen from the opposite riverbank; this constitutes an important element of the formation of spatial order.
- Concept of land development takes into account the existing vistas on a local and urban scale.
- Study area was developed with the preservation of its unique, on a European scale, character, in the form of a commonly accessible waterfront space called the Vistula Boulevards (Bulwary Wiślane).

4.5. Spatial and Functional Solutions

Thanks to its new building development, a post-industrial character of Zabłocie area is being transformed into a fashionable place for living and spending free time. The residents of the apartment buildings will also co-form and co-use this local centre. The figure below shows a diagram of the formation of the body of the designed multifunctional complex (Figure 23).

The ground floor of the designed complex houses rooms and facilities enabling community integration (Item a, Figure 24, orange colour), including activities such as: craftwork, organisation of presentations and training sessions, after school activities for children. The design provides for a large multifunctional room. The ground floor area also includes: two entrance zones to the office part (Item a, Figure 24, blue colour), the hotel part (Item a, Figure 24, green colour), an aperitif-bar and restaurant with cooking facilities, and the space of the museum exhibitions with the entrance zone (Item a, Figure 24, yellow colour).

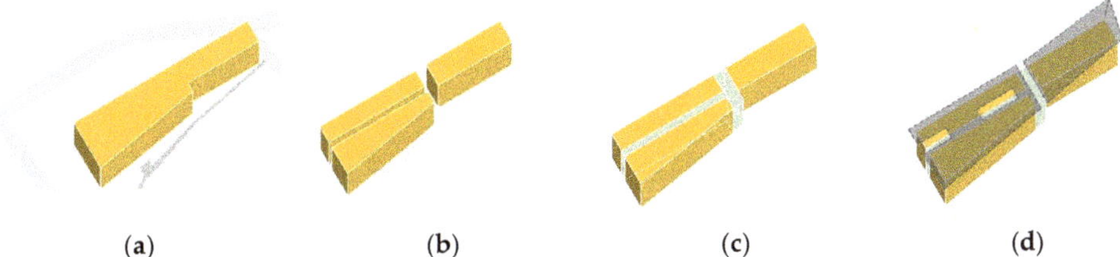

Figure 23. Diagram of the building body formation: (**a**) Adjustment of the object to the building plot after the delineation of new roads; (**b**) Division of the complex into museum, hotel and office functions; (**c**) Structural isolation of the 'wedge' with circulation and big groups of greenery; (**d**) Application of a flat roof and a spatial cover reminding of an aircraft wing (elaborated by E. Latusek).

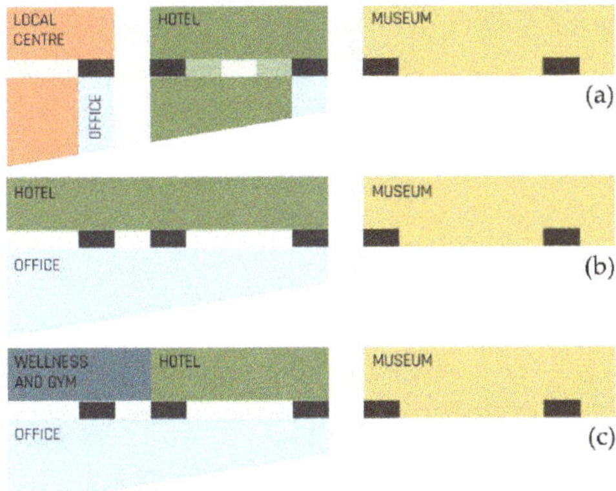

Figure 24. Functional diagram of the designed multifunctional complex: (**a**) ground floor; (**b**) repeatable storeys +1, +2; (**c**) storey +3 (elaborated by E. Latusek).

Storeys +1, +2 and +3 were designed in accordance with the division into three chief functions: office, hotel and museum. The last floor was supplemented with the space dedicated to biological renewal and bodybuilding to meet the requirements of the four-star hotel. The above-mentioned functions are connected by means of a structurally isolated "wedge" with big clusters of greenery and circulation. The roof reminds of an aircraft wing. It has openings providing natural lighting to the single-space "wedge" and the rooms on lower storeys. The main emphasis, however, was put on the proper design of the museum space. In its central part there is space for a large exhibit–the replica of a Liberator aircraft. There are also exhibition rooms connected with the local history.

4.6. Comparative Analysis with the Existing European Examples of Development–Regeneration of Degraded Riverside Areas

To undertake a discussion with the existing examples of riverside development, an analysis of several projects was attempted, such as Paris Rive Gauche, Oslo Fjord City & Akerselva River, Refshaleoen in Copenhagen. The above-mentioned areas considerably differ in their scale from the study area discussed herein; however, they introduce similar functional and spatial solutions.

4.6.1. Paris Rive Gauche

Paris Rive Gauche constitutes the new 13th municipal district of Paris located south of the river Seine. The area is an example of a high-quality urban-development space (including 10 hectares of green areas) and the integration with sustainable transport. The vicinity of Austerlitz Nord contributed to the construction of a series of office buildings with shops and catering establishments on their ground floors in order to meet needs of the local community for the creation of a local centre. Similarly in Kraków-Zabłocie, there is a railway station adjacent to the study area. The analyses conducted showed that there is also a need for the creation of an active local centre. Therefore, it was proposed that the coworking office centres should be built there (as a type of space most often used by freelancers), including a museum, a culture zone for the inhabitants and a hotel with conference and catering facilities.

Also, similarly to Cracow (Kraków), the development of Austerlitz Nord takes advantage of the difference in the terrain altitude (~9 m) using it for the passages for pedestrians from alleys to the waterfront. In the design of the Vistula Riverside development, a pedestrian and cycling footbridge (in a north–south direction) is planned over a heavy traffic artery located in a terrain depression. The footbridge is to facilitate traffic and circulation between the designed multifunctional centre and the riverside boulevard.

The National Library in the district Paris Rive Gauche is located in a similar urban-development context to the designed multifunctional complex in Kraków-Zabłocie. A massive body of the library is separated from the river by municipal infrastructure and a row of high trees. On the other hand, in Kraków-Zabłocie, due to the development of the terrain and introduction of changes in the road and pedestrian infrastructure, the safety of free circulation was improved between the designed building and vast green areas on the Vistula (Wisła) waterfront. Thanks to that, the space in front of the multifunctional complex freely blends with the riverside boulevard.

4.6.2. Oslo Fjord City & Akerselva River

Oslo Fjord City is one of the most interesting concepts of regeneration and creation of space of urban areas in the riverside part of Oslo downtown. Former harbour space was converted into the "salon" of the city with residential buildings interwoven with commercial objects. The area of Kraków-Zabłocie is undergoing similar transformations. Significant historic post-industrial sites were built with new residential and public utility objects resulting in the gradual increase in the number of users and inhabitants of this area in recent years.

The example of Oslo is important due to the way of merging public buildings with the waterfront. The central point is the Opera House—the largest building devoted to culture that has been built in Norway over the span of 700 years. This is a building whose roof was made available to users. It plays a function of a viewpoint and constitutes an integral part of the public space areas. The designed multifunctional complex in Cracow is supposed to play a similar role. Having conducted historical analyses, it was noticed that there was a need for the creation of an architectural dominant in the form of a multifunctional complex (including a museum, local centre, hotel and offices). The museum commemorating the catastrophe of a Liberator aircraft and the history of the Allied Forces Pilots is to serve the local community for the promotion of significant events connected with the history of Kraków-Zabłocie. Moreover, a multifunctional centre located in the area of the meeting point of three central districts of Cracow (Old Town, Grzegórzki, Podgórze) may be a magnet drawing new users to the southern district of Cracow (Kraków).

Another solution having similar features to the one applied in Kraków-Zabłocie is the development of the waterfront of the river Akerselva, which flows entirely within the boundaries of the city of Oslo. Upon this river, meandering through the municipal park areas, there is an object proving an interesting transformation of a degraded post-industrial silo into a students' residence hall. This exceptional building became the landmark of the city of Oslo and was awarded a prize in 2002. This shows how important it is for the space users to preserve characteristic features of the place. Similarly to the

Akerselva riverside, the Vistula (Wisła) riverfront witnesses architectural and urban planning changes. Historic housing development of the Podgórze district is being subjected to gradual regeneration: a museum was created in the former Oskar Schindler's Enamel Factory, the MOCAK museum was constructed, post-industrial objects were transformed into a students' residence hall 'Livinn Kraków' and an office centre, the area of the former railway station was transformed into a park 'Park Stacja Wisła'. The study area encompasses objects under conservation policy, i.e., a restaurant 'Zabłocie 13' and an independent culture and community centre 'Warsztat' (Workshop), which are included in the context of the new development as representing historic and social values.

4.6.3. Refshaleoen in Copenhagen

Refshaleoen is a flourishing place where its users may spend leisure time in an attractive way. Once degraded, the warehouses and space remaining after historic shipyard Burmeister & Wain are nowadays filled with private business which has brought in fresh commercial energy and serves the local community. This new fashionable district of Copenhagen is located just 15 min away by bicycle from the city centre. The example of the thriving island of Refshaleoen confirms that areas located in some distance from the city centre may be sufficiently attractive to appeal to many users. A similar distance must be covered to reach the area of Zabłocie from the centre of Cracow. The proposed concept of the Vistula riverside development aims to introduce solutions which will regenerate the degraded areas in terms of economic, social and pro-ecological issues.

5. Conclusions

In many European cities there is a noticeable trend to build compact multifunctional complexes and public buildings in degraded areas. Not only does this help to save costs and urban space, but also enables mixing of functions and the community of users. As it turns out, local urban centres may be created not only in open public space but also in the form of 'city salons' within the structure of objects. Multifunctionality makes it possible to mix different groups of users having various needs and interests, as for instance in the Lucerne Culture and Convention Centre KKL or the Oslo Opera House.

Attractiveness of the regenerated spaces is increased by objects having characteristic appearance. Similarly to the case of converting a silo into a students' residence hall in Oslo, in the study area in Kraków-Zabłocie, the buildings under conservation policy (Restaurant 'Zabłocie 13' as well as Culture and Community Centre 'Warsztat'–'Workshop') were included as historically valuable objects in the context of the development.

The concept of the development of Podolski Boulevard (Bulwar Podolski) in Kraków-Zabłocie and turning it into a multifunctional complex meets the needs of the local community and enables further development of tourism, which is so important to Cracow. The analysis of the diversified design and social issues related to the study area aimed at the creation of an interesting programme as well as attractive functional and spatial solutions. In addition, historic events connected with the study area made it possible to assign a special character to this place. The concept of the development of the Vistula (Wisła) riverside including the design of a multifunctional complex having pro-social, museum, hotel and coworking functions resulted from the conducted holistic research. This made it possible to achieve expected results in compliance with vital urban planning documents for this area and the idea of sustainable development in terms of ecological, economic and social aspects.

Author Contributions: Conceptualization, B.M.-P.; Data curation, E.L. and B.M.-P.; Formal analysis, B.M.-P.; Funding acquisition, E.L. and B.M.-P.; Investigation, E.L.; Methodology, E.L.; Project administration, E.L.; Resources, E.L.; Software, E.L.; Supervision, B.M.-P.; Validation, B.M.-P.; Visualization, E.L.; Writing—original draft, E.L.; Writing—review & editing, B.M.-P. All authors have read and agreed to the published version of the manuscript.

Funding: This research received no external funding.

Conflicts of Interest: The authors declare no conflicts of interest.

References

1. Cracow Municipal Office-Urząd Miasta Krakowa; The Institute of Ecology of Industrialised Areas in Katowice-Instytut Ekologii Terenów Uprzemysłowionych Katowice. *Local Programme of Zabłocie Regeneration-Lokalny Program Rewitalizacji Zabłocia*; Cracow Municipal Office-Urząd Miasta Krakowa: Cracow, Poland; The Institute of Ecology of Industrialised Areas in Katowice-Instytut Ekologii Terenów Uprzemysłowionych Katowice: Katowice, Poland, 2008; pp. 1–71.
2. Minor Poland's Infrastructure of Spatial Information-Małopolska Infrastruktura Informacji Przestrzennej. Available online: www.miip.geomalopolska.pl (accessed on 9 October 2018).
3. Municipal Council of District 13 Podgórze-Rada Dzielnicy XIII Podgórze. Resolution Concerning the Petition to the Mayor of the City of Cracow to Give a Section of Podolski Boulevard (Bulwar Podolski) the Name of the Boulevard of the Allied Forces' Pilots (Bulwar Lotników Alianckich). Publisher Municipal Office: Cracow, Poland, 2014; p. 1.
4. Przybyła, M. *Regeneration of the Post-Industrial Area of Zabłocie-Rewitalizacja Poprzemysłowego Obszaru Zabłocie*; Kołodziejczyk, J., Ed.; Zarządzanie Publiczne Numer 1 (21); Publisher Jagiellonian University in Cracow: Cracow, Poland, 2013; pp. 103–113.
5. Kuboś, R. *Zeszyty Naukowe Politechniki Śląskiej*, 1st ed.; Wydawnictwo Politechniki Śląskiej: Gliwice, Poland, 2012; pp. 267–274.
6. Tischner, J. Pojęcia Wyrosłe z Idei. Oficyna Wydawnicza Impuls: Crakow, Poland, 2000; pp. 95–120.
7. Tymkiewicz, J. Metody badań w architekturze-Methods of Research in Architecture. In *Lecture in the Subject: Methodology of Research Work*; Faculty of Architecture at the Silesian University of Technology in Gliwice-Wydział Architektury Politechniki Śląskiej: Gliwice, Poland, 2018.
8. Przegon, W. *Zmiany Użytkowania Ziemi na Przykładzie Miasta Podgórza i Zamościa w Świetle Materiałów Kartograficznych–Changes in the Use of Land Discussed on the Example of the Towns of Podgórze and Zamość on the Basis of Cartografic Rcords*, 1st ed.; Wydawnictwo Naukowe Akapit: Crakow, Poland, 2011; pp. 46–64.
9. Cracow Municipal Office-Urząd Miasta Krakowa. *The List of Individual Buildings and Building Complexes under the Legal, Conservation and Urban Planning Protection-Wykaz Obiektów i Zespołów Objętych Prawną Ochroną i Opieką Konserwatorską Oraz Ochroną Ustaloną Planem, Annex no 3 to the Resolution*; Publisher Cracow Municipal Office: Cracow, Poland, 2017; p. 2.
10. Tatara, P. *Mosty Krakowa–The Bridges of Cracow, Mosty 2010*; Publisher Mota-Engil Central Europe SA: Cracow, Poland, 2010; pp. 58–60.
11. Resolution of the Cracow City Council-Uchwała Rady Miasta Krakowa. *Update of the Municipal Programme of Cracow Regeneration-Aktualizacja Miejskiego Programu Rewitalizacji Krakowa*; Publisher Cracow Municipal Office: Cracow, Poland, 2014; pp. 12–16.
12. Resolution of the Cracow City Council-Uchwała Rady Miasta Krakowa. *Resolution Concerning the Passing of the Local Plan of Spatial Development of the Area of Zabłocie-Uchwała Dotycząca Miejscowego Planu Zagospodarowania Przestrzennego Obszaru Zabłocie*; Publisher Cracow Municipal Office: Cracow, Poland, 2006; pp. 1–94.
13. Millennials at Work, Reshaping the Workplace. Available online: www.pwc.com (accessed on 17 December 2018).
14. Chimczak, P. *Mieszkania Adresowane do Generacji Y Jako Sposób na Przyciąganie Talentów–Flats Addressed to Generation Y as a Way of Attracting Talents*; Level: Completed; The Warsaw University of Technology-Politechnika Warszawska: Warsaw, Poland, 2014.
15. Boulevard of the Allied Forces Pilots-Bulwar Lotników Alianckich. Available online: www.podgorze.pl (accessed on 2 November 2018).
16. Liberator from Zabłocie–Liberator z Zabłocia. Available online: www.muzeumlotnictwa.pl (accessed on 9 October 2018).
17. A Memorial Place has been Created in Zabłocie, the Fallen Pilots have been Commemorated-Stworzono Miejsce Pamięci na Zabłociu, Polegli Piloci Zostali Uczczeni. Available online: www.dziennikpolski24.pl (accessed on 17 December 2018).
18. Liberator Above Zabłocie, a Walk and Unveiling the Mural–Liberator nad Zabłociem, Spacer i Odsłonięcie Muralu. Available online: www.krakow.pl (accessed on 9 October 2018).
19. Cracow will Pay Tribute to the Allied Forces Pilots-Kraków Odda Hołd Lotnikom Alianckim. Available online: www.krakow.pl (accessed on 9 October 2018).

20. Resolution of the Cracow City Council-Uchwała Rady Miasta Krakowa. *Resolution Concerning the Passing of the Local Plan of Spatial Development of the Area of Vistula Boulevards; Uchwała w Sprawie Uchwalenia Miejscowego Planu Zagospodarowania Przestrzennego Obszaru Bulwary Wisły*; Publisher Cracow Municipal Office: Cracow, Poland, 2013; pp. 1–37.
21. Szalewska, K. *Fabryka Schindlera-Tektonika Dyskursów–Schindler's Factory–Tectonics of Discourses*; Jednak Książki: Cracow, Poland, 2015; pp. 69–88.
22. Sobotka, A.; Radziejowska, A. *Budowa Cricoteki projekt–The Construction of Cricoteka-Design*; Builder: Cracow, Poland, 2015; pp. 70–74.
23. Results of the Competition for the Centre of Literature and Language-Planeta Lem. Available online: www.krakow.pl (accessed on 1 December 2019).
24. Zabłocie Business Park. Available online: www.gsbk.pl (accessed on 1 December 2019).
25. Zabłocie Business Park A Sold-Zabłocie Business Park A sprzedany-. Available online: www.rp.pl (accessed on 1 December 2019).
26. Park Vistula Station Nominated-Park Stacja Wisła Nominowany. Available online: www.podgorze.pl (accessed on 12 December 2018).
27. ATAL Residence Investment Description. Available online: www.atal.pl (accessed on 1 December 2019).
28. Marina in Zabłocie for 50 Million-Marina na Zabłociu za 50 Milionów. Available online: www.dziennikpolski24.pl (accessed on 9 October 2018).

 *buildings*

Article

Living Environment Quality Determinants, Including PM$_{2.5}$ and PM$_{10}$ Dust Pollution in the Context of Spatial Issues—The Case of Radzionków

Rafał Blazy

Faculty of Architecture, Cracow University of Technology, 31-155 Cracow, Poland; rblazy@pk.edu.pl

Received: 3 February 2020; Accepted: 11 March 2020; Published: 14 March 2020

Abstract: This article discusses living environment determinants in Central and Eastern Europe. It is based on a case study of the city of Radzionków, which has 16 thousand inhabitants and is located in the Silesian agglomeration in southern Poland. Hard coal has been mined in this area for almost two hundred years, and it is the main fuel used for central heating. A total of 360 buildings, divided into groups of 60 buildings each, were investigated in the selected city. Three distinct areas were distinguished in terms of living environment quality, depending on building technical condition, heating method and location. These qualities were found to be largely determined by site-specific spatial and geophysical conditions. A significant portion of the literature was found to ignore the spatial factors mentioned in this paper, instead focusing primarily on statistical data concerning pollution. This study examines site-specific variables and presents differences in air pollution levels as examined in relation to the morphological structure of development, the degree of building modernisation and heating system types.

Keywords: living environment quality; spatial location conditions; air pollution; urban ventilation; EU subsidies targeting environmental quality improvement

---

1. Introduction

At present, the problem of air pollution caused by inefficient heat sources has become one of the most important urban problems of Central and Eastern Europe. On the macro scale, air pollution resulting from the use of fossil fuels for heating is extremely harmful to the Earth's climate due to greenhouse gas emissions, while on the local and the regional scale, it has a direct, negative impact on human health (e.g., PM$_{2.5}$ and PM$_{10}$ pollution) [1–3]. Particulates, due to their dimensions, can be transported over considerable distances even by weak winds. It was found that solid particles (0.1–1 μm) can be transported over a distance of up to several thousand kilometres [4]. The latest research [5] shows that a significant amount of PM$_{10}$ emitted in Silesia can be transported several hundred kilometres to eastern or northern Poland, and even to Scandinavia. Air pollution emitted in this area of Poland can be considered an external source of air pollution, affecting, among others, the eastern regions of the Czech Republic [6]. Poland, along with Bulgaria, is one of the countries with the highest levels of PM$_{10}$ air pollution (above 50 μg/m^3) in Europe [7–9]. Both high-altitude and low-altitude emissions have a significant impact on measurable and perceptible air quality in living areas [10–15]. Emissions at an altitude of 40 m above ground level are considered to be high altitude emissions. Low altitude emissions include all sources of particle pollution up to a height of 40 m. In Poland, the primary sources of low altitude emissions are domestic furnaces and boilers, with road transport also being a significant but smaller contributor.

Particle pollution caused by fine and coarse particulates (PM$_{2.5}$ and PM$_{10}$, respectively) is presently one of the most important scientific problems [16]. PM$_{2.5}$ is particulate matter that consists of particles smaller than 2.5 μm in diameter, while PM$_{10}$ is composed of particles that range between 2.5 μm and 10 μm in diameter. Both types are extremely dangerous. Inhaling harmful substances can lead to various respiratory and cardiovascular disorders [17]. Epidemiological studies confirm that permanent exposure to $PM_{2.5}$ significantly contributes to a higher incidence of cardiopulmonary disease and complications that can lead to increased mortality rates [18,19]. Therefore, a better and more comprehensive understanding of all conditions related to $PM_{2.5}$ and PM_{10} pollution is necessary. This can significantly contribute to preventing air pollution and protecting human health. A substantial amount of research has been conducted on $PM_{2.5}$ and PM_{10} pollution around the world. The main areas of focus include spatial issues and the time of occurrence of pollution [5–10], particle pollution sources [20–23], effects on human health [24–27] and estimation studies [28–31].

Studies have shown that on the macro scale, meteorological conditions affect $PM_{2.5}$ and PM_{10} levels [32–35]; on the micro scale, $PM_{2.5}$ pollution is strongly associated with land use [36–39]. Some researchers suggest that land use can be optimised to reduce $PM_{2.5}$ and PM_{10} levels on an urban level [40–42]. However, there is a noticeable lack of research into the link between spatial development, urban composition and topography in the context of $PM_{2.5}$ and PM_{10} emissions [37,43]. Therefore, the author made an attempt to determine the impact of land use on $PM_{2.5}$ and PM_{10} pollution levels and to present pollutant distribution across different functional and spatial zones. This can be considered necessary and significant even in the case of Radzionków, a city located in southern Poland, which has very distinct climatic, topographic and spatial conditions.

There is insufficient available data to carry out research of land use impact on the degree of $PM_{2.5}$ and PM_{10} pollution, which is a major challenge. Several attempts to study this subject have been made during the current decade. Pollutant concentrations should be interpolated using dense monitoring grids, while in reality, monitoring sites are usually rare and sparsely placed, with less than ten locations in large cities, while in small towns there is often only one. Dispersion models that simulate pollutant behaviour may be useful in some cases, but their dependence on many spatial variables requires very accurate input data [44,45].

In Poland, energy and climate policy, particularly concerning air pollution prevention, has been steadily increasing in prominence in public debates, political decisions and academic research. However, it is more commonly associated with problems of the mining and energy sector rather than the problems of the direct pollution of urban areas. Popular opinion among Polish citizens holds that the air in small-town and rural areas is cleaner and healthier than in large cities and metropolises. However, scientific publications in this field prove that the situation is quite the opposite—air quality in small towns and rural areas is often observed to be much poorer [46–49]. The problem of air pollution in small urbanised areas in Poland is largely connected with the use of conventional, old and inefficient energy carriers and central heating systems [50]. The effects of using low-calorie coal include the emission of harmful compounds into the atmosphere. These compounds include sulphur dioxide (SO_2), nitrogen oxides (NO_x), carbon oxides (CO_x) and harmful particulates. Particulates with a diameter below 10 μm, including $PM_{2.5}$ and PM_{10}, are considered to be the most harmful [51]. Direct inhalation of particulates ($PM_{2.5}$) is associated with an increased incidence of cardiovascular disease [52]. These particulates often include other impurities such as arsenic, cadmium, nickel and polycyclic aromatic hydrocarbons, which are considered mutagenic, such as benzo(a)pyrene, as well as substances that contribute to carcinogenicity. They are also one of the most dangerous air pollutants [53,54]. The burning of conventional energy carriers such as hard coal contributes to the degradation of the natural environment via the greenhouse effect and water and soil pollution. In Polish small towns and rural areas, the problem is additionally compounded by obsolete power grids.

Gas distribution grids are also poorly developed and are absent in many areas. The quality of the living environment depends on many factors [55]. It is largely determined by a location's existing spatial conditions. The topography and development spatial structure of an area can be considered significant when analysing air quality and pollution. Many reports on the living environment do not take spatial factors into account, as they focus solely on pollution statistics [56–58]. In this study, the author, taking into account site-specific variables, shows differences in air pollution measurements for designated locations relative to the morphological structure of their development. The study aimed at answering two fundamental questions.

- What are the key factors that contribute to $PM_{2.5}$ and PM_{10} pollution under the conditions of a small Polish town whose development has historically been based on coal mining?
- Which structures generate the most pollution in this situation?

2. Object and Method of Research

Radzionków is a city and municipality in south-western Poland, in Tarnowskie Góry County in the Silesian Voivodeship. It is located in the northern part of the Upper Silesian Industrial District (GOP), which is one of the most important cultural, academic and economic centres in Poland. The oldest mentions of Radzionków are dated to between 1326 and 1357. The subject of the study was the northern and central part of the municipality of Radzionków. In terms of morphology, the central part of its structure is a surviving oval system with a historical trail. Most of the buildings in the area are located along streets that are delineated along plot boundaries. The buildings located in the city's core form frontages and have a regional, classicist character, with stone featured in their facades. Their architecture is typical of the end of the nineteenth and the start of the twentieth centuries. Historical architecture is complemented by Modern and Functionalist buildings. Most residential buildings were built after the Second World War, either in a Modern or Postmodern style. In the eastern part of the town, there is a railway line that divides its area into separate zones, with circulation between them provided only by historical tunnels or viaducts above the railway tracks. The eastern part of the city also includes the Silesian Insurgents Park and the Silesian Botanical Garden.

Three types of city heating zones were determined after first outlining six characteristic zones. These areas were selected based on their location, similarities within their development structure in terms of age and architectural form, as well as similar topography. Building insulation and central heating systems were inspected on site. Within the six areas, 60 buildings were randomly selected, and their heating systems and insulation were analysed. In total, the analysis covered 360 buildings divided into 6 groups composed of 60 buildings each. Based on the findings of the analysis, the author delineated 3 major urban zones that displayed distinct heating and insulation systems (Figure 1).

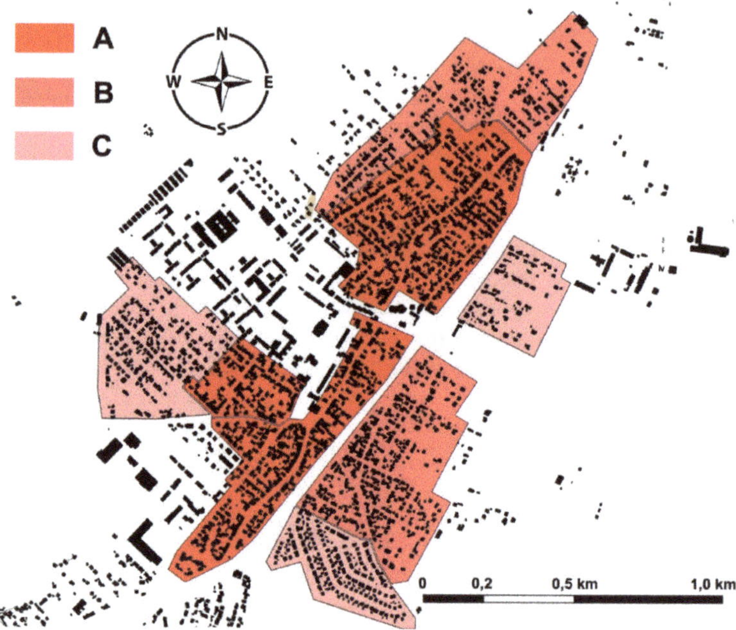

Figure 1. Three zones (A, B and C) with different morphological and environmental conditions in the city of Radzionków, for which different spatial policies should be pursued in terms of protecting the quality of the living environment. Zone A: 53% of buildings were heated with coal; zone B: 39% of buildings were heated with coal; zone C: 35% of buildings were heated with coal.

To better explain and illustrate the problem, 4 terrain profiles were presented, showing the valley-like character of the development structure (Figures 2–5). Slightly smaller height differences between the valley floor and its edges are present in the northern tip of the valley (Figures 2 and 3). However, the most urbanised part shows the greatest terrain height differences, as here the valley floor is surrounded on both sides by hills with a height of 30 metres (Figures 4 and 5). This adversely affects the natural ventilation of the central zone. One of the presented terrain profiles shows the rise of the terrain opposite to the direction from which fresh air is supplied, along with the prevalent wind directions, i.e., from the south-west (Figures 6–8). The area primarily sees south-westerly winds, and this dominant wind direction was presented against the background of the existing urban fabric. The figure also presents streets that are aligned with the city's direction of ventilation and the streets that form barriers and obstacles to ventilation (Figures 7 and 8). Figure 1 presents the main zones that were delineated after the analysis, designated A, B and C, respectively. Detailed tests and observations of $PM_{2.5}$ and PM_{10} levels were performed for zone A (Figure 9). The city has one official air pollution monitoring station. Measurements in each zone were conducted using a certified portable manual measuring device—Steinberg 10030389 SBS-PM2.5/EX10030389. Tests for individual zones were conducted during the winter period in 2018 and 2019. The inspection of the individual zones was performed by moving the particulate matter measuring device to each location and then comparing its readings with results from the official air quality measurement station. The measurements were carried out at 3 locations in each of the following zones: A1, A2, A3, B1, B2, C1, C2 and C3. A total of 25 measuring points were designated in the city. The measurements were performed on days when air pollution readings were high, medium and very low. The test was repeated to verify the results. Sample measurement results will be presented in Figures 10 and 11 The tests were verified and compared with official documents [56,57].

Figure 2. Cross-section through the northern section of Radzionków's urban zone—the difference in elevation between the central and valley part is around 32 m; the terrain has a northwest-facing slope with an incline of around 7%. (Original work based on data from https://mapy.geoportal.gov.pl).

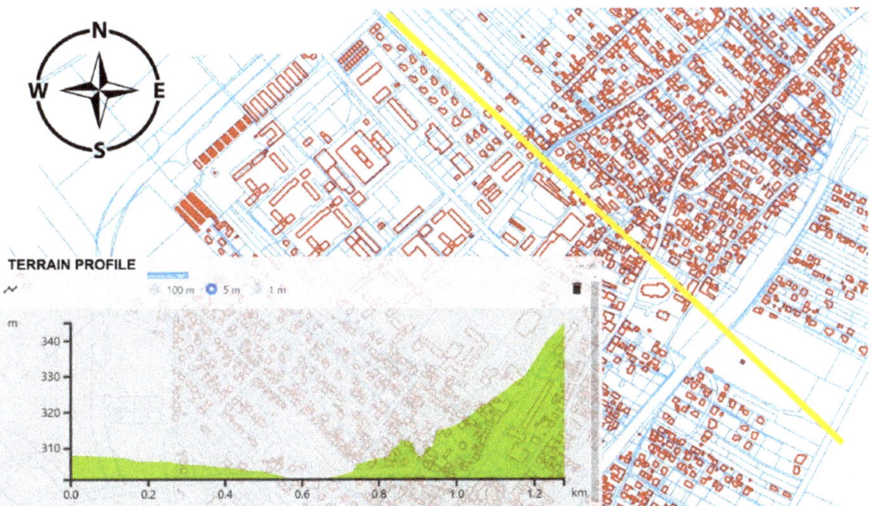

Figure 3. Cross-section through the central part of Radzionków's urban area—the difference in elevation between the valley part and the south-eastern side is 45 m, with a 7% incline. (Original work based on data from https://mapy.geoportal.gov.pl).

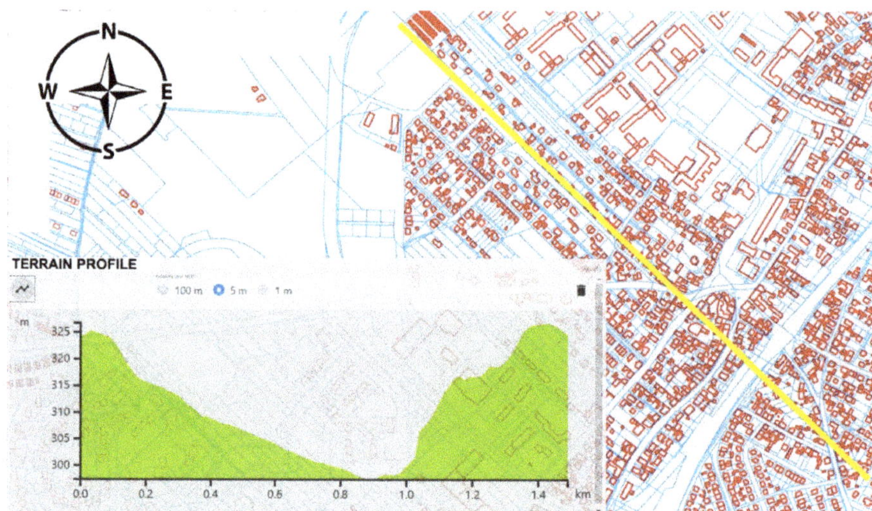

Figure 4. Cross-section through the valley. There are differences in elevation of 30 m on the north-west side and of 35 on the south-west side. The valley has a width of 1400 m in this area. There is an incline of 3.3% in the north and 5% in the south. (Original work based on data from https://mapy.geoportal.gov.pl).

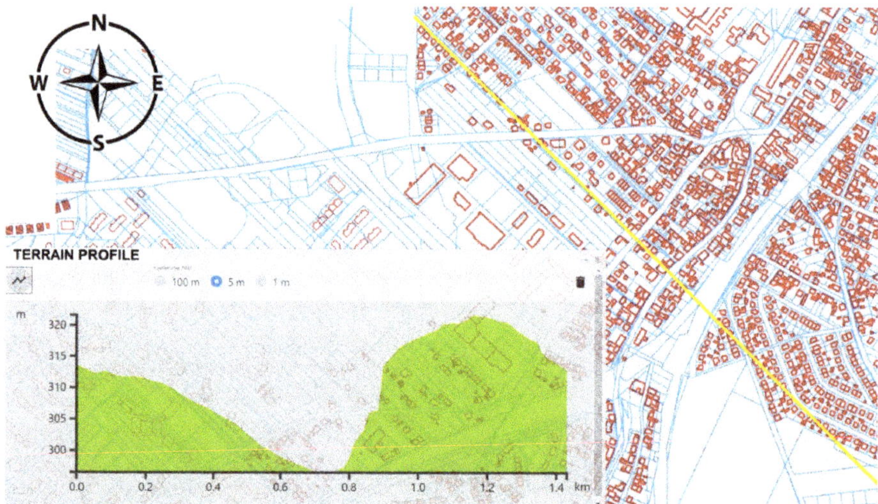

Figure 5. Cross-section of the urbanised area of Radzionków, the difference in elevation between the valley floor and the surrounding hills is up to 35 m; the lowest and highest points are less than 500 m apart, and the terrain has an incline of 7%. (Original work based on data from https://mapy.geoportal.gov.pl).

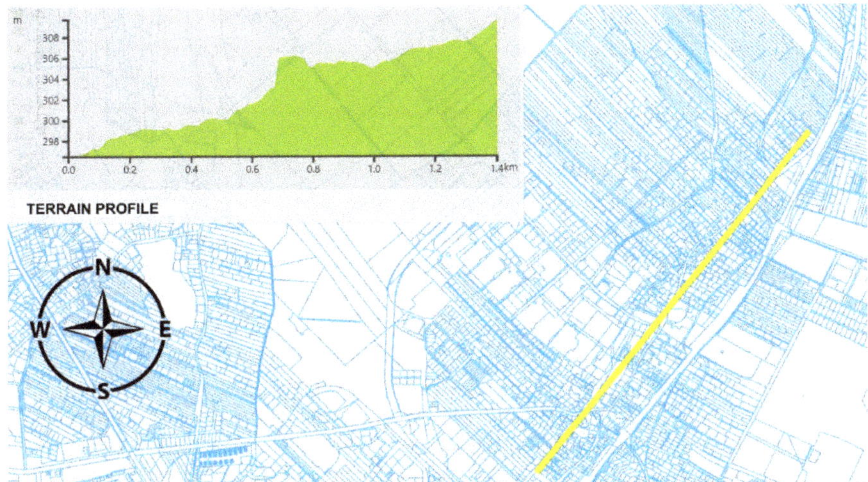

Figure 6. Longitudinal cross-section of Radzionków's urban zone—a difference in elevation of 14 m over a length of 1400 m has been observed. (Original work based on data from https://mapy.geoportal.gov.pl).

POLAND SLASKIE VOIVODESHIP

Figure 7. Dominant wind directions for the Silesian Voivodeship. The illustration shows the strength and direction of the wind at two altitudes: at 10 m and at 50 m. Differences in wind force are clearly noticeable, with significantly weaker wind forces at low altitudes. This leads to there being very little air movement inside the valley, as most of it moves at higher altitudes along with south-westerly winds. (source: https://globalwindatlas.info/area/Poland/%C5%9Al%C4%85skie).

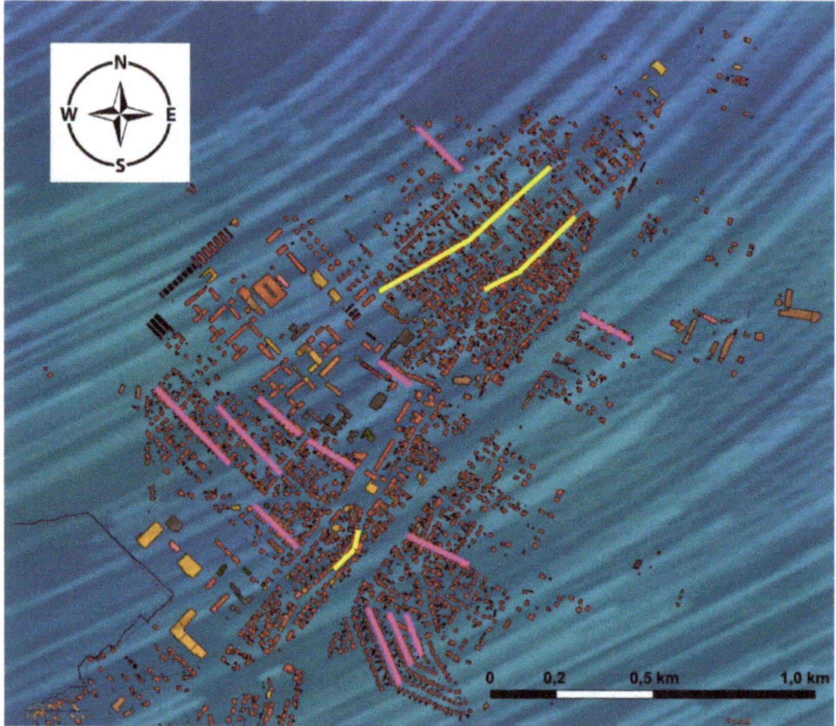

Figure 8. Radzionków's urban development structure compared with the dominant wind direction. The figure presents the roads that are perpendicular to the direction of wind flow (coloured magenta) and form barriers that inhibit the city's ventilation, as well as the low number of roads aligned with the directions of the dominant number of winds (marked yellow).

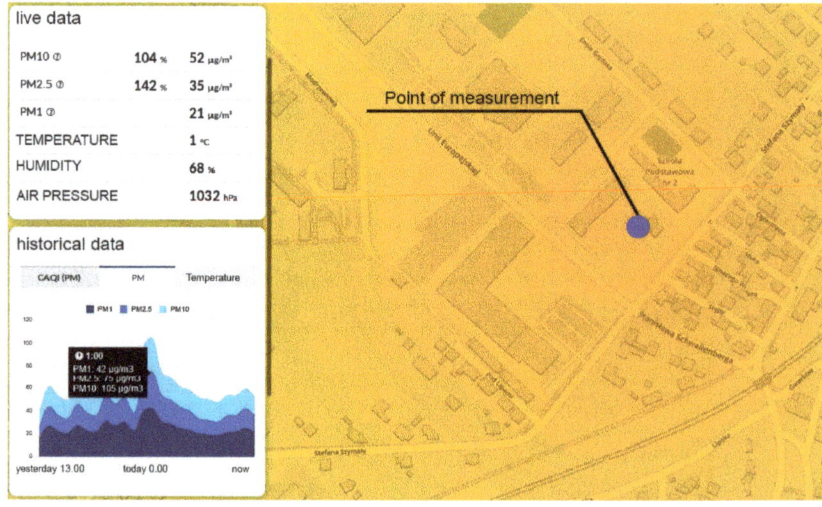

Figure 9. Air pollution at the measurement station at Szymały Street. (https://airly.eu/map).

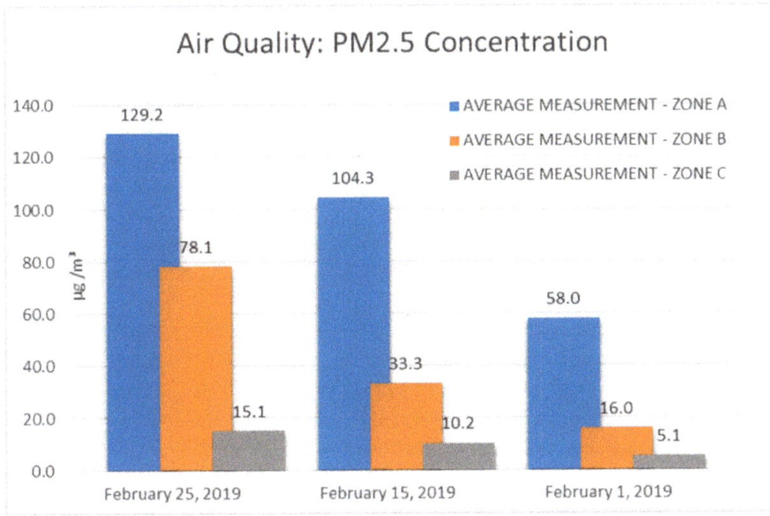

Figure 10. Results of average $PM_{2.5}$ level measurements performed in zones A, B, C for 3 different days, one with very high (25 February 2019), high (15 February 2019) and low pollution, respectively (1 February 2019).

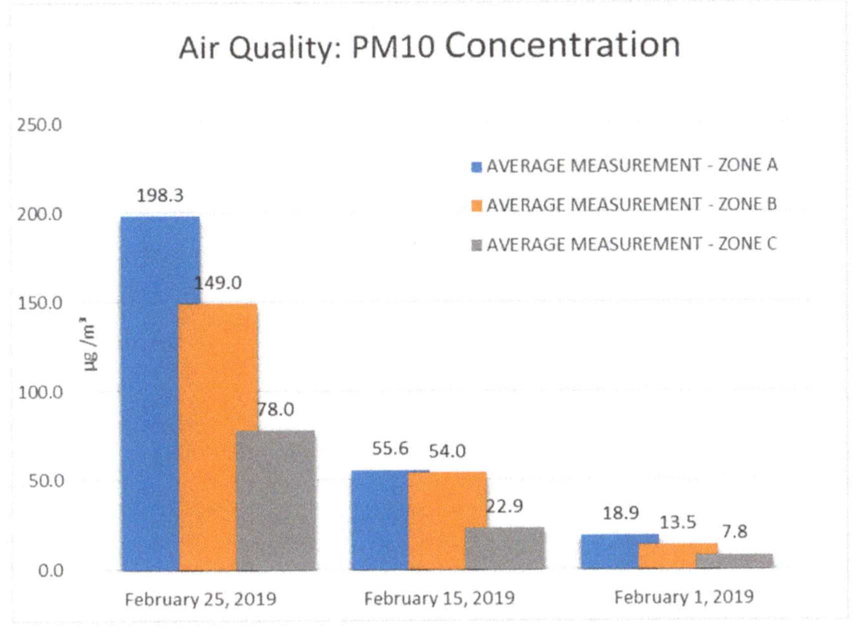

Figure 11. Results of average PM_{10} level measurements in zones A, B, C for 3 different days, one with very high (25 February 2019), high (15 February 2019) and low pollution, respectively (1 February 2019).

According to official documents, during the formulation of the environmental quality policy, surveys were conducted for 991 buildings/dwellings (780 detached buildings, 116 semi-detached buildings, 76 terraced houses and 19 apartments in housing blocks or tenement houses), 24 businesses and 14 public buildings. In housing, the average house/flat floor area was 129.7 m^2 [56,57]. The oldest residential building was dated to 1860, the newest was dated to 2014, while the average building completion date was 1961. Some citizens reported having modern boilers (the latest was from 2015), but a large group of respondents used outdated central heating systems (the oldest being from 1960). The average boiler production year was 2003, and the average power was 20.3 kW. On average, there were 4 people per household. In businesses (firms), the average heated area was reported as 247.9 m^2, and the average year of building completion was 1951 (the oldest being from 1800 and the newest from 2011). On the other hand, for used boilers, the average power was 27.5 kW, and the average production year was 2005. For public buildings covered by the survey, the average heated area was 579.0 m^2, and the year of building completion was 1960 (the oldest being from 1885 and the newest from 1986). The average boiler power was reported to be only 5 kW, as most public buildings were reported to be connected to the heating grid, and boilers were used only as an additional source of domestic hot water. The average production year reported for these boilers was 2003. The municipal authorities have a plan to replace existing heating installations and co-finance boiler replacement with RES heating media. However, as these changes are being introduced very slowly, they were not observed to have a significant impact on the environment of Radzionków. An additional survey regarding interest in replacing the heat sources conducted on a group of 701 people showed that 81.31% were interested in replacing their heat sources or in purchasing renewable energy sources or a more eco-friendly energy source [56,57]. Another survey showed that only 9.13% of the respondents planned to postpone the replacement of their heating systems to an unspecified point in the future.

3. Results and Discussion

Radzionków's urban area cannot be treated as a morphologically homogeneous structure. As evidenced by the attached terrain profiles (Figures 2–6), Radzionków's entire urban area is located in a small river valley located in the Szarlejka river catchment, with an incline of 1% from the north-east to south-west is 1%. The attached 4 transverse terrain profiles (Figures 2–5) and one longitudinal terrain profile (Figure 6) show the valley-like character of the town's layout. In addition, the historical layers of the town's urban tissue development indicate its expansion from the valley floor to areas located at a higher elevation. In order to examine specific environmental zones, 6 areas were initially delineated, selected on the basis of their location, similar development structure, the age and type of local buildings and similar topography. In these areas, 60 buildings were randomly selected and analysed in terms of their heating and insulation systems. A total of 360 buildings were analysed, divided into 6 zones with 60 buildings each. The findings of the analysis enabled the delineated of 3 major urban zones that were distinct in terms of dwelling heating and insulation solutions (Figure 1). Considering the most important environmental conditions from the point of view of urban tissue ventilation, westerly and south-westerly winds were observed to encounter a natural barrier in the form of a 35-metre tall hill on the eastern side of the city (Figures 2–5 and Figure 7). Additional obstacles include groups of buildings placed parallel to the roads running along the northwest–southeast direction (Figure 8). In the entire spatial arrangement of the city, only 19% of roads were observed to be oriented along the southwest–northeast direction, the remaining 81% being roads that are mostly perpendicular and thus not intended to ventilate the city (Figure 8). The road layout outside the city centre was mostly created via delineating them along the original outline of fields—perpendicularly to the slopes of hills. Considering the topography of the city and the dominant south-westerly winds, the city's location can be considered highly unfavourable in terms of its ventilation across the entirety of its territory.

Against the background of existing development tissue, when the age of the buildings is considered, three zones, named A, B and C, respectively, can be clearly distinguished on the basis of their distinct environmental conditions (Figure 1). They provide a basis for formulating conclusions concerning necessary actions related to the high PM$_{2.5}$ and PM$_{10}$ levels observed in the city (Figure 9).

In zone A, 53% of buildings were heated with coal; in zone B, 39% of buildings were heated with coal; in zone C, 35% of buildings were heated with coal. Outside these zones, new housing estates were heated with gas and renewable energy sources, public buildings were heated using the heating grid.

In zone A, there were 44% insulated buildings; in zone B, 48% of buildings were insulated; in zone C, 68% of buildings were insulated. Outside of these zones, contemporarily built buildings constructed today were either all insulated or met thermal insulation standards (Figure 1) [58].

3.1. Characteristics of Zone A

Zone A (Figure 1) was observed to have the worst environmental conditions since 53% of buildings were heated with coal. It is also the zone located at the lowest elevation on the valley floor, ranging between 296 m above sea level to 306 m above sea level, with an average 1% incline in the opposite direction to that of the area's natural ventilation. This zone is hemmed in from the southeast and the northwest by 35-metre hills. In this zone, only 6 streets were observed to run parallel to the direction of the area's dominant winds. The remaining roads, i.e., approx. 39, were oriented perpendicular to the dominant wind direction. The ventilation of this zone was observed to be the worst because it had the most physical barriers—most of the buildings were oriented parallel to the roads. In zone A, 44% of the buildings were thermally insulated, although the average was skewed upwards by the smallest, sometimes new residential buildings. The largest percentage of the buildings were not thermally insulated. In general, small buildings emit much less pollution than large multi-family buildings, which were reported to be heated by either several individual installations or central former coal installations.

It is also the oldest and the most intensively built-up zone. The development density (the floor area to plot area ration) of this zone ranges between 0.3 and 2.0. In order to improve the quality of the environment, it is necessary to increase expenditure on new building heating systems This can be done by introducing subsidies for gas heating or by taking advantage of existing potential in the form of connection to the heating grid that is linked to the Jerzy Ziętek housing estate. The largest stone buildings, often under heritage protection, require special expenditures.

3.2. Characteristics of Zone B

In zone B (Figure 1), 48% of buildings were insulated, with the average inflated by infill buildings after 1980. The closer to the historical layout of the town centre, the more buildings without insulation and that used coal for heating were observed. On the other hand, insulated buildings were observed to be in breach of the building code (Ordinance concerning the technical conditions to be met by buildings and their placement of 2014), specifying the maximum energy intensity parameter at 120 kWh/(m^{2}annum). To meet these parameters, common traditional brick wall technology requires a layer of mineral wool or polystyrene sheets with a thickness of 15 cm and a lambda factor of 0.031 W/(mK) (Figure 12) [58]. Typically, the insulation thickness ranged between 5 and 10 cm. In zone B, 39% of the buildings were heated with coal, which significantly contributed the deterioration of the quality of the environment in this area but at the same time affected the deterioration of the environment in zone A, which is located on the valley floor. In area B, on the slopes of hills, streets perpendicular to the direction of ventilation become natural pollution funnels that direct it towards the valley floor. According to Office of the City of Radzionków, the average age of coal boilers in Radzionków area were boilers installed 17 years ago—it can be assumed that the average boiler age in this area is compliant with official data. Most of the buildings in this zone are dated to the post-war period and were built mostly between 1945 and 1990.

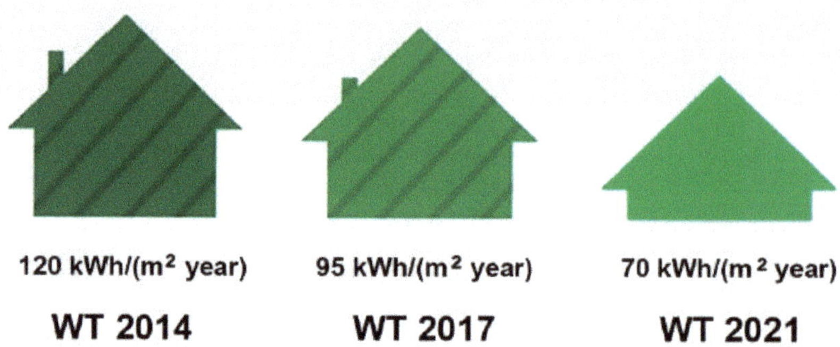

Figure 12. Change of standards related to thermo-modernisation in Polish legislation in accordance with the Ordinance concerning the technical conditions to be met by buildings and their placement of 2014—120 (kWh/m² annum); after 2017, the maximum value decreased to 95 (kWh/m² annum), while after 2021, it is to decrease further down to 70 (kWh/m² annum) [58].

3.3. Characteristics of Zone C

In zone C (Figure 1), 68% of the buildings were insulated, as they were mostly relatively new buildings built on the outskirts of the city. The closer to the centre, the more the number of insulated buildings decreased and their age increased. Most of the insulated buildings were observed to be in breach of the building code (Ordinance concerning the technical conditions to be met by buildings and their placement of 2014), specifying the maximum energy intensity parameter at 120 kWh/(m²annum). This situation was similar to the one observed in zone B [58]. The thickness of buildings insulation layers varied between 5 and 10 cm (Figure 12). In zone C, 35% of buildings were heated with coal, which significantly contributed to the deterioration of the quality of the environment in this area, similarly affecting zone A, located at the valley floor. According to the Municipal Office, the average age of coal-powered boilers in Radzionków was 17 years—it can be assumed that the average age of boilers in this area was slightly lower than listed in statistical data. Most of the buildings in this zone are dated to the post-war period and were erected between 1960 and 2010.

3.4. Measurements of Air Pollution in the Urban Area of Radzionkow

Currently, there is only one air quality measurement station (Figure 9) in Radzionków. It is located on Szymały Street outside of the designated zones. Its readings seem to be unreliable, due to the open and flat space that surrounds it, which facilitates the area's ventilation. Furthermore, there are no buildings that use coal for heating up to a distance of 1 km to its west. The measurement station is actually located on the edge of Radzionków's urban area, where environmental conditions are highly favourable. The measuring equipment is located at a height of approx. 6.5 m. It should, therefore, be assumed that its indications show the most favourable situation in the city, at a site located outside the designated reliable location (A, B, C). On a sunny, windy day, e.g., 30 December 2019, the sensor's indicators showed a PM_{10} level of 52 $\mu g/m^3$. The acceptable total is 50 $\mu g/m^3$ per day. The readings did not fall below 50 $\mu g/m^3$ during the entire day, at times reaching 110 $\mu g/m^3$, with the average hourly reading being approx. 65 $\mu g/m^3$. Calculated for all hours of the day, this resulted in a combined PM_{10} amount of 1440 $\mu g/m^3$ [55,59]. This demonstrates that the readings exceeded standards by as much as 2280% for the entire day, with the daily average exceeding standards by 130%, while overall it exceeded 360 $\mu g/m^3$ per day—under very favourable weather conditions and measured at a convenient location. According to WHO recommendations, the permissible daily concentration should not be exceeded for more than 35 days in a year, with levels of 300 $\mu g/m^3$ (daily average) being cause for alarm [55,59].

In terms of permissible $PM_{2.5}$ standards, European Union Member States have set their maximum levels at 25 $\mu g/m^3$ [55]. Measurements taken for the purposes of this study showed a value of 30 $\mu g/m^3$, while in zone A, measured with a hand-held pollution measurement device, they reached 120 µg/m^3. Inside a gas-heated building, the readings were as high as 91 µg/m^3. In zone A, the test results were four times worse than the readings of the official measurement station. At the same time, this demonstrates that the standard was exceeded by 480%. Daily tests performed during the period with the highest concentrations of PM$_{2.5}$ at the measurement station, which reported a level of 75 µg/m^3, produced readings at a level of 300 µg/m^3 on the hand-held device in zone A, exceeding the standard by 1200%. On less windy days, in zone A, readings taken with the hand-held device exceeded standards by 2000%.

4. Conclusions

In recent years, mathematical LUR (Land Use Regression) models have become an alternative to conventional research approaches [60]. They are used to predict the concentration of atmospheric pollution at a given location by establishing statistical relationships between pollution measurements and potential predictive variables, e.g., land use, traffic and physical features of land use [60,61]. High PM$_{2.5}$ and PM$_{10}$ levels observed in a mostly residential and service-oriented zone did not confirm frequently reported findings in which industrial zones have higher air pollution levels than residential zones. This is mainly related to PM$_{2.5}$ and PM$_{10}$ emission sources, which are more closely associated with restrictive regulations and inspections of zones used by industrial entities under Polish conditions. On the other hand, in the zones under study, differences in pollution around new multi-family housing and service areas were clearly noticeable. The study did not confirm the findings presented in the simulation model described in [62] and focused on the topographic and functional specificity of areas in Central Europe, rooted in their historical, economic and spatial conditions. As a result of political and economic changes that took place in the twentieth century, Poland has engaged in measures aimed at promoting and supporting renewable energy only relatively recently. The first significant reduction of air pollutant emissions took place in Poland in the 1990s. This was due to the abandonment of old industrial technologies that caused significant environmental pollution. At that time, biomass and coal combustion technology in existing domestic boilers was developed quite successfully [63]. Unfortunately, changes in domestic heating technologies affected less urbanised areas to a smaller extent. In the municipal and multi-family housing sector in Poland, they are regulated and modernised in many places. In the private sector, low-efficiency furnaces with no dedusting systems are used the most often. The purpose of this paper is to highlight the problem of reducing emissions at low altitudes with a particular emphasis on topography. The paper discusses development challenges that are often faced by selected municipal authorities in the area under study.

The research showed that:

- The location of air pollution measurement stations is crucial as it can affect their readings, and well-ventilated sites at higher altitudes can underreport pollution values;
- The topographic altitude of urban development has a very strong impact on PM$_{2.5}$ and PM$_{10}$ air pollution readings;
- Development layouts that inhibit ventilation of the spatial structure adversely affect air quality, even in cases of quite favourable altitude (part of zone C in the southern part of Radzionków);
- Built-up basins and valleys arranged perpendicularly (corridor) to the dominant wind direction, wherein most streets are oriented perpendicular to the most common wind directions, generate very unfavourable initial conditions in terms of air pollution (the shape and form of the Radzionków basin in relation to the wind rose);
- There is evidence in support of the argument that the period during which the Silesian agglomeration and its land ownership structure developed determined their impact on the environment;
- Residential buildings built before 1980 (zone A and B) were observed to generate the greatest air pollution,

- The smallest air pollution was generated by buildings owned by municipalities and education authorities, which rely on the municipal heating grid.

Several postulates can be formulated for the case under study:

- Using coal for domestic heating should be prohibited in zones A and B of the Radzionków basin on account of their extremely unfavourable topographic conditions, with a requirement to achieve building energy consumption levels as stipulated in applicable Polish construction regulations for 2017, i.e., 95 kWh/(m^{2}annum);
- In zone A, designating ventilation corridors should be a priority, while in zone C, all buildings should be fitted out with insulation;
- The co-financing planned for the insulation of buildings and supporting the replacement of coal-fuelled boilers with gas-powered ones or other heating devices using renewable energy planned in the municipality should apply to zones A and B to a greater extent. Municipal funds should be allocated appropriately for each zone; efficient results can be achieved as a measure of environmental policy in some quantitative value, for example: 55% of funds for zone A, 35% of funds for zone B, 15% of funds for zone C; in addition, the allocated funds should include a clause on the necessity of spending them within a year of receipt;
- The planned measures intended to reduce $PM_{2.5}$ and PM_{10} levels should first focus on the centre and then shift outwards from there.

Funding: This research received no external funding.

Conflicts of Interest: The author declares no conflict of interest.

References

1. Bilgen, S.; Sarıkaya, I. Energy for environment, ecology and sustainable development. *Renew. Sustain. Energy Rev.* **2015**, *51*, 1115–1131. [CrossRef]
2. Kampa, M.; Castanas, E. Human health efects of air pollution. *Environ. Pollut.* **2008**, *151*, 362–367. [CrossRef] [PubMed]
3. Hendryx, M.; Fedorko, E. The relationship between toxics release inventory discharges and mortality rates in rural and urban areas of the United States. *J. Rural Health* **2011**, *27*, 358–366. [CrossRef] [PubMed]
4. QUARG. *Airborne Particulate Matter in the UK*; Third Report of the Quality of Urban Air Review, Group; Harrison, R.M., Ed.; Institute of Public and Environmental Health, University of Birmingham: Birmingham, UK, 1996.
5. Grewling, Ł.; Bogawski, P.; Kryza, M.; Magyar, D.; Šikoparija, B.; Skjøth, C.A.; Udvardy, O.; Wermer, M.; Smith, M. Concomitant occurrence of anthropogenic air pollutants, mineral dust and fungal spores during long-distance transport of ragweed pollen. *Environ. Pollut.* **2019**, *254*, 112948. [CrossRef] [PubMed]
6. Kozáková, J.; Pokorná, P.; Vodicka, P.; Ondráčková, L.; Ondrácek, J.; Krumal, K.; Mikuška, P.; Hovorka, J.; Moravec, P.; Schwarz, J. The influence of local emissions and regional air pollution transport on a European air pollution hot spot. *Environ. Sci. Pollut. Res. Int.* **2019**, *26*, 1675–1692. [CrossRef] [PubMed]
7. Adamczyk, J.; Piwowar, A.; Dzikuc, M. Air protection programmes in Poland in the context of the low emission. Environ. *Sci. Pollut. Res.* **2017**, *24*, 16316–16327. [CrossRef]
8. Naydenova, I.; Petrova, T.; Velichkova, R.; Simova, I. PM10 exceedance in Bulgaria. In Proceedings of the CBU International Conference Proceedings, Prague, Czech Republic, 21–23 March 2018; Volume 6, pp. 1129–1138.
9. Beloconi, A.; Chrysoulakis, N.; Lyapustin, A.; Utzinger, J.; Vounatsou, P. Bayesian geostatistical modelling of PM_{10} and $PM_{2.5}$ surface level concentrations in Europe using high-resolution satellite-derived products. *Environ. Int.* **2018**, *121*, 57–70. [CrossRef]
10. Gehrsitz, M. The efect of low emission zones on air pollution and infant health. *J. Environ. Econ. Manag.* **2017**, *83*, 121–144. [CrossRef]
11. Ellison, R.B.; Greaves, S.P.; Hensher, D.A. Five years of London's low emission zone: Effects on vehicle fleet composition and air quality. *Transp. Res. D Transp. Environ.* **2013**, *23*, 25–33. [CrossRef]

12. Holman, C.; Harrison, R.; Querol, X. Review of the efficacy of low emission zones to improve urban air quality in European cities. *Atmos. Environ.* **2015**, *111*, 161–169. [CrossRef]

13. De Marco, A.; Proietti, C.; Anav, A.; Ciancarella, L.; D'Elia, I.; Fares, S.; Fornasier, M.F.; Fusaro, L.; Gualtieri, M.; Manes, F.; et al. Impacts of air pollution on human and ecosystem health, and implications for the National Emission Ceilings Directive: Insights from Italy. *Environ. Int.* **2019**, *125*, 320–333. [CrossRef] [PubMed]

14. Cai, B.; Liang, S.; Zhou, J.; Wang, J.; Cao, L.; Qu, S.; Xu, M.; Yang, Z. China high resolution emission database (CHRED) with point emission sources, gridded emission data, and supplementary socioeconomic data. *Resour. Conserv. Recycl.* **2018**, *129*, 232–239. [CrossRef]

15. Alyuz, U.; Alp, K. Emission inventory of primary air pollutants in 2010 from industrial processes in Turkey. *Sci. Total Environ.* **2014**, *488*, 369–381. [CrossRef] [PubMed]

16. Han, L.; Zhou, W.; Li, W.; Li, L. Impact of urbanization level on urban air quality: A case of fine particles (PM2.5) in Chinese cities. *Environ. Pollut.* **2014**, *194*, 163–170. [CrossRef]

17. Makkonen, U.; Hellén, H.; Anttila, P.; Ferm, M. Size distribution and chemical composition of airborne particles in south-eastern Finland during different seasons and wildfire episodes in 2006. *Sci. Total Environ.* **2010**, *408*, 644–651. [CrossRef]

18. Pope, C.A., III; Dockery, D.W. Health effects of fine particulate air pollution: Lines that connect. *J. Air Waste Manag.* **2006**, *56*, 709–742. [CrossRef]

19. Pope, C.A.; Turner, M.C.; Burnett, R.; Jerrett, M.; Gapstur, S.M.; Diver, W.R.; Krewski, D.; Brook, R.D. Relationships between fine particulate air pollution, cardiometabolic disorders, and cardiovascular mortality. *Circ. Res.* **2015**, *116*, 108–115. [CrossRef]

20. Behera, S.N.; Sharma, M. Reconstructing primary and secondary components of PM2.5 composition for an urban atmosphere. *Aerosol Sci. Technol.* **2010**, *44*, 983–992. [CrossRef]

21. Wang, Z.; Hu, M.; Wu, Z.; Yue, D.; He, L.; Huang, X.; Liu, X.; Wiedensohler, A. Long-term measurements of particle number size distributions and the relationships with air mass history and source apportionment in the summer of Beijing. *Atmos. Chem. Phys.* **2013**, *13*, 10159–10170. [CrossRef]

22. Wu, D.W.; Fung, J.C.; Yao, T.; Lau, A.K. A study of control policy in the Pearl River Delta region by using the particulate matter source apportionment method. *Atmos. Environ.* **2013**, *76*, 147–161. [CrossRef]

23. Wang, Y.; Li, L.; Chen, C.; Huang, C.; Huang, H.; Feng, J.; Wang, S.; Wang, H.; Zhang, G.; Zhou, M.; et al. Source apportionment of fine particulate matter during autumn haze episodes in Shanghai, China. *J. Geophys. Res.* **2014**, *119*, 1903–1914. [CrossRef]

24. Brook, R.D.; Rajagopalan, S.; Pope, C.A.; Brook, J.R.; Bhatnagar, A.; Diez-Roux, A.V.; Holguin, F.; Hong, Y.; Luepker, R.V.; Mittleman, M.A. Particulate matter air pollution and cardiovascular disease an update to the scientific statement from the American Heart Association. *Circulation* **2010**, *121*, 2331–2378. [CrossRef]

25. Bell, M.L. *Assessment of the Health Impacts of Particulate Matter Characteristics*; Research Report, No 161; Health Effects Institute: Boston, MA, USA, 2012; pp. 5–38.

26. Fann, N.; Lamson, A.D.; Anenberg, S.C.; Wesson, K.; Risley, D.; Hubbell, B.J. Estimating the national public health burden associated with exposure to ambient PM2.5 and ozone. *Risk Anal.* **2012**, *32*, 81–95. [CrossRef]

27. Kim, S.Y.; Peel, J.L.; Hannigan, M.P.; Dutton, S.J.; Sheppard, L.; Clark, M.L.; Vedal, S. The temporal lag structure of short-term associations of fine particulate matter chemical constituents and cardiovascular and respiratory hospitalizations. *Environ. Health Perspect.* **2012**, *120*, 1094. [CrossRef]

28. Beckerman, B.S.; Jerrett, M.; Serre, M.; Martin, R.V.; Lee, S.-J.; van Donkelaar, A.; Ross, Z.; Su, J.; Burnett, R.T. A hybrid approach to estimating national scale spatiotemporal variability of PM2.5 in the contiguous. United States. *Environ. Sci. Technol.* **2013**, *47*, 7233–7241. [CrossRef] [PubMed]

29. Geng, G.; Zhang, Q.; Martin, R.V.; van Donkelaar, A.; Huo, H.; Che, H.; Lin, J.; He, K. Estimating long-term PM2.5 concentrations in China using satellite-based aerosol optical depth and a chemical transport model. *Remote Sens. Environ.* **2015**, *166*, 262–270. [CrossRef]

30. Just, A.C.; Wright, R.O.; Schwartz, J.; Coull, B.A.; Baccarelli, A.A.; Tellez-Rojo, M.M.; Moody, E.; Wang, Y.; Lyapustin, A.; Kloog, I. Using high-resolution satellite aerosol optical depth to estimate daily PM2.5 geographical distribution in Mexico City. *Environ. Sci. Technol.* **2015**, *49*, 8576–8584. [CrossRef] [PubMed]

31. Zhang, T.; Gong, W.; Wang, W.; Ji, Y.; Zhu, Z.; Huang, Y. Ground level PM2.5 estimates over China using satellite-based geographically weighted regression (GWR) models are improved by including NO2 and enhanced vegetation index (EVI). *Int. J. Environ. Res. Public Health* **2016**, *13*, 1215. [CrossRef]

32. Arain, M.A.; Blair, R.; Finkelstein, N.; Brook, J.R.; Sahsuvaroglu, T.; Beckerman, B.; Zhang, L.; Jerrett, M. The use of wind fields in a land use regression model to predict air pollution concentrations for health exposure studies. *Atmos. Environ.* **2007**, *41*, 3453–3464. [CrossRef]

33. Madsen, C.; Carlsen, K.C.L.; Hoek, G.; Oftedal, B.; Nafstad, P.; Meliefste, K.; Jacobsen, R.; Nystad, W.; Carlsen, K.-H.; Brunekreef, B. Modeling the intra-urban variability of outdoor traffic pollution in Oslo,Norway—A GA 2 LEN project. *Atmos. Environ.* **2007**, *41*, 7500–7511. [CrossRef]

34. Wilton, D.; Szpiro, A.; Gould, T.; Larson, T. Improving spatial concentration estimates for nitrogen oxidesusing a hybrid me teorological dispersion/land use regression model in Los Angeles, CA and Seattle, WA.Sci. *Total Environ.* **2010**, *408*, 1120–1130. [CrossRef] [PubMed]

35. Li, X.; Liu, W.; Chen, Z.; Zeng, G.; Hu, C.; León, T.; Liang, J.; Huang, G.; Gao, Z.; Li, Z.; et al. The application of semicircular-buffer-based land use regression models incorporating wind direction in predicting quarterly NO2 and PM10 concentrations. *Atmos. Environ.* **2015**, *103*, 18–24. [CrossRef]

36. Lam, T.; Niemeier, D. An exploratory study of the impact of common land-use policies on air quality. *Transp. Res. D Transp. Environ.* **2005**, *10*, 365–383. [CrossRef]

37. Bandeira, J.M.; Coelho, M.C.; Sá, M.E.; Tavares, R.; Borrego, C. Impact of land use on urban mobility patterns, emissions and air quality in a Portuguese medium-sized city. *Sci. Total Environ.* **2011**, *409*, 1154–1163. [CrossRef] [PubMed]

38. Zhang, R.S.; Pu, L.J.; Liu, Z. Advances in research on atmospheric environment effects of land use and land cover change. *Area Res. Dev.* **2013**, *32*, 123–128.

39. Chen, L.D.; Sun, R.H.; Liu, H.L. Eco-environmental effects of urban landscape pattern changes: Progresses, problems and perspectives. *Acta Ecol. Sin.* **2013**, *33*, 1042–1050. [CrossRef]

40. Briggs, D.J.; de Hoogh, C.; Gulliver, J.; Wills, J.; Elliott, P.; Kingham, S.; Smallbone, K. A regression-based method for mapping traffic-related air pollution: Application and testing in four contrasting urban environments. *Sci. Total Environ.* **2000**, *253*, 151–167. [CrossRef]

41. Jerrett, M.; Arain, A.; Kanaroglou, P.; Beckerman, B.; Potoglou, D.; Sahsuvaroglu, T.; Morrison, J.; Giovis, C. A review and evaluation of intra-urban air pollution exposure models. *J. Expo. Sci. Environ. Epidemiol.* **2005**, *15*, 185–204. [CrossRef]

42. Liu, C.; Henderson, B.H.; Wang, D.; Yang, X.; Peng, Z.-R. A land use regression application into assessing spatial variation of intra-urban fine particulate matter (PM2.5) and nitrogen dioxide (NO2) concentrations in City of Shanghai, China. *Sci. Total Environ.* **2016**, *565*, 607–615. [CrossRef]

43. Sun, L.; Wei, J.; Duan, D.H.; Guo, Y.M.; Yang, D.X.; Jia, C.; Mi, X.T. Impact of Land-Use and Land-Cover Change on urban air quality in representative cities of China. *J. Atmos. Sol.-Terr. Phys.* **2016**, *142*, 43–54. [CrossRef]

44. Henderson, S.B.; Beckerman, B.; Jerrett, M.; Brauer, M. Application of land use regression to estimate long-term concentrations of traffic-related nitrogen oxides and fine particulate matter. *Environ. Sci. Technol.* **2007**, *41*, 2422–2428. [CrossRef] [PubMed]

45. Liu, W.; Li, X.; Chen, Z.; Zeng, G.; León, T.; Liang, J.; Huang, G.; Gao, Z.; Jiao, S.; He, X.; et al. Land use regression models coupled with meteorology to model spatial and temporal variability of NO_2, and PM_{10}, in Changsha, China. *Atmos. Environ.* **2015**, *116*, 272–280. [CrossRef]

46. Olszowski, T.; Tomaszewska, B.; Góralna-Włodarczyk, K. Air quality in non-industrialised area in the typical Polish countryside based on measurements of selected pollutants in immission and deposition phase. *Atmos. Environ.* **2012**, *50*, 139–147. [CrossRef]

47. Błaszczyk, E.; Rogula-Kozłowska, W.; Klejnowski, K.; Kubiesa, P.; Fulara, I.; Mielżyńska-Švach, D. Indoor air quality in urban and rural kindergartens: Short-term studies in Silesia, Poland. *Air Qual. Atmos. Health* **2017**, *10*, 1207–1220. [CrossRef]

48. Kobza, J.; Geremek, M.; Dul, L. Characteristics of air quality and sources affecting high levels of PM 10 and PM 2.5 in Poland, Upper Silesia urban area. *Environ. Monit. Assess.* **2018**, *190*, 515. [CrossRef]

49. Zajusz-Zubek, E.; Mainka, A.; Korban, Z.; Pastuszka, J.S. Evaluation of highly mobile fraction of trace elements in PM10 collected in Upper Silesia (Poland): Preliminary results. *Atmos. Pollut. Res.* **2015**, *6*, 961–968. [CrossRef]

50. Dzikuc, M.; Kułyk, P.; Urban, S.; Dzikuc, M.; Piwowar, A. Outline of ecological and economic problems associated with the low emission reductions in the Lubuskie Voivodeship. *Pol. J. Environ. Stud.* **2019**, *28*, 65–72. [CrossRef]

51. Nguyen, T.; Park, D.; Lee, Y.; Lee, Y.C. Particulate matter (PM10 and PM2.5) in subway systems: Health-based economic assessment. *Sustainability* **2017**, *9*, 2135. [CrossRef]

52. Dabass, A.; Talbott, E.O.; Venkat, A.; Rager, J.; Marsh, G.M.; Sharma, R.K.; Holguin, F. Association of exposure to particulate matter (PM2.5) air pollution and biomarkers of cardiovascular disease risk in adult NHANES participants (2001–2008). *Int. J. Hyg. Environ. Health* **2016**, *219*, 301–310. [CrossRef]

53. Błaszczyk, E.; Rogula-Kozłowska, W.; Klejnowski, K.; Fulara, I.; Mielżyńska-Švach, D. Polycyclic aromatic hydrocarbons bound to outdoor and indoor airborne particles (PM2.5) and their mutagenicity and carcinogenicity in Silesian kindergartens, Poland. *Air Qual. Atmos. Health* **2017**, *10*, 389–400. [CrossRef]

54. Kujda, Ł.; Kozacki, D.; Pociech, D.; Hryniewicz, M. Effect of the renewable energy resources on the reduction of pollution emissions from the rural areas. *Probl. Agric. Eng.* **2016**, *93*, 59–67.

55. Directive 2008/50/EC of the European Parliament and of the Council of 21 May 2008 on Ambient Air Quality 192 and Cleaner Air for Europe. Available online: https://eur-lex.europa.eu/legal-content/en/ALL/?uri=Celex (accessed on 15 December 2020).

56. Low-Emission Economy Plan for the Radzionków Municipality for the Years 2015–2020. Update of the Draft Assumptions to the Plan for the Supply of Heat, Electricity and Gas Fuels for the Radzionków Commune. July 2018. Available online: http://www.bip.radzionkow.pl/?p=documentaction=showid=27696bar_id=17663 (accessed on 15 December 2020).

57. Municipal Subsidies for Boiler Replacement in Radzionkow. Available online: http://www.radzionkow.pl/ochrona-zasobow-srodowiska/powietrze/440-dotacje-do-modernizacji-systemu-ogrzewania-z-budzetu-gminy (accessed on 15 December 2020).

58. Ordinance of the Minister of Infrastructure of 12th April 2002 Concerning the Technical Conditions to be Met by Buildings and Their Placement. Available online: http://prawo.sejm.gov.pl/isap.nsf/DocDetails.xsp?id=WDU20020750690 (accessed on 15 December 2020).

59. Health Effects of Particulate Matter. Policy Implications for Countries in Eastern Europe, Caucasus and Central Asia. Available online: http://www.euro.who.int/data/assets/pdf_file/0006/189051/Health-effects-of-particulate-matter-final-Eng.pdf (accessed on 15 December 2020).

60. Olvera, H.A.; Garcia, M.; Li, W.-W.; Yang, H.; Amaya, M.A.; Myers, O.; Burchiel, S.W.; Berwick, M.; Pingitore, N.E., Jr. Principal component analysis optimization of a PM2.5 land use regression model with small monitoring network. *Sci. Total Environ.* **2012**, *425*, 27–34. [CrossRef]

61. Briggs, D.J.; Collins, S.; Elliott, P.; Fischer, P.; Kingham, S.; Lebret, E.; Pryl, K.; Reeuwijk, H.V.; Smallbone, K.; Van Der Veen, A. Mapping urban air pollution using GIS: A regression-based approach. *Int. J. Geogr. Inf. Sci.* **1997**, *11*, 699–718. [CrossRef]

62. He, Z.J.; Yuan, S.L.; Xiao, M. Pollution levels of airborne particulate matter PM10 and PM2.5 in summer inNanchang City. *J. Anhui Agric. Sci.* **2009**, *38*, 1336–1338.

63. Dzikuc, M.; Piwowar, A. Ecological and economic aspects of electric energy production using the biomass co-firing method. The case of Poland. *Renew. Sustain. Energy Rev.* **2016**, *55*, 856–862. [CrossRef]

 *buildings*

Article

The Role of an Architect in Creating the Image of an Elderly-Friendly Sustainable Smart City

Joanna Tymkiewicz

Faculty of Architecture, Silesian University of Technology, 44-100 Gliwice, Poland; joanna.tymkiewicz@polsl.pl; Tel.: +48-322-372-418

Received: 22 September 2019; Accepted: 17 October 2019; Published: 21 October 2019

Abstract: The idea of sustainable smart city has extensive scientific literature where the architects' role in designing built environments, being a physical platform for implementing "elderly-friendly" solutions, is poorly referenced. The main objective of the article is to define the role of architects in creating the image of sustainable smart cities, focusing on senior citizens. The paper surveys the available literature on the subject and describes pilot studies carried out at the indicative level among the students of one of architecture faculties in Poland, based on the design thinking method. The studies demonstrate how students imagine intelligent elderly-friendly cities in the future from the architects' perspective. In addition, examples of other studies with the students of that faculty are presented. Following the analyses combining the conclusions of research and pilot studies with the students, a tabular summary of the architects' tasks and roles were provided—these were divided into six building blocks of a smart city and as a reference to the elements shaping the image of cities, districts and buildings. This is a new, innovative classification of architectural issues. The perspectives for further desk research and field participatory research were indicated, which should, in the future, translate into a novel holistic approach to the problem.

Keywords: sustainable smart city; architect; image of the city; participatory design; body of the building; facades; roofs; built environment; design thinking method

1. Introduction

The sustainable development of cities is one of the key challenges of the modern world. There are authors who believe that smart cities are just an example of another concept of the ideal city [1,2]. It should be also noted that, in the literature, the concepts of sustainable city and smart city are sometimes investigated separately. Thomas L. Saaty and Pierfrancesco De Paola treat the sustainable city, smart city and compact city as separate models. According to the authors—taking into account the process of urban sprawl, the transformation of buildings and economic impact on the environment—the choice of a compact city model is the best solution for future urban design and planning [3]. Alessio Russo and Giuseppe T. Cirella point out that the features of a modern compact city support sustainable development [4]. In turn, Matthew E. Kahn claims that the improvements of information technology, and advances in the know-how on the reduction of pollution, and following it health benefits for residents imply that more cities will be striving for "smart" sustainable development [5]. Such a viewpoint is presented in the present paper.

A sustainable smart city is an interdisciplinary concept, and as such, it has many definitions, emphasizing various aspects and proposing different assessment indicators [6]. An attempt to systematize the concepts existing in that area was made by Mattias Höjer and Josefin Wangela [7], and [8], as well as Rasha F. Elgazzar and Rania F. El-Gazzar [9]. The latter of the aforementioned authors has been extensively discussing and clarifying the meaning of the words/concepts such as "smart", "sustainability", "sustainable development", and also the meaning of complex concepts such as "sustainable cities", "smart cities" and "smart sustainable cities". They also quote the definition of smart sustainable cities developed by the focus group on smart sustainable cities adopted in October 2015 by the International Telecommunication Union (ITU—T Study Group 5), which reads as follows: "A smart sustainable city is an innovative city that uses information and communication technologies (ICTs) and other means to improve the quality of life, efficiency of urban processes and services, as well as competitiveness ensuring at the same time that that it meets the needs of present and future generations with respect to economic, social, environmental as well as cultural aspects" [10].

The above definition represents a holistic character of the reflections on sustainable smart cities. In the discussion on such a concept of urban development, the following notions are taken into account: "sustainability, quality of life, urban aspects, and the main topics comprise: society, economy, environment, and governance" [6]. As it was mentioned above, the knowledge about sustainable smart cities was created by the representatives of various scientific disciplines, who often worked in interdisciplinary teams and describe the problem from various perspectives, such as: social [11,12], economic [13], information technology (IT) and telecommunications [14–16], environmental [17], health [18–21], legal [22], transport and mobility [23,24], urban and spatial planning [25,26], etc. It should be emphasized that the recurring feature of the publications is that they do not concentrate on the discussion of one problem, but they show it in a broader context (therefore, the above pairing of the authors with a given aspect only informs that a given topic occurs in a given publication). The market offer involving the segmentation of the smart city is dedicated to an even wider group of recipients, including:

- "local government officials: presidents and mayors of cities, village mayors,
- directors and heads of investment departments (roads, cubature investments),
- management staff of municipal companies (heating plants, combined heat and power plants, city cleaning, road management),
- representatives of the private sector (developers, investors, designers)" [27].

In the vast and diverse group of stakeholders, the architectural aspect, and the role of an architect in the development of sustainable smart cities, are getting lost. For example, in the above-mentioned, very comprehensive report [6] (pp. 6–7), the Table presents the definitions of "sustainable smart cities", assigning keywords to each of them. Further, there is no keyword of "architecture" among them. A very indirect reference can be seen in the definitions quoted below:

- "Hitachi's vision of a smart sustainable city seeks to show concern for the global environment and lifestyle safety and convenience through the coordination of infrastructure. Smart sustainable cities realized through the coordination of infrastructures consist of two infrastructure layers that support consumers' lifestyles together with the urban management infrastructure that links these together using information technology (IT)" [28];
- "Replacing the actual city infrastructures is often unrealistic in terms of cost and time. However, with recent advances in technology, we can infuse our existing infrastructures with new intelligence. By this, we mean digitizing and connecting our systems, so they can sense, analyze and integrate data, and respond intelligently to the needs of their jurisdictions. In short, we can revitalize them so they can become smarter and more efficient. In the process, cities can grow and sustain quality of life for their inhabitants" [29];

Furthermore, in the document of the Economic Commission for Europe Committee on Housing and Land Management [30] (pp. 7–9), from among the mentioned 72 smart sustainable city indicators, only the following ones have indirect reference to architecture:

- "Topic: Physical infrastructure—buildings;
- Indicator: Integrated management in public buildings" [30]
- "Topic: Environmental quality:
- Indicator: Perception on environmental quality
- Indicator: Green areas and public spaces" [30].

The problem of overlooking architectural issues when considering sustainable cities was noted by Emile Mardacana who pointed out that, "The definition of a smart city based on six key smart elements, including economy, management, people, science and technology, life and the environment, ignores such a basic component as the built environment, which is a physical platform of a smart city" [31]. To provide an example, we can refer to publication [32], in which each of the six pillars (building blocks of a smart city) was assigned from 10 to 22 features. In the presented set there was only one, very general reference to architecture, in the context of the latest research conducted by universities "for cultural heritage, architecture, planning, development, and the like" [32], (p. 13). Therefore, it should be noted that: "Architecture is all around us. From our homes to our offices, our stations to our skylines, the built environment defines the world we live in [33]". It should be added here that even the concept of "smart city architecture" has been taken over by computer science and it does not apply to the built environment, but to the structure of software [34]. In turn, in the literature on the subject of architecture and urban planning, even if there are problems at the intersection of architecture and sustainable development and smart city technology, the authors focus on environmental or technological aspects, and the reference to architectural aspects is missing. Architecture appears only in general keywords such as "smart buildings" and "urban infrastructure", but with no specific information, for example, how the idea of sustainable smart city translates into the form or facades of buildings. Furthermore, yet the environment of sustainable smart cities is designed by architects in terms of usability, but also in terms of form. Architects provide a physical form for the entire ecosystem of solutions created by smart and sustainable residential houses, smart and sustainable public utility buildings, which in turn create smart and sustainable housing districts [35]. In this way, the external image (appearance) of architecture is created, which in turn translates into the image of districts and cities together with their recreation and rest spaces furnished with small architecture. On this point, it should be noted that the literature on the subject includes a few publications referring to a more general level, i.e., to the urban form of smart cities [36]. Yet, the image at the architecture level, i.e. the form of buildings, facades, roofs, details and what is happening in the space around these buildings—in terms of the development of sustainable smart cities has not been a popular research topic so far.

This particular aspect - the external image of sustainable smart cities and the role of the architects in their creation—is discussed in the article. It is the main objective. Furthermore, the indirect objective of the article involves the synthesis of knowledge and an attempt to define the elements of sustainable smart cities, whereof a designing process is within the competence of the architect. Due to the extensive subject matter, the article provides a limited scope of analyses, attempting to define the elements that shape the image of a sustainable, senior friendly city, and to identify the role of an architect in this context. We must add that in line with the accepted assumption, an elderly-friendly city should be understood as one that meets the following criteria set by WHO:

- "it takes into account the diversity of older people,
- it prevents exclusion and promotes the contribution of seniors to all areas of life,
- it respects the choices, decisions and way of life of older people,
- it anticipates and flexibly responds to the needs of people growing old" [37].

The world and Polish resources of scientific publications contain works that combine the issues of smart cities and senior-related issues in the following aspects:

- security [37],
- intelligent technologies offering amenities for seniors [38],
- models of city management allowing for the role of people aged 65+ as a creative class [39],
- implemented amenities for seniors [40],
- a system for collecting and managing data on daily routines of seniors [41],
- problems of seniors in urban areas, mainly in terms of transport [42],
- technological skills and computer competences of seniors [43],
- very general recommendations, also in relation to the built environment [44].

At that stage of research, when searching through the resources of scientific publications published in the open access, as journals, books or conference proceedings, no scientific publication could be found that would combine the following aspects:

- issues of sustainable smart cities,
- senior issues,
- architectural issues.

It seems that this may be a new approach to research problems.

It should be clearly emphasized that due to the presumed novelty of the undertaken topic, the research is in the initial phase, referred to as the indicative level. It consists in observation and general review on the investigated topic, during which we can identify some dependencies, but without diagnosing their origin. The rise to higher levels, i.e.;

- investigative (application of scientific methods, an attempt to explain the problem and its cause) and;
- diagnostic ones (comparative tests that diagnose and indicate recurring problems that should be eliminated) [45],

requires in-depth research using scientific methods, techniques and research tools.

Thus, the paper:

- sums up only a certain scope of knowledge, recommending that further desk researches should be carried out,
- shows a sample of new research, with the participation of students of the Faculty of Architecture of the Silesian University of Technology,
- quotes research studies conducted earlier in which students of the Faculty of Architecture of the Silesian University of Technology also participated,
- indicates further research directions, placing them in Tables related to six building blocks of a smart city, which seems to be an unprecedented form of classification of architectural problems.

As to the scope of the work, it should be added that due to the scope of the undertaken topic and that of the article, we had to omit some relevant and interesting issues, such as:

- examples of smart and sustainable buildings, districts and cities friendly to seniors worldwide and in Poland (a catalog of good practices),
- various types of reports, guides, guidebooks, guidelines for the implemented innovations in the senior-friendly built environment,
- online platforms dedicated to the subject of smart cities, constituting the basis of current information on the programs being implemented,
- social (e.g., exclusion, isolation) and medical context (various forms of physical and intellectual disability of seniors) that can affect the perception of the architectural environment.

These issues will be the subject of further research and publications.

2. Materials and Methods

Nowadays, architects propose solutions not found before and they use technological novelties to create innovative building concepts. It is the first of the architects' roles: creating innovative building concepts which make use of technological achievements and presenting them in attractive visualizations and films used to promote the idea of sustainable smart cities.

Of course, only a few architects become famous innovators and visionaries. Despite this, a lot of emphasis is put on the development of creativity in the education process of architects at the Silesian University of Technology in Gliwice (Poland) [46]. As an example, we can provide a study conducted by the author of the article with second-cycle students at the Faculty of Architecture. The topic was formulated as follows: "How will seniors be participating in the life of a future smart city (in the perspective of 10 years) and what amenities will it be offering them?" It combined two very important issues, i.e., the problem of designing a friendly space for seniors and the problem of a smart city.

The layout of the research work is presented in Figure 1. It shows the duality of the research approach, which has been also demonstrated in the structure of the article, i.e., the "gray path" (on the left) presents the pilot study with the participation of students, and the "blue" path (on the right) presents the expert research documented by the author's own publications, which are listed in the bibliography. Both sources combine to obtain the synthesis of knowledge.

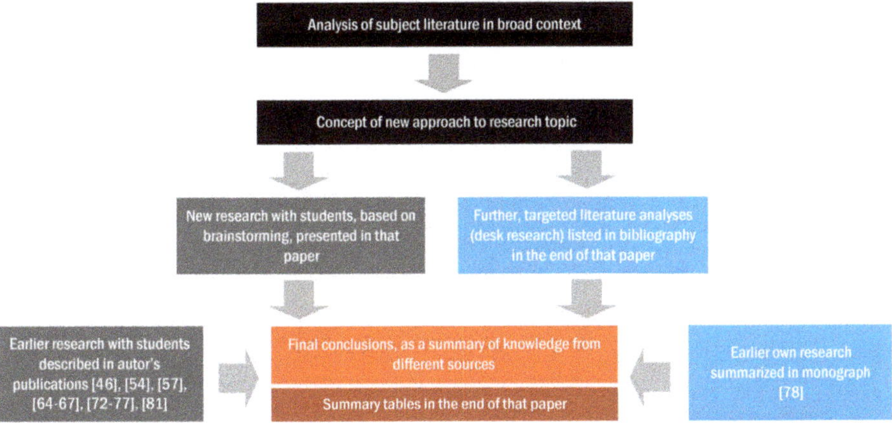

Figure 1. Layout of research procedures used in the article. It refers to the indicative level and illustrates the author's pursuit of the synthesis of knowledge and holistic approach to the investigated problem, i.e., "the role of the architect in creating the image of a sustainable smart city friendly to seniors" (author's own study).

The study was conducted in two groups of 12 students, which were then divided into teams of 4. In total, 24 students divided into 6 teams took part in the study. It should be noted that the first group had in-depth knowledge of the needs of the elderly, acquired during the previous research studies. This research concerned a different topic and was conducted under the guidance of Anna Szewczenko [47], associated professor an expert in the field of architectural solutions friendly to seniors. In the course of the study, the group developed, among others, "personae" or characteristics of fictitious people of senior age, who were given names, their health condition was described, scenarios of their functioning and spending free time was conceived, including passions and interests, as well as their social roles, fears and worries. The "personae" characterized in this way were used in the further part of the study described here. The second group of students had only general knowledge in the field of senior related issues.

The scientific objective of the research was to find out the vision of future architects about the life of seniors in a smart city of the future in the perspective of 10 years, and to collect inspiring, innovative ideas. The study had also a didactic goal—stimulating students' creativity, familiarizing them with brainstorming techniques and sensitizing future architects to the problems of older people in cities of the future. The study was conducted using brainstorming, which is an important element of the design thinking method. According to the definition of Willemien Visser: "Design Thinking refers to design-specific cognitive activities that designers apply during the process of designing" [48]. Items explaining design thinking principles are presented in the following publications:

- Brown, T. "Change by Design. How Design Thinking Transforms Organizations and Inspires Innovation" [49],

- Brown, T., "Design thinking" [50],
- Eleutheriou, V.; Depiné, Á.; de Azevedo, I.; Teixeira, C. "Smart Cities and Design Thinking: sustainable development from the citizen's perspective" [51],
- Thoring, K., Müller, R. M., "Understanding Design Thinking: a process model based on method engineering" [52],

Examples involving the application of Design Thinking in the discipline of architecture and urban planning, are presented in the following articles:

- Stangel, M., Witeczek, A., "Design thinking and role-playing in education on brownfields regeneration. Experiences from Polish-Czech cooperation" [53];
- Tymkiewicz, J.; Bielak-Zasadzka, M. "The design thinking method in architectural design, particularly for designing senior homes" [54];
- Stangel, M., Szóstek, A., "Empowering citizens through participatory design: a case study of Mstów, Poland" [55].

The brainstorming was carried out with a classic division into two stages (as shown in Table 1):

- in the first stage, the students freely submitted ideas and wrote them down on post-it notes (with the provision that there were no restrictions on self-expression and no criticism);
- in the second stage, the students looked at the results written on the notes, discussed them and tried to select ideas that in their opinion suited them best.

The collection of ideas for each of the partial topics took five minutes. Ideas were written on post-it notes and stuck on boards assigned to each team.

During the brainstorming the students were to imagine that in 10 years, cities would change, becoming smart cities. There was an auxiliary research question related to this was: "What facilities for seniors represented by the three "personae" should be found in the smart city in the future?" It was clearly emphasized that the smart city had evolved since the creation of that concept—from smart city 1.0 focused on technology to human smart city 4.0 focused on people [56]. Importantly, the problem was to be considered by the students from the architect's perspective.

The diagram of the research methodology (Figure 2) and Table 1 presents the scenario of the research below.

Table 1. Research scenario with the participation of students of the Faculty of Architecture of the Silesian University of Technology involving the following research problem: "How will seniors be participating in the life of a future smart city (in the perspective of 10 years) and what amenities will it be offering them?".

	Introduction to research	Duration
organizational activities	• selection of team leaders and compiling the students into particular groups; • explaining the course of research;	2 min
information explaining the undertaken research problem	• reading out the definition of a smart city by the leader and a brief explanation of its development stages; • reading out the characteristics of the three "personae" prepared earlier by students—potential residents / users of a smart city representing the following age ranges: 60–65 years, 71–75 years and 81–85 years; • short discussion, answers to questions;	10 min

Table 1. *Cont.*

Introduction to research		Duration
Brainstorming part 1–unrestrained presentation of ideas		
brainstorming problem_1	• urban information,	
brainstorming problem_2	• communication and transport - public and individual,	
brainstorming problem_3	• forms of residential housing,	
brainstorming problem_4	• accessibility of buildings,	
brainstorming problem_5	• availability of services,	55 min in total
brainstorming problem_6	• medical services (as city-wide service),	
brainstorming problem_7	• ensuring safety,	
brainstorming problem_8	• recreation and free time,	
brainstorming problem_9	• new urban functions,	
brainstorming problem_10	• ecology and sustainable development,	
brainstorming problem_11	• aesthetics and the appearance of buildings and urban space,	
Brainstorming part 2: summary and conclusions.		
analysis of ideas	• brief discussion and the evaluation of ideas,	
	• selection of the best solutions,	10 min
	• summary and conclusions.	

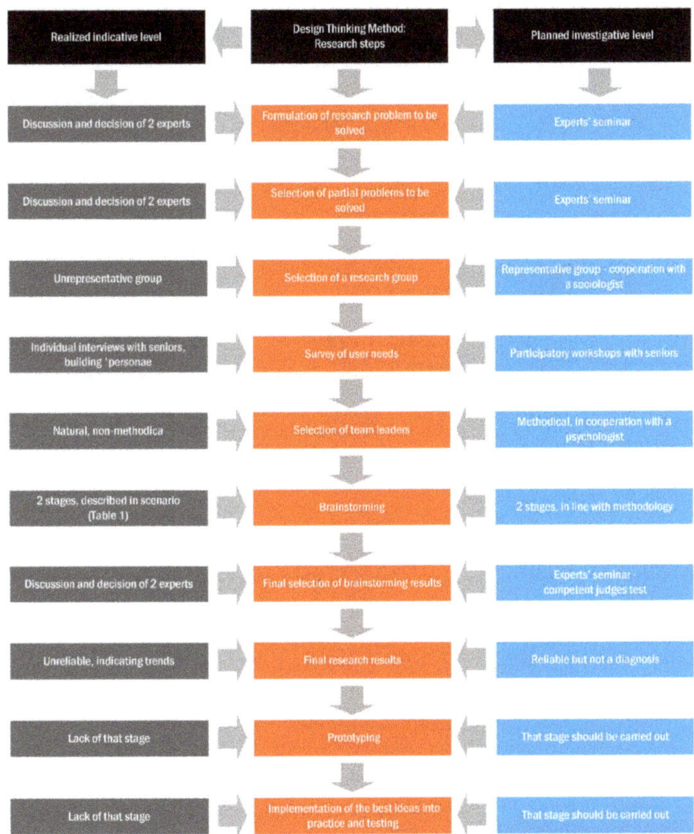

Figure 2. Diagram of the research methodology presents the design thinking method step by step (red path). Additional explanations are presented on left and right side. On the right (grey path) you can see how research activities have been realized on indicative level, and on the right (blue path)—how it should be conducted on investigative level, when the full design thinking methodology in cooperation with sociologist and psychologist should be applied.

3. Results

As presented in Figure 2, the research at the indicative level was conducted on an unrepresentative group, and therefore its results could not be generalized, as they refer only to the opinion of the group participating in the research. In the future, the study can be repeated in cooperation with sociologists, which will allow to select a methodologically correct research sample and to make the generalization of results. It seems that brainstorming in terms of the number of generated ideas is an effective method [46,57]. In the study, 24 people submitted 348 ideas on a given topic within 55 min. The team working most effectively in terms of quantity generated 70 ideas, and the smallest number of ideas submitted by a 4-person team was 42. The situation was different when assessing to the quality of the generated ideas. The ideas which satisfied the key condition of the research, i.e., a reference to architecture, could be assessed as rather conservative, and replicating what is known from reality. They are quoted and discussed below.

3.1. Results of Research

The results of the research will be discussed merely as some tendencies in the formulation of answers to the stated research problem.

3.1.1. Urban Information

With regard to urban information, the students proposed "information points as urban accents in stopover places; horizontal signs—arrows, changing surface textures for the blind; simple, legible and consistent visual messages; maps of the city with our location; city districts identified by colors; avoiding excessive visual information on digital displays; avoiding sound reverberation, echo, understandable information transfer (in relation to hearing)"; as part of the "smart" solutions, the idea of "a communication assistant: senior robot" was proposed.

3.1.2. Communication and Transport-Public and Individual

In the field of communication and public or individual transport, the following was proposed: "roofed, lit, closable stops (protection against wind), with ergonomic seats and places for shopping bags, walking sticks, crutches, walkers, pets; moving walkways in popular public places, or quiet traffic paths separated from faster traffic paths (separating people being in a hurry from people who walk)"; in the area of "smart" solutions, the idea of "drones distributing mail or medicines ordered from the pharmacy" was proposed.

3.1.3. Forms of Residential Housing

The proposed forms of housing for seniors in a smart city are identical to the already existing solutions: "old age homes connected to an orphanage or kindergarten (preventing loneliness), multi-generational housing estates; ground floors for seniors, multifunctional facilities, a residential complex containing services, medical and entertainment functions; community centers, gardens for common cultivation; common areas; places encouraging intergenerational integration; all flats should be flexible (for able-bodied and disabled people); color marking of buildings-better orientation; elevators, ramps in every building; senior cohousing"; "smart" solutions comprise: "fingerprint-activated door opening; flats equipped with buttons to activate assistance (neighborly, medical)".

3.1.4. Accessibility of Buildings

In the field of building accessibility, students reported conservative ideas such as: "no architectural barriers; clear marking of entrances; designing kitchen systems which are mobile and adapted for people in wheelchairs (e.g., the possibility of rolling the wheelchair under the worktop)", and more audacious ones: "antigravity domes, teleporting; mobile buildings".

3.1.5. Availability of Services

The availability of services would be ensured by: "a large number of benches in the infrastructure of public space; stops along the routes of seniors; appropriate functional zoning; creating interactive community work centers for seniors (earning money); creating special senior zones in stores (lower shelves, wider passageways); activation of seniors with children and with animals"; the ideas for "smart" solutions include: "mobile shops in housing estates for the elderly; shopping via mobile applications with home delivery; automatic shopping assistant (a robot); drones for seniors (a shop); self-cooking and self-cleaning robots; printing of clothes".

3.1.6. Medical Services (as City-Wide Services)

With regard to medical services, the students indicated a number of solutions which are currently regarded as technological novelties, namely: "a symptoms-interpreting machine (in bed); artificial intelligence at home as a physician; a hologram and interactive meetings with specialist doctors; individual medical scanners". The students did not in any way refer those technological proposals to the solutions in the field of architectural space in which they would be applied.

3.1.7. Ensuring Safety

The ideas offered by the students on security issues for seniors in a smart city were as follows: "urban lighting with a motion sensor; robotic guards on the street, transhumanism and technological implants to improve the fitness of seniors". The students overlooked the issue of spatial morphology, which has a significant impact on safety. It is not only about the visibility of areas for surveillance cameras, but it also involves the elimination of places in terms of spaces which are conducive to crime [58,59]. Recently, the problem of counteracting terrorism has been added, which is also reflected in urban space projects [60]. This problem was not perceived by the students.

3.1.8. Recreation and Free Time

In the next subtopic "recreation and free time", there were repeated proposals that had been made earlier-regarding the activation and integration of seniors into society (social gardens, activities with children), and only the following was added: "prestigious loges for seniors in cultural institutions; attractive meeting centers for seniors; a larger number of senior architecture points, e.g., graduation towers"; the "smart" solutions involved: "VR (Virtual reality) walking, safe boxes where walking is possible in optimal conditions (temperature, humidity, wherever you want)".

3.1.9. New Urban Functions

With respect to new urban functions, the students once again proposed traditional solutions, some of which are already known: "city markets, where you can sell your products (fruit, vegetables, crafts, art); milk bars (places with cheap food); community and intergenerational cafes and workshops; training places for the elderly in the field of interactive services (IT); intergenerational dialogue centers; spaces for exchanging skills and passing on traditions; drama series cinemas for seniors; mutual reading parlors (for the poor sighted)"; the "smart" ideas included: "modern agora: a random selection of topics related to real problems, and brainstorming ideas for their solutions; Flyspot salons (wind tunnels) for people with mobility difficulties—stimulation of movement".

3.1.10. Ecology and Sustainable Development

In the described study, ecology and sustainable development were a separate partial problem. The submitted ideas contained individual references to important problems, such as: energy ("cheap, ecological heating, energy sources; power stations for seniors for electricity/energy generation; personal, garden, public solar umbrellas; solar clothing"), pollution ("cleaning air in every house"); waste ("self-segregation of wastes; biodegradable packaging; eco-medicines"); green infrastructure ("gardens in housing estates—mini vegetable farms, orchards; sensory gardens for the activation of seniors; less paved surfaces and more greenery"). Yet, there is a feeling of paucity about the above ideas, both in terms of quantity and the lack of more inquisitive insight into the problem in the context of architecture.

3.1.11. Aesthetics and the Appearance of Buildings and Urban Space

In regards to the aesthetics and appearance of buildings and urban space in a smart city friendly to seniors, the students attached great importance to wayfinding issues, proposing such solutions as: "urban functions-characteristic and easy to recognize; automatic signaling informing about the location of, e.g., the entrance zone; cleaning up visual information, in particular from advertising; clear description of streets and districts; identification of individual residential buildings-color-coded staircases and building entrances; murals—characteristic and acceptable". They also reported the need "to keep the spaces open, to maintain private and semi-private areas; to use calm colors; bright, well-lit rooms; natural materials, cozy interiors and balconies, and to use balustrades that give a sense of security, not fear." All in all, the ideas were very conservative, already known and rather ordinary. More innovative and "smart" thinking was represented by such ideas as: "personalized building facade-automatic signaling sensors; goggles-lenses that allow you to adjust the style of the facade to your taste; paint that changes color; changing eco-facades made of nanorobots".

4. Discussion

When analyzing the research results, we could observe that it was quite difficult for students to relate the problems of designing a smart city friendly to seniors to architecture. It seems that it was also difficult for them to empathize with the role of an architect in this aspect. There were no differences in the quality of the submitted ideas in any of the groups (the level was very even). The image of the city of the future emerging from the students' brainstorms was not futuristic, but real, while the proposed "smart" solutions were frequently not associated with architecture. There were several inspiring ideas, but also ideas that dangerously approached the exclusion or stigmatization of seniors as a social group (separate pavements, separate zones in shops, robotic guards on the streets). On the positive side, it should be noted that students pay attention to extremely important issues: ergonomics, accessibility of buildings and space for everyone, wayfinding and the need for places of social integration. Yet, there were no proposals to solve such important issues as: climate change and what it means for architecture. In addition, the problems of environmental pollution or waste utilization were treated marginally, although they can be very creatively related to architectural design. As an example, we can refer to the heat and power waste-to-energy plant building in Amager, Copenhagen (Denmark), designed by the Bjarke Ingels Group architectural office. This facility proves that innovations in the field of architecture are possible. In the film "BIG Architecture" [61], Bjarke Ingels passionately described how the concept of this—in principle, very utilitarian building—was being developed. It was known that it would be a large cubature, towering over the surroundings. Then, during the brainstorming, the idea of a "mountain building" with a ski slope was proposed. In this way, an incinerator was created, producing electricity and energy, producing more water than it consumes, with a sloping roof, on which a year-round "artificial ski slope, hiking slope and climbing wall" were created [62]. The problem of the chimney, which had to be placed on the body of the building, was also attempted to be solved from the architect's point of view: it was supposed to emit smoke (because it could not be eliminated), but in the shape of circles (smoke rings) [63]. In that way, the most desirable features in the contemporary architecture had been combined: innovation for the benefit of the environment and people. Such innovative ideas directly relating to architectural design, relating to buildings: to their outer shells, roofs, facades, functions and needs of the natural environment and people were definitely missing among the ideas submitted by the students.

When analyzing the above results of the study, it is worth referring to two earlier studies conducted by the author with the participation of the students of the Faculty of Architecture. The first study concerned the notion of students about retirement homes for seniors in the future (but more distant future, when they are seniors themselves). In this case, the full design thinking methodology was used with prototyping in various ways. The results of the research were more satisfying in terms of the innovation of the proposed solutions. The description of the research and the results were published in the articles of Joanna Tymkiewicz and Maria Bielak-Zasadzka [54,64].

The second of the related studies only indirectly concerned the subject of this article, i.e., the image of sustainable smart cities and the role of the architect in shaping them. In this case, students - young architects faced the following research question: "Would everyone want to live in futuristic-looking buildings?" Young people probably would, but would seniors too? Such doubts arose from the conclusions of the pilot studies regarding, "The evaluation of facade solutions and their impact on the aesthetic quality of the external image of buildings" (2016). The multithreaded scientific and didactic project was realized in cooperation of the Faculty of Architecture with the University of the Third Age in Gliwice and the Laboratory of Architecture 60+ Foundation (Lab 60+). As part of the research, a focus meeting was arranged with seniors who were asked to express their opinions on various aesthetic solutions for the facades of residential buildings intended for the elderly. The examples of such buildings were presented on large photographs and slides. The course of the research and the conclusions have been described in the publications [65,66]. It turned out that seniors criticized the façades of buildings developed in an avant-garde way. Of course, the research was not conducted on a representative group and the results are not authoritative, but for reference, it is worthwhile quoting some examples of seniors' opinions about the highly rated by architects Home for Elderly People in Alcácer do Sal, Portugal (Aires Mateus Arquitectos, 2010), the finalist of the "European Union Prize for Contemporary Architecture—Mies van der Rohe Award 2013". They read as follows: "it looks good in the picture, but it is not for living", "cubist shape, too modern", "an unsightly box, a block" [66]. In the quoted studies—in the further creative part of the workshop with seniors, some assumptions regarding the aesthetics and functionality of façade solutions had been developed. They favored the tendencies advocating "moderation, traditional aesthetic solutions, friendly scale of buildings, natural materials". Favorite elements of the façade comprised "large, shared terraces with the opportunity of socializing, intimate balconies, fenced off from one another, but in a way ensuring contact with neighbors" [66]. These completely preliminary guidelines inspire further, in-depth research on this topic and allow us to conclude that if cities of the future are to be focused on people and their needs, and seniors are becoming an increasing part of the society, facade solutions must be consulted with them, and the public should be educated to enhance the understanding and acceptance of innovative, novel and avant-garde solutions—also in terms of aesthetics. Here, architects have a lot of scope to act: as educators, convincing to new architectural solutions and as the liaison between various groups of stakeholders (e.g., between investors and users).

This role of architects is associated with another one—initiating and conducting pre-project participatory research in which representatives of future, potential users would take part. An example of this type of research carried out by the author with the participation of students can be illustrated by the venture, "The Experimental Project on 'Soft' Intervention Aimed at Enlivening the Academic Zone" (2017). The said research project aimed to gather ideas from the academic community to revive the newly designed campus space of the Silesian University of Technology in Gliwice (Figure 3). The research was described and published in the article [67]. We can also add here that among many offered proposals, there were also ones that would certainly improve the image of the academic district in terms of space characterized by the "sustainable" and "smart" features which take into account the needs of people with disabilities. In the project, one of the student teams proposed for the academic district a network of elements improving the functionality and accessibility of the campus using the so-called small architecture and information technology. The network of connections would consist of: pylons with interactive information, an application, typhlographic signs, i.e., tactile information for the blind and a 3D campus model, as well as multifunctional boxes. The boxes would offer: "a charging station for electric carts/wheelchairs, wireless charging of mobile phones, access to electricity from renewable sources, access to Wi-Fi and to academic library resources, a working station for a student, a place for books, available through the application". The box would be equipped with "devices used to read the resources of digital library, e.g., a tablet, e-book reader, and it would ensure access to the equipment for listening to music files, as well as to the text-to-speech software. The information would be also written in Braille." Such solutions would facilitate the use of the boxes for the visually impaired and hearing impaired. The elements of the concept were presented in the article [67], (p. 29). In that case, the students-future architects played a triple role: researchers, potential users and designers. The presented student concept fits into a frequently adopted strategy which says that smart solutions should be first implemented in a city quarter or district, and not in the whole city. In this aspect, the academic campus is a very good place to initiate smart solutions, with the care for sustainable development [68].

Figure 3. Fragment of the central part of the space of the academic campus of the Silesian University of Technology in Gliwice (Upper Silesia, Poland), as of today (photograph taken by J. Tymkiewicz).

In this article, we have already mentioned the architects' role in designing the image of elderly-friendly sustainable smart cities. These are roles which are located on two opposite poles and which define two attitudes of the architect: as an innovator-visionary (1) as well as a researcher and educator (2). The tasks reflecting these two attitudes are as follows:

1. setting the direction of development, creating new ideas, visions, innovations in the field of the architecture of sustainable smart cities that are not always possible to implement, using the achievements of science and technology as well as digital technologies (architect innovator-visionary);

2. initiating and carrying out pre-project participatory research allowing to diagnose the real needs of users, and developing guidelines for design works, which take into account such needs; conducting research on buildings and urban spaces which expand knowledge on their functioning (architect-researcher); educating the society (including senior citizens) on new solutions (architect-educator).

The above roles should be supplemented with the most important one, being in the middle and combining both of the above roles, namely: designing the image of cities which implement the idea of elderly-friendly sustainable smart city in real conditions. Cities are currently struggling with the problems mentioned at the beginning of this article, i.e., the increasing number of inhabitants, including seniors, energy consumption and environmental pollution, and the consequences of climate change. Cities have also the potential contained in the development of digital technologies that allow them to diagnose and to some extent solve the mentioned problems. Architects, through their designs and implementations supply a form for the conceptual solutions pertaining to the above problems. They also create a physical space for virtual digital technologies (e.g., for various types of applications) that are supposed to make life easier for users (including the elderly) living in the real built environment. The number of applications proposed for residents is constantly increasing. A review of the European classification of smart city applications has been presented in the article [69]. In addition, in Polish cities various types of "smart" and "sustainable" as well as "pro-senior" applications have already been implemented as, e.g., "active senior card", mobile navigation for visually impaired people, and a smart public transport system [70,71].

There are many more implemented solutions, but is the real world, or the built environment, following the development of the virtual world? The above systems and applications operate in city spaces representing different qualities. On the one hand, these are organized, architecturally valuable, historical or modern city centers. Their attractive image is consistent with the image of a city aspiring to the title of "smart" and "sustainable". But in the same city there are also neglected, devastated districts, or developing districts-but in a chaotic way. Even if there are applications or other smart facilities operating in such places, it is difficult to feel the idea of sustainable smart city. There is an unpleasant gap between technological advancement in the virtual world and low quality of the real world. What is missing is the appropriate quality of the built environment mentioned at the beginning, as a physical platform for implementing the idea of smart city in a sustainable way. There is also no deeper reflection on the design projects being implemented in the field of "smart" and "sustainable" and no reference to architecture.

It seems that an architect-designer should not only use elements of the language of global architecture of sustainable smart cities (listed in Chapter 2), but also take into account real problems of cities (including local problems), buildings and the needs of their users, as well as legal regulations and economic potential. It is too early to make conclusions and summarize the discussed topic as well as to start discussion with other researchers. More research is needed, but now we can at least create a list of tasks related to the role of the architect in shaping the image of a sustainable smart city. They comprise such elements: outer shells/bodies of buildings, facades with details, roofs, and urban spaces in their vicinity (such as: recreational areas, squares, etc.). The diagnosed tasks and roles have been presented in a tabular form, taking into account the division into six building blocks of a smart city (the division has been taken from the already cited publication [32] (pp. 12–16)). It should be added that the partial problems presented below have in many cases been already investigated, but only as general problems, without reference to the issues of designing sustainable smart cities friendly to seniors. For some of them—examples of bibliographic references were given as the nucleus of desk studies. A broader approach requires further literature review and in-depth analyses. At this stage, we could state that knowledge is dispersed and there is a need to consolidate it.

The photographs below are presenting facades and public spaces as the architectural elements of image of Gdańsk—the city in Poland, which develops the idea of an elderly-friendly sustainable smart city (Figures 4–8).

Figure 4. Gdańsk—Museum of the Second World War, main façade (photograph taken by J. Tymkiewicz).

Figure 5. Gdańsk—Old Town District (photograph taken by J. Tymkiewicz).

Figure 6. Gdańsk—modern facades on the banks of the River Motława (photo by J. Tymkiewicz).

Figure 7. View of public space of the "Forum Gdańsk" (photograph taken by J. Tymkiewicz).

Figure 8. Modern public space of the "Forum Gdańsk" (photo by J. Tymkiewicz).

The following tables (Tables 2–7) contain the content selected and formulated by the author, supported by the author's experience of research carried out for many years on the functions of facade in the holistic aspect, summarized, among others, in the conference proceedings [72,73], articles [74–77], and in the monograph [78]. The tables contain also—students' inspiring ideas for the author, voiced during the brainstorming described in this article.

It was the author's intention that the tables presented below—ordering architectural issues and presenting architects' roles in a new way, analogously to the issue of sustainable smart cities—could become a starting point to undertake research at the investigative and diagnostic levels in this area, focused on the needs of the residents of sustainable smart cities, especially senior citizens.

Table 2. "Smart people"—in the context of the architect's role in designing the image of a sustainable smart city (author's study).

Tasks in designing the image of a sustainable smart city	Role of architect
Identifying users' needs prior to the designing work (pre-design studies);	researcher
Consulting architectural projects with the local community in terms of the body and facades of buildings, and offering conclusions that are taken into account in the design process (participatory design);	researcher, liaison between different stakeholder groups
Educating the society and persuading to new innovative solutions so that they are understood and accepted;	educator
Creating attractive places in the city (urban spaces and / or buildings) in the "human scale", having good proportions, serving as places of social integration, centers of creativity and activity of residents, where meetings, workshops, brainstorming can take place, for example: "city markets, where you can sell your products (fruit, vegetables, crafts, art); milk bars (places with cheap food); community and intergenerational cafes and workshops; training places for the elderly in the field of interactive services (IT); intergenerational dialogue centers; spaces for exchanging skills and passing on traditions; drama series cinemas for seniors; mutual reading parlors (for the poor sighted)"*;	designer
Respecting the opinions of residents and using the ideas and innovative solutions they propose in the designing process;	designer
Integrating universities with the city and designing them in such a way so that the buildings evoke positive emotions, so that they do not intimidate, but in a friendly way invite residents to want to stay there and learn throughout their lives;	educator
Designing in the way ensuring that architectural solutions prevent the exclusion of specific social groups and are friendly to all user groups (design for all, "no architectural barriers; clear marking of entrances", "automatic signaling informing about the location of, e.g., the entrance zone" *, universal design);	designer

* The ideas generated during the brainstorming described in the article.

Table 3. "Smart economy" in terms of the role of the architect in designing the image of a sustainable smart city (author's study).

Tasks in designing the image of a sustainable smart city	Role of architect
Designing visually attractive, avant-garde buildings that have the potential to become tourist attractions affecting the economic development of the city (Bilbao effect);	visionary and designer
Designing buildings which have the shapes which exclude natural ventilation, air-conditioned, with windows which do not open /facades only in places where it is economically justified (buildings maintained from the budget often have problems, because the costs of air conditioning are too large for them) [78,79];	designer and researcher
Designing architecturally impressive double-skin facades only in places where it is economically justified and the facade can be an efficiently functioning element of the building's ventilation system which does not generate substantial costs;	designer and researcher

Table 3. *Cont.*

Tasks in designing the image of a sustainable smart city	Role of architect
Designing facades as one of the building layers, with shorter durability than the structure (building life cycle analysis), and thus easily replaceable after the period of technical or aesthetic wear [80];	designer
Designing façades with the appropriate proportions of glazed surfaces, thanks to which natural lighting of the interior will be provided, without additional costs of illuminating it with artificial light during the day;	designer
Incorporation of sunshade systems (the most effective are external ones) into the architectural design of the facade, which can reduce the costs of air conditioning of the interior and improve the comfort of users [78,81];	designer
Proposing "green facades" and "green roofs" only in places where the owner can afford the costs of their maintenance (infrastructure, gardening services, water, fertilizers, etc.);	designer
In the design of greenery, choosing native species that grow and develop well in the climate and conditions prevailing in the area and require minimal irrigation and fertilization (which generates additional costs) [4];	designer
In the design of facades, taking into account the economic aspects of electricity consumption in night light scenography and by media facades [82];	designer

Table 4. "Smart mobility" in terms of the role of the architect in designing the image of a sustainable smart city (author's study).

Tasks in designing the image of a sustainable smart city	Role of architect
Designing buildings and building complexes with good proportions, in human scale, harmoniously fitting into the existing architectural and urban context, with interesting facades, respecting cultural values, emphasizing the identity of the place and genius loci ("Smart heritage" [83], space morphology [4]), with active ground floor services ("ground floors for seniors" *), roofed arcades—which encourages walking and supports the mobility of residents (pedestrian-friendly cities);	designer
Integrating the city's green infrastructure into a network of connections, which facilitates the creation of interesting walking and cycling routes, not only in green areas, but also along architecturally attractive frontages, or through interesting urban interiors;	designer
Designing facades—hallmarks, city icons that support orientation in space (wayfinding);	visionary, designer
Designing buildings and safe spaces, well-lit and having a form which supports their monitoring and security, but also ensures privacy (where it is desirable), with a friendly appearance that encourages residents to walk [84,85];	designer
Interesting design of the space around stops, stations, underground passages, good graphic signage (also for blind or visually impaired people)—as elements supporting pedestrian traffic; "roofed, lit, closable stops (protection against wind), with ergonomic seats and places for shopping bags, walking sticks, crutches, walkers, pets""cleaning up visual information, in particular from advertising; clear description of streets and districts" *;	designer

* The ideas generated during the brainstorming described in the article.

Table 5. "Smart environment" in terms of the role of the architect in designing the image of a sustainable smart city (author's study).

Tasks in designing the image of a sustainable smart city	Role of architect
Parametric designing, which takes into account various factors affecting the shape of building body and facades, such as wind (aerodynamics of the body, alleviation of drafts, but also ensuring good ventilation of building interiors and urban interiors), sun (insolation, shading, overheating of interiors, glare caused by reflected light), noise (specific forms of the façade can strengthen or weaken sound waves), snow (build-up on roofs and sloping facades) [86,87];	designer
Taking into account a place for the infrastructure related to new technologies in the architectural design so that it does not disturb the aesthetics of the facade or roof;	designer

Table 5. *Cont.*

Tasks in designing the image of a sustainable smart city	Role of architect
The use of ecological materials, but taking into account all aspects of the problem, including production, transport, durability and disposal;	designer
Integrating renewable energy sources with building facades (photovoltaic cells are more real, and wind turbines are less real) and educating the public in this respect;	designer, educator
Designing the arrangement of solar panels and/or solar collectors on roofs in such a way that they do not spoil the aesthetics of the building (e.g., additional mounting frames protruding above the roofs);	designer

Table 5. *Cont.*

Tasks in designing the image of a sustainable smart city	Role of architect
Designing greenery on facades—including: -local atmospheric conditions (in Polish conditions, these are not technologically advanced vertical gardens like Patrica Blanca's designs, but rather climbing plants resistant to weather conditions), -real advantages (CO_2 absorption and oxygen production, absorption of pollutants), -and disadvantages (moisture retention and possible destruction of wall surfaces if climbing plants do not climb along suitable support frames);	designer
Designing vegetable or herb gardens on roofs, or separate buildings adapted for cultivation, so-called urban farming; designing "gardens in housing estates—mini vegetable farms, orchards; sensory gardens for the activation of seniors" *	designer, visionary
Preventing the formation of "urban heat islands"—wherever possible, preservation of the existing plant cover in design projects: lawns, trees and shrubs; "a larger number of senior architecture points, e.g. graduation towers"*	designer
In the design of greenery, careful selection of plant species, taking into account the local ecosystem and the danger of its disturbance by the introduction of new, invasive species (currently a fashionable trend is to give up mowing grass in cities, which saves energy, reduces the amount of exhaust from mowers, allows to create an "urban meadow");	researcher, designer
In the design projects, taking into account underground waste collection and segregation systems which do not spoil the aesthetics of the city;	designer

* The ideas generated during the brainstorming described in the article.

Table 6. "Smart living" in terms of the architect's role in designing the image of a sustainable smart city (author's study).

Tasks in designing the image of a sustainable smart city	Role of architect
Designing so that people, their needs, health, well-being and satisfaction from being in the built environment are always in the center; architecture (also through its image) should support human development and people's activity throughout their entire life;	researcher, designer
In the design of green infrastructure of the city, taking into account the impact of plants on human health; the positive aspect is represented by medicinal plants, herb gardens, therapeutic horticulture, sensory gardens; negative impact involves pollen shedding harmful to allergy sufferers, the presence of insects dangerous for health (ticks, mosquitoes) [4];	designer
Including elements liked by users in facade designs, such as terraces and balconies, properly protected from the sun, ensuring very desirable privacy to residents (properly selected distances between balconies, and balcony covers), but also providing a visual connection between the interior and the outside [81];	designer

Table 6. *Cont.*

Tasks in designing the image of a sustainable smart city	Role of architect
Conducting pre-project studies—research on the perception of the facade by residents, taking into account the impact of new technologies and social media [88]—the assessment of what facade solutions (detail, color, texture, composition of facade elements) are accepted by the residents and best express the ambition, development and creativity of the local community;	researcher, designer
Establishing cooperation with artists, and integrating art with architecture (e.g., murals—"characteristic and acceptable" *, permanent or temporary installations) sending out a message: this place is inhabited by creative people;	designer, liaison between different stakeholder groups
Including a graphic information system in the facade design that is friendly to the elderly and to people with disabilities;"information points as urban accents in stopover places; horizontal signs—arrows, changing surface textures for the blind; simple, legible and consistent visual messages; maps of the city with our present location; city districts identified by colors; avoiding excessive visual information on digital displays"*	designer
Designing media facades, taking into account the nuisance that they can generate for nearby residents (pulsating light);	designer
The presence of the infrastructure in the design projects related to the availability of buildings and space (lifts, ramps, ground floors accessible from ground level), building entrances easy to find, panes (in shop windows or doors) properly marked to avoid collisions; compliance with the principles of universal design;	designer

* The ideas generated during the brainstorming described in the article.

Table 7. "Smart governance" in terms of the role of the architect in designing the image of a sustainable smart city (author's study).

Tasks in designing the image of a sustainable smart city	Role of architect
Undertaking participatory research to define the features of the building's form and facade (e.g., a government building), which can evoke desired associations that reflect the idea of a sustainable smart city, e.g., intelligence, respect for nature, ecology, modernity, creativity, efficient management, democracy, openness to citizens;	researcher and designer
Designing facades that correctly inform about the purpose of the building (denoting the function of the object), evoking positive emotions and connotations [89]; "urban functions—characteristic and easy to recognize"*;	designer
Anticipating advertising space and city information displays in façade designs—not disturbing the city's aesthetics;	designer

*The idea generated during the brainstorming described in the article.

5. Conclusions

The above elements presented in the tables are not, "in principle", smart, but they can successfully create a "physical" sustainable built environment, oriented on human needs (including the needs of seniors), which can be entwined by a network of "virtual" connections, offering various "smart" facilities. Such a more "analogue" approach should prevent the exclusion of some social groups, e.g., seniors due to their lower proficiency in the digital world, or poorer citizens—due to insufficient access to new technologies (e.g., smartphones). The most important role of an architect in this context is:

- the role of a designer,
- the role of a visionary
- the role of a researcher,
- the role of an educator,
- the role of a liaison between different stakeholder groups.

Creating innovative ideas, but also researching and recognizing user needs, reliable, knowledge-based co-creation and co-designing with local communities, educating and persuading to new solutions—these are the architect's main roles and tasks in the context of contemporary challenges of the architecture of elderly-friendly sustainable smart cities. The undertaken topic is very broad and the content presented is only a voice in the discussion on the development of modern cities and the role of the architect in this aspect. The list of tasks involving the creation of an image anew is certainly not closed. It presents a holistic approach and may be the nucleus of further in-depth research on each constituent. In the future, it could provide an opportunity to develop design guidelines for architects, indicating at the same time their important role, which has been so far insufficiently exposed in the designing process of elderly-friendly sustainable smart cities.

Funding: This research has received no external funding.

Conflicts of Interest: The author declares no conflicts of interest.

References

1. Yigitcanlara, T.; HoLeeb, S. Korean ubiquitous-eco-city: A smart-sustainable urban form or a branding hoax? *Technol. Forecast. Soc. Chang.* **2014**, *89*, 100–114. [CrossRef]

2. Karbowniczek, A. Smart City-Next Step to the Ideal City. Available online: www.ejournals.eu/pliki/art/13374/ (accessed on 20 August 2019). [CrossRef]

3. Saaty, T.L.; De Paola, P. Rethinking Design and Urban Planning for the Cities of the Future. *Buildings* **2017**, *7*, 76. [CrossRef]

4. Russo, A.; Cirella, G.T. Modern Compact Cities: How Much Greenery DoWe Need? *Int. J. Environ. Res. Public Health* **2018**, *15*, 2180. [CrossRef] [PubMed]

5. Kahn, M.E. Sustainable and Smart Cities. The World Bank Sustainable Development Network Urban and Disaster Risk Management Department. May 2014. Available online: https://openknowledge.worldbank. org/bitstream/handle/10986/18748/WPS6878.pdf?sequence=1&isAllowed=y (accessed on 20 August 2019).

6. ITU-T Focus Group on Smart Sustainable Cities. *Smart Sustainable Cities: An Analysis of Definitions*; Focus Group Technical Report; International Telecommunication Union: Genève, Switzerland, 2014; pp. 1–63.

7. Höjer, M.; Wangel, J. Smart Sustainable Cities Definition and Challenges. In *ICT Innovations for Sustainability, Advances in Intelligent Systems and Computing*; Hilty, L.M., Aebischer, B., Eds.; Springer International Publishing: Zurich, Switzerland, 2014; pp. 333–349. [CrossRef]

8. Trindade, E.P.; Hinnig, M.P.F.; Moreira da Costa, E.; Marques, J.S.; Bastos, R.C.; Yigitcanlar, T. Sustainable development of smart cities: A systematic review of the literature. *J. Open Innov. Technol. Mark. Complex.* **2017**, *3*, 11. [CrossRef]

9. Elgazzar, R.F.; El-Gazzar, R.F. Smart Cities, Sustainable Cities, or Both? A Critical Review and Synthesis of Success and Failure Factors. In *Conference on Smart Cities and Green ICT Systems*; SCITEPRESS–Science and Technology Publications, Lda: Setúbal, Portugal, 2017; pp. 250–257. ISBN 978-989-758-241-7. [CrossRef]

10. ITU 2015. Available online: https://www.itu.int/en/ITU-T/focusgroups/ssc/Pages/default.aspx (accessed on 25 August 2019).

11. Monfaredzadeha, T.; Krueger, R. Investigating Social Factors of Sustainability in a Smart City. *Procedia Eng.* **2015**, *118*, 1112–1118. [CrossRef]

12. Castelnovo, W.; Misuraca, G.; Savoldelli, A. Citizen's Engagement and Value co-Production in Smart and Sustainable Cities. Available online: https://www.ippapublicpolicy.org/file/paper/1433973333.pdf (accessed on 25 August 2019).

13. Anand, P.B.; Navío-Marco, J. Governance and economics of smart cities: Opportunities and challenges. Elsevier. *Telecommun. Policy* **2018**, *42*, 795–799. [CrossRef]

14. Alam, M.T.; Porras, J. Architecting and Designing Sustainable Smart City Services in a Living Lab Environment. *Technologies* **2018**, *6*, 99. [CrossRef]

15. Sánchez, L.; Gutiérrez, V.; Galache, J.A.; Sotres, P. Engaging individuals in the smart city paradigm: Participatory sensing and augmented reality. *Interdiscip. Stud. J.* **2014**, *3*, 1–14, Laurea University of Applied Sciences.

16. Mazhar Rathore, M.; Awais, A.; Anand, P.; Seungmin, R. Urban planning and building smart cities based on the Internet of Things using Big Data analytics. *Comput. Netw.* **2016**, *101*, 63–80. [CrossRef]

17. Jong, M.; Joss, S.; Schraven, D.; Zhan, C.; Weijnen, M. Sustainable–smart–resilient–low carbon–eco–knowledge cities: Making sense of a multitude of concepts promoting sustainable urbanization. *J. Clean. Prod.* **2015**, *109*, 25–38. [CrossRef]

18. Boulos, M.N.K.; Al-Shorbaj, N.M. On the Internet of Things, smart cities and the WHO Healthy Cities. *Int. J. Health Geogr.* **2014**, *13*, 10. [CrossRef] [PubMed]

19. Cook, D.J.; Duncan, G.; Sprint, G.; Fritz, R. Using Smart City Technology to Make Healthcare Smarter. *Proc. IEEE* **2018**, *106*, 708–722. [CrossRef] [PubMed]

20. Sprint, G.; Cook, D.; Fritz, R.; Schmitter-Edgecombe, M. Using Smart Homes to Detect and Analyze Health Events. *Computer* **2016**, *49*, 29–37. [CrossRef]

21. Trencher, G.; Karvonen, A. Stretching 'Smart': Advancing Health and Wellbeing through the Smart City Agenda. *Local Environ.* **2017**, *24*, 610–627. [CrossRef]

22. Ferrara, R. The Smart City and the Green Economy in Europe: A Critical Approach. *Energies* **2015**, *8*, 4724–4734. [CrossRef]

23. Zawieska, J.; Pieriegud, J. Smart city as a tool for sustainable mobility and transport decarbonisation. *Transp. j.tranpol.2017.11.004*CrossRef]

24. Pawłowska, B. Intelligent transport as a key component of implementation the sustainable development concept in smart cities. *Transp. Econ. Logist.* **2018**, *9*. [CrossRef]

25. Hajer, M.A. On Being Smart about Cities: Seven Considerations for a New Urban Planning and Design. January 2015, pp. 50–62. Available online: https://www.researchgate.net/publication/283873474_On_being_smart_about_cities_Seven_considerations_for_a_new_urban_planning_and_design (accessed on 26 August 2019).

26. Mora, L.; Deakin, M. *Untangling Smart Cities-From Utopian Dreams to Innovation Systems for a Technology-Enabled Urban Sustainability*; Elsevier: Amsterdam, The Netherlands, 2019; ISBN 978-0-12-815477-9.

27. Available online: https://www.smartcityexpo.pl (accessed on 26 August 2019).

28. Hitachi. Smart Sustainable City Overview. Available online: http://www.hitachi.com/products/smartcity/vision/concept/overview.html (accessed on 3 October 2019).

29. IBM. "India Needs Sustainable Cities." IBM SMARTER PLANET, Web. Available online: http://www.ibm.com/smarterplanet/in/en/sustainable_cities/ideas/ (accessed on 2 October 2019).

30. The UNECE-ITU Smart Sustainable Cities Indicators, Economic Commission for Europe Committee on Housing and Land Management Seventy-Sixth Session Geneva, 14–15 December 2015 Item 6 (b) of the Provisional Agenda Review of the Implementation of the Programme of Work 2014–2015 Sustainable Urban Development. Available online: http://www.unece.org/fileadmin/DAM/hlm/projects/SMART_CITIES/ECE_HBP_2015_4.pdf (accessed on 26 August 2019).

31. Mardacany, E. Smart cities characteristics: Importance of built environment components. In Proceedings of the Conference: IET Conference on Future Intelligent Cities, London, UK, 4–5 December 2014. [CrossRef]

32. Vinod Kumar, T.M.; Dahiya, B. Smart Economy in Smart Cities. In *Smart Cities, Local Community and Socio-Economic Development: The Case of Bologna*; Vinod Kumar, T.M., Ed.; Springer: Berlin, Germany, 2017; pp. 3–76. [CrossRef]

33. Available online: https://www.arcadis.com/en/global/what-we-do/our-capabilities/design/architecture/ (accessed on 26 August 2019).

34. Gaura, A.; Scotneya, B.; Parra, G.; McCleana, S. Smart City Architecture and its Applications based on IoT. *Procedia Comput. Sci.* **2015**, *52*, 1089–1094. [CrossRef]

35. Gorynski, B.; Mikolajczyk, P.; Muller, T.; Gelsin, A. Smart City, Smart Region, Smart City Guidebb Smart City Guide. 2019, pp. 2–56. Available online: https://hub.beesmart.city/hubfs/04-insights/02-landing-pages/lp-smart-city-atlas-de/beesmartcity_Handlungsleitfaden_SmartCity_Smart%20Region_web.pdf?utm_campaign=%23smart-city-atlas&utm_source=hs_automation&utm_medium=email&utm_content=70904257&_hsenc=p2ANqtz-8Z7KWa5NDomLPE62lWo4gAy_SZMBqiui0KX9m16hYipsSAtcBJ-JWgtTdgABE14iVAUQlcY0h95yDqdWdxRtsenSa6KocQcYRGuLRunCz4tMOGibM&_hsmi=70904257 (accessed on 15 July 2019).

36. Gorgol, N.K. The Analysis of the Relationship between the Idea of Smart City and the Urban Form on the Example of Oslo and Vienna. p. 41. Available online: yadda.icm.edu.pl (accessed on 28 August 2019). [CrossRef]

37. Trzpiot, G.; Szołtysek, J. *Safety of the Elderly in Smart City*; Research Papers of Wrocław University of Economics nr 483; Publishing House of Wrocław University of Economics: Wrocław, Poland, 2017. [CrossRef]
38. Skouby, K.E.; Kivimäki, A.; Haukipuro, L.; Lynggaard, P.; Windekilde, I. Smart Cities and the Ageing Population, OUTLOOK Visions and Research Directions for the Wireless World. 2014. No 1. Available online: https://pdfs.semanticscholar.org/5bf3/050e0a44322a6d5cccb633d53bffcb7a8a1f.pdf (accessed on 27 August 2019).
39. Fazlagić, J. Koncepcja Smart Cities w Kontekście Produktywności Pracowników Wiedzy 65 Plus; (The Concepof Smart Cities in the Context of Productivity of Knowledge Workers Aged 65+) ZNUV 2016; Volume 46. Available online: https://www.google.com/url?sa=t&rct=j&q=&esrc=s&source=web&cd=2&cad=rja&uact=8&ved=2ahUKEwiBstXW3qXlAhWObVAKHVCgAFoQFjABegQIAxAC&url=http%3A%2F%2Fcejsh.icm.edu.pl%2Fcejsh%2Felement%2Fbwmeta1.element.desklight-86c950b2-cac4-471d-b2ab-2b9e4ea8f467%2Fc%2FVistula-Zeszyty-naukowe-46_2016.79-90.pdf&usg=AOvVaw1UlYrtdWABPmDm0jf3udsb (accessed on 26 August 2019).
40. Tomczyk, Ł.; Klimczuk, A. Inteligentne Miasta Przyjazne Starzeniu Się-Przykłady z Krajów Grupy Wyszehradzkiej (Smart, Age-Friendly Cities: Examples in the Countries of the Visegrad Group (V4). *Rozwój Regionalny i Polityka Regionalna* **2016**, *34*, 79–97.
41. Martínez, R.M.; Azkune, G.; Jiménez, P.A.; Almeida, A. An IoT-Aware Approach for Elderly-Friendly Cities. *IEEE Access* **2018**, *6*, 7941–7957. Available online: https://www.researchgate.net/publication/322844110_An_IoT-aware_Approach_for_Elderly-Friendly_Cities (accessed on 29 August 2019). [CrossRef]
42. Brdulak, A. The concept of a smart city in the context of an ageing population. *Transp. Econ. Logist.* **2017**, *68*. [CrossRef]
43. Černá, M.; Poulová, P.; Svobodová, L. The Elderly in SMART Cities, Chapter January 2019. *Smart Education and e-Learning*. 2018, pp. 224–233. Available online: https://www.researchgate.net/publication/325426946_The_Elderly_in_SMART_Cities (accessed on 29 August 2019). [CrossRef]
44. Smart Age-friendly Cities|Age-friendly Smart Cities! by Willeke van Staalduinen (Age-friendly Nederland), Rodd Bond (NetwellCASALA), Carina Dantas (Caritas Coimbra), Ana Luísa Jegundo (Caritas Coimbra). Available online: https://ec.europa.eu/eip/ageing/sites/eipaha/files/library/smart_age-friendly_cities_age-friendly_smart_cities.pdf (accessed on 30 August 2019).
45. Preiser, W.F.E.; Rabinowitz, H.Z.; White, E.T. *Post Occupancy Evaluation*; Van Nostrand Reinhold Company: New York, NY, USA, 1988.
46. Tymkiewicz, J. Creative methods and techniques in didactics at the Faculty of Architecture. In Proceedings of the ICERI 12th Annual International Conference of Education, Research and Innovation, Seville, Spain, 11–13 November 2019.
47. Szewczenko, A. Enhancing the students' competences using the action research method: architecture education towards ageing society. In Proceedings of the 11th International Technology, Education and Development Conference. INTED 2017, Valencia, Spain, 6–8 March 2017; Gomez Chova, L., Lopez Martinez, A., Candel Torres, I., Eds.; IATED Academy, 2017; pp. 3495–3501, (INTED Proceedings; 2340-1079).
48. Available online: http://dschool.stanford.edu/dgift (accessed on 20 February 2016).
49. Brown, T. *Change by Design. How Design Thinking Transforms Organizations and Inspires Innovation*; Harper Collins Publishers Inc.: New York, NY, USA, 2011.
50. Brown, T. Design Thinking. *Harvard Business Review*, 23 October 2008; Volume 86, 84–92.
51. Eleutheriou, V.; Depiné, Á.; de Azevedo, I.; Teixeira, C. Smart Cities and Design Thinking: Sustainable development from the citizen's perspective. In Proceedings of the IV Regional Planning Conference, Aveiro, Portugal, 23–24 February 2017.
52. Thoring, K.; Müller, R.M. Understanding Design Thinking: A process model based on method engineering. In *Proceedings of the International Conference on Engineering and Product Design Education, London, UK, 8–9 September 2011*; City University: London, UK, 2011; Available online: https://www.designsociety.org/publication/30932/Understanding+Design+Thinking%3A+A+Process+Model+based+on+Method+Engineering (accessed on 30 August 2019).
53. Stangel, M.; Witeczek, A. Design thinking and role-playing in education on brownfields regeneration. Experiences from Polish-Czech cooperation. *Archit. Civ. Eng. Environ. (ACEE)* **2015**, *8*, 19–28.

54. Tymkiewicz, J.; Bielak-Zasadzka, M. The design thinking method in architectural design, particularly for designing senior homes. *Archit. Civ. Eng. Environ. (ACEE)* **2016**, *9*, 43–48. [CrossRef]

55. Stangel, M.; Szóstek, A. Empowering citizens through participatory design: A case study of Mstów, Poland. *Archit. Civ. Eng. Environ. (ACEE)* **2015**, *8*, 47–58.

56. Campolargo, M. From Smart Cities to Human Smart Cities. In Proceedings of the 48th Hawaii International Conference on System Sciences (HICSS), Kauai, HI, USA, 5–8 January 2015. [CrossRef]

57. Tymkiewicz, J. Team work efficiency in finding innovative solutions. Experience with the design thinking method implemented into teaching at the Faculty of Architecture. In Proceedings of the ICERI 2015 8th International Conference of Education, Research and Innovation, Seville, Spain, 16–18 November 2015; Chova, L.G., Martinez, A.L., Torres, I.C., Eds.; IATED Academy: Valencia, Spain, 2015; pp. 5894–5902, ISBN 978-84-608-2657-6.

58. Design out Crime. Designing out Crime a Designers' Guide. Available online: https://www.designcouncil. org.uk/sites/default/files/asset/document/designersGuide_digital_0_0.pdf (accessed on 27 August 2019).

59. Schuilenburg, M.; Peeters, T. Smart cities and the architecture of security: Pastoral power and the scripted design of public space. In *City, Territory and Architecture*; Springer: Berlin, Germany, 24 October 2018; Available online: https://link.springer.com/article/10.1186/s40410-018-0090-8 (accessed on 29 August 2019).

60. Eckes, A. Landscape architecture in protection of pedestrian zones against acts of terrorism. *Archit. Civ. Eng. Environ. (ACEE)* **2018**, *11*, 7–12. [CrossRef]

61. *BIG Architecture*; Kaspar Astrup Schroeder: Copenhagen, Denmark, 2017.

62. Available online: https://en.wikipedia.org/wiki/Amager_Bakke (accessed on 30 August 2019).

63. Available online: https://www.archdaily.com/601952/here-s-how-big-s-power-plant-ski-slope-will-blow-smoke-rings (accessed on 30 August 2019).

64. Tymkiewicz, J.; Bielak-Zasadzka, M. Senior homes of the future in the eyes of students of architecture. Didactic experience from the application of the design thinking method. *Archit. Civ. Eng. Environ. (ACEE)* **2016**, *9*, 49–56. [CrossRef]

65. Tymkiewicz, J. The collaboration with external entities and problem-based learning at the Faculty of Architecture. In Proceedings of the ICERI 2017 10th Annual Conference of Education, Research and Innovation, Seville, Spain, 16–18 November 2017; Chova, L.G., Martinez, A.L., Torres, I.C., Eds.; IATED Academy: València, Spain, 2017; pp. 786–794.

66. Tymkiewicz, J. Elewacje dla seniorów-badania jakościowe preferencji estetycznych i funkcjonalnych (Facades for seniors-quality research of aesthetic and functional preferences). In Proceedings of the Post Conference Monograph of the 2nd Conference on Interdisciplinary Research in Architecture, Gliwice, Poland, 20–21 April 2017; Tymkiewicz, J., Ed.; 2017; Volume 4, pp. 25–40. Available online: http://delibra.bg.polsl.pl/dlibra/doccontent?id=44113 (accessed on 30 August 2019).

67. Tymkiewicz, J.; Winnicka-Jasłowska, D.; Fross, K. The campus space in research and student projects. In *Advances in Human Factors, Sustainable Urban Planning and Infrastructure, Proceedings of the AHFE 2018 International Conference on Human Factors, Sustainable Urban Planning and Infrastructure, Orlando, FL, USA, 21–25 July 2018*; Charytonowicz, J., Falcao, C., Eds.; Loews Sapphire Falls Resort at Universal Studios: Orlando, FL, USA; Springer International Publishing: Berlin, Germany, 2019; pp. 24–35, (Advances in Intelligent Systems and Computing; Volume 788, pp. 2194–5357). [CrossRef]

68. Ravesteyn, P.; Plessius, H.; Mens, J. Smart Green Campus: How IT can Support Sustainability in Higher Education. In Proceedings of the 10th European Conference on Management Leadership and Governance, Conference proceedings, Zagreb, Croatia, 13–14 November 2014.

69. Zubizarreta, I.; Seravalli, A.; Arrizabalaga, S. Smart City Concept: What It Is and What It Should Be. Agricultural Information Institute, 07/18/16. Available online: http://agri.ckcest.cn/ass/NK006-20160801005. pdf (accessed on 24 August 2019).

70. Available online: http://seniorzy.bialystok.pl/ (accessed on 25 August 2019).

71. Available online: https://www.forbes.pl/gospodarka/ranking-forbesa-najbardziej-innowacyjne-miasta-w-polsce/we22dl4 (accessed on 25 August 2019).

72. Tymkiewicz, J. The advanced construction of facades. The relations between the quality of facades and the quality of buildings. In *Advanced Construction 2010, Proceedings of the 2nd International Conference, Kaunas, Lithuania, 11–12 November 2010*; Kaunas University of Technology: Kaunas, Lithuania, 2010; pp. 274–281.

73. Tymkiewicz, J. Quality analyses of facades based on post occupancy evaluation. Research experience with students of architecture participation. The Silesian University of Technology. In Proceedings of the ICERI 2016 9th International Conference of Education, Research and Innovation, Seville, Spain, 14–16 November 2016; Chova, L.G., Martinez, A.L., Torres, I.C., Eds.; IATED Academy: Valencia, Spain, 2016; pp. 8831–8838, ISBN 978-84-617-5895-1. (ICERI Proceedings; 2340-1095).

74. Tymkiewicz, J. Guidelines for programming and modernising facades as a follow-up of users' needs analyses. *Archit. Civ. Eng. Environ. (ACEE)* **2008**, *1*, 37–46.

75. Tymkiewicz, J. Facades and problems in correct recognition of the functions that buildings perform. *Archit. Civ. Eng. Environ. (ACEE)* **2012**, *5*, 15–22.

76. Tymkiewicz, J. The sun, wind and water in designs of exterior walls and facades-natural forces potential in shaping the architecture of sustainable development. *Archit. Civ. Eng. Environ. (ACEE)* **2012**, *5*, 31–40.

77. Tymkiewicz, J. Technological aesthetics of modern facades. *Czas. Tech.* **2014**, *111*, 257–263.

78. Tymkiewicz, J. *Funkcje Ścian Zewnętrznych w Aspektach Badań Jakościowych. Wpływ Rozwiązań Architektonicznych Elewacji na Kształtowanie Jakości Budynku (Functions of the Exterior Walls of Buildings in View of Quality Analyses; The Impact of Architectural Design Solutions of Facades on the Quality of Building)*, Gliwice; Wydawnictwo Politechniki Śląskiej: Gliwice, Poland, 2012; p. 304. Available online: http://delibra.bg.polsl.pl/dlibra/doccontent?id=17876 (accessed on 24 August 2019).

79. Baborska-Narozny, M.; Bać, A. Preliminary Evaluation of Design and Construction Details to Maximize Health and Well-Being in a New Built Public School in Wroclaw, book Sustainability in Energy and Buildings. In Proceedings of the 4th International Conference on Sustainability in Energy and Buildings (SEB'12), Stockholm, Sweden, 15 May 2012; pp. 581–590. [CrossRef]

80. Brand, S. *How Buildings Learn; What happens after they're built*; Penquin Books: New York, NY, USA, 1994.

81. Tymkiewicz, J. The architect vs. users-the problem of balconies in residential buildings. Conclusions from student research. In Proceedings of the ICERI 2019 12th International Conference of Education, Research and Innovation, Seville, Spain, 11–13 November 2019.

82. Adonina, A.; Akhmedova, E.; Kandalova, A. Realization of smart city concept through media technology in architecture and urban space: From utopia to reality. *MATEC Web Conf.* **2018**, *170*. [CrossRef]

83. Vattano, S. European and Italian experience of Smart Cities: A model for the smart planning of city built. In *Techne*; Firenze University Press: Firenze, Italy, 2013; pp. 110–116. [CrossRef]

84. Rodriguez, J.A.; Fernandez, F.J.; Arboleya, P. Study of the Architecture of a Smart City. *Proceedings* **2018**, *2*, 1485. [CrossRef]

85. Elmaghraby, A.S.; Losavio, M.M. Cyber security challenges in Smart Cities: Safety, security and privacy. *J. Adv. Res.* **2014**, *5*, 491–497. [CrossRef] [PubMed]

86. Szolomicki, J.; Golasz-Szolomicka, H. Technological Advances and Trends in Modern High-Rise Buildings. *Buildings* **2019**, *9*, 193. [CrossRef]

87. Coates, C. Dimming Disney Hall; Gehry's Glare Gets Buffed. Los Angeles Downtown News (21 March 2005). Available online: www.downtownnews.com/articles/2005/03/21/news/news02.txt (accessed on 28 August 2019).

88. Bagnolo, V.; Manca, A. Image beyond the form. Representing perception of urban environment. *Archit. Civ. Eng. Environ. (ACEE)* **2019**, *12*, 7–16. [CrossRef]

89. Rostański, K.M. *Connotations in Architecture. On the Art of Observation of Associations Drawn from Culture*; Wydawnictwo Politechniki Śląskiej: Gliwice, Poland, 2018; p. 179. ISBN 978-83-7880-595-3.

 *buildings* 

Article

A SWOT Analysis of the Use of BIM Technology in the Polish Construction Industry

Krzysztof Zima *, Edyta Plebankiewicz and Damian Wieczorek

Faculty of Civil Engineering, Cracow University of Technology, 31-155 Krakow, Poland; eplebank@l7.pk.edu.pl (E.P.); dwieczorek@l7.pk.edu.pl (D.W.)
* Correspondence: kzima@L7.pk.edu.pl

Received: 28 November 2019; Accepted: 16 January 2020; Published: 20 January 2020

Abstract: The present paper presents a SWOT analysis, the aim of which is to evaluate the strategic implementation of BIM technology in the construction industry in Poland. The authors created a SWOT matrix presenting strengths, weaknesses, opportunities, and risks associated with the use of BIM. Using literature analyses, own experience, and market reports, all elements of the SWOT matrix are described in detail. Basic indicators characterizing the strategic position of BIM on the Polish construction market are calculated. Finally, the matrix of strategic tasks and actions that should be applied in order to promote and develop BIM in Poland are defined.

Keywords: building information modeling; BIM; construction management; SWOT

1. Introduction

At present in the world, as well as in Poland, building information modeling (BIM) constitutes the fastest developing concept in the field of construction management. In many countries it is gradually becoming the standard for construction projects. In Poland, this technology is still being discussed and considered as an alternative to traditional planning and execution of construction projects. There exist considerable concerns about the introduction of a mandatory use of BIM technology in public works contracts. On the construction market one may notice a great deal of misunderstanding regarding this technology, lack of experience of the majority of participants in the construction process, and significant caution in its implementation.

However, the global literature emphasizes the major advantages of BIM. Approaches for acquiring 3D building information rely on using digital building information models and simplifying them (geometrically and semantically). BIM are object-oriented, semantically-rich, and up-to-date, thus allowing a query of necessary building parts in views [1]. This means that, according to the BIM idea, it is not only important to obtain information but, above all, to simplify the information contained in the model, systematize it, and use the information contained in the model. In contrast to standard models CAD, BIM models are now able to contain both geometric and semantic information as they develop during all stages of the life cycle of a building.

Due to the advantages and popularity of BIM globally, the authors of this paper decided to analyze the current situation in Poland, with the aim to examine the strengths and weaknesses of the introduction of the BIM technology in Poland, as well as the opportunities and risks it offers. In their analysis, the authors of the paper used a SWOT analysis, which is a tool employed mainly in the process of strategy building, but also in other areas of management. The challenges faced by both the Polish public administration and the participants of construction projects were presented. The authors also attempted to analyze the future trends in the development of BIM technology in Poland.

2. Methods—SWOT Analysis

The SWOT analysis is commonly used in strategic management when building the strategy of a given organization. It is a kind of diagnostic tool, often used at the very beginning of the process of defining future strategic plans. The SWOT analysis is a simple but powerful tool for sizing up an organization's resource capabilities and deficiencies, its market opportunities, and the external threats to its future [2]. SWOT is the acronym created from the initial letters of words that describe the characteristics of an organization's resources and its environment. The analysis has two dimensions: internal and external. The internal dimension includes organizational factors, also strengths and weaknesses, while the external dimension includes environmental factors, as well as opportunities and threats [3]. The SWOT analysis allows to obtain information on the possibilities of using strengths and enhancing the weaknesses in order to take advantage of the opportunities the environment offers and limit the risks that the environment can bring about. Strengths and weaknesses are of an internal nature, while opportunities and threats are of an external nature. Thus, the two dimensions generate four categories of factors:

- External positive (Opportunities)—opportunities (development opportunities in the environment).
- Internal positive (Strengths)—strengths of the organization.
- External negative (Threats)—environmental hazards.
- Internal negative (Weaknesses)—weaknesses of the organization.

The analysis performed by the authors aims to assess the opportunities offered by the introduction of the BIM technology in Poland, to evaluate risks, as well as analyze the strengths and weaknesses of support for the construction and the investment process using the BIM technology. The combination of strengths and weaknesses with opportunities and threats in the analysis strives for finding the best use of the potential that the BIM method offers in Polish conditions. The measure of the internal strength of the BIM technology, its strategic attractiveness and probability of strategic success will also be determined.

Advantages and disadvantages of the BIM technology will be presented in the analysis as strengths and weaknesses; moreover, the opportunities and risks of implementing BIM will be identified. The analysis will be performed in three stages:

- STAGE I—Identification of factors related to the implementation of BIM as positive or negative for the construction project and its environment, and assessment of these factors on a numerical scale from 1 to 5 where: 1—very weak factor influence, 5—very strong factor influence.
- STAGE II—Assessment of the strategic situation of the BIM method.
- STAGE III—Identification of the strategic tasks and actions by combining and analyzing the strengths and weaknesses of the BIM technology with opportunities and threats by using the strengths of BIM to obtain maximum benefits from the opportunities offered by the environment, overcoming the weaknesses by using the opportunities that exist in the environment, using the strengths of BIM to avoid threats from the environment, and minimizing the effects of the existence of BIM weaknesses to avoid risks from the environment.

3. BIM Technology

According the United States National Institute of Building Sciences (NIBS) "A BIM is a digital representation of physical and functional characteristics of a facility. As such, it serves as a shared knowledge resource for information about a facility forming a reliable basis for decisions during its lifecycle from inception onward" [4].

The task of the BIM technology is, therefore, to support the activities performed during the entire life cycle of a building by providing information on the geometry of the building as well as descriptive information on the building and its individual elements. The development of BIM and other digital technologies supporting the construction industry allows to improve the construction process and accelerate the implementation of buildings. Along with the development of such technologies as building information modeling (BIM), augmented reality (AR), virtual reality (VR), and the Internet of Things (for instance, near-field communication (NFC) and radio-frequency identification (RFID) sensors), new hardware and software tools have been introduced into the construction industry. These technologies allow the automation of construction processes, monitoring of construction works and management of information flow, as well as quality inspections [5].

The main advantage of BIM is the possibility to collect data in one place, namely in the BIM model, together with the method of geometric presentation of the building structure in a three-dimensional view. Collecting data through the design, construction, and operational phases would allow for further analysis of these data, generating new insights and simulations to identify clashes and interdependencies. Moreover, creating new methods of data visualization using visual and mixed reality improves communication and provides on-site information [6].

What should be emphasized is that, in comparison to the traditional construction process, not only the way of presenting geometric data (3D view) and collecting all the necessary data in the BIM model differ, but also the view and mentality of the participants in the construction process should also be changed. For example, the process of cost analysis benefiting from the availability of the BIM model differs in terms of quick extraction of information and data necessary as input for the developed predictive models [7]. However, every participant in a construction project has access to the data throughout the entire construction cycle. The cost data can, therefore, be adjusted on an ongoing basis. Moreover, each participant has access to the model and the data it contains, so they can, for instance, verify the data on an ongoing basis or have up-to-date information about the sample cost analysis and its changes.

Another advantage of BIM is the possibility to check the geometry and information included in the model easily and quickly. Revising the possible clash detection which can make works difficult to implement in order to discuss other solutions or remove errors occurring already at the pre-execution stage, is a significant advantage of the BIM technology. BIM is at the forefront of digital transformation in the AEC industry, [...] with a view toward streamlining a number of operations, such as collaboration and design review while addressing issues such as speed, cybersecurity, and data exchange integrity [8].

Considering the evolution imposed by the use of BIM, Table 1 summarizes the basic differences between the possibilities of using the BIM model in the construction process and the possibilities of CAD software supporting the traditional construction process.

Table 1. Comparison of CAD and BIM software capabilities (own study).

CAD Environment	BIM Technology
No link between drawings, any changes require manual correction	Parametric design allows automatic change of object and element parameters
2D view, no visualization of the 3rd dimension (height of the object and its elements)	3D, 4D, 5D, ... , xD
The features of the elements are given by the designer. The drawing consists of 2D lines which are later interpreted as objects	Ready-made elements that already have their properties can be used. Specific relationships between the elements are established
Possibility of describing the properties of an element in a drawing or in a technical description for a project	The properties of the elements are correlated with the element and can be elicited at any time by selecting the element
Industry documentation is often created independently	The BIM model is a data source for all industries and integrates them
Collisions between industries are often detected only at the construction stage	Collisions between industries are detected at the design stage

Based on their research of publications on BIM published since 2004, Liu et al. [9] concluded that, in the recent decade, the research in the field of BIM was developing continuously, which had completely subverted the traditional operation mode of the AEC industry and attracted more and more researchers' attention at the same time. Considering the advantages of BIM technology and the pace of its development in many countries around the world, but also taking into account the disadvantages of this technology, the authors decided to investigate the trends in its development in Poland.

4. BIM Adoption in the World

The United States is the world leader in the field of BIM adoption, so it is difficult to be surprised by the dominance of the North American continent. Nevertheless, Europe should be put in second place, which, using American models, but primarily creating its own dynamically developments in the use of BIM technology. Australia and Oceania were considered the next most advanced and were strong in the design modeling. Asia is followed by strong leaders in the form of Korea, Hong Kong, China, and Japan. In South and Central America, despite close patterns flowing from the leaders from the USA and Canada, BIM adoption is much slower, only individual countries decide to introduce BIM in public procurement within a few years and there are not such large countries as Argentina and Brazil. BIM adoption in the Middle East and Africa is low. Private investors are eager to implement projects using BIM technology, while public investors only use BIM for infrastructure investments involving the construction of airports or passenger service terminals (e.g., Istanbul Grand Airport, Bahrain International Airport, and Abu Dhabi Midfield Terminal). The list and brief characteristics of BIM adoption by regions are shown in Table 2.

Table 2. BIM adoption by regions (own study).

Region/Continent	BIM Adoption Brief Characteristic
North America	In the US, BIM is mandatory since 2008—a definite leader in BIM adoption. In Canada, the BIM adoption program has been running from 2014 to today. North America apparently ranked as the most advanced continent in every approach.
South and Central America	It is planned to introduce BIM obligatory in government projects in the years 2020–2022 in a few countries.
Europe	Open BIM standards and mandate in a few countries (especially in Scandinavian countries and Great Britain), many countries are preparing to introduce BIM standards or to make BIM obligatory in public procurement).
Asia	Korea and Hong Kong are becoming leaders in the region, China and Japan have great government support in implementing BIM standards. The level of adoption on the continent, however, must be assessed low, despite these four strong leaders.
The Middle East	Unlike many Far East countries, BIM adoption in the Middle East is low. Individual projects are implemented, but there are no specific actions for the adoption of BIM in individual countries of the Middle East.
Africa	South African government's growth targets in respect of technology usage in South Africa would be beneficial to BIM stakeholders. However, there is a lack of efforts to adopt BIM on the continent. Only representatives of architecture, engineering and construction (AEC) recommended the adoption of BIM techniques in Egypt, starting with the acquisition of full awareness of the BIM framework, different levels and BIM stages also in the perspective of the entire life cycle of construction projects.
Australia and Oceania	Mandate in place in Australia, in New Zealand there is government support and significant BIM promotion.

In Europe, the prevailing trend can be seen that western and northern countries have adopted BIM earlier and some of them, primarily the United Kingdom and Scandinavian countries are leaders on the BIM market. Southern and eastern countries are taking BIM at a slower pace. The situation is similar if we look at government initiatives, where especially Finland, Norway, Great Britain, France, and Italy dominate in the scope of solutions introduced in public procurement. Poland belongs to the second group of countries that are just planning to introduce standards or an obligation to use

BIM in public procurement and the adoption of BIM is slightly slower. In Poland, there is a lack of both generally used BIM standards and government support and mandatory solutions introduced in public procurement. It seems, however, that market awareness, especially of designers, is growing and dozens of public procurement, mainly related to the BIM model designing, have already been announced and carried out.

5. The SWOT Analysis of the Use of BIM Technology in Poland

The subsequent steps of the SWOT analysis described in chapter 2 are presented below. The analysis concerns the assessment of the strategic use of the BIM in the Polish construction industry.

5.1. Stage I—Identification of Factors

Factor identification was based on the analysis of both Polish and foreign literature, authors' own experience in implementing BIM in Poland and reports published by Autodesk:

- In October 2015 [10] on behalf of Autodesk, the Millward Brown Institute conducted a study in a group of 350 companies from the architectural and construction industry (architectural studios, construction design and installation companies, and development companies), called "BIM—Polish perspective".
- In October 2019 [11] on a sample of 287 companies, the Kantar Poland Institute (commissioned by Autodesk) prepared and published a study "BIM, cooperation, cloud in Polish construction".

Table 3 presents strengths and weaknesses, opportunities and threats of the BIM technology implementation in Polish construction projects. The Table also includes the evaluation of elements in the SWOT matrix on a numerical scale from 1 to 5, where: 1—very weak impact of the element and 5—very strong impact of the element. The assessment was made subjectively by the authors on the basis of market data, reports, and their own experience.

Table 3. Strengths and weaknesses, as well as opportunities and threats of BIM technology implementation in Polish construction projects (own study).

STRENGTHS (S)		WEAKNESSES (W)	
Better documentation	5	No universal software platform	2
Reduction of costs of the construction project	5	High labor consumption of the correct BIM model	5
Reduction of construction material waste	3	Errors in reflecting the true form of the building	4
Automation of drawing execution	4	High costs of BIM implementation in a company	1
TOTAL	17	TOTAL	12
OPPORTUNITIES (O)		**THREATS (T)**	
High interest of the leaders of the construction market	4	Lack of legal regulations and binding standards concerning BIM in Poland	5
Implementation of the BIM technology in many countries	3	Lack of qualified and experienced staff	5
Developing higher awareness among all stakeholders	5	Unwillingness of the contractors/clients/users to employ BIM	3
Educating students in BIM	2	-	-
TOTAL	14	TOTAL	13

The elements of the SWOT matrix which were qualified to its individual components are briefly described below.

5.1.1. STRENGTHS (S)

Better Quality of Documentation

Due to the greater transparency of the documentation presented in a 3D view and the requirements concerning the degree of model development (Level of Development, LOD), it is possible to obtain documentation of better quality and adapted to the requirements at a given stage of investment (for example, requirements in accordance with the LOD 100, ..., LOD 500 classification). Additionally, BIM tools allow to detect spatial collisions of elements in the BIM model, the so-called "clash detection". Thus, it is possible to detect early irregularities and design mistakes and, thus, reduce cost-generating errors during construction.

According to [12], clash detection with the use of BIM caused a significant reduction in RFIs (Requests For Information) on all surveyed projects. In case studies presented by [13], RFIs were reduced by 34% on a small tilt-wall project, 68% on a three-story assisted living facility, and 43% on a midrise commercial condominium project. The number of change orders was reduced by 40%, 48%, and 37%, respectively. In accordance with reports [10,11], the greatest benefit of implementing the BIM technology was the creation of better quality projects (2015—61.4%; 2019—69.4%) and the possibility of minimizing errors (60.5%, 51.0%), both in terms of design and implementation.

Final rating of the feature (on a scale of 1–5): **5** (due to over 60% the highest rating in surveys).

Reduction of Construction Project Costs

Information entered into the model and accuracy of take-off calculations have a considerable impact on calculated costs of a construction project [14]. By gathering information in one place, granting access to this information for all investment participants, improving the information flow process and clash detection, BIM allows to reduce the cost of the entire investment. In case studies presented by [13], the ROI (Rate On Investment) of BIM varied greatly from 16% to 1.654%. According to the report [10] the greatest savings are generated by the stage of redundancy and cost estimation (as stated by 70% of respondents), implementation (about 55% of respondents) and use (about 50% of respondents). In 2019, a similar hierarchy was maintained. The greatest controversy was caused by the architectural design stage, since about 55% of respondents declared that BIM reduced the costs of this process, while 30% claimed that it increased the costs. Generally speaking, it can be said that BIM definitely has a positive impact on the reduction of costs of a construction project in the whole process of the construction of a building.

Final rating of the feature (on a scale of 1–5): **5** (due to 70% of respondents' support for opinions on cost savings when using BIM).

Reduction of Construction Material Waste

The BIM technology helps to reduce building material waste by making accurate measurements based on the BIM model. Therefore, BIM allows to control the amount of material that should be used at a given stage of a construction project. With flexible purchasing and delivery management, waste and unused materials can be reduced by both contractors and subcontractors. However, environmental benefits are not prioritized in Poland, hence, the much lower assessment of this element.

Final rating of the feature (on a scale of 1–5): **3** (due to the relatively low assessment of environmental benefits in Poland).

Automation of the Drawing Process

The drawing process based on the BIM technology is automated. Changes made to the model immediately generate corrected drawings, thus saving time that can be used to refine the model. At the moment, these design methods are regarded to be an area of professional skills, connected with computer techniques [15]. The use of parametric design allows to create even very complex shapes. It uses advanced algorithms which are based on the parameters entered into the computer. Parametric design allows to set parameters for certain element sizes, when appropriate values are entered and, thus, the shape of the element is updated. In this way a series of types is created without having to draw all the changes.

Kapliński in [16] includes BIM modeling and the integrated BIM process in construction as construction trends shaping the industry in 2016 and beyond. In report [12] the respondents highly appreciated the possibility of improving the way of designing, as well as the possibility of creating more efficient projects (2015—38.4%; 2019—34.7%).

Final rating of the feature (on a scale of 1–5): **4** (due to the high position in the respondents' opinions in the report, but lower than creating better quality projects).

5.1.2. WEAKNESSES (W)

No Universal Software Platform

Cooperation of models created in programs from one manufacturer is not a problem; however, when there is a need to exchange data between software from different manufacturers, smooth data exchange and information loss issues may occur. Figure 1 shows the loss of data in the export and import of models to the various stages of investment. The bottom graph presents problems arising from data loss in the model due to the lack of a universal data exchange platform and a lack of interoperability between the software used in the construction process. The downward spikes at the end of each project phase means loss of geometrical and/or non-geometrical data. These losses usually occur when the project is exported from BIM to 2D CAD format, but also during the export to IFC format or from the IFC format the model is converted to native formats. The top line represents a practically ideal situation in which data and knowledge are gradually increased throughout the duration of the project (the model is successively updated and supplemented with new information, no data loss).

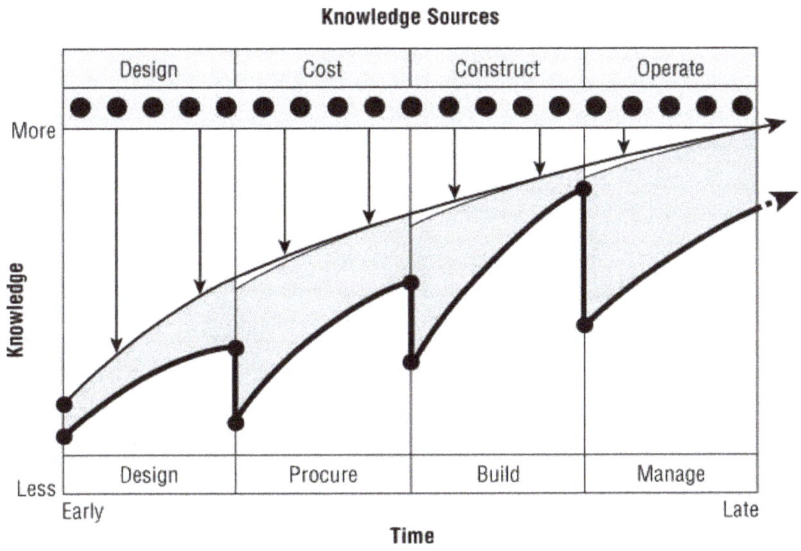

Figure 1. The BIM curve shows data loss without interoperability at project milestones (After: [17]).

Appropriate interoperability helps to eliminate the problem of the lack of a universal information exchange platform. It seems that the activities of the buildingSmart organization striving to provide a universal foundation for sharing information and improving processes in the design and construction industry. This goal is implemented through actions to develop standards, norms, and tools supporting the exchange of information regardless of the IT platform used. Authors think that focusing on one format—IFC—and its development is the right thing to do. Working on common standards, e.g., within the EU, and a common dictionary of building terms and the BIM dictionary (which Bilal Succar does creating the new BIM Dictionary platform) and promoting such activities can lead to solutions to many problems.

Final rating of the feature (on a scale of 1–5): **2** (subjective assessment).

High Labor Consumption of Developing the Correct BIM Model

Creating a correct BIM model which will be properly made and filled with information required at a given stage of the project is strongly laborious according to the designers. Research results in Poland [10,11] reveal the respondents' mentioning the most frequent barrier in connection with the implementation of BIM in Polish companies, namely extremely low prices of projects, which is directly related to the labor consumption of their implementation (2015—83.9%; 2019—67.4%).

Final rating of the feature (on a scale of 1–5): **5** (due to the highest rated BIM development barrier in Poland by respondents in reports).

Errors in Reflecting the True Form of a Building

The availability of design software on the market is relatively high; however, there are few specialists able to properly design a building in accordance with world standards promoted by the AIA (American Institute of Architects). In the research [10,11] competence gaps were mentioned (few BIM specialists): shortage of qualified staff (as stated in [10]—71.4%, according to [11]—58%) and a low level of knowledge about BIM in Poland (60.2% and 68.4%, respectively).

Final rating of the feature (on a scale of 1–5): **4** (due to the second most important barrier to BIM development in Poland).

High Implementation Costs of BIM in a Company

The requirements for hardware supporting BIM applications are much higher than for CAD applications, due to the large size of BIM files and the higher graphic demands. Implementing BIM often involves the purchase of new computers and software. Examples of the costs of implementing BIM in a construction company are presented in Table 4.

Table 4. Costs of implementing BIM in a construction company.

Software Type	Example Software	Number (unit)	Unit cost (PLN/unit)	Cost (PLN)
Design	Archicad/Revit	1	10,000	10,000
Cost estimates	BIMestiMate	3	2190	6570
IFC browser	SMV/BIM Vision	3	0	0
	TOTAL			16,570

Moreover, there are additional costs related to training. Sending three employees to post-graduate studies means a cost of approximately 6000 PLN per person.

Final rating of the feature (on a scale of 1–5): **1** (due to relatively low costs for large and medium companies)

5.1.3. OPPORTUNITIES

High Interest of the Construction Market Leaders

Large companies (both investors and construction contractors) that are leaders in the construction industry are highly interested in the implementation of the BIM technology due to the wish to reduce the costs of construction investments and to increase the attractiveness of their company. The interest of large public procurers is also slowly growing, as proved by the first pilot programs performed by infrastructure procurers, such as the Polish State Railways, the General Directorate for National Roads and Motorways and the Polish Power Grids.

Final rating of the feature (on a scale of 1–5): **4** (subjective assessment).

Implementation of the BIM Technology in a Number of Countries

The implementation of BIM technology is currently taking place all over the world. Some countries already have legislation in place and in some countries BIM is mandatory in public procurement. The state of global adoption of BIM for 2017 is depicted in Figure 2.

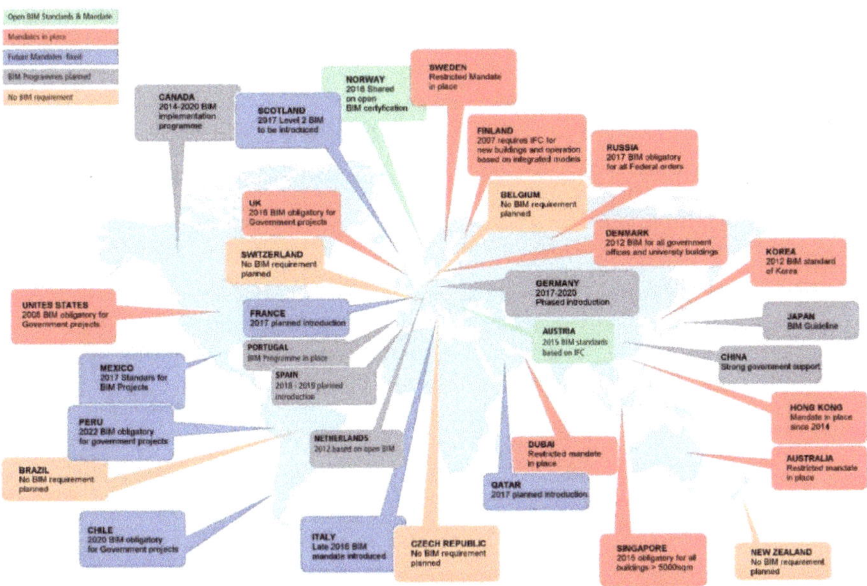

Figure 2. Overview of global BIM adoption [18].

Final rating of the feature (on a scale of 1–5): **3** (compared to other European countries, Poland is only on the path of planned BIM adoption)

Developing Higher Awareness among All Stakeholders

Increasing stakeholder awareness regarding both the benefits and losses associated with running an enterprise using BIM technology will cause less distrust on the market. More investors, knowing the benefits and potential threats, will decide to impose the use of BIM by designers and contractors in the construction process.

Developing higher awareness was placed in reports [10,11] as the first BIM activities that should be implemented in Poland so that the architectural and construction industry could fully use BIM

(2015—40.4% and 2019—55.3% respondents believe that BIM awareness is currently the most important activity that should be carried out in Poland).

Final rating of the feature (on a scale of 1–5): **5** (due to the most important activity that should be carried out in Poland).

Students' Education about BIM

The awareness of future engineers and architects about BIM is increasing. The universities offer master's and post-graduate courses to develop knowledge about the BIM technology. Therefore, there is a chance that, in a few years' time, the current lack of personnel in the field of BIM knowledge will be filled with young professionals.

Final rating of the feature (on a scale of 1–5): **2** (subjective assessment).

5.1.4. THREATS

No Legal Regulations and Binding Standards Regarding BIM in Poland

Poland lacks regulations, or at least standards, in force, regarding the use of the BIM technology. In public procurement, investors use the provisions of the British specification PAS 1192-2:2013 (in 2017 replaced by BS EN ISO 19650) or LOD (level of development) specified by the AIA (American Institute of Architects). In the reports [10,11] it is proposed to create Polish BIM standards as the second most important action to be taken for the development of BIM in Poland, immediately after building investor awareness (2015—36.0% and 2019—51.1% respondents believe in that).

Final rating of the feature (on a scale of 1–5): **5** (due to the one of most important barrier to BIM development in Poland).

Lack of Qualified and Experienced Staff

An efficient use of the BIM technology requires new skills and experience and, thus, time, which in turn prolongs the construction project when creating the first BIM projects. A survey conducted on 30 September 2016 for the Ministry of Infrastructure and Construction shows that the main problem in the implementation of investments in Poland with the use of BIM is the lack of competence of the staff (Figure 3).

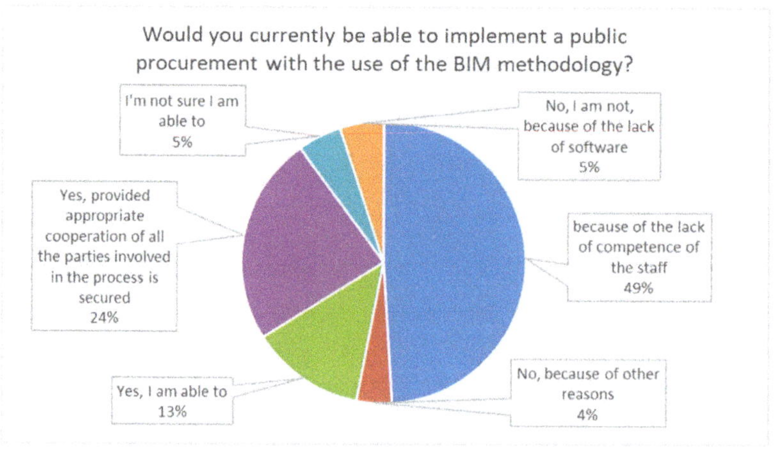

Figure 3. Survey on the implementation of BIM [19].

Final rating of the feature (on a scale of 1–5): **5** (due to lack of competence of the staff—49% respondents of survey point out on that problem).

Unwillingness of the Contractors/Clients/Users to Use BIM

Most of the participants of the construction market in Poland do not see or are not aware of the benefits of BIM. The participants do not want to change the way they work because of their habits. Internal barriers are quite important, that is: reluctance to change or no openness to new solutions, which is a common and typical behavior in the case of all novelties.

Final rating of the feature (on a scale of 1–5): **3** (subjective assessment).

5.2. Stage II—Assessment of the Strategic Situation of the BIM Method

Table 2 shows the total points for each of the four forces (*S*, *W*, *O*, *T*). It is important to note that:

$$S = \sum_{i=1}^{4} S = 17 > W = \sum_{j=1}^{4} W = 12$$

$$O = \sum_{k=1}^{4} O = 14 > T = \sum_{l=1}^{3} T = 13$$

This means a maxi-maxi strategic position, which gives BIM a privileged position in the market because of the advantage of the strengths over the weaknesses and opportunities over the threats. The strategy of further proceedings concerning the BIM technology on the Polish market should aim at maintaining its current position.

Moreover, using the data from Table 2 for the SWOT analysis, it is possible to determine the attractiveness of the environment (*AS*) of the BIM technology, which is a function of opportunities and threats *AS* = f (*O*,*T*), calculated from the following formula:

$$AS = \frac{O}{O+T} = \frac{14}{14+13} = 0.519$$

The market position of the BIM technology can also be determined (SP—internal strength), as well as the likelihood of strategic success (PSS—probability of a strategic success):

$$SP = \frac{S}{S+W} = \frac{17}{17+12} = 0.586$$

$$PSS = \frac{SP+AR}{2} = \frac{0.59+0.52}{2} = 0.552$$

Assuming that the PSS limit value is 0.5, it can be concluded that the BIM technology has a chance to develop, though probably slowly and not without problems.

5.3. Stage III—Defining Strategic Tasks and Actions

Each analyzed pair created by the strong or weak side of the use of BIM technology in Poland and by the opportunities and threats related to the environment, delimits a certain framework of a specific problem. In order to solve it and to indicate the ways of further proceedings, the before-mentioned SWOT matrix and associative techniques can be used.

The matching technique may be a specific heuristic method, such as brainstorming, as well as direct or symbolic analogies. Combining strengths and weaknesses with opportunities and threats in order to make the best use of the potential of the BIM method is illustrated in Table 5.

Table 5 shows the matrix of tasks and strategic actions that should support the development of the BIM technology in Poland. The application of heuristic methods results in the proposed strengthening of strategic activities undertaken in Poland, for example S1O3, S2O1, and W2O4, proposals to implement actions not yet taken, such as W4O2, and showing the dangers, for instance W1T1, W2T3.

Table 5. Matrix of strategic tasks and activities.

	STRENGTHS (S)	**WEAKNESSES (W)**
	1. Creating better quality projects 2. Decreasing the construction project costs 3. Decreasing the amount of construction waste 4. Automation of drawings	1. No common software platform 2. High labour consumption of creating a correct BIM model 3. Errors in reflecting the real form of the building 4. High costs of BIM implementation in a company
OPPORTUNITIES (O) 1. High interest of the construction market leaders 2. Mandatory implementation of BIM in public procurement in many countries 3. Developing higher awareness among all stakeholders 4. Education of students in BIM	S1O3—higher quality of projects should contribute to developing higher awareness of all participants in the project S2O1—reducing the cost of construction projects will increase interest of market leaders in BIM technology S4O4—automation of work should increase students' interest in BIM technology	W1O1—the interest of market leaders can help to create universal tools and software platforms W2O4—education of students on BIM will allow to introduce to the market people able to quickly perform the correct BIM model W4O2—implementation of BIM in public procurement in Poland may result in increased investment in the implementation of BIM in the company
THREATS (T) 1. No regulations or binding standards concerning BIM in Poland 2. No qualified or experienced staff 3. Unwillingness of the contractors/clients/users to use BIM	S2T3—reduction of construction project costs should convince all the reluctant to introduce this technology S4T2—design automation should encourage young engineers in particular to learn the BIM technology	W1T1—the lack of a universal software platform may discourage the government from creating legal regulations and hinder the creation of the standards of conduct W2T3—high labor consumption and low wages may discourage designers from using BIM technology W4T2—high costs for software and staff training may discourage employers from training their staff

6. Summary and Conclusions

The SWOT analysis shows that the implementation of BIM in Poland currently has a favorable position on the market, resulting from the existence of strengths over weaknesses and opportunities over threats. However, it is difficult to count on a fast dynamics of changes in Poland in terms of the implementation of BIM in construction.

The best strategic solution for the implementation of the BIM technology seems to be an aggressive development strategy, which is recommended for "maxi-maxi" situations. Such a strategy is based on maximizing the use of strengths and opportunities to streamline the dynamic implementation of BIM for everyday use. The strengths of BIM should be fully exploited when the environment provides an opportunity to do so.

The promotion of BIM should take advantage of the interest of companies that are leaders in the construction market, who should be reminded about reducing investment costs, which should encourage the use of BIM.

Introduction of fields of study specializing in BIM at universities should, in turn, allow to fill in the gaps in the staff able to proficiently use various BIM applications in a few years.

Author Contributions: Conceptualization: K.Z.; formal analysis: K.Z., E.P., and D.W.; investigation: K.Z. and D.W.; methodology: E.P. and D.W.; supervision: E.P.; writing—original draft: K.Z. All authors have read and agreed to the published version of the manuscript.

Funding: This research received no external funding.

Conflicts of Interest: The authors declare no conflict of interest.

References

1. Isikdag, U.; Zlatanova, S. A SWOT analysis on the implementation of Building Information Models within the Geospatial Environment. In *Urban and Regional Data Management—UDMS Annual 2009*; Krek, A., Rumor, M., Zlatanova, S., Fendel, E.M., Eds.; Taylor & Francis Group: Boca Raton, FL, USA, 2009; pp. 15–30.
2. Thompson, A.A.; Strickland, A.J.; Gamble, J.E. *Crafting and Executing Strategy-Concepts and Cases*, 15th ed.; McGraw-Hill/Irwin: Boston, MA, USA, 2007.
3. Gürel, E.; Tat, M. SWOT analysis: A theoretical review. *J. Int. Soc. Res.* **2017**, *10*, 994–1006. [CrossRef]
4. US National Institute of Building Sciences. *National Building Information Modelling Standard*; Version 1—Part 1: Overview, Principles, and Methodologies, Glossary; US National Institute of Building Sciences: Washington, DC, USA, 2007.
5. Ratajczak, J.; Riedl, M.; Matt, D.T. BIM-based and AR Application Combined with Location-Based Management System for the Improvement of the Construction Performance. *Buildings* **2019**, *9*, 118. [CrossRef]
6. Elagiry, M.; Marino, V.; Lasarte, N.; Elguezabal, P.; Messervey, T. BIM4Ren: Barriers to BIM Implementation in Renovation Processes in the Italian Market. *Buildings* **2019**, *9*, 200. [CrossRef]
7. Juszczyk, M.; Zima, K.; Lelek, W. Forecasting of sports fields construction costs aided by ensembles of neural networks. *J. Civ. Eng. Manag.* **2019**, *25*, 715–729. [CrossRef]
8. Nawari, O.N.; Ravindran, S. Blockchain and Building Information Modeling (BIM): Review and Applications in Post-Disaster Recovery. *Buildings* **2019**, *9*, 149. [CrossRef]
9. Liu, Z.; Lu, Y.; Peh, L.C. A Review and Scientometric Analysis of Global Building Information Modeling (BIM) Research in the Architecture, Engineering and Construction (AEC) Industry. *Buildings* **2019**, *9*, 210. [CrossRef]
10. BIM–polska perspektywa. Raport z badania / BIM - Polish Perspective. Research Report. Available online: http://damassets.autodesk.net/content/dam/autodesk/www/campaigns/bim-event/BIM_raport_final.pdf (accessed on 14 October 2019).
11. BIM, współpraca, chmura w polskim budownictwie / BIM, Cooperation, Cloud in Polish Construction. Available online: https://www.autodesk.pl/campaigns/aec/bim-report-2019#form-section (accessed on 31 October 2019).
12. Fan, S.L.; Skibniewski, M.J.; Hung, T.W. Effects of Building Information Modeling During Construction. *J. Appl. Sci. Eng.* **2014**, *17*, 157–166.
13. Giel, B.K.; Issa, R.R.A. Return on Investment Analysis of Using Building Information Modeling in Construction. *J. Comput. Civ. Eng.* **2013**, *27*, 511–521. [CrossRef]
14. Zima, K. Impact of information included in the BIM on preparation of Bill of Quantities. *Procedia Eng.* **2017**, *208*, 203–210. [CrossRef]
15. Kapliński, O. An important contribution to the discussion on research methods and techniques in designing. *Eng. Struct. Technol.* **2015**, *7*, 50–53. [CrossRef]
16. Kapliński, O. Innovative solutions in construction industry. Review of 2016–2018 events and trends. *Eng. Struct. Technol.* **2018**, *10*, 27–33. [CrossRef]
17. Read, P.; Krygiel, E.; Vandezande, J. *Mastering Autodesk®Revit®Architecture 2013*; John Wiley & Sons: Ottawa, ON, Canada, 2012.
18. McAuley, B.; Hore, A.; West, R. *BICP Global BIM Study—Lessons for Ireland's BIM Programme*; Construction IT Alliance (CitA) Limited: Dublin, Ireland, 2017. [CrossRef]
19. KPMG Advisory Sp. z o.o. Sp. k. Building Information Modeling. In *Ekspertyza Dotycząca Możliwości Wprowadzenia BIM w Polsce*; Ministerstwo Infrastruktury i Budownictwa/Expert Opinion on the Possibility of introducing BIM in Poland; Ministry of Infrastructure and Construction, KPMG Public: Amstelveen, The Netherlands, 30 September 2016.

Article

BIM Management Measure for an Effective Green Building Project

Reuven Maskil-Leitan [1,*], Ury Gurevich [2] and Iris Reychav [1]

[1]   Industrial Engineering & Management Department, Ariel University, Ariel 40700, Israel; irisre@ariel.ac.il
[2]   Government Office, Petach Tikva 4959253, Israel; ury.gurevich@gmail.com
*   Correspondence: ruven.maskille@msmail.ariel.ac.il

Received: 29 June 2020; Accepted: 25 August 2020; Published: 27 August 2020

Abstract: In light of the gap in research and practice, with regard to achieving the sustainability goals of green building, while maximizing combination with building-information-modeling (BIM) as a social system—a gap that is expressed in the absence of integration of all stakeholders—a managerial measure is proposed to integrate them and promote sustainable green building. By using a framework for implementing BIM as a social system, and through network analysis, an index is developed to assess its integration into the green building—the Green BIM Index. This measure consists of comparing a social benchmark for optimal implementation with the actual implementation, in a given project. The index is intended to help score the BIM integration level in a green building. Comparing the BIM management measure results with social benefit assessments, and the effectiveness of BIM in nine case-studies enables to understand project outcomes in terms of schedules, budgets, and quality. The paper demonstrates the index applicability, pointing to possible significant economic improvements through the implementation of BIM social capabilities. BIM management benchmarking is helpful for the comparative evaluation of similar projects incorporating green building with BIM, indicating the level of integration to improve benefits.

Keywords: BIM; green building; benefits of BIM; public construction clients; project outcomes

1. Introduction

The use of building-information-modeling (BIM) as a promoter of green processes has received considerable attention among practitioners of the construction industry [1]. Given the momentum of BIM and green building applications, many construction companies have sought to leverage green building projects through BIM, in order to realize the synergies between them while achieving sustainability through them [2]. Green building is 'a holistic concept that starts with the understanding that the built environment can have profound effects, both positive and negative, on the natural environment, as well as the people who inhabit buildings' [3] (p. 1). It is 'an effort to amplify the positive and mitigate the negative of these effects' [3] (p. 1). BIM is a 'digital representation of physical and functional characteristics of a facility. A BIM is a shared knowledge resource for information about a facility forming a reliable basis for decisions' [4] (p. 1). The construction industry has been driven to adopt green building strategies from sustainability considerations, such as reducing CO_2 emissions and energy dependency on fossil fuels. BIM has been regarded by many to be an opportunity for making the best use of the available design data for sustainable design and performance analysis [5]. The convergence of these separate trends into emerging practice has been referred to as green BIM [1].

BIM's technical advantages were soon joined to help facilitate more effective processes related to budget control, schedules, and environmental data, in an effort to increase green building effectiveness [6]. Green BIM has been perceived as a combination of green building, required to address environmental issues [7], with BIM as a technical tool [8], which serves it. However, incorporating BIM, as a socio-technical system, into green building, as sustainable construction, also requires consideration of the social component involved in achieving effectiveness.

Green building integration with BIM can be presented as one that necessitates coordination among many involved, using sophisticated modeling and system analysis to bring about a sustainable project. This combination requires managerial capabilities to improve efficiency [9]. However, alongside measurement systems—mostly environmental—for green building, and alongside mostly technical indicators—for BIM, there is no benchmarking system for examining these capabilities in green BIM.

In light of the gap in research and practice, with regard to achieving the sustainability goals of green building, while maximizing combination with BIM as a social system—a gap that is expressed in the absence of integration of all stakeholders—a managerial measure is proposed to integrate them and promote sustainable green building. The measure for this implementation of the social system—the integration of BIM in the interactions between all those involved in the project—emphasizes the reliance on an intra-firm organizational structure for the realization of a sustainable purpose. Presenting green BIM as a combination that requires high levels of interaction between participants and complex technology systems, the paper highlights the need for social integration through stakeholder management. The purpose of the research is to explain the importance of addressing this need to achieve effectiveness and to offer it an appropriate response.

In order to refer to the green BIM in this social context, this study uses a corporate-social-responsibility (CSR) model for BIM application as a benchmark for evaluating this integration and promoting its benefits. From this social point of reference to the agreements and working relationships between parties to a given project, the participation of stakeholders in the organization is examined, and their connection to the construction process is assessed. Considering that industrial practice scarcely includes reference to social components of sustainability, this study suggests bridging the gap using a CSR-based BIM index: the Green BIM Index. This metric refers to the basic question of whether and how social sustainability can be measured [10], by considering a green building project as a means of achieving sustainable benefits, and by presenting practical BIM-based indicators for assessing social sustainability in green building projects. According to the proposed measure, the BIM implementation is calculated by the CSR benchmark for a given project as a standard, and a match is made between its results and the actual BIM implementation results, using social-network-analysis (SNA). The need for collaboration is evident in times of crisis, such as the COVID-19 pandemic, when work is done remotely. A socially-based BIM may help achieve quality as a result of conducting a proper collaborative process.

The index is examined and applied in nine case-studies to confirm its validity and to examine its effectiveness by comparing the actual use of BIM methodology, the social benefits, and the objective effectiveness of the project. In this study, a special emphasis is placed on data quality in order to establish the index. Accordingly, green public projects of the owner-occupier type are carefully chosen and assessed at various project stages, to allow for an appropriate comparison of all model criteria and evaluation. The attempt is to raise awareness of the planning method and the concept of sustainability to achieve success [11]. This demonstration of social integration expands the understanding of CSR as required to implement BIM's social role in the industry while presenting practical means to promote a sustainable green building project.

2. Background

2.1. Effective Green BIM—Social Characteristics and Requirements for Sustainable Benefits

The integration of the green building with BIM is a combination of a highly complex project-based organization and sophisticated environmental modeling and analysis systems, which require managerial responsibility for achieving broad sustainable benefits. Its characteristics and requirements, as indicated by the literature, are presented below.

In contrast to the traditional work methods, the management of design information and processes integration in green building planning involves a wider range and a larger number of consultants, using sophisticated environmental modeling and analysis systems. A comprehensive understanding of the multi-level interconnections between technologies, people, project phases, processes, and systems is needed to address the green BIM requirements [12]. Green BIM requires consideration of processes and technologies, as well as an information management strategy that supports inter-organizational collaboration for a sustainable project. Hence, the effective management of information is likely to require an extensive dialogue with stakeholders to meet these green BIM requirements. Moreover, in a detailed examination of the integration of BIM technology with green practices, in order to achieve the sustainable environmental benefit, the need to achieve social benefit is revealed. It turns out that alongside the technical issues—which include references to software [13], technical skills [14], and the technological process [15]—there is a need for access [16] and awareness of all the parties involved in the project [17]. Moreover, in evaluating the success factors that can increase the connection between BIM and green building, it has been found that stakeholders have an important part in this, in light of the fact that among the first factors is their level of awareness and involvement [18]. The full application of green BIM thus requires reference to social integration alongside technical integration.

The literature review provides some important insights on optimizing the effective adoption of BIM for sustainability, which involves the need for an appropriate collaborative practice [19]. It turns out that the overall level of collaboration in the common data environment is not at a threshold level enough to realize BIM's full potential [20]. Moreover, conventional contractual measures do not appear to be compatible with the characteristics of BIM. Thus, some studies have begun to examine an integrative approach for addressing social issues. Integrated-project-delivery (IPD) has been proposed to improve communication and collaboration, enabling sustainable achievement [14]. Against this backdrop, the need for empirical research has been raised to learn the best practice of providing BIM for better social sustainability [21].

It is, therefore, possible that the prevailing combination of green buildings and BIM, a combination that has received considerable coverage in the literature [22], does not fully fulfill its purpose. Although BIM has been proposed as a solution to common obstacles in green construction, which include cost overruns and delays related to increasing design and construction complexity, the proposals are required to address the fact that this solution is a function of the full realization of its social dimension. In view of this, a means is needed to examine the realization of BIM's social capacity in this combination.

2.2. Means of Implementing Social Integration through BIM for Sustainable Green Building

In order to address the need to integrate stakeholders in green BIM and to make full use of the socio-technical integration, appropriate guidance and evaluation measures are required. The following lines have reviewed various possible measures to promote social sustainability in the construction industry, as part of an attempt to achieve, through BIM, a sustainable green building project.

The green building assessment tools have been developed with the aim of assisting in the implementation of sustainable development in the construction industry. However, they lack a detailed analysis of the social aspect of sustainable development [23]. Furthermore, BIM integration can provide information to support the calculation of a number of credit points to define goal levels of sustainability associated with green building rating systems [24]. It allows the evaluation of multiple design scenarios simultaneously, environmentally, and financially [25]. However, there is no comprehensive assessment and measurement tool to promote social sustainability through BIM in green building.

In an attempt to promote social sustainability in the construction industry, it has been suggested to use social-network-analysis [26] and apply it to construction—where a project-based organization is prevalent [27]. While identifying the status of the stakeholders within the social network of a project indeed allows an assessment of the social value obtained from it, there is no standard for achieving this value in a green building project. Therefore, the study has used this view to present a normative approach for the application of BIM in green building by SNA. Based on a socio-technical perspective that presents the implementation of BIM as an influencing factor for the project-based organization, and considering the importance of BIM's social application for achieving sustainable green building, this study presents a standard for its application, as well as an SNA measure for its assessment.

In addition, the need to promote sustainable development in the construction industry has led to the development of a framework to assess the performance of the corporate-social-responsibility of the construction corporation [28]. Different indicators should provide guidance for the implementation of social responsibility in the construction industry and allow organizations to build and assess the performance of social responsibility, which, in turn, could assist in achieving sustainable business development. However, it seems that for purposes of presenting CSR indicators, a transparent weight system is required, as well as an examination of stakeholders' interactions [28].

CSR is based on the premise that organizations need to behave in a socially responsible manner [29]. It is also possible to point out the complementarity of CSR and stakeholder theory [30]. This theory recognizes that organizations have commitments not only to shareholders but also to other interest groups, such as customers, employees, suppliers, and the wider community [31,32]. Sustainable development, corporate sustainability, and CSR are closely related to stakeholder relationship management, but at different levels of performance. While sustainable development is a guiding model at the society level, and corporate sustainability is a sustainable development model at the corporate level, CSR is a management approach to business contribution for sustainable development [29]. On this basis and for the purpose of this paper, CSR is defined as a corporate management approach that addresses all stakeholders involved in a construction project in an attempt to realize their sustainable benefits within business processes that include the use of BIM. Mapping of information transfer through BIM may facilitate the examination and evaluation of interactions between the corporate stakeholders. Therefore, this study uses the BIM application to promote sustainable green building while presenting a standard and CSR-based SNA index for its evaluation.

3. Formulation and Application of Standard and Index for Social Integration in Green BIM

3.1. Proposed Standard for Social Integration in Green BIM

In response to the need for social integration to achieve sustainable goals of green BIM, a CSR-based BIM application model is presented. In terms of relevant guidelines for the application of BIM as a social system—the CSR-based model includes elements for achieving social sustainability, which, according to this paper hypothesis, enable effective green building. Social sustainability components include; fairness, which provides equal opportunities for all; awareness, which fosters alternative consumption habits; participation, which relates to the inclusion of as many groups as possible in decision-making; cohesion, which strengthens community integration [33]. In addition, in terms of suitable project delivery conditions, the integrated-project-delivery (IPD) method includes contractual components that reinforce the BIM application model for achieving sustainable green building. IPD is a method that attempts to align interests by a group-based attitude. The main group participants consist of the owner, architect, general contractor, and major consultants. Beyond IPD principles according to industry definitions—including multi-party agreement, early involvement of all parties, and shared risk and reward—a survey has found that 'good leadership is required to encourage a collaborative team environment' [34]. Moreover, the IPD vision includes the involvement of end-users at the beginning of the planning process [35]. This method gains different levels of detail and application [36]. The broad definition refers to many owners, mainly public owners, who are not authorized to enter multi-party agreements and to bring subcontractors into the planning process. However, to take advantage of some of the key benefits of IPD-type delivery, many contractual provisions and project procedures can be modified. These include bringing the construction manager (CM) to the project at the beginning of the process, co-location of the team, and establishing a team decision-making process and structure [37]. The classification of social sustainability components—in the combination of BIM and the interrelations between those involved in a project—mirrors relationships with stakeholders of a responsible construction corporation in terms of CSR interpretation. This enables improved management of all involved, professionals and non-professionals alike, throughout all phases of the building. The carefully shaped project-based organization can be quantified and evaluated as a basis for comparison in terms of social networks. A social network is based on a set of actors and the relationships between them [38]. With the use of social network theory and social-network-analysis, it is possible to describe and analyze interactions between participants in construction projects [39–43]. Five levels of social sustainability are identified and classified according to CSR for a green building project [44]. These levels are adapted to IPD and are translated into SNA indices [45]. The criteria for this model, with their indices, are becoming a benchmark—The BIM Integrated Application Standard—for evaluating the socio-technical integration and promoting its benefits. Given that standardization is a key enabler for advancing BIM implementation [46], this specific standard is proposed to promote the implementation of BIM's social potential in green building. Standard components, which include (1) stakeholder management—through BIM manager centrality, (2) use of a BIM-based social network, (3) benefits through connecting all stakeholders, (4) and tenants, to the BIM manager, are presented below in describing the benchmark index (A detailed conceptual description is given in Appendix A).

3.2. Green BIM Index

The proposed examination is consistent with the trend in literature to introduce formal and informal institutions that influence a project-based organization [47]. As a result, the use of BIM is presented as another factor shaping this organization. However, unlike the tendency in studies to examine formal institutions, such as project delivery contracts, alongside informal institutions, such as work practices, the proposed examination is in relation to an external reference point. In view of the gap reflected in the BIM guidelines and the various valuation methods, in terms of achieving the sustainable goals of green building, and given the importance of BIM's social integration to achieve them, it is proposed to compare its actual combination with its optimal one; i.e., a standard combination in a given project. This comparison by standard criteria is conducted using the components of the Green BIM Index.

3.2.1. Index Components

1. Stakeholder management ratio—weighted degree centrality index (X1). This measure constitutes a means for assessing the implementation of the stakeholder management criterion, by presenting the ratio between the actual weighted degree and a standard weighted degree, in relation to the BIM manager in each project. In this way, the degree of centrality of the BIM manager, as a management implementer for stakeholders, relative to the required level, is reflected. The use of this index component enables the realization of the full potential of BIM leadership in a given project (Equation (A2) in Appendix B).
2. Stakeholder participation ratio—BIM-based social networking cluster index (X2). This measure constitutes a means for assessing the implementation of the stakeholder participation criterion by presenting a ratio between the number of project participants connected to a BIM-based social network and the optimal number of connections. In this way, the participation of stakeholders, relative to nodes identified, as required in a BIM-based network, is reflected. This index component enables the examination of the promotion of collaboration in a given project (Equation (A3) in Appendix B).
3. The ratio of professional involvement—ego network cluster index of BIM manager (X3). This measure serves as a means to assess the implementation of the engagement criteria of all professionals by presenting a ratio between the number of direct links to the BIM manager and the optimal number of links. The index component enables examination of the connectivity of the professional team, including the contractor or the construction manager, already at the design phase of the building (Equation (A4) in Appendix B).
4. The ratio of tenant involvement—ego network cluster index of tenant representative (X4). This measure serves as a means to examine the implementation of the engagement criterion of all

non-professionals by presenting a ratio between the number of their direct connections to their representative and the optimal number of connections. The ego network cluster index enables the examination of connectivity, in terms of tenants and end-users, through their representatives, from the design phase of the building (Equation (A5) in Appendix B).

5. The ratio by stage (X5). As part of comparing the actual use to the standard, which reflects a project reference, the index enables the application of the criteria to be considered throughout the phases of the building. This examination is based on the calculations of the indices at each phase within a project (Equation (A6) in Appendix B).

3.2.2. Demonstrating the Way to Use Index Components

The way to use the components of the index is illustrated below, in a specific example of a project that combines BIM in green building. The examination of the integration between the two is performed using three tables: SNA results table for the actual implementation of BIM in green building (Table 1), SNA results table for optimal implementation of BIM in green building, according to standard assumptions (Table 2), and table of findings of the Green BIM Index (Table 3). This project relates to a detailed design stage of a cafeteria. Although the standard assumption in the given project refers to 12 required stakeholders, in practice, two of them did not participate in the sample (the tenant and the landscape architect).

Table 1. SNA results for the actual implementation of BIM in green building.

ID	Label	Actual Implementation Results		
		Weighted Degree	Weighted Out-Degree	Weighted In-Degree
1	AC engineer	43	13	30
2	Architect	55	20	35
3	BEM Specialist	28	3	25
4	Tenant	0	0	0
5	Constructor	37	10	27
6	Electrical engineer	38	10	28
7	Cafeteria consultant	67	25	42
8	Landscape architect	0	0	0
9	Plumbing Consultant	42	12	30
10	Project Manager	22	0	22
11	Regulatory advisor	28	3	25
12	BIM manager	228	198	30

Table 2. SNA results for optimal implementation of BIM in green building.

ID	Label	Optimal Application Results		
		Weighted Degree	Weighted Out-Degree	Weighted In-Degree
1	AC engineer	27	5	22
2	Architect	27	5	22
3	BEM Specialist	23	1	22
4	Tenant	23	1	22
5	Constructor	27	5	22
6	Electrical engineer	27	5	22
7	Cafeteria consultant	23	1	22
8	Landscape architect	23	1	22
9	Plumbing Consultant	27	5	22
10	Project Manager	23	1	22
11	Regulatory advisor	23	1	22
12	BIM manager	273	242	31

Table 3. Green BIM Index findings.

BIM Integrated Application Standard	Project Results		Green BIM Index
Components	Actual	Optimal	Components
Participants connected to a BIM-based social net	9	11	0.8
Connections with professionals	9	10	0.9
Connection with a tenant representative	0	1	0
BIM for stakeholder management-weighted degree	228	273	0.84
Score			0.64

The assumption for a BIM manager is that he/she sends information to everyone daily, receives information from key participants on a daily basis, receives information from the secondary to them on a weekly basis while receiving from other participants on a monthly basis. In the example, the BIM manager did send information as required but did not receive a standard response (some sent on a weekly basis, and some did not send at all). The assumption results in a weighted degree of 273 compared to 228 in practice, resulting in an index finding of 0.84 (X1).

In addition, according to the standard, all involved, professionals and tenants alike, should be connected, directly or indirectly, to a BIM-based social network. This is reflected in 11 optimal connections versus nine actual connections, resulting in an index finding of 0.8 (X2).

Besides, the standard assumption is that there is no interaction but through a BIM-based network and with a BIM manager or agents. This is shown in the result of 10 optimal connections of professionals compared to nine of their actual connections and in an index finding of 0.9 (X3). This is in addition to one optimal connection of a tenant representative compared to 0 connections of an actual tenant representative and to an index finding of 0 as a result (X4).

The Green BIM Index score in relation to the BIM application at the detailed design stage in the green building project is 0.64 (X5). This index can be used as a means of improving capabilities in additional stages. Figure 1 graphically shows the performance metrics at the given stage.

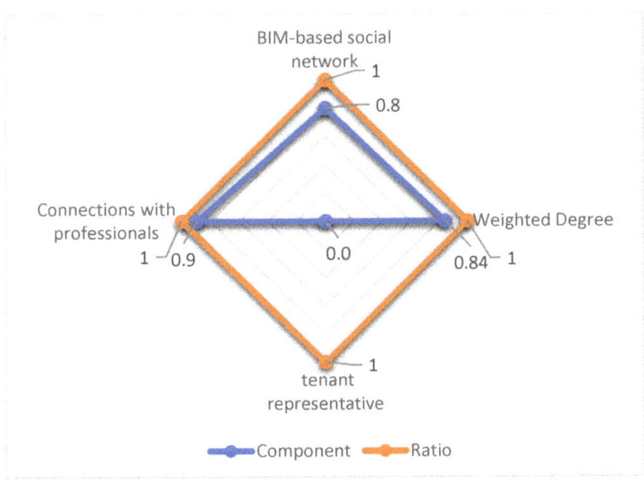

Figure 1. Graphical display of the Green BIM Index components.

4. Research Method

4.1. Applying CSR for Sustainable Benefits of Green BIM Using SNA

It turns out, then, that the green BIM is not only a combination with a purely technical environmental dimension but also a combination with a complex human aspect, which requires an overall social reference. This includes managing a large number of participants, including end-users with real environmental interests, as well as other interests associated with their connection to the building. Implementing green BIM through CSR may facilitate the social paradigm shift in the construction industry to achieve effectiveness for all involved. Reflecting green BIM performance on quantitative measures to assess the various interactions of a BIM combination in a green building project may be a significant tool for evaluating this application.

The research workflow outlined in Figure 2 presents the measures taken across the various parts of the research to present an applicable benchmark for broad social integration of BIM with green building. The study begins with a comprehensive review of the literature on green BIM to identify social characteristics and requirements for sustainable benefits. Different tools are examined in terms of social integration in order to achieve sustainable green building. The results illustrate the absence of effective integration between BIM and green building and the lack of suitable means for its evaluation (top row in Figure 2). This serves as the basis for the implementation of a CSR-based model for the BIM application, adapted to the IPD method and quantified by SNA. Based on the SNA indices for comparing actual results and optimal results—according to the standard—of the BIM system, the possibility of maximal integration of BIM with the green building is proposed. In addition, this systemic perception is examined using a participant questionnaire to assess the social sustainability benefits alongside effectiveness in a given project (middle row in Figure 2). Comparison and validation of this measure are conducted in nine case-studies, reflecting deferent types and stages of a project (the bottom line in Figure 2).

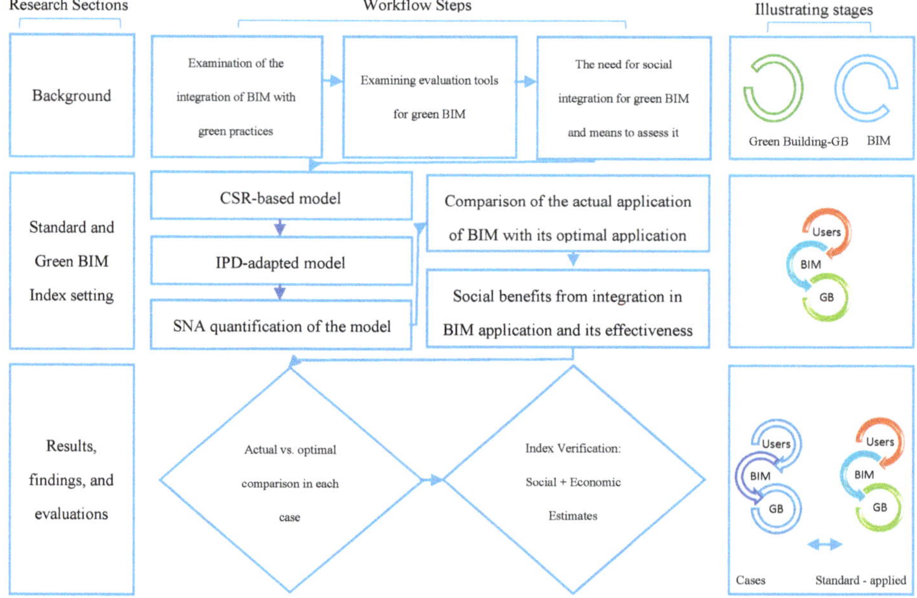

Figure 2. Outline of the research workflow.

4.2. The Case-Study Approach

The index proposed in this paper is intended to serve as a means of measuring social integration in a green BIM in an attempt to realize its benefits. In order to demonstrate and evaluate the implementation of the Green BIM Index, the case-study approach is selected. This approach is considered most appropriate for this research inquiry, as it tries to construct realistic representations of project-related communications in uncontrolled surroundings [48]. In the absence of a comprehensive study on the social perception of BIM in green building projects, and in order to attain certainty [49], green building cases that incorporate BIM are selected to examine the possibility of achieving effectiveness in this building through this socio-technical system.

For a thorough examination of the model implementation, we have chosen to focus on public projects. In Israel, public projects are taking the main share of green BIM. The public sector is often characterized by the publication of project auctions in which it is owner-occupier—allowing accessibility to examine the implementation of all parts of the model, including involving end-users. The urban public sector usually allows the end-user community to be located, so we have chosen two municipal projects, which incorporate office dwellers or end-users at the planning stage. The two selected projects are located in cities that belong to the 15 cities forum that incorporates most of the largest cities in Israel, which has taken on the integration of green building. Both are designed by the same architect and by the same green building consultant, with the desire to integrate the constructor manager, so the selection has allowed a comparison on a similar basis. The government public sector is a major client in the construction industry that uses BIM and assimilates the green building principles through threshold requirements for tenders. This sector allows the examination of large-scale projects, so we have chosen to compare seven projects from it—two projects of each type/budget, stage, and even architectural teams at the same stage in a project. In this way, the selection allows for a broad comparison of BIM social implementation on an equal basis. The types of projects and the clear separation between the design phase and the execution phase in this sector have resulted in adapting the model to the different number of stakeholders in each. Table 4 shows the key characteristics of the carefully chosen projects. In the various cases selected, the attempt is to validate the findings of the proposed index with the results of the social benefit assessment and the effectiveness of incorporating BIM in green building. The following sections describe how data is collected and analyzed in an attempt to show how measurement is performed and confirmed.

Table 4. Characteristics of the projects.

No.	Project	Stage	Purpose	Budget	Size	Features
1	Municipality project	Conceptual design	City Hall	$24,548,913	4151 m^2	Downtown area
2	Municipality project	Conceptual design	Central station	$67,789,183	27,620 m^2	Downtown area
3	Government project	Conceptual design	Catering	$2,857,143	530 m^2	Rural area
4	Government project	Conceptual design	Campus	$42,857,143	20,900 m^2	Suburban area
5	Government project	Detailed design	Catering	$10,000,000	3400 m^2	Rural area
6	Government project	Detailed design	Campus	$10,000,000	3000 m^2	Rural area
7	Government project	Construction	Catering	$6,285,714	1686 m^2	Rural area
8	Government project	Construction	Campus—Part 1	$52,571,429	13,400 m^2	Rural area
9	Government project	Construction	Campus—Part 2	$58,000,000	16,700 m^2	Rural area

4.3. Network Properties and Data Collection

4.3.1. SNA Questionnaire

Setting up network properties is necessary to mirror the actual interactions and to display the relevant components for examining BIM applications in each project. Network boundaries are defined by the BIM communication platform for professionals and by all types of communications for tenants. Because the optimal model is implemented according to a given project, the boundaries include several tenants' representatives in urban projects—to express personnel involvement, as well as a closed list of professionals required at different stages and types of government projects—to compare them. Data is collected through SNA questionnaires for actors, selected by case documents and interviews, which are senior representatives of the parties to the projects and are directly involved in their communication. The questionnaire presents a detailed list of actors from which actors are asked to choose the ones they communicate with in exchange for information. They are also requested to indicate their communication frequencies with them. The urban projects require extensive preparation work to find relevant representatives of office users. For this purpose, visits and background interviews are held. In one municipality, representatives of office users, selected by the municipality's management, are guided by feedback from all municipal employees. In a second municipality, the planning team includes elected representatives from departments and offices. The questionnaires are filled out via email correspondence and through telephone interviews. In the government projects, the questionnaires are filled out by telephone interviews with representatives of the AEC (Architecture, Engineering and Construction) industry, and with the owner-occupier representatives who worked with them (the questionnaire is presented in Appendix C). Data collected is transformed into the weighting of contacts (Table in Appendix D).

4.3.2. Participant Questionnaire

In order to determine the validity of the measure, questionnaires for project stakeholders are set. The questionnaires are based on frameworks and models presented in the literature for assessing social sustainability and benefits from BIM. In urban projects, access is given to various personnel teams during the planning stage. Their representatives, selected by them as having access to BIM, fill out the questionnaires (five from the City Hall project, three from the station project). In the government projects, representatives of all stakeholders fill the questionnaires. For comparing them—in terms of type and stage—at the conceptual planning stage and at the detailed planning stage, 12 stakeholders are required for a cafeteria-type project and 11 for a campus-type project. These include the BIM manager, architect, constructor, air conditioning, electricity and plumbing consultants, landscape architect, project manager/owner representative, energy modeling consultant, green building consultant, and tenant representative. A cafeteria-type project also requires a special consultant. At the construction stage, a contractor is added (13, 12, respectively). To cover all involved, experienced and inexperienced, in working with BIM, both subjective and objective assessments are combined. As part of their assessments, representatives are asked to rate, according to the Likert scale, the social benefits and effectiveness of BIM integration (1–5). In addition, objective project data is requested. Four questions deal with social aspects, including fairness, awareness, participation, and cohesion [33]. These elements are chosen because of the important conceptual presentation that identifies four general social concepts and links them to environmental imperatives. Five to seven questions, depending on the project stage, are related to subjective evaluation of a BIM implementation, which includes project quality improvement, better cost forecasting, quick client approval cycles, reduction of construction disputes, improvement of collective understanding of planning intent, reduction of construction changes, and reduction of RFI—request for information [50]. Two questions, intended for the project manager, are related to an objective evaluation of the BIM implementation and include highly mentioned metrics in the literature, schedules, and cost changes [51]. These calculation frameworks for analyzing the benefits of BIM are chosen because they summarize major subjective and objective parameters in the literature. In urban projects, where the reference is to the benefit of office users, the questionnaires are filled by a broad representation on their behalf. In the government projects, where the reference is to the benefit of involvement in various stages and types of projects, the questionnaires are filled out by all the representatives (the questionnaire is presented in Appendix E).

4.4. Data Analysis

The social networks are analyzed using Gephi software [52]. Gephi is an open-source network analysis and visualization software package that has been used in a number of research projects [53].

Weighted-degree-centrality and cluster are used to specify connections, as well as the most connected nodes, especially with respect to key actors according to the IPD-adapted model: clients—including owners and tenants, architects and contractors, or alternatively, construction managers. In the government projects, the model is adapted to the project-delivery-method used, the design-bid-build method, which involves the contractor only during the construction stage. For each project, an optimal SNA model and an actual SNA model are prepared. The optimal models are based on the forms of information transfer, which make the most of BIM in terms of sharing and engagement. The preparation of the actual models requires data filling in two tables, one representing the frequency of information giving, and the other the frequency of receiving information. After specifying the information transfers, adjustments are made according to the frequency weight, and the data is entered in Gephi software. The results of the different SNA models are used for the index equations. The index is based on the understanding that each project has unique characteristics and different social composition. Therefore, it is formulated as a tool for quantitatively assessing the performance of each green building project in relation to its optimal workability with the BIM system. Along with these findings, the evaluations of the benefits and effectiveness from the integration into the BIM application are calculated. The findings and evaluations are compared to confirm the validity of the index in an attempt to emphasize the importance of promoting social integration in green BIM to achieve its effectiveness. Figure 3 visually shows the comparisons being conducted to examine whether the social benefit level has implications for green building effectiveness and whether this is reflected in the proposed index.

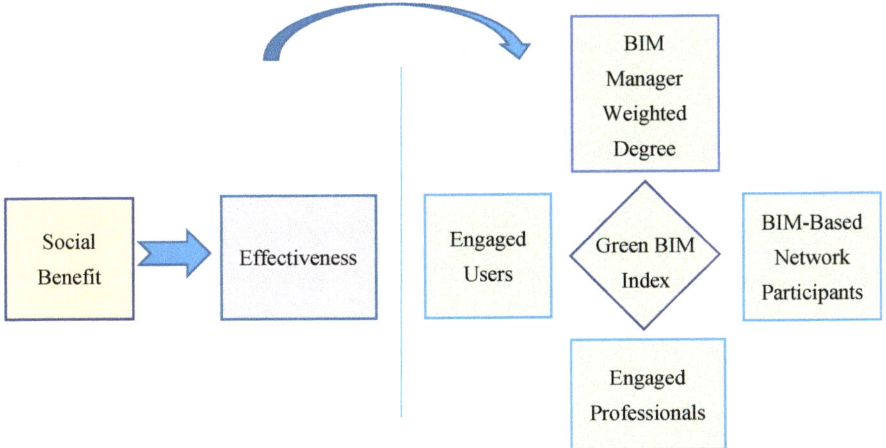

Figure 3. BIM application in terms of social benefit, effectiveness, and the proposed measure.

5. Evaluations, Index Findings, and Their Fitting

Table 5 presents the evaluations of social benefit and effectiveness alongside the findings of the index in each project. Each of the project parameters is averaged, sorted by levels—high (H), medium (M), low (L)—for comparison. For this purpose, level H—indicates a rating from 4/0.8 and higher, level M—indicates a range that reaches 2.75/0.55, while L—indicates a lower rate. Likewise, an objective estimate for the change in effectiveness is rated H—for performance unchanged and below, M—for a change up to 80%, and L—for change beyond that. The fit reflects identical results in the parameters.

Table 5. Comparison of evaluations and index findings in the various projects *.

| No. | Project Purpose | Stage | Social Benefit | | | | | Effectiveness | | | | | | | | | | | | Green BIM Index | | | | | Fit |
| | | | | | | | | Subjective | | | | | | | | Objective | | | | | | | | | |
			Fairness	Awareness	Participation	Cohesion	Average	Quality Improve	Cost Forecasting	Quick Approval	Reduces Dispute	Understanding	Reduces Changes	Reduces RFI	Average	Schedule Changes	Cost Changes	Average	Weighted Degree	BIM-Based Net	Engaged Pro	Engaged Users	Average	
1	City Hall	Conceptual design	4.6	4.2	4.8	4.4	4.5 H	5	5	5	-	5	-	5	5 H	No	No	H	0.43	0.96	0.95	1	0.84 H	✓
2	Central station	Conceptual design	5	4	4.6	4	4.4 H	5	3	5	-	5	-	3	4.2 H	No	No	H	0.34	0.94	0.93	1	0.8 H	✓
3	Catering	Conceptual design	3.37	2.37	2.5	2	2.56 L	2.62	1.62	1.25	-	2.87	-	2	2.07 L	+100%	+2%	L	0.07	0.63	0.7	0	0.35 L	✓
4	Campus	Conceptual design	4.77	4.88	4.66	4.33	4.66 H	4.1	4.1	3.88	-	4	-	4.21	4.05 H	No	No	H	0.84	0.8	0.78	1	0.85 H	✓
5	Catering	Detailed design	3.9	3.4	3.3	2.5	3.28 M	3.5	2.2	2.5	-	3.8	-	3.1	3.02 M	+71%	+16%	M	0.83	0.81	0.9	0	0.63 M	✓
6	Campus	Detailed design	4.22	2.33	2.56	2.22	2.83 M	3.44	1.78	2.22	-	4.1	-	2.55	2.81 M	+50%	+10%	M	0.9	0.8	0.88	0	0.64 M	✓
7	Catering	Construction	4.33	3.22	3.33	3.11	3.5 M	4.11	3.11	3.22	4.44	4.78	3.78	3.22	3.81 M	No	No	H	0.83	0.67	0.64	1	0.78 M	✓
8	Campus—Part 1	Construction	4.11	3	3.33	3.22	3.42 M	4	2.89	3.22	4.22	4.78	3.67	3.44	3.75 M	No	No	H	0.78	0.7	0.7	1	0.79 M	✓
9	Campus—Part 2	Construction	5	4.44	4.78	4.67	4.72 H	4.67	4.33	4.89	4.56	5	4.22	3.44	4.44 H	−10%	−1.3%	H	0.9	0.73	0.7	1	0.83 H	✓

* level H—indicates a rating from 4/0.8 and higher, level M—indicates a range that reaches 2.75/0.55, while L—indicates a lower rate. Likewise, an objective estimate for the change in effectiveness is rated H—for performance unchanged and below, M—for a change up to 80%, and L—for change beyond that.

Project #7 – Catering – Construction – Comparison per project

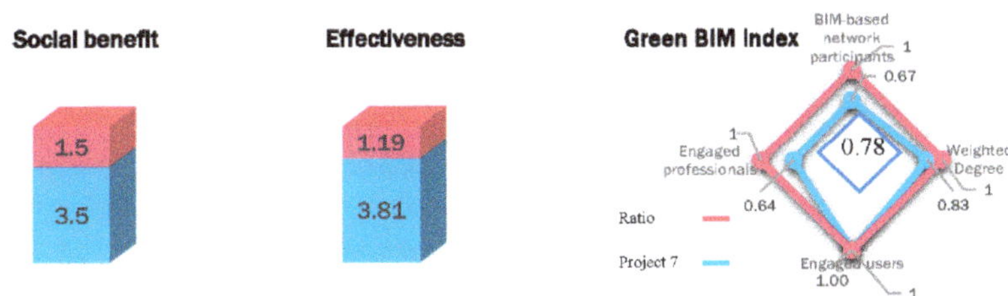

Project #5 vs. Project #3 – Catering – Design – Comparison per stage

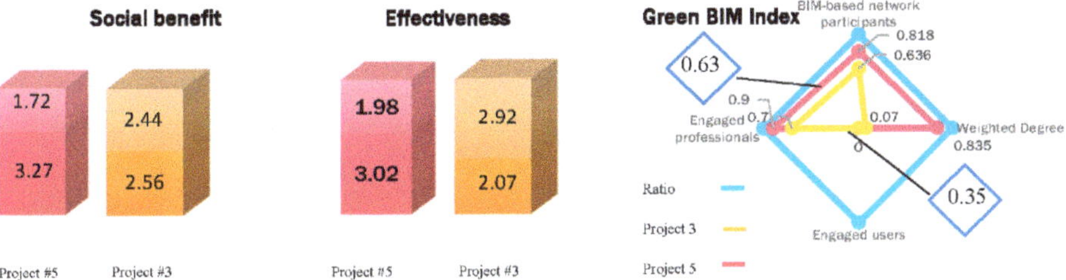

Project #4 vs. Project #6 – Campus – Design – Comparison per stage

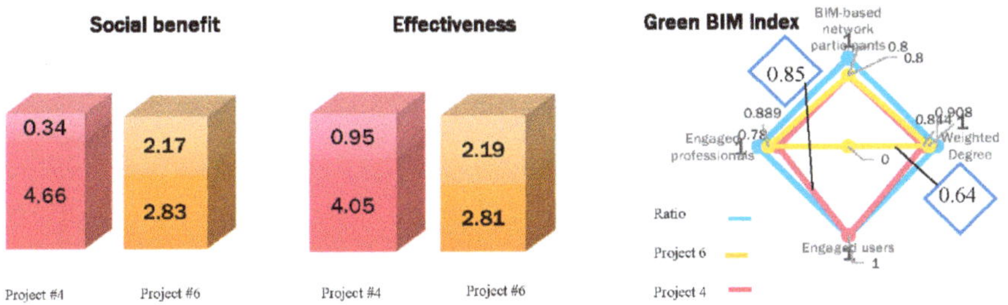

Project #9 vs. Project #8 – Campus – Construction – Team comparison

Figure 4. Comparison of evaluations and index findings for projects by stages and teams.

In addition to this comparison of parameters for each project, Figure 4 presents a comparison of projects. The figure shows in detail the process of comparison conducted in the index between optimal implementation and actual implementation in a given project. It also presents a classification of projects with the same purpose, by stages and by teams, in an attempt to examine the possibility of overlap in trends between the three types of parameters.

6. Discussion

In this paper, the optimal BIM implementation models are characterized, for all stages and project types, to compare projects, with their social and economic implications, using BIM management measures. The tables above clearly show the relationship between social benefit and effectiveness, alongside the ability to reflect this in the proposed index.

Examining the projects reveals a fit for each project, as well as between projects—in terms of linear changes by stages and by teams in the same project. This applies to both professional and non-professional populations in projects. Social and economic background data back up the findings of the index. For example, in urban projects, there is a social attempt to integrate selected representatives of employees and municipalities into the planning team, which results in great efficiency and a high index. Projects 3 and 5 are characterized by the involvement of catering consultants, and, therefore, a special optimal BIM implementation model has been prepared. These consultants are not involved in communications, and the other professionals are not adequately skilled, resulting in unnecessary delays and expenses. However, in project number 5, the BIM manager is more involved, which has made the index different. In contrast to project 4, in project 6, the end-user is not involved, which has led to a late revision of the campus plan, change of schedules, and huge costs, a difference that is well reflected in the index. Projects 7,8,9 belong to the same compound and are deliberately separated for research purposes. Project 7 requires the preparation of a BIM implementation model for the construction stage, which also includes the contractor teams. Projects 8 and 9 are carried out by various teams and require, in terms of research, the preparation of a BIM implementation model for the campus in the construction stage. Although, in terms of objective effectiveness data, projects 7 and 8 are the same, and the comparison with the optimal model reveals the great work that the BIM manager has done in dealing with the lack of skills and social connection. The same BIM manager has gained greater collaboration in project number 9. This means that according to industry-standard project management parameters, the work processes required at various points of construction could not be discovered. The difference between the results is due to the resolution of the index, which is measured point-wise, compared to parameters measured over a period of the project. The index score thus allows evaluating performance and improving it further during the work, while providing a targeted response to the needs and differentiation in the project. It can be seen, therefore, that there is a direct relation between the level of social benefit and the subjective and objective levels of effectiveness per project, which is reflected in the index.

The examination validates the concept presented by the research model regarding the relationship between BIM social system and the social and economic benefits of a green building project. In examining the suitability between the management of the BIM system and the social benefit and effectiveness of its combination, it emerges that: the closer the BIM performance index is to 1, the greater the social benefits of those involved in the project and the effectiveness. These conclusions are supported by literature on the positive effects of teamwork on project performance [54] and extend it by presenting the effect of social integration on sustainability in a green building project.

7. Conclusions

This study expands the understanding of the importance of social integration through BIM to achieve sustainable green building while introducing a benchmark for BIM socio-technical application to increase green building effectiveness. This importance is presented in a gradual manner through three stages, summarized below.

1. Presentation of a socio-technical model for defining the integration of BIM with green building for sustainable benefit;
2. Presentation of a CSR-based BIM system application standard for achieving a socially sustainable effective green building project;
3. Presentation of a confirmed green BIM index as a useful tool for evaluating this application.

First, this study outlines a broad concept of green BIM, with reference to the possibility of social integration between green building and BIM. The definition of this reference is made by clarifying the social purposes of this building and the socio-technical capabilities of this system as a background for examining their actual integration. Given the gap, expressed in the absence of effective social-involvement, a BIM management application model is presented based on the CSR concept.

Second, in order to address the need, which has emerged from the literature, to integrate stakeholders into green BIM, and to use it fully in socio-technical integration, the paper has proposed a means of guiding and assessing this integration as standard. This reference is based on adapting the BIM application model to the preferred method for a green BIM project. Following the conceptualization of a project-based organization as a social network by literature, SNA is used to examine the combination of BIM and green building.

Third, through the concrete realization of the application model, this study presents a BIM system index in an attempt to bridge the gap and promote effectiveness in the green building industry. By comparing the actual application of BIM with the criteria of The BIM Integrated Application Standard, using the Green BIM Index, an option is presented to improve the flow of information and assist in the processing of information for all those involved in a green building project. This index is implemented and evaluated in nine case-studies, reflecting a variety of different types and stages of projects. The examination confirms the validity of the index and provides evidence that the more the actual use of BIM fits the proposed standard, the greater the social benefit and effectiveness of the project.

The conclusions of the study, regarding the importance of social involvement through BIM for the success of the green building, may form the basis for further research in the field. Possible research directions may include the assessment of social impact through social networks [55], as well as the use of email log [56] as empirical evidence from industry to validate the Green BIM Index and present its implications in various projects. By using a primary data source for mapping project communication networks, it is possible to compare BIM index-based projects in different industries.

This study presents for the first time a practical and feasible tool for examining BIM integration with green building, in an effort to promote the effectiveness of this combination. The presentation of this index to examine the application of building information management enables the construction industry to examine the social responsibility of green building companies in relation to their stakeholders along the supply chain. As a result, this study may have implications for the perception of a green building project as a sustainable project. In the green BIM projects, which are more socially complex and technologically sophisticated than regular projects, implementing the Green BIM Index, which facilitates stakeholder integration, may be of great benefit to the industry.

Author Contributions: R.M.-L. conceived the research, collected and analyzed the data, drafted, and edited. U.G. collected objective data from government projects and reviewed the draft. I.R.—guidance of data analyses, literature review support, revision of manuscript for important intellectual content, supervision. All authors read and agreed to the published version of the manuscript.

Funding: This research received no external funding.

Acknowledgments: The authors would like to thank the experts who reviewed the manuscript.

Conflicts of Interest: The authors declare no conflict of interest.

Appendix A. The Conceptual Description for BIM Integrated Application Standard Components

	Criteria	SNA Indices	Basic Assumptions
1	Stakeholder management	Weighted degree centrality. In order to examine the centrality of the BIM manager as a management implementer in terms of stakeholders, the study uses the means of weighted degree centrality. The BIM manager is important if there are many other nodes—actors that link to him/her or if he/she links to many other nodes. Weighted degree centrality of a node is 'the sum of weights of the ties of the node with the other nodes' (Kapoor et al., 2013). It follows that the higher the centrality, the greater the BIM management leadership.	The standard assumption regarding the BIM manager, based on preliminary research, is that he/she sends information to everyone daily, receives information from key participants on a daily basis, receives information from the secondary to them weekly while receiving from other participants on a monthly basis. These transitions are required for optimal application. To manage stakeholders, it is suggested that the BIM manager should have a high level of weighted degree centrality, as standard.
2	Stakeholder participation	Cluster. In order to examine the participation of stakeholders in a specific workgroup, the research is based on a cluster approach. Cluster nodes are 'more connected with nodes of this cluster than with nodes outside the cluster' (Cuvelier and Aufaure, 2012). Finding a cluster is based on how they are connected to one another, and therefore, according to that, it is possible to describe a shared connection or access to joint resources in a project.	It is proposed that all involved, professionals and tenants alike, should be connected, directly or indirectly, to a BIM-based social network. The reference in this BIM indicator is to participants in an IPD-based cross-border coordination mechanism. This technological accessibility, as an expression of the possibility of equal participation as a conscious group, is required for optimal application in a given project.
3	All professional stakeholders are engaged	Ego net cluster. In order to examine the way stakeholders are involved and their connectivity as a group, the research is assisted by means of ego nets clusters. Ego nets represent the connections among the neighbors of a given node (Epasto et al., 2015). In an attempt to locate a community of direct contacts with the BIM manager and the tenant representative, it is suggested that the size measure be used as algorithms for identifying communities or clusters, thus allowing specifying the number of participants in the BIM manager and the tenant representative groups. Since the size of an ego-network is considered as the amount of neighboring nodes plus the ego (Hanneman and Riddle, 2005), it is possible to specify the direct links to them.	The participation can be divided into sub-clusters. The references in these BIM indicators are in relation to the manner in which the different participants are involved. With regard to the engagements of all professional and non-professional, it is proposed to define the ego nets of the BIM manager and the tenant representative, equal to the maximum possible size for each. The standard assumption, based on the need for management of stakeholders through the BIM manager, to promote information flow and full processing, is that there is no interaction except through a BIM-based network and with a BIM manager or agents.
4	Tenants are engaged		
5	Management, participation, engagements at all phases, within a project	The examination of the management of stakeholders—their participation and engagement—is required throughout the project.	The basic assumptions above regarding the indices in each criterion are applied at the design, construction, and maintenance phases, according to a given project.

Appendix B. The Equations of the Green BIM Index

The suitability between the results of the actual implementation of BIM and the results of the optimal application of the BIM standard proposed by the study is expressed in the BIM application index, which can be presented in the equation (Equation (A1)):

$$BIM\ Application\ Index\ (X) = \frac{The\ Actual\ Application\ of\ BIM}{Standard\ for\ Optimal\ Application\ of\ BIM} \tag{A1}$$

The comparison is applied in a number of formulas for examining the suitability in the various BIM standard parts:

1. **Stakeholder Management (Equation (A2))**

$$Weighted\ Degree\ Centrality\ Index\ (X1)\ = \frac{Weighted\ Dedree}{Standard\ Weighted\ degree} \tag{A2}$$

2. **Stakeholder Participation (Equation (A3))**

$$BIM-Based\ Social\ Net\ Clustering\ Index(X2) = \frac{\sum Stakeholders\ Connected}{Stakeholder\ Standard} \tag{A3}$$

3. **Involvement of Professionals (Equation (A4))**

$$Ego-Net\ Cluster\ Index\ of\ BIM\ Manager\ (X3) = \frac{\sum BIM\ Manager\ Connections}{BIM\ Manager\ Connections\ Standard} \tag{A4}$$

4. **Involvement of Tenants (Equation (A5))**

$$Ego-Net\ Cluster\ Index\ of\ Tenant\ Representative\ (X4) = \frac{\sum Tenants\ Rep.Connections}{T.\ Rep.\ Connections\ Standard} \tag{A5}$$

5. **The Application of BIM System at one of the building phases (Equation (A6))**

$$BIM\ Index\ by\ phase\ (X5) = \frac{X1 + X2 + X3 + X4}{4} \tag{A6}$$

The higher the suitability, the more the result of the index is close to 1.00. This means that as the BIM management team is more centralized in the network, as network sharing becomes apparent, for the connectivity of all involved, professionals and end-users alike, at any phase of the building, the performance of the project may be relatively more social.

Appendix C. SNA Questionnaire

Receiving Information		Frequencies			
	Mark the Respondent	**Daily**	**Weekly**	**Monthly**	**No interaction**
BIM manager					
Architect					
Constructor					
AC engineer					
Electrical engineer					
Plumbing engineer					
Landscape architect					
Project manager					
BEM consultant					
GB consultant					
Tenant					
* Contractor					
* Cafeteria consultant					

Giving Information		Frequencies			
	Mark the Respondent	**Daily**	**Weekly**	**Monthly**	**No Interaction**
BIM manager					
Architect					
Constructor					
AC Engineer					
Electrical engineer					
Plumbing engineer					
Landscape architect					
Project manager					
BEM consultant					
GB consultant					
Tenant					
* Contractor					
* Cafeteria consultant					

* By project type or project stage.

Appendix D. Weighting of Information Flow

Level of Information Flow	Weighting of Connection	Reasoning
Never	0	
Monthly	1	At least once a month
Weekly	5	At least 4 per month
Daily	22	At least 5 per week

Appendix E. Participant Questionnaire

Appendix E.1. Social Benefits

1	Have you been given open or equal access, relative to the other participants, to the project's BIM forum?	1–5
2	Did you get the opportunity to promote awareness, non-material ethical preferences, in the project's BIM forum?	1–5
3	Were you involved, independently or through delegates, in the BIM forum, so that your preferences were reflected in the decision-making process?	1–5
4	Have you been involved in the BIM forum, in a way that allowed you to work as a team or feel part of teamwork?	1–5

Appendix E.2. Effectiveness

1	How has BIM's specific use/application improved the quality of the project?	1–5
2	How has the specific use/application of BIM improved the predictability of project costs?	1–5
3	How has BIM's specific use/application improved client approval speed?	1–5
4	How has the specific use/application of BIM reduced conflicts during construction? *	1–5
5	How has the specific use/application of BIM enhanced the collective understanding of planning intentions?	1–5
6	How has the specific use/application of BIM reduced changes during construction? *	1–5
7	How has the specific use/application of BIM reduced the number of information requests on your part?	1–5

* This question is for the construction stage.

References

1. McGraw Hill Construction. Green BIM: How Building Information Modeling Is Contributing to Green Design and Construction. Available online: http://construction.com/market_research/FreeReport/GreenBIM/MHC_GreenBIM_SmartMarket_Report_2010.pdf (accessed on 26 June 2020).
2. Wu, W.; Issa, R. Integrated process mapping for BIM implementation in green building project delivery. In Proceedings of the 13th International Conference on Construction Applications of Virtual Reality, London, UK, 30–31 October 2013; pp. 30–39.
3. USGBC. What Is Green Building? U.S. Green Building Council. Available online: https://www.usgbc.org/articles/what-green-building (accessed on 26 June 2020).
4. NIBS. What Is BIM? National Institute of Building Sciences. Available online: https://web.archive.org/web/20141016190503/http://www.nationalbimstandard.org/faq.php#faq1 (accessed on 26 June 2020).
5. Lu, Y.; Wu, Z.; Chang, R.; Li, Y. Building Information Modeling (BIM) for green buildings: A critical review and future directions. *Autom. Constr.* **2017**, *83*, 134–148. [CrossRef]
6. Azhar, S.; Carlton, W.A.; Olsen, D.; Ahmad, I. Building information modeling for sustainable design and LEED® rating analysis. *Autom. Constr.* **2011**, *20*, 217–224. [CrossRef]
7. Wong, J.K.W.; Zhou, J. Enhancing environmental sustainability over building life cycles through green BIM: A review. *Autom. Constr.* **2015**, *57*, 156–165. [CrossRef]
8. El-Diraby, T.; Krijnen, T.; Papagelis, M. BIM-Based collaborative design and socio-technical analytics of green buildings. *Autom. Constr.* **2017**, *82*, 59–74. [CrossRef]
9. Ayman, R.; Alwan, Z.; McIntyre, L. BIM for sustainable project delivery: Review paper and future development areas. *Archit. Sci. Rev.* **2020**, *63*, 15–33. [CrossRef]
10. Stender, M.; Walter, A. The role of social sustainability in building assessment. *Build. Res. Inf.* **2019**, *47*, 598–610. [CrossRef]
11. Bonenberg, W.; Kapliński, O. The architect and the paradigms of sustainable development: A review of dilemmas. *Sustainability* **2018**, *10*, 100. [CrossRef]
12. Gandhi, S.; Jupp, J.R. Characteristics of Green BIM: Process and information management requirements. In *IFIP International Conference on Product Lifecycle Management*; Springer: Berlin, Germany, 2013; pp. 596–605.
13. Gourlis, G.; Kovacic, I. Building information modelling for analysis of energy efficient industrial buildings—A case study. *Renew. Sustain. Energy Rev.* **2017**, *68*, 953–963. [CrossRef]
14. Wong, K.; Fan, Q. Building information modelling (BIM) for sustainable building design. *Facilities* **2013**, *31*, 138–157. [CrossRef]
15. Gerrish, T.; Ruikar, K.; Cook, M.; Johnson, M.; Phillip, M. Using BIM capabilities to improve existing building energy modelling practices. *Eng. Constr. Archit. Manag.* **2017**, *24*, 190–208. [CrossRef]
16. Wong, J.; Wang, X.; Li, H.; Chan, G.; Li, H. A review of cloud-based BIM technology in the construction sector. *J. Inf. Technol. Constr.* **2014**, *19*, 281–291.

17. Antón, L.Á.; Díaz, J. Integration of LCA and BIM for sustainable construction. *Int. J. Soc. Behav. Educ. Econ. Bus. Ind. Eng.* **2014**, *8*, 1378–1382.

18. Olawumi, T.O.; Chan, D.W.M. Critical success factors for implementing building information modeling and sustainability practices in construction projects: A Delphi survey. *Sustain. Dev.* **2019**, *27*, 587–602. [CrossRef]

19. Harding, J.; Suresh, S.; Renukappa, S.; Mushatat, S. Do building information modelling applications benefit design teams in achieving BREEAM accreditation? *J. Constr. Eng.* **2014**. [CrossRef]

20. Raouf, A.M.I.; Al-Ghamdi, S.G. Building information modelling and green buildings: Challenges and opportunities. *Archit. Eng. Des. Manag.* **2018**, *15*, 1–28. [CrossRef]

21. Chong, H.-Y.; Lee, C.-Y.; Wang, X. A mixed review of the adoption of building information modelling (Bim) for sustainability. *J. Clean. Prod.* **2017**, *142*, 4114–4126. [CrossRef]

22. Liu, Z.; Lu, Y.; Peh, L.C. A Review and Scientometric Analysis of Global Building Information Modeling (BIM) Research in the Architecture, Engineering and Construction (AEC) Industry. *Buildings* **2019**, *9*, 210. [CrossRef]

23. Atanda, J.O.; Öztürk, A. Social criteria of sustainable development in relation to green building assessment tools. *Environ. Dev. Sustain.* **2018**, *22*. [CrossRef]

24. Maltese, S.; Tagliabue, L.C.; Cecconi, F.R.; Pasini, D.; Manfren, M.; Ciribini, A.L. Sustainability assessment through green BIM for environmental, social and economic efficiency. *Procedia Eng.* **2017**, *180*, 520–530. [CrossRef]

25. Acampa, G.; Ordóñez García, J.; Grasso, M.; Díaz-López, C. Project sustainability: Criteria to be introduced in BIM. *J. Valori Valutazioni* **2019**, *23*, 119–128.

26. Almahmoud, E.; Doloi, H.K. Assessment of social sustainability in construction projects using social network analysis. *Facilities* **2015**, *33*, 152–176. [CrossRef]

27. Wang, H.; Zhang, X.; Lu, W. Improving social sustainability in construction: Conceptual framework based on social network analysis. *J. Manag. Eng.* **2018**, *34*, 05018012. [CrossRef]

28. Zhao, Z.Y.; Zhao, X.J.; Davidson, K.; Zuo, J. A corporate social responsibility indicator system for construction enterprises. *J. Clean. Prod.* **2012**, *29*, 277–289. [CrossRef]

29. Asif, M.; Searcy, C.; Zutshi, A.; Fisscher, O.A.M. An integrated management systems approach to corporate social responsibility. *J. Clean. Prod.* **2013**, *56*, 7–17. [CrossRef]

30. Roberts, R.W. Determinants of corporate social responsibility disclosure: An application of stakeholder theory. *Account. Organ. Soc.* **1992**, *17*, 595–612. [CrossRef]

31. Carroll, A.B. The pyramid of corporate social responsibility: Toward the moral management of organizational stakeholders. *Bus. Horiz.* **1991**, *34*, 39–48. [CrossRef]

32. Freeman, R.E. *Strategic Management: A Stakeholder Approach*; Pitman: Boston, MA, USA, 1984.

33. Murphy, K. The social pillar of sustainable development: A literature review and framework for policy analysis. *Sustain. Sci. Pract. Policy* **2012**, *8*, 15–29. [CrossRef]

34. Kent, D.C.; Becerik-Gerber, B. Understanding construction industry experience and attitudes toward integrated project delivery. *J. Constr. Eng. Manag.* **2010**, *136*, 815–825. [CrossRef]

35. AIA. *Integrated Project Delivery: A Guide*; The American Institute of Architects: Washington, DC, USA, 2007.

36. Yee, L.S.; Saar, C.C.; Yusof, A.M.; Chuing, L.S.; Chong, H.Y. An empirical review of integrated project delivery (IPD) System. *Int. J. Innov. Manag. Technol.* **2017**, *8*, 1–8. [CrossRef]

37. NASFA; COAA; APPA; AGC; AIA. *Integrated Project Delivery for Public and Private Owners*; National Association of State Facilities Administrators (NASFA); Construction Owners Association of America (COAA); APPA: The Association of Higher Education Facilities Officers; Associated General Contractors of America (AGC); American Institute of Architects (AIA): Austell, GA, USA, 2010.

38. Wasserman, S.; Faust, K. *Social Network Analysis: Methods and Applications*; Cambridge University Press: Cambridge, UK, 1994.

39. Kapoor, K.; Sharma, D.; Srivastava, J. Weighted Node Degree Centrality for Hypergraphs. In Proceedings of the 2013 IEEE 2nd Network Science Workshop (NSW), West Point, NY, USA, 29 April–1 May 2013; pp. 152–155.

40. Cuvelier, E.; Aufaure, M.A. Graph mining and communities detection. In *Business Intelligence*; Aufaure, M.A., Zimanyi, E., Eds.; Springer: Berlin, Germany, 2012; Volume 96, pp. 117–138.

41. Epasto, A.; Lattanzi, S.; Mirrokni, V.; Sebe, I.O.; Taei, A.; Verma, S. Ego-Net Community Mining Applied to Friend Suggestion. *Proc. VLDB Endow.* **2015**, *9*, 324–335. [CrossRef]

42. Hanneman, R.A.; Riddle, M. *Introduction to Social Network Methods*; University of California: Riverside, CA, USA, 2005.
43. Pryke, S. *Social Network Analysis in Construction*; Wiley: Hoboken, NJ, USA, 2012.
44. Reychav, I.; Maskil Leitan, R.; McHaney, R. Sociocultural sustainability in green building information modeling. *Clean Technol. Environ. Policy* **2017**, *19*, 2245–2254. [CrossRef]
45. Maskil-Leitan, R.; Reychav, I. A sustainable sociocultural combination of building information modeling with integrated project delivery in a social network perspective. *Clean Technol. Environ. Policy* **2018**, *20*, 1017–1032. [CrossRef]
46. Sacks, R.; Gurevich, U.; Shrestha, P. A review of building information modeling protocols, guides and standards for large construction clients. *J. Inf. Technol. Constr. (ITcon)* **2016**, *21*, 479–503.
47. Wang, H. The Interplay of Formal and Informal Institutions for Procurement Innovation: A Social Network Approach. Ph.D. Thesis, The University of Hong Kong, Hong Kong, China, 2015.
48. Yin, R.K. *Case Study Research: Design and Methods*, 5th ed.; Sage: Thousand Oaks, CA, USA, 2014.
49. Simons, H. *Case Study Research in Practice*; Sage: London, UK, 2009.
50. Yang, J.B.; Chou, H.Y. Subjective benefit evaluation model for immature BIM-enabled stakeholders. *Autom. Constr.* **2019**, *106*, 102908. [CrossRef]
51. Barlish, K.; Sullivan, K. How to measure the benefits of BIM—A case study approach. *Autom. Constr.* **2012**, *24*, 149–159. [CrossRef]
52. Bastian, M.; Heymann, S.; Jacomy, M. Gephi: An open source software for exploring and manipulating networks. In Proceedings of the Third International Conference on Weblogs and Social Media (ICWSM 2009), San Jose, CA, USA, 17–20 May 2009.
53. Malisiovas, A.; Song, X. Social Network Analysis (SNA) for construction projects' team communication structure optimization. In Proceedings of the Construction Research Congress 2014, American Society of Civil Engineers (ASCE), Atlanta, GA, USA, 19–21 May 2014; pp. 2032–2042.
54. Franz, B.; Leicht, R.; Molenaar, K.; Messner, J. Impact of team integration and group cohesion on project delivery performance. *J. Constr. Eng. Manag.* **2016**, *143*, 04016088. [CrossRef]
55. Acampa, G.; Grasso, M.; Marino, G.; Parisi, C.M. Tourist flow management: Social impact evaluation through social network analysis. *Sustainability* **2020**, *12*, 731. [CrossRef]
56. Du, J.; Zhao, D.; Issa, R.R.A.; Singh, N. BIM for improved project communication networks: Empirical evidence from email logs. *J. Comput. Civ. Eng.* **2020**, *34*, 04020027. [CrossRef]

 *buildings* 

Article

Multi-Objective Optimizing Curvilinear Steel Bar Structures of Hyperbolic Paraboloid Canopy Roofs

Jolanta Dzwierzynska

Department of Architectural Design and Engineering Graphics; Rzeszow University of Technology, 35-084 Rzeszow, Poland; joladz@prz.edu.pl

Received: 10 January 2020; Accepted: 24 February 2020; Published: 28 February 2020

Abstract: The paper concerns shaping curvilinear steel bar structures that are hyperbolic paraboloid canopy roofs by means of parametric design software Rhinoceros/Grasshopper and Karamba 3D. Hyperbolic paraboloid shape has found applications in various solutions of building roofs, mainly as reinforced concrete or steel coverings made of bent sheets. The hyperbolic paraboloid as a ruled surface can be a good base surface for forming bar grids. However, there are few studies on the effect of its division and the obtained topology of bar structures on their load-bearing capacity. In order to fill this gap, the aim of the presented research was to compare the effectiveness of various curvilinear steel bar structures of hyperbolic paraboloid canopy roofs covering the same plane, as well as defining both the most effective pattern of their structural grids and the optimal supporting system. This analysis was carried out thanks to the application of genetic algorithms enabling the free flow of information between geometrical and structural models, as well as thanks to the obtained result of multi-objective optimizations of the shaped structures for given boundary conditions. Minimal mass of the structure as well as minimal deflection of the structural members were assumed as the optimization criteria.

Keywords: a curvilinear structure; a hyperbolic paraboloid; shaping structures; structural optimization; parametric design; genetic algorithms; multi-objective optimization; topology; Grasshopper; FEM

1. Introduction

In general, steel bar structures are determined as spatial structures made of slender members which are directly connected in order to carry loads. Historically, curvilinear steel bar structures, mostly in the form of cylindrical lattice structures, began to be created in the mid-nineteenth century. However, due to serious difficulties in both calculating and constructing from repeatable elements, they began to be used on a larger scale only in the 1940s. During this period, the beginning of steel mass production and the invention of many devices influenced the great development of various manufacturing technologies of steel roof structures. The most popular were layered geodesic domes, which were shaped using the procedures of sphere division into triangles elaborated by Buckminster Fuller [1,2]. Due to this fact, the problem of the most regular subdivision of the spherical surface was one of the major challenges for scientists in the steel structures field. Various ways of dividing a sphere have been developed over the years, in order to achieve different types of grid, like Lamell's lattice and Schwedler's lattice [3]. However, combining different parts of the sphere into larger forms was one of the ways of obtaining new shapes of grid shells [3]. The broad review of various types of spatial grid structures and their development is described in [4], whereas broad analytical approaches concerning plane bar grids and double layer trusses are given in [5]. The method of forming steel bar structures, placing their vertices on the so-called base surfaces which are Catalan surfaces—has been presented in [6]. On the other hand, the shaping of bar structures based on minimal surfaces, especially the Enneper surfaces, is presented in [7,8].

A hyperbolic paraboloid constitutes an especially interesting and important basic shape for various single or complex architectural roof forms. The use of a hyperbolic paraboloid shape for constructing thin shells was pioneered in the post-war era as the result of the combination of modern architecture with structural engineering. The hyperbolic paraboloid as the element for creating complex forms was used widely by F. Candela for the implementation of lightweight shell concrete structures, constituting coverings that are free of intermediate supports [9]. The great interest in this shape was caused by its positive static properties, allowing the creation of shells with a large span, as well as a great possibility of various arrangements of single shells in compound ones.

Hyperbolic paraboloids are exceptionally stiff, due to their double curvature [10]. They exhibit membrane action, wherein internal forces are efficiently transmitted through the surface, which is the subject of various publications [11,12]. Most of the research concerns theoretical, experimental and constructional problems related to hyperbolic paraboloid concrete or reinforced concrete shells [9,13]. The method of shaping freeform buildings, roofed with profiled steel sheets effectively transformed into strips of screw ruled surfaces, is presented in [11,14]. However, the behaviour of gabled hyperbolic paraboloid shells is studied in [15]. Although a hyperbolic paraboloid as a ruled surface constitutes a good basis for creating lattice grids, there is little research into the effect of the division of this surface, as well as received grid pattern, on the bearing capacity of the bar structure created. However, the variety of compound roof structures that can be obtained by combining several hyperbolic paraboloid grid modules is presented in [12]. The canopy roofs of hyperbolic paraboloid shape are worth considering due to both their interesting form and relative simplicity of construction.

At present, the use of curvilinear steel bar structures is increasing thanks to advanced technology in the field of steel bar structures. Thus, more and more curvilinear steel bar structures are created, with a great variety of geometric forms and technical solutions. Therefore, a rational and effective attempt to shaping of this type of structures is important. The shaping phase is the design phase preceding all subsequent stages of the design process, which is why it is the most creative phase, as well as having a significant impact on the final form of the structure [9,16]. Due to this fact, it is very important to consider as many aspects of the future project as possible in the early stage. The principles of the rational shaping of steel bar structures are presented in many publications [17,18]. However, shaping the curvilinear steel bar structures can sometimes be a challenging task.

The possibilities of rational shaping depend not only on creativity and practical skills, but also on the design tools used. In the last twenty years, the progress of digital technologies has affected the entire field of both architectural design and structural engineering. That is due to the fact that digital tools greatly facilitate the creation of complex geometry, as well as performing advanced structural calculations [19]. The practical application of the digital design tools by European design studios is presented in the research reported in [20]. Various computer-aided design (CAD) tools enable both the creation of two-dimensional documentation and the creation of three-dimensional models based on two-dimensional drawings [21–23]. CAD technology enabled the generation of digital models, their geometry visualization and, finally, analysis of their structural behavior. Moreover, the progress in design caused by development of computer technology and integration of digital modeling systems has facilitated cooperation in various design areas, such as architecture and structural engineering [24,25]. Especially, in the field of steel bar structures, whose shaping is accompanied by a number of issues, the interdisciplinary approach is often required.

Recently, development of the modeling process based on Non-Uniform Rational B-Splines (NURBS) has had a great impact on forming the structures' shapes. NURBS can be controlled during modeling, and therefore they can constitute a base for the generation of various digital changeable forms with diverse topologies. Moreover, digital environment, especially algorithmic-aided shaping structures, has created new possibilities for performing various simulations, which further enable structural optimization [26]. In 1977, the idea of solving evolutionary optimization problems was introduced by means of a computer simulation of evolutionary transformations [27]. The evolutionary algorithms for such a simulation are the stochastic search methods that mimic natural biological evolution.

These algorithms have been developed in order to arrive at near-optimum solutions to large-scale optimization problems for which traditional mathematical techniques might fail [28]. A comparison of the formulation and results of five recent evolutionary-based algorithms—genetic algorithms, particle swarm algorithms, ant-colony systems, and shuffled frog leaping—is presented in [28].

The algorithms presented in the paper are genetic algorithms inspired by Darwin's theory of evolution, which mimic natural selection and gene mutation. In each genetic algorithm, the optimization process of the given problem begins with creating a set of random solutions called individuals. The set of variables is treated as chromosomes. As in nature, the whole set of possible solutions is considered a population. The most efficient and strong chromosomes are selected in order to create the next population. The evolutionary principles of genetic algorithms allow the generation of well-performing instances and search for the solution closest to the optimal one in the given space [29]. In order to solve the optimization problem, the parameters and constraints of the problem should be identified. Depending on the nature of the objective function applied, the optimization problem can be classified into either single objective or multi objective. In the literature of the subject, the term multi-objective optimization refers to problems with up to four objectives [27].

The optimization of bar structures can deal with many aspects: the weight of the structure, appropriate support method, or topology related to both ultimate limit states (ULS) and serviceability limit states (SLS) [6,8,30,31]. In the case of a steel structure analyzed in the paper, the ultimate limit state referring to internal failure involves the resistance of cross sections and the resistance of the structure and its members. If the design value of effect of actions E_d does not exceed design value of corresponding resistance R_d, then this should be verified. The design value of the effects of actions E_d is determined by combining the various values of actions that are considered to occur simultaneously.

However, verification of SLS primarily aims at preventing excessive movements or vibrations of structures [31]. Whether the design value of the effects of actions specified in the serviceability criterion E_d does not exceed limiting design value of the relevant serviceability criterion (e.g., design value of displacement) should be verified.

In the paper, the effects of displacements and deformations are mostly taken into account, assuming deflection limits equal to

$$f = L/250 \tag{1}$$

where L–span of the structure [31].

The application of evolutionary structural optimization (ESO) for the shaping of steel bar structures is a new field of research, which can lead to obtaining effective structures.

Referring to the above conditions, the article attempts the multi-objective optimization of curvilinear steel bar structures forming roofs of hyperbolic paraboloid shape. Although the hyperbolic paraboloid as a ruled surface may be a convenient base for forming bar grids, there are few studies on the effect of its division and the topology of the obtained bar structures on their load-bearing capacity. In order to fill this gap, the aim of the presented research is the comparison of the effectiveness of canopies—curvilinear steel bar structures formed based on hyperbolic paraboloids covering the same plane. The research goal is to determine the most effective pattern of grids and the optimal supporting system, as well as the mass of the structure.

2. Materials and Methods

The research was conducted with the application of modern digital tools working in Rhinoceros 3D software developed by Robert McNeel and Associates [32]. These tools are: Grasshopper plug-in for parametric modeling and Karamba 3D plug-in developed to predict the behaviour of structures under external loads [33]. The active use of Rhinoceros 3D/Grasshopper software in the architectural design process is becoming increasingly popular in the world, mainly as a tool for generating models with complex geometry. Moreover, interactive structural evolutionary optimization has recently gained some popularity for optimization in structural design [6,8,9,34]. New methods of design solutions based on genetic optimization are analyzed in [35]. However, algorithmic structural shaping, which is the process in which both the geometric model and structural analysis are carried out using multi-objective interactive structural evolutionary optimization algorithms, is a new field of research. Therefore, the approach presented in the paper to shape curvilinear steel bar structures of hyperbolic paraboloid canopy roofs is innovative.

During the tests, in order to generate geometric models, Rhinoceros 5.0 version was used in combination with Grasshopper. This enabled the creation of complex generative algorithms and the parallel exploration of the shaped geometric models in the Rhinoceros 3D viewport. The shaping strategy presented in the paper consisted of forming of curvilinear steel bar structures by placing their structural nodes on the so-called base surfaces, which were hyperbolic paraboloids. However, the structures' geometries were generated algorithmically using a set of various specified input parameters.

Then, on the basis of the created geometric models of the analyzed structures, as well as the adopted boundary conditions concerning the supporting systems (loads), as well as the type of joins and material properties, the structural models were established. The integration of geometrical shaping and structural analysis took place by the Karamba 3D. The topology and cross-sections of the structures' bars were optimized taking into account the minimal structure's self-weight, as well as minimal deflection, as the optimisation criteria.

The general scheme of the conducted analysis dealing with shaping hyperbolic paraboloid canopy roof is presented in Figure 1. However, a more detailed description of the individual steps is provided in the following sections.

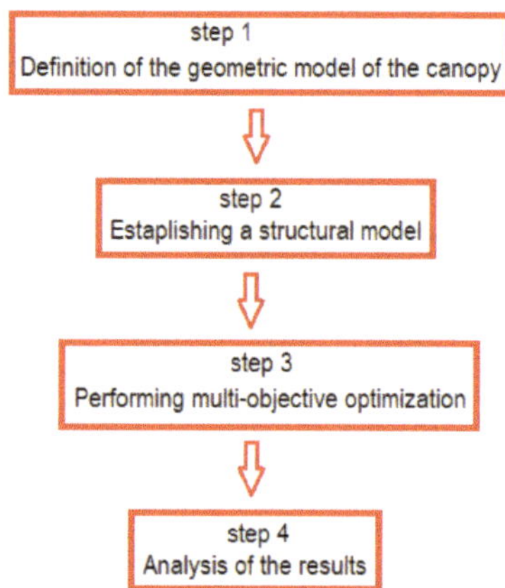

Figure 1. The general scheme of the conducted analysis.

2.1. Definition of the Geometric Model of the Canopy

Canopy roofs of hyperbolic paraboloid shape were chosen as a case study. It was assumed that the roofs covered a square plan of an area of 100 square meters. Each roof was supported by four columns placed symmetrically, whereas each column was joined with the grid by four branches, Figure 2. The positions of the columns have been set as parametric variables, as well as the locations of the branches' nodes.

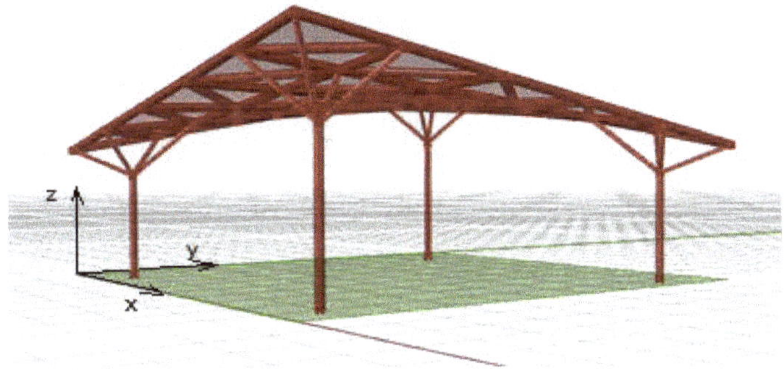

Figure 2. The view of the considered structure.

The hyperbolic paraboloid roof surface as the ruled surface was established by two skew lines—the directrix lines and a director plane to which all surface's rulings are parallel, Figure 3. This surface constituted the base surface to form a grid of bars.

Figure 3. A hyperbolic paraboloid with the directrix lines expressed.

The Grasshopper's algorithm, composed of the connected block components, was created in such a way that two skew lines defined parametrically by two pairs of various points were distinguished as its input. Each of the lines were next divided into the same number of elements to establish a series of points on them. These series of points were next joined by lines to define a hyperbolic paraboloid, which constituted a base surface for structural grid creation. Therefore, the obtained surface was discretized by dividing it into the same number of parts in two directions. Thanks to this a three-dimensional quadrate grid was obtained, whose vertices lay on the base surface. The Grasshopper's block script for roof's base surface creation is presented in Appendix A, Figure A1.

Next, each spatial polygon of the obtained grid was divided into two triangles to form a triangular bar grid. Depending on the division direction of each of the quadrangles, which can be done along shorter or longer diagonals, and depending on the number of subdivisions of the base surface as well as its type, various patterns of bar grids can be obtained. In Figure 4, the examples of grid patterns obtained due to eight-fold division of the hyperbolic paraboloid surface are shown. The structure with the grid pattern split along a short diagonal is further called the structure of type a in the paper, whereas the structure with the grid pattern split along a long diagonal is called the structure of type b.

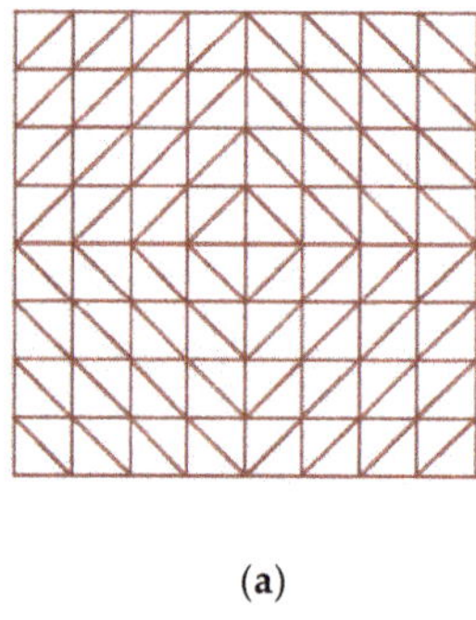 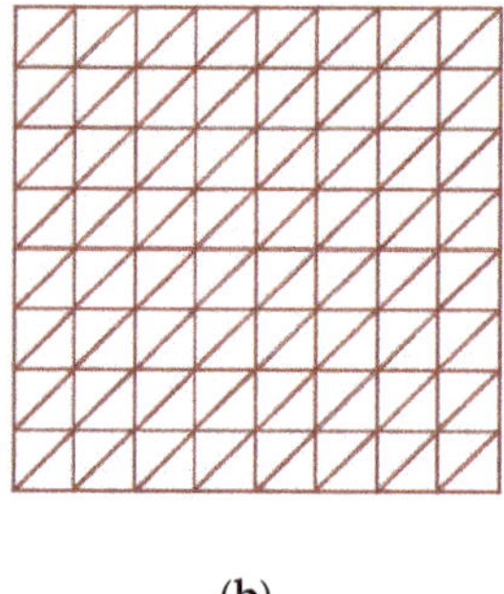

(a) **(b)**

Figure 4. Rectangular projections of considered grid patterns: (**a**) the pattern—split along a short diagonal; (**b**) the pattern—split along a long diagonal.

The geometry of each considered structure was determined using a block algorithm with variable parameters. However, during the simulations carried out, the following variables were adopted:

- The amount of parts into which the surface was divided or the lengths of grid bars, which determined the division;
- The distances of the branches' nodes from the ground;
- The locations of the supports expressed by the distances of the column bases from the borders of the covered square, Figure 5.

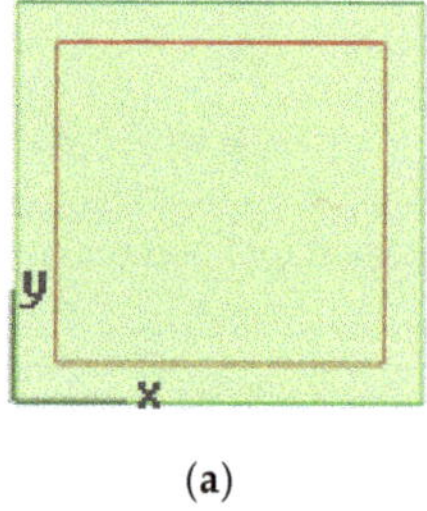 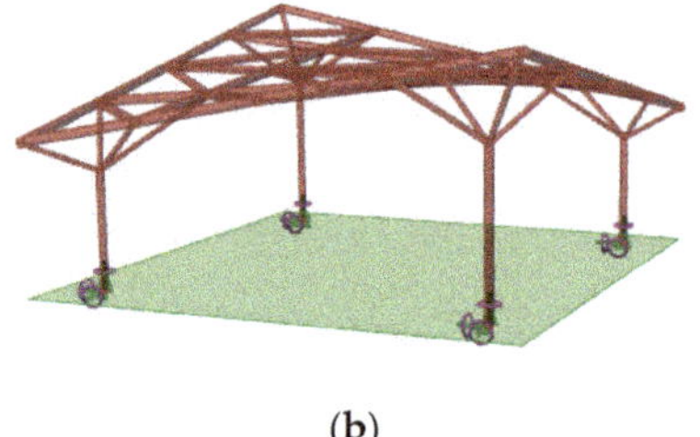

(a) **(b)**

Figure 5. Presentation of the allowed area of supports' positions: (**a**) a horizontal projection; (**b**) a perspective view.

Moreover, it was assumed that each branch node was placed at the column's end point, whereas the column length was equal to a distance of between sixty and eighty percent of the distance of the ground support from the roof surface. However, the columns were assumed to be located within the rectangular plan, but no further than one meter from the place's border (offset from the edge of the square in both x and y directions of 0.0–1.0 m), Figure 5.

2.2. Establishing of the Structural Model of the Canopy and Assumptions for the Evolutionary Optimization

Due to the fact that the geometry of the structure plays a crucial role in any optimization problem, the scripts developed to achieve the geometric forms of the roof structures were used as the part of the scripts defining the structural models for optimization performed by Karamba 3D. For that reason, grid lines were changed into beams, whereas grid vertices were changed into structural nodes. The assumed boundary conditions regarding the means of support, as well as both joints and material properties, were specified too. The structure was assumed to consist of round steel pipes. The structural nodes of the grid were assumed to be rigid, while the branches' joins with the grid as pinned joins. For a roof covering, polycarbonate plastic panels with a thickness of 10 mm were chosen.

The optimization was performed by Octopus which is the Grasshopper's plug-in for applying evolutionary principles to parametric design and problem solving by multi-objective optimization. Octopus as an evolutionary simulator can approach optimal solution sets through iterative tests and constant self-adaptation. It possesses the ability to cross-reference multiple parameters simultaneously. However, it requires multiple objectives to be input.

The goal of the performed optimization was to determine the best structure in terms of bar grid topology, the location of supports, and the locations of branches' nodes. However, the optimization objectives were as follows:

- Minimize total mass m;
- Minimize deflection d.

Structural constrains resulting from general structural principles presented in [31–36] were as follows:

- Due to ULS, the structure should be able to bear acting loads, but this verification was carried out automatically;
- Due to SLS, the deflection limit for any load case should fulfill the condition $f = L/250$, where L is a span of the structure, so for considered structures, $f \leq 40$ mm;
- Kind of structural material applied: steel S235.

Established variables:

- Dimensions of the rectangular plan—10 m × 10 m;
- Number of supports—four;
- The height of the whole structure—5 m;
- The height of the roof's surface—2 m;
- Elements' cross sections—circular hollow, walls' thickness not less than 3.2 mm.

Variables for optimization:

- Location of the supports—within the rectangular plan, however, no further than one meter from the place's border;
- Bars' length: 1.0–3.0 m;
- Location of the column branching node in the scope of 60–80%d, where d is the distance of the column's base to the roof's surface.

During simulations it was assumed that the structures were composed of round tubes with cross-sections, as expressed in Table 1.

Table 1. Division of the structures' bars and their cross sections.

Lattice Bars Cross-Section Radius/Wall Thickness [cm/mm]	Branches Bars Cross-Section Radius/Wall Thickness [cm/mm]	Columns Cross-Section Radius/Wall Thickness [cm/mm]
16.83/5.0	16.83/5.0	24.45/7.1

Moreover, it was assumed that each structure is loaded by its self-weight, as well as environmental loads from snow and wind. The wind load was applied locally to the grid structure whereas snow wind was applied globally. These loads were calculated in the form of pressure coefficients acting over the surface of the roof assuming the structure's location in Rzeszow, Poland [34,36]. Several combinations of loads have been considered, including asymmetric ones, which, in the case of canopy roofs, can be crucial when shaping the structures. However, in this case, the worst case scenario was achieved for the combination when the drifted snow load is a main load and the wind load is an associated load acting from above on the structure.

Due to the symmetry of each roof structure and its shape, some simplifications were proposed; that is, the snow load could be calculated similarly, as in the case of a butterfly roof, whereas roof inclination angle was determined according to Figure 6.

Figure 6. Determination of the roof inclination angle.

3. Result

The simulation was carried out four times: twice for structures with the grid pattern of type a and twice for structures with the grid pattern of type b, presented in Figure 4. The first simulation for both structures was carried out assuming bar lengths within the scope of 1.5–2.0 m, while the second one assumed bar lengths within the scope of 1.0–3.0 m.

Due to the fact that the optimization objectives indicated previously—the minimization of total mass and minimization of deflection—are conflicting, several results of each simulation for both structures with pattern a and b have been chosen. The graph of the Pareto front with the best solutions for the structure of type **a** received during the first simulation is presented in Figure 7. The individuals that are displayed closest to the origin are equally optimal for all three objectives, however, in our case solutions which meet both ULS and SLS were chosen. As was mentioned earlier, ULS and SLS are verified automatically. However, deflection cannot exceed 0.04 m. The generated solutions are characterized by the fact that the greater the mass of the structure, the smaller the deflection. Therefore, several solutions were chosen for which the deflections are close to but not exceeding 0.04 m. This guarantees the minimum mass of the structure.

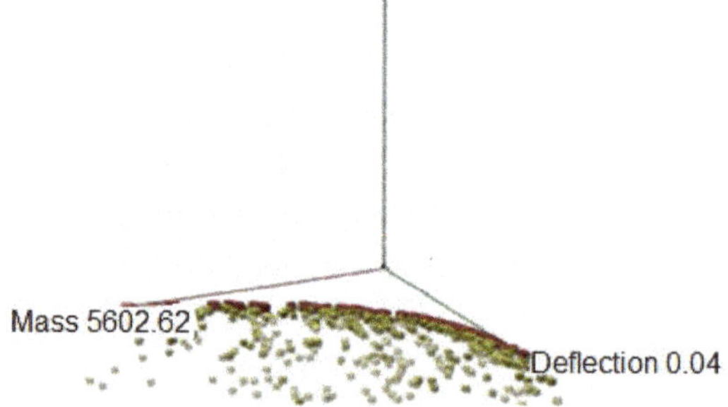

Figure 7. The 2D graph of the Pareto front for the hyperbolic paraboloid canopy structure with pattern **a** (split along the short diagonal—division into eight parts).

The chosen results of the first simulation performed for the structure of type a (grid split along a short diagonal) are given in Table 2. However, the results of the simulation performed for the structure of type b (grid split along a long diagonal) are given in Table 3.

Table 2. The results of the simulation assuming bar lengths within the scope of 1.5 m–2.0 m.

Number of the Canopy Roof	Distances of Columns from Place's Boarder in x, y Directions [m]	Distance of the Brunches' Node from the Surface [m]	Maximum Displacement [m]	Total Mass [kg]
a1	0.83	0.80	0.039	5552.33
a2	0.95	0.80	0.036	5554.83
a3	1.00	0.80	0.035	5557.13

Table 3. The results of the simulation assuming bar lengths within the scope 1.5–2.0 m.

Number of the Canopy Roof	Distances of Columns from Place's Boarder in x, y Directions [m]	Distance of the Brunches' Node from the Surface [m]	Maximum Displacement [m]	Total Mass [kg]
b1	0.87	0.80	0.040	5560.4
b2	0.95	0.80	0.034	5563.3
b3	1.00	0.80	0.034	5565.6

It is worth noticing that the achieved bar grids during the first simulation were obtained as a result of the division of the surface into eight parts in both directions. Due to the smallest mass, the structure a1 is the most efficient, Figure A2.

During the second simulation, a maximum bar length of 3 m was allowed and bar grids were obtained as a result of the division of the surface into four parts in both directions. The Pareto front for the structure type b is presented in Figure 8. However, several solutions of the simulation assuming bar lengths within the scope of 1.0–3.0 m are presented in Tables 4 and 5.

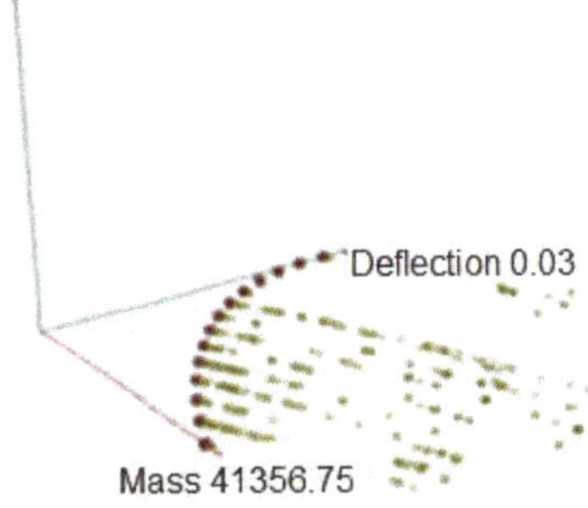

Figure 8. The 2D graph of the Pareto front for the hyperbolic paraboloid canopy structure with pattern b (split along long diagonal—division into four parts).

Table 4. The results of the simulation assuming bar lengths within the scope of 1.0–3.0 m.

Number of the Canopy Roof	Distances of Columns from Place's Boarder in x, y Directions [m]	Distance of the Brunches' Node from the Surface [m]	Maximum Displacement [m]	Total Mass [kg]
a1	0.95	0.71	0.026	4338.54
a2	0.98	0.71	0.026	4339.27
a3	1.00	0.73	0.026	4339.54

Table 5. The results of the simulation assuming bar lengths within the scope of 1.0–3.0 m.

Number of the Canopy Roof	Distances of Columns from Place's Boarder in x, y Directions [m]	Distance of the Brunches' Node from the Surface [m]	Maximum Displacement [m]	Total Mass [kg]
b1	1.00	0.66	0.022	4347.58
b2	1.00	0.74	0.027	4343.73
b3	1.00	0.73	0.026	4343.61

The analysis of the above results suggests that there was a certain reserve in the amount of deflections in the structure, which created the possibility of reducing bars' cross-sections. Therefore, another simulation was performed assuming bars' cross-sections as in Table 6.

Table 6. Division of the structures' bars and their cross sections.

Lattice's Bars Cross-Section Radius/Wall Thickness [cm/mm]	Branches' Bars Cross-Section Radius/Wall Thickness [cm/mm]	Columns Cross-Section Radius/Wall Thickness [cm/mm]
16.0/5.0	16.0/5.0	22.45/5.0

The results of this simulation for structures of type a and type b are given in Tables 7 and 8, respectively. However, the optimum results generated due to performed simulation are presented in Figures A3 and A4.

Table 7. The results of the simulation assuming bar lengths within the scope of 1.0–3.0 m.

Number of the Canopy Roof	Distances of Columns from Place's Boarder in x, y Directions [m]	Distance of the Brunches' Node from the Surface [m]	Maximum Displacement [m]	Total Mass [kg]
a1	1.00	0.71	0.038	3255.75
a2	1.00	0.72	0.039	3254.73
a3	1.00	0.73	0.040	3253.84

Table 8. The results of the simulation assuming bar lengths within the scope 1.0–3.0 m.

Number of the Canopy roof	Distances of Columns from Place's Boarder in x, y Directions [m]	Distance of the Brunches' Node from the Surface [m]	Maximum Displacement [m]	Total Mass [kg]
b1	1.00	0.75	0.040	3256.79
b2	1.00	0.74	0.039	3257.40
b3	1.00	0.73	0.038	3258.15

On the basis of the above, the results of the most efficient structure resulted in the structure a1, presented in Figure 9.

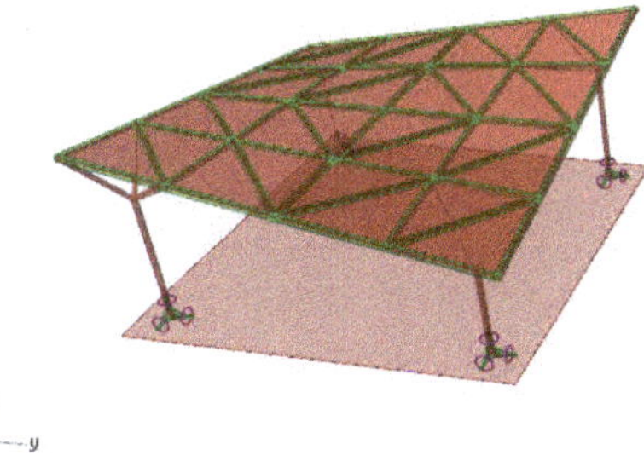

Figure 9. Perspective view of the optimal structure; the structure of pattern a, received due to surface division into four parts along its edges (structure a3, T7).

4. Discussion

An original algorithmic-aided method of shaping the hyperbolic paraboloid canopy roof structures has been proposed. This method verifies simulation structures' geometries with respect to structural requirements. The performed analyses found several solutions that meet ULS at given boundary conditions and, at the same time, meet SLS, i.e., do not exceed the allowable deflection of 40 mm. Based on the results compiled in the tables T2, T3, T4, T5, T7, T8, it can be concluded that the mass of the considered structure depends on the grid topology. In turn, this topology is dependent on the method of surface division, as well as the number of divisions applied. The curvilinear steel bar structures resulting from the division of the hyperbolic paraboloid into four parts are much lighter than the structures resulting from the division into eight parts. Due to the fact that the weight of the structure significantly affects its cost, these structures are more effective.

As previously mentioned, triangular grids of curvilinear steel bar structures analyzed in the study are obtained by dividing spatial quadrate grids, and the division can take place along the longer diagonals or shorter ones. The research revealed large differences in the masses of the structures depending on the shaping of triangular grids based on quadrangular grids. The analysis of the structures carried out showed that the grid structures obtained by dividing quadrangles along the longer diagonals are much heavier than the grid structures formed when dividing them along shorter diagonals.

The location of the supporting columns is another aspect that has a significant impact on the efficiency of the structure. The simulations showed that the further the columns are moved away from the edge of the covered square, the larger the mass of the structure. Moreover, the research found the optimal branch node positions, and thus the optimal column lengths for each structure.

5. Conclusions

The optimal design of structures is an important direction for the development of research. The presented work is a contribution to the research conducted in the field of design optimization by modern digital tools. The studies have shown that the multi-objective optimization does not give one unambiguous optimal solution, especially when the assumed criteria are contradictory. However, it can pre-estimate solutions and select the most favourable ones. The solutions selected due to multi-objective optimisation should be subjected to further analysis and selection in order to choose the most optimal result. However, this kind of optimization at the initial stage of shaping has a justification when it is difficult to assess the behaviour of the structure and choose the right solution.

Due to its properties, a hyperbolic paraboloid always constitutes an important basic shape for various interesting single or complex architectural forms of different types; therefore, this study should be continued. It is very important to take into account other optimization criteria of the structure, such as the unification of elements and the method of connection, which will be a goal of the author's further research.

Funding: This research received no external funding.

Conflicts of Interest: The authors declare no conflict of interest.

Appendix A

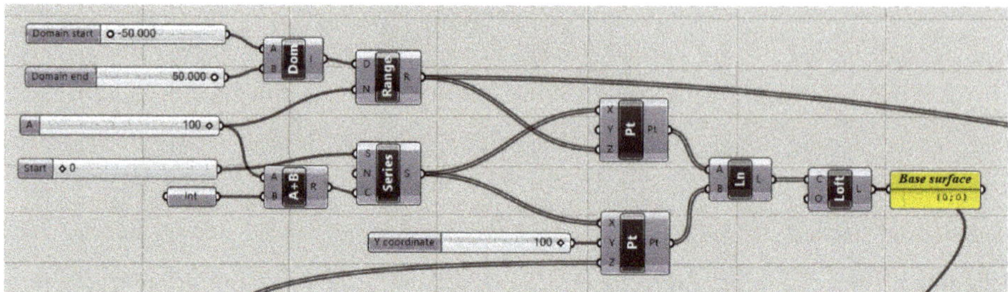

Figure A1. Block script for roof's base surface creation.

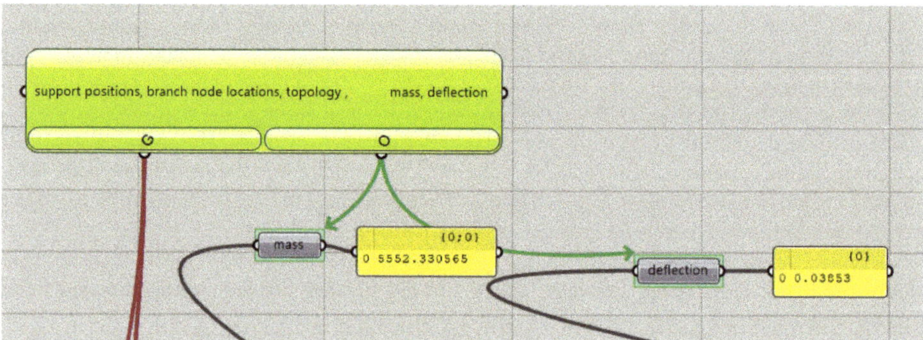

Figure A2. Result of the first simulation for a structure with pattern a.

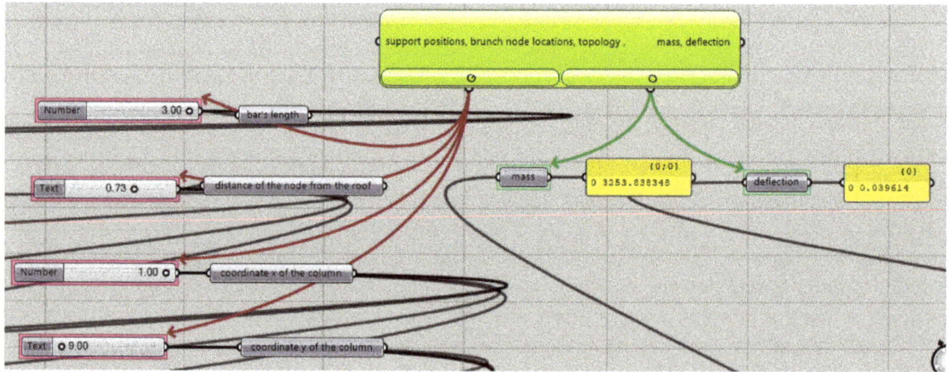

Figure A3. Result of the simulation for a structure with pattern a.

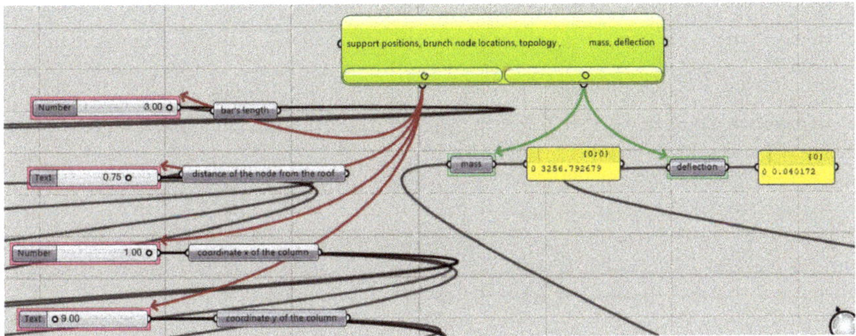

Figure A4. Result of the simulation for a structure with pattern b.

References

1. Pilarska, D. Geodesic Bars' Domes—Propositions of Large Area Coverings. *J. Civ. Eng. Environ. Archit.* **2016**, *63*, 447–454. [CrossRef]
2. Tarnai, T. Spherical Grids of Triangular Network. *Acta Tech. Acad. Sci. Hung.* **1974**, *76*, 307–338.
3. Lee, I.; Bae, E.; Hinton, E. Shell topology optimization using the layered artificial material model. *Int. J. Numer. Methods Eng.* **2000**, *47*, 843–867. [CrossRef]
4. Rebielak, J. Bar space structures—Rules of Shaping. In Proceedings of the Third Interdisciplinary Symmetry Symposium and Exhibition, Washington, DC, USA, 14–20 August 1995; 6, pp. 442–445.
5. Obrębski, J.B. Review of own complex researches related to bar structures. In *Lightweight Structures in Civil Engineering-Contemporary Problems, Proceedings of the Local Seminar Organized by Polish Chapter of IASS, Warsaw, Poland, 5 December 2008*; MICRO-PUBLISHER C-P Jan B. Obrębski Wydawnictwo Naukowa + SB: Warsaw, Poland, 2008; pp. 87–128.
6. Dzwierzynska, J. Integrated Parametric Shaping of Curvilinear Steel Bar Structures of Canopy Roofs. *Buildings* **2019**, *9*, 72. [CrossRef]
7. Dźwierzynska, J. Shaping curved steel rod structures. *Tech. Trans. Civ. Eng.* **2018**, *8*, 87–98.
8. Dzwierzynska, J. Rationalized Algorithmic-Aided Shaping a Responsive Curvilinear Steel Bar Structure. *Buildings* **2019**, *9*, 61. [CrossRef]
9. Dzwierzynska, J.; Prokopska, A. Pre-Rationalized Parametric Designing of Roof Shells Formed by Repetitive Modules of Catalan Surfaces. *Symmetry Spec. Issue Civ. Eng.* **2018**, *10*, 105. [CrossRef]
10. Pottman, H.; Asperl, A.; Hofer, M.; Kilian, A. *Architectural Geometry*, 1st ed.; Bentley Institute Press: Exton, PA, USA, 2007; pp. 35–194. ISBN 978-1-934493-04-5.
11. Abramczyk, J. *Shell Free Forms of Buildings Roofed with Transformed Corrugated Sheeting*, 1st ed.; Publishing House of Rzeszow University of Technology: Rzeszow, Poland, 2017.
12. Dzwierzynska, J. Shaping of Spatial Steel Rod Structures Based on a Hyperbolic Paraboloid. *Arch. Civ. Eng.* **2018**, *64*, 309–320. [CrossRef]
13. Vafai, A.; Farshad, M. A Study of Hyperbolic-Paraboloidal Concrete Shell Structures. *Archit. Sci. Rev.* **1980**, *23*, 90–95. [CrossRef]
14. Reichhart, A. *Geometric and Structural Shaping of Corrugated Sheet Shells*, 1st ed.; Publishing House of Rzeszow University of Technology: Rzeszow, Poland, 2002. (In Polish)
15. Chang-Soon, R.; Seung-Nam, K.; Eunjong, Y. Behavior of Gabled Hyperbolic Paraboloid Shells. *J. Asian Archit. Build. Eng.* **2015**, *14*, 159–166. [CrossRef]
16. Elango, M.; Devadas, M.D. Multi-Criteria Analysis of the Design Decisions In architectural Design Process during the Pre-Design Stage. *Int. J. Eng. Technol.* **2014**, *6*, 1033–1046.
17. Urbańska-Galewska, E.; Łukowicz, A. On the possibilities of optimizing steel structures. *Tech. Technol. Mod. Halls* **2011**, *4*, 26–30.
18. Bródka, J.; Broniewicz, M. *Design of Steel Structures According to Eurocodes*, 1st ed.; Polskie Wydawnictwo Techniczne: Warsaw, Poland, 2013. (In Polish)
19. Oxman, R. Theory and design in the first digital age. *Des. Stud.* **2006**, *27*, 229–265. [CrossRef]

20. Bonenberg, W. Digital Design Tools in National Architectural Practice in the Background of the Developed European Countries. In Proceedings of the 65 Scientific Committee for Civil Engineering of the Polish Academy of Sciences and Science Committee of the Polish Association of Civil Engineers (PZITB), Krynica Zdrój, Poland, 15–20 September 2019.

21. Dzwierzynska, J. Computer-Aided Panoramic Images Enriched by Shadow Construction on a Prism and Pyramid Polyhedral Surface. *Symmetry* **2017**, *9*, 214. [CrossRef]

22. Dzwierzynska, J. Reconstructing Architectural Environment from a Panoramic Image. *IOP Conf. Ser.-Earth Environ. Sci.* **2016**, *44*, 042028. [CrossRef]

23. Dzwierzynska, J. Single-image-based Modelling Architecture from a Historical Photograph. *IOP Conf. Ser. Mater. Sci. Eng.* **2017**, *245*, 062015. [CrossRef]

24. Luo, Y.; Dias, J.M. Development of a Cooperative Integration System for AEC Design. In *Cooperative Design, Visualization, and Engineering (CDVE 2004)*; Luo, Y., Ed.; Springer: Berlin/Heidelberg, Berlin, 2004; Volume 3190.

25. Wang, J.; Chong, H.-Y.; Shou, W.; Wang, X.; Guo, J. BIM—Enabled Design Collaboration for Complex in Cooperative Design, Visualization, and Engineering Building. In Proceedings of the 9-th International Conference, CDVE 2013, Alcudia, Mallorca, Spain, 22–25 September 2013; Luo, Y., Ed.; Springer: Heidelberg/Berlin, Germany; New York, NY, USA; Dordrecht, The Netherlands; London, UK, 2013; ISBN 978-3-642-40840-3.

26. Turrin, M.; von Buelow, P.; Stouffs, R. Design explorations of performance driven geometry in architectural design using parametric modeling and genetic algorithms. *Adv. Eng. Inform.* **2011**, *25*, 656–675. [CrossRef]

27. Mirjalili1, S.Z.; Mirjalili, S.; Saremi, S.; Faris, H.; Aljarah, I. Grasshopper optimization algorithm for multi-objective optimization problems. *Appl. Intell.* **2018**, *48*, 805–820. [CrossRef]

28. Elbeitagi, E.; Hegazy, T.; Grierson, D. Comparison among five evolutionary-based optimization algorithms. *Adv. Eng. Inform.* **2005**, *19*, 43–53. [CrossRef]

29. Toutou, A.; Fikry, M.; Mohamed, W. The parametric based optimization framework daylighting and energy performance in residential buildings in hot arid zone. *Alex. Eng. J.* **2018**, *57*, 3595–3608. [CrossRef]

30. Tajs-Zielińska, K.; Bochenek, B. Topology Optimization—Engineering Contribution to Architectural Design. *IOP Conf. Ser. Mater. Sci. Eng.* 2017. [CrossRef]

31. PN-EN 1990: 2004 Eurocode. *Basis of Structural Design*; PKN: Warsaw, Poland, 2004. (In Polish)

32. Available online: https://www.rhino3d.com/ (accessed on 8 January 2020).

33. Preisinger, C. Linking Structure and Parametric Geometry. *Archit. Des.* **2013**, *83*, 110–113. [CrossRef]

34. Dźwierzyńska, J. *Algorithmic-Aided Shaping Curvilinear Steel Bar Structures*, 1st ed.; Publishing House of Rzeszow University of Technology: Rzeszow, Poland, 2019; ISBN 978-83-7934-300-3.

35. Bonenberg, W.; Giedrowicz, M.; Radziszewski, K. *Contemporary Parametric Design in Architecture*, 1st ed.; Wydawnictwo Politechniki Poznańskiej: Poznan, Poland, 2019; ISBN 978-83-7775-548-8.

36. PN-EN 1991-1-1:2004 Eurocode 1. *Actions on Structures. Part 1–3: General Actions—Snow Loads*; PKN: Warsaw, Poland, 2004.

Article

Multi-Criteria Analysis of Design Solutions in Architecture and Engineering: Review of Applications and a Case Study

Karolina Ogrodnik

Faculty of Civil Engineering and Environmental Sciences, Bialystok University of Technology, 15-351 Bialystok, Poland; k.ogrodnik@pb.edu.pl

Received: 24 November 2019; Accepted: 13 December 2019; Published: 17 December 2019

Abstract: The primary goal of this paper is to present the application potential of MCDM/MCDA (multi-criteria decision-making/multi-criteria decision analysis) methods in the field of architecture and urban planning and in energy efficient construction, especially in the context of sustainable development paradigm. The first part of this paper is devoted to literature studies pertaining to multi-criteria decision-making support in the selected fields. On the basis of the delivered review, it was demonstrated that the most popular methods belonging to the MCDM/MCDA group that have been used so far for the purpose of resolving selected decision-making challenges, is the AHP (analytic hierarchy process) method with modifications, TOPSIS (technique for order of preference by similarity to ideal solution) method, as well the up-and-coming COPRAS (complex proportional assessment) method. In addition, by reviewing the literature, it was found that MCDM/MCDA methods constitute an effective support tool at the stage of evaluating and selecting project solutions, and are especially helpful in framing various social, economic, environmental criteria that are permanently linked to the rule of sustainable development. The empirical section of this paper, through a case study, presents a comparative analysis of the classical AHP method with its extension onto fuzzy sets. The case study pertained to the criteria for the location of single-family residential buildings with solar installations.

Keywords: multi-criteria decision-making; multi-criteria decision analysis; fuzzy AHP; sustainable development; design solutions

---

1. Introduction

The concept of sustainable development serves as the answer to the current civilization shifts and the relentless growth of the population, economy, and technology [1]. Many definitions of sustainable development exist, but the most popular one can be found in Brundtland's Report, "Our Common Future" [2]. The concept of sustainable development constitutes, without a doubt, an important research subject from a theoretical standpoint and—from a practical point of view—it determines the direction of development in multiple fields. The concept of sustainable development is increasingly frequently seen in construction engineering, which includes all stages of a building's lifecycle, starting from the design phase, through the construction work, its planning and management, all the way to its usage [3]. In the light of the growing climate changes and legal requirement, energy-efficient construction plays a key role.

Contemporary architectural design and urban planning are also based on the sustainable development paradigm. Establishing sustainable city structures is one of the challenges of modern urban planning [4]. In city policy-making, the assumptions of sustainable models and developmental concepts such as compact city, smart city, or green city are more frequently taken into account [5].

Architecture, urban planning, and construction, including energy-efficient construction are all connected fields that shape our common space and meaningfully impact the environment and climate. Additionally, everyday decision problems in the field of architecture and urban planning and construction, such a selecting an architectural design [6], a revitalization scenario [7], or a construction solution [8], require a multi-criteria approach, the inclusion of various, often contradictory decision factors as well as taking into account the preferences of the stakeholders. Here, methods from MCDM/MCDA (multi-criteria decision-making/multi-criteria decision analysis) groups can serve as important supporting tools.

Multi-criteria decision-making/multi-criteria decision analysis (with multiple-criteria decision-making/multiple-criteria decision analysis being the alternative terms also in use) is a well-known and a well-developed branch of operational research that encompasses various techniques and mathematical tools, which all facilitate the analysis and selection of decision-making alternatives against the pre-defined criteria. What's important, it is an interdisciplinary branch that is based not only on mathematics, but also takes advantage of the theory of economics and IT [9].

The subject matter literature describes various methods and their classification. In general, MCDM methods can be divided into two categories: discrete MADM (multi-attribute decision-making) and continuous MODM (multi-objective decision-making) [10,11]. According to Dytczak, the two fundamental trends should be identified within MCDM: MCDA and MODM. MODM allows for the creation of a set of decision-making alternatives by using mathematical programming. MCDA methods can be divided into aggregation (most well-known methods include the AHP method and its extensions and the MAUT (multi-attribute utility theory) method), as well as surpassing methods (ELECTRE (elimination and choice expressing the reality) and PROMETHEE (preference organization method ranking for enrichment evaluations) families of methods). Because of the diversity and continuous development of the methods, a third group of the so-called remaining methods can be distinguished. These group encompasses the following methods: geometric distance methods (mostly TOPSIS and VIKOR (VlseKriterijuska Optimizacija I Komoromisno Resenje)), interactive (e.g., RUBIS) or methods of verbal decision analysis (e.g., ZAPROS) [12]. A similar classification was also proposed by Kobryń, who added simple ranking methods, such as Bordy, Arrow-Raynaud, or Copeland methods [13].

As proven by overview works e.g., [9,14,15], MCDM/MCDA methods can constitute a universal tool for supporting the decision-making processes in various fields of life and science. It is worth mentioning that the selected methods are increasingly used in architectural design and urban planning, as well as in construction (see examples of the applications in Tables 1 and 2 in the next part of the work). The interest in this subject is growing systematically, which can be confirmed by the already-delivered general overview work, as well as work focused solely on decision-making challenges in the field of construction [11,16].

In this paper, an overview of the MCDM/MCDA methods applied to the selected decision-making challenges in the field of architecture and urban planning as well as construction was carried out. The analysis was narrowed down to issues pertaining to energy-efficient construction. Special attention was paid to decision-making challenges connected to sustainable development paradigms in the selected fields. Moreover, the second part of the work was devoted to AHP method and its modification (FAHP), which most frequently appeared in the overview. A comparative analysis of these methods was carried out. The numerical example concerned the location of single-family residential building with renewable energy sources, and three scales were used in the calculations, including two fuzzy triangular scales.

The primary goal of this paper is to present the application potential of MCDM/MCDA methods in the field of architecture, urban planning, and in energy-efficient construction, while indicating the most popular methods and research problems. This paper consists of four fundamental parts: introduction, literature overview, and a case study presenting the comparative analysis of AHP and Fuzzy AHP. The paper is concluded with the discussion and conclusions that take into consideration the future research directions.

Table 1. Examples of the applications of selected MCDM/MCDA (multi-criteria decision-making/multi-criteria decision analysis) methods in architecture and urban planning.

Author(s) (year)	The Place of the Case Study	The Selected Method(s)	The Main Subject of the Research
Tamosaitiene J., Sipalis J., Banaitis A., Gaudutis E. (2013) [17]	Vilnius (Lithuania)	SAW (Simple Additive Weighting)	Model of location assessment for high-rise buildings in the city (8 alternatives and 12 criteria were taken into consideration).
Bielinskas V., Burinskiene M., Palevicius V. (2015) [18]	Vilnius (Lithuania)	COPRAS	Identifying assessment indicators for neglected areas in the city (finally, a set of 15 indicators related to the selected economic, social, spatial, and environmental aspects was proposed).
De Toro P., Iodice S. (2016) [19]	Cava De' Tirreni in Salerno (Italy)	PROMETHEE	Multi-criteria decision-making support for the selection of Operational Plans (3 alternatives assessed against 29 defined criteria).
Ogrodnik K. (2017) [20]	Bialystok Chorzow Czestochowa Lublin (Poland)	PROMETHEE	Assessment of sustainable development of the selected Polish cities against 66 quantitative indicators.
Zinatizadeh S., Azmi A., Monavari S.M., Sobhanardakani S. (2017) [21]	Kermanshah (Iran)	SAW ELECTRE TOPSIS	Assessment of sustainable development in selected urban areas (6 selected areas were assessed against 44 indicators divided into 3 groups: social and welfare, economic growth, environmental protection).
Pujadas P., Pardo-Bosch F., Aguado-Renter A., Aguado A. (2017) [22]	Barcelona (Spain)	MAUT AHP	Assessment of public investments (15 heterogeneous public investment projects were included in the case study).
Chen C.S., Chiu Y.H., Tsai L.C. (2018) [23]	Sun Yat-Sen Historical Museum in Taipei (Taiwan)	ANP (Analytic Network Process)	Selection of secondary uses for historic building (the decision model includes four alternative functions of building utilization which were assessed against selected economic, social, environmental and historical aspects).
Harputlugil T. (2018) [6]	the design studios of Department of Architecture of Cankaya University (Turkey)	AHP	Assessment of students' architectural designs (the basis for the assessment were the main decision criteria: functionality, presentation, process, build quality, innovation and impact, and sub-criteria defined within them).
Tian G.D., Zhang H.H., Feng Y.X., Wang D.Q., Peng Y., Jia H.F. (2018) [24]	China	AHP GC-TOPSIS (Grey correlation TOPSIS)	Multi-criteria support for selecting green decoration materials (ten kinds of solid woods were tested against six criteria).
Masoumi Z., Genderen J.V. (2019) [25]	Zanjan (Iran)	AHP TOPSIS	Determining the direction of the city's sustainable development, taking into account social, economic and environmental perspectives (eventually, the assessment of the usefulness of urban areas in terms of future development was performed).
Della Spina L., Giorno C., Galati Casmiro R. (2019) [7]	Catanzaro (Italy)	ANP	Identifying the preferred urban revitalization scenarios (the decision model consisted of the following levels: objective, strategic criteria included in four groups, three urban revitalization scenarios and three selected urban areas).
Ribera F., Nesticò A., Cucco P., Maselli G. (2019) [26]	Palazzo Genovese in Salerno (Italy)	AHP	Identification of the best way to utilize historical buildings, taking into account selected economic, social, cultural and historical-architectural aspects.
Pons O., Franquesa J., Amin Hosseini S.M. (2019) [27]	The Polytechnic University of Catalonia (Spain)	MIVES (Modelo Integrado de Valor para una Evaluación Sostenible), AHP, Knapsack algorithm	Selection of the best set of educational activities to be used in lectures on architecture.

Source: author's own work.

Table 2. Examples of using the selected MCDM/MCDA methods in energy-efficient constructions.

Author(s) (year)	The Place of the Case Study	The Selected Method(s)	The Main Subject of the Research
Yang Y., Li B., Yao R. (2010) [28]	China	group AHP	Developing a method for identifying and weighing indicators for the energy efficiency assessment of residential buildings (eventually, 17 indicators were proposed).
Kuzman M.K., Grošelj P., Ayrilmis N., Zbašnik-Senegačnik M. (2013) [29]	Slovenia	AHP	A comparative analysis of the selected types of passive house construction along with the identification of key decision-making factors (the study included: solid wood, wood-frame, aerated concrete and brick).
Siozinyte E., Antucheviciene J. (2013) [30]	the vernacular dwelling from Aukštaitija region (Lithuania)	AHP COPRAS TOPSIS WASPAS (Weighted Aggregates Sum Product Assessment)	Multi-criteria analysis of insolation improvements in a vernacular building (three variants and six criteria were included).
Ruzgys A., Volvaciovas R., Ignatavicius C., Turskis Z. (2014) [31]	residential buildings in Vilnius and Siauliai (Lithuania)	SWARA (Step-wise Weight Assessment Ratio Analysis) TODIM (an acronym in Portuguese of Interactive and Multi-criteria decision-making)	Evaluation of six cases of residential building modernization along with identifying key factors affecting the efficiency of this process.
Siozinyte E., Antucheviciene J., Kutut V. (2014) [32]	the vernacular dwelling from Aukštaitija region (Lithuania)	TOPSIS Grey AHP	Selection of the best modernization method for an old vernacular building (nine variants and ten criteria included).
Medineckiene M., Dziugaite-Tumeniene R. (2014) [33]	an existing low energy individual family house in Vilnius (Lithuania)	AHP WASPAS TOPSIS	Selection of the optimal combination of technologies for a building energy system, taking into account economic and ecological aspects (five decision alternatives were subject to analysis).
Chen L., Pan W. (2015) [34]	high-rise commercial buildings in Hong Kong (China)	Fuzzy PROMETHEE	Development of a fuzzy MCDM model integrated with BIM for the selection of Low-Carbon Building (LCB) indicators (five criteria and nine alternatives identified).
Sedláková A., Vilčeková S., Krídlová Burdová E. (2015) [35]	single-family house in Slovakia	CDA (concordance discordance analysis) IPA (ideal points analysis) WSA (the weighted sum approach) TOPSIS	Assessment of materials for designing the construction details of the foundations, walls and floors in the light of environmental and thermophysical criteria.
Motuziene A., Rogoza A., Lapinskiene V., Vilutiene T. (2016) [36]	energy efficient single-family house in Lithuania	AHP COPRAS	Development of an algorithm for choosing the most rational design solution, using LCA (life cycle assessment), LCC (life cycle cost) and the selected MCDA methods (three types of envelopes: masonry, log, and timber frame of an energy efficient building were subject to this analysis).
Vujošević M.L., Popović M.J. (2016) [37]	the atrium type hotel buildings in Belgrade (Serbia)	PROMETHEE Borda	Energy performance comparative analysis for atrium-type hotel buildings (four alternatives were considered, looking for the most energy-efficient atrium hotel model for a selected climate zone).
Tomczak K., Kinash O. (2016) [38]	detached family house in Cracow (Poland)	AHP	Technical and economic analysis of selected construction and building variants of an energy-efficient building (three alternatives were analyzed).
Seddiki M., Anouche K., Bennadji A., Boateng P. (2016) [39]	neoclassical colonial collective building in Oran (Algeria)	Delphi Swing PROMETHEE	Multi-criteria decision support of a thermal renovation project for a masonry building (selected renovation solutions were subject to analysis).

Table 2. *Cont.*

Author(s) (year)	The Place of the Case Study	The Selected Method(s)	The Main Subject of the Research
Zavadskas E.K., Antucheviciene J., Kalibatas D., Kalibatiene D. (2017) [40]	apartments in brick dwelling houses in Vilnius (Lithuania)	WASPAS ARAS (Additive Ratio Assessment) TOPSIS	Assessment of the condition of buildings along with comparative analysis with the optimal alternative, i.e., nearly zero-energy building (NZEB) (the case study pertained to 13 apartments).
Cortes J.P.R., Ponz-Tienda J.L., Delgado J.M., Gutierrez-Bucheli L. (2017) [41]	a new University's facility construction in Colombia	CBA (Choosing By Advantages)	Selection of a structural contractor (four alternatives were subject to analysis).
Guzman-Sanchez S., Jato-Espino D., Lombillo I., Diaz-Sarachaga J.M. (2018) [42]	Valencia Santander Madrid (Spain)	AHP TOPSIS	Assessment of the selected roof types in terms of sustainable development requirements (analyzed in the following order: self-protected, gravel finishing, floating flooring and green; the analysis was performed for three climate scenarios).
Moghtadernejad S., Chouinard L.E., Mirza M.S. (2018) [43]	Canada	AHP Choquet TOPSIS	Multi-criteria sustainable facade design (four facade panel alternatives were assessed: concrete, aluminum-glazing, fiber cement composite and sandwich panels against eight criteria).
Zolfani S.H., Pourhossein M., Yazdani M., Zavadskas E.K. (2018) [44]	five-star hotel in Tehran (Iran)	SWARA COPRAS	Assessment of hotel projects in terms of sustainable development (five different project alternatives were subject to a multi-criteria analysis).
Arroyo P., Mourgues C., Flager F., Correa M.G. (2018) [45]	hotel building in Orlando, Florida (USA)	CBA	Selection of a design solution (the study included 1000 different alternatives based on different combinations of design variables).
Jalilzadehazhari E., Vadiee A., Johansson P. (2019) [8]	Sweden	AHP	Selection of a construction solution (the study included 375 different alternatives based on different combinations of design variables).
Beltran R.D., Martinez-Gomez J. (2019) [46]	social dwelling in Ecuador	COPRAS-G (COPRAS-Gray) TOPSIS VIKOR	Phase change materials (PCM) analysis for the purpose of constructing wall panels and roofs (nine alternative building materials included).

Source: author's own work.

2. Materials and Methods

2.1. MCDM/MCDA in Architecture, Urban Planning and Energy-Efficient Construction: Literature Review

As a result of the undertaken literature studies, Tables 1 and 2 present examples of previous applications of the MCDM/MCDA methods for the selected decision-making challenges in the field of architecture, urban planning, and energy-efficient constructions. What is important, selected scientific papers which are indexed in the Web of Science and Scopus databases were assumed as the basis for this elaboration. The developed list is the result of a several-stage analysis of scientific papers in the abovementioned databases. In the first stage, a preliminary selection of articles available in databases was made, using keywords: "MCDM/MCDA in architecture," "MCDM/MCDA in urban planning," "MCDM/MCDA in energy-efficient construction." Then, selected scientific papers were analyzed in terms of criteria which result from the main purpose of the work. Only the research in which the research problem and the methods applied were precisely determined and the territorial delimitation of research was defined, was selected.

The research is presented chronologically, divided into examples of applications in architecture and urban planning (Table 1) and in energy-efficient construction (Table 2). The literature review included 33 publications.

Based on the literature review, it can be noticed that MCDM/MCDA methods are universal tools, which is confirmed by the variety of decision-making challenges presented in Tables 1 and 2. In architecture and urban planning, the selected methods are applied—for example—at the stage of selecting urban revitalization scenarios and for renovation of historical buildings. In addition, assessing sustainable development of urban areas is a popular research topic. MCDM/MCDA methods are also gaining popularity in energy-efficient construction, in which the selection of a design solution often requires one to take into consideration many criteria of different nature. In addition to technical criteria, economic and environmental criteria often occur; a phenomenon that stems from the sustainable development paradigm.

It is also worth mentioning the effectiveness of selected MCDM/MCDA methods. Based on the literature review, it can be stated that the selected methods facilitate decomposition of the decision problem, improve the transparency of decision processes, facilitate comparison of various decision alternatives, identify their strengths and weaknesses [19,29]. However, in addition to the advantages, some authors also noted the weaknesses and limitations of selected MCDM/MCDA methods. For example, Moghtadernejad et al. among the limitations of the AHP and TOPSIS methods (method which were used in their case study) indicate the lack of consideration of interactions between various design criteria [43]. In contrast, Zinatizadeh et al. compared three selected methods: SAW, ELECTRE and TOPSIS, where the TOPSIS method came out best. Other methods, SAW and ELECTRE, were rated lower, e.g., in terms of ability in pair comparison or ability to manage low quality input data [21]. In the work of multi-criteria decision support of a thermal renovation project for a masonry building, the authors used the Delphi, Swing, and PROMETHEE methods, indicating their several limitations, e.g., the method did not take into account the various uncertainties (regarding the assessment of criteria or decision makers' preferences) that could have influenced the final ranking of decision variants [39]. More about the pros and cons of selected MCDM/MCDA methods can be found in works [47,48]. It is worth adding that because of the limitations of individual methods, the hybrid approach is becoming increasingly popular, Zavadskas et al. state that "to have comprehensive assessment, it is better to use two or three different MCDM" [40].

MCDM/MCDA methods are not only universal, but also well-known tools, as evidenced by territorial delimitation of the conducted research. Figure 1 presents places where research focused on the subject of multi-criteria decision support was carried out. The major research centers in this field are Europe (primarily Lithuania, followed by Spain, Italy, and Poland), Asia, and both the Middle and the Far East (Iran and China) (compare: [3]).

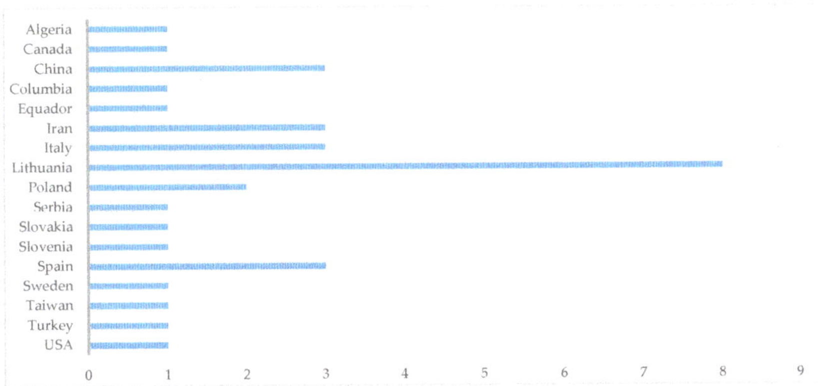

Figure 1. Countries in which selected studies on MCDM/MCDA were carried out with the focus on architecture and urban planning, as well as in the area of energy-efficient construction. Source: author's own work.

In addition, when analyzing the methods used in the selected papers, it can be seen that the most commonly used method was AHP and its extensions. The AHP method appeared in 16 papers included in the literature overview. Importantly, the AHP method was most often used to calculate the weighted criteria, which were then used to rank decision variants using other methods, e.g., TOPSIS. In addition, research conducted to-date shows that MCDM/MCDA methods can be combined by using the so-called hybrid approach, which allows one to increase the efficiency of the adopted methods. The AHP and TOPSIS hybrid is the most popular combination. A conjunction with the COPRAS method is also increasingly applied. It is worth adding that MCDM/MCDA methods can be integrated with other methods and systems, such as the dynamically developing BIM (building information modeling) system or the GIS (geographic information system) system.

The AHP method, which—for many years now—has been subject to significant interest in many areas around the world see for example: [49,50], is also being considered in terms of its advantages and disadvantages. Among the advantages of the AHP method the following can be mentioned: decomposition of a decision problem using a hierarchical structure tree, the possibility of making comparisons through element pairs that are located at given levels of the structure, the possibility of assessing the consistency of the comparisons by using consistency ratio. The AHP algorithm is also subject to criticism, for example: the independence of the analyzed elements, excessive subjectivity, or difficulty of accounting for uncertainties associated with judgments [43,51].

MCDM/MCDA methods are constantly evolving, and that is why fuzzy multi-criteria methods are becoming an important and equally popular group of computational tools. The most commonly used method is Fuzzy AHP, which dates back to 1983 [52]. The next part of this paper presents this method's algorithm that is subsequently used to develop a case study.

2.2. Fuzzy AHP—Algorithm and Selected Scales

The fuzzy sets theory was first presented in the 1960s by Zadeh, as a mathematical method to facilitate the framing of uncertainty and imprecision, which often accompanies human assessments. In the 1970s, the fuzzy sets theory also appeared within the scope of decision-making challenges. As mentioned in the previous section, the most commonly used method is the Fuzzy AHP method (abbreviated FAHP). Among the researchers who contributed to the development of FAHP are: van Laarhoven and Pedrycz, Buckley, and Chang [53,54]. Importantly, this work uses a known algorithm proposed by Buckley. Individual FAHP calculation steps are shown in Figure 2.

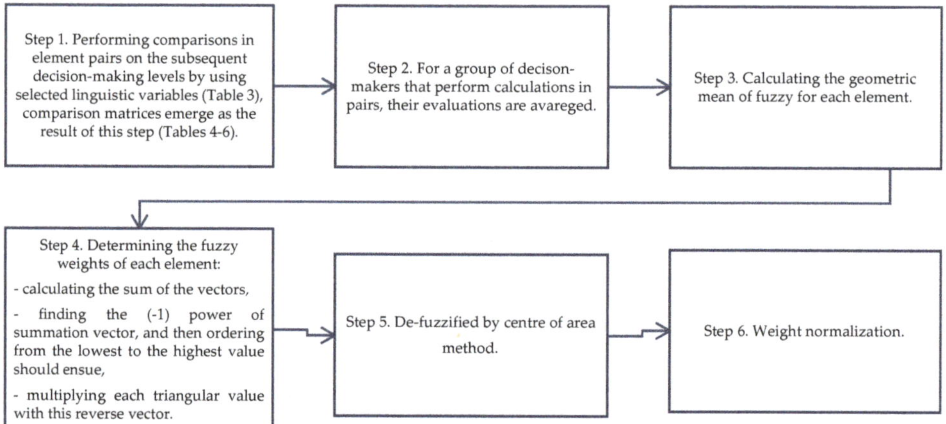

Figure 2. Main calculation stages of Fuzzy AHP. Source: author's own work based on [54].

Importantly, the literature on the subject matter quotes various scales used in FAHP (e.g., in online output software application, the user has a choice of nine different scales) [59]. In this paper, in addition to the classic Saaty scale, two fuzzy triangular scales (Table 3) were used. Calculation results and comparative analysis of weights obtained by using individual scales are presented in the subsequent part of this paper.

Table 3. Classic Saaty scale and selected Fuzzy triangular scales.

Definition	Classic Saaty Scale	Fuzzy Triangular Scale—Variant I	Fuzzy Triangular Scale—Variant II
Equal Importance	1	1,1,1	1,1,1
Weak or slight	2	1,2,3	1,2,4
Moderate importance	3	2,3,4	1,3,5
Moderate plus	4	3,4,5	2,4,6
Strong importance	5	4,5,6	3,5,7
Strong plus	6	5,6,7	4,6,8
Very strong	7	6,7,8	5,7,9
Very, very strong	8	7,8,9	6,8,9
Extremely strong	9	9,9,9	7,9,9

Source: [54–58].

3. Results

The subject of this research is a comparative analysis of the weights for the location criteria of residential area with solar installations and obtained via the classical AHP method [60] with the weights of these criteria, calculated by means of the Fuzzy AHP method (two fuzzy scales included). The assessment of the suitability of the area for residential development with solar installations can be considered as a multi-criteria decision problem. Selecting the location for this type of investment is the net sum of not only spatial, environmental, or legal conditions, but the inhabitants' preferences also play a key part in that decision-making process. For that reason, the excessive subjectivity (of which AHP and FAHP methods are often accused), may be its advantage in the case of such decision-making challenges. Importantly, the decision-making criteria used for the case study are the result of research described in the following papers [60,61].

3.1. Hierarchical Structure Tree

The basis of the hierarchical structure tree (Figure 3) was a set of criteria proposed in the papers on assessment of area in terms of the location of single-family residential buildings with solar installations [60,61]. Importantly, the groups of criteria and main criteria were developed on the basis of the subject matter literature review i.e., [62–64] and selected legal provisions. Fifteen main criteria were proposed, taking into account the distance from selected facilities, properties of plots. Moreover, selected climatic factors likely to affect the efficiency of solar installations have been taken into consideration. The last decision level, i.e., sub-criteria have local character and depends on the research area, the subcriteria of these figure were proposed on the basis of spatial analysis of Bialystok.

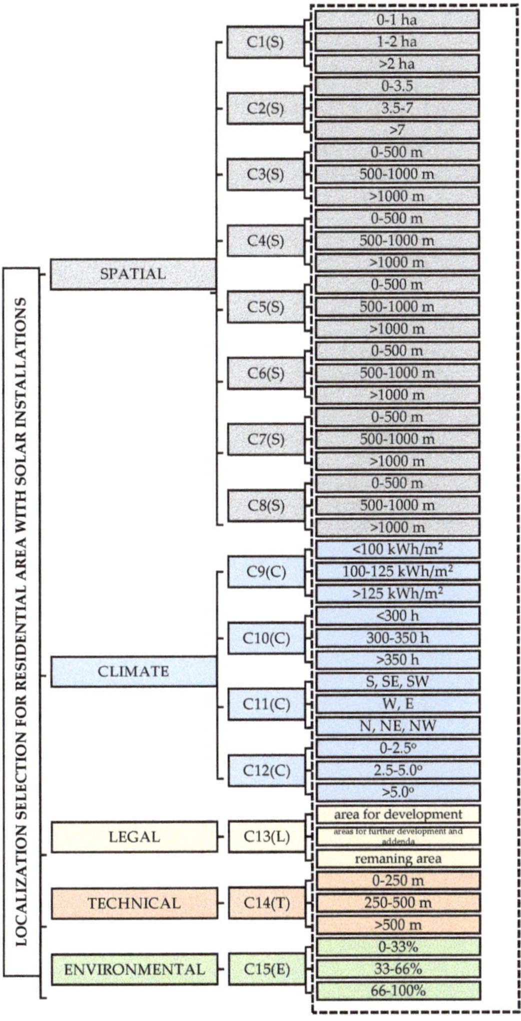

Figure 3. Hierarchical structure tree. * The last decision level (in frame) depends on the research area; the subcriteria of these figure were proposed on the basis of spatial analysis of Bialystok. Source: author's own work based on [60,61].

A hierarchical structure tree was developed, in which the following levels were distinguished:

- Purpose: location of residential buildings with solar installations;
- Groups of criteria: spatial (S), climate (C), legal (L), technical (T), environmental (E);
- Main criteria: plot area C1(S), plot shape C2(S), distance from education facilities C3(S), distance from healthcare facilities C4(S), distance from commercial and service facilities C5(S), distance from bodies of forest C6(S), distance from sports and recreation areas C7(S), distance from places of worship C8(S), total insolation C9(C), duration of sunlight exposure C10(C), exposure of the area C11(C), sloping C12(C), purpose in urban planning C13(L) (in the case of Poland it is a study of conditions and directions of spatial development), accessibility of communication infrastructure C14(T), degree of plot forestation C15(E);
- Sub-criteria-constituting the ranges defined within individual main criteria, they are local in character and are determined by the conditions of the given area.

Importantly, the groups of criteria presented in Figure 3, main criteria and sub-criteria, the hierarchical structure tree presents only the so-called soft criteria, i.e., factors enabling the suitability assessment of a given area in terms of the location of the selected investment.

3.2. Comparison Matrices

A series of pairwise comparisons of individual elements was subsequently made at each decision-making level (Tables 4–6 show comparison matrices for groups of criteria developed by using three different scales).

Table 4. Comparison matrix for criteria groups (classic Saaty's scale).

	Spatial	Climate	Legal	Technical	Environmental
Spatial	1	3	5	7	9
Climate	1/3	1	3	5	7
Legal	1/5	1/3	1	3	5
Technical	1/7	1/5	1/3	1	3
Environmental	1/9	1/7	1/5	1/3	1

Source: author's own work.

Table 5. Comparison matrix for criteria groups (Fuzzy triangular scale—variant I).

	Spatial			Climate			Legal			Technical			Environmental		
Spatial	1	1	1	2	3	4	4	5	6	6	7	8	9	9	9
Climate	1/4	1/3	1/2	1	1	1	2	3	4	4	5	6	6	7	8
Legal	1/6	1/5	1/4	1/4	1/3	1/2	1	1	1	2	3	4	4	5	6
Technical	1/8	1/7	1/6	1/6	1/5	1/4	1/4	1/3	1/2	1	1	1	2	3	4
Environmental	1/9	1/9	1/9	1/8	1/7	1/6	1/6	1/5	1/4	1/4	1/3	1/2	1	1	1

Source: author's own work.

Table 6. Comparison matrix for criteria groups (Fuzzy triangular scale—variant II).

	Spatial			Climate			Legal			Technical			Environmental		
Spatial	1	1	1	1	3	5	3	5	7	5	7	9	7	9	9
Climate	1/5	1/3	1	1	1	1	1	3	5	3	5	7	5	7	9
Legal	1/7	1/5	1/3	1/5	1/3	1	1	1	1	1	3	5	3	5	7
Technical	1/9	1/7	1/5	1/7	1/5	1/3	1/5	1/3	1	1	1	1	1	3	5
Environmental	1/9	1/9	1/7	1/9	1/7	1/5	1/7	1/5	1/3	1/5	1/3	1	1	1	1

Source: author's own work.

3.3. Comparative Analysis

Because of the volume of data, the next part of the work presents only the results of pairwise comparisons, i.e., the weight of all the elements at subsequent decision-making levels in three variants (Table 7 and Figures 4 and 5).

Table 7. Weights of groups of criteria, main criteria and sub-criteria calculated by the AHP and FAHP (Fuzzy AHP) methods.

		Classic Saaty's Scale						Fuzzy Triangular Scale—Variant I							Fuzzy Triangular Scale—Variant II						
Group	Weights	Criteria	Local Weights	Global Weights	Sub-Criteria	Local Weights	Global Weights	Weights	Criteria	Local Weights	Global Weights	Sub-Criteria	Local Weights	Global Weights	Weights	Criteria	Local Weights	Global Weights	Sub Criteria	Local Weights	Global Weights
Spatial	0.5128	C1(S)	0.2277	0.1168	0–1 ha	0.0953	0.0111	0.5034	C1(S)	0.2206	0.1111	0–1 ha	0.0975	0.0108	0.4745	C1(S)	0.2195	0.1042	0–1 ha	0.1067	0.0111
					1–2 ha	0.2499	0.0292					1–2 ha	0.2550	0.0283					1–2 ha	0.2776	0.0289
					>2 ha	0.6548	0.0765					>2 ha	0.6475	0.0719					>2 ha	0.6157	0.0641
		C2(S)	0.2277	0.1168	0–3.5	0.0953	0.0111		C2(S)	0.2206	0.1111	0–3.5	0.0975	0.0108		C2(S)	0.2195	0.1042	0–3.5	0.1067	0.0111
					3.5–7	0.2499	0.0292					3.5–7	0.2550	0.0283					3.5–7	0.2776	0.0289
					>7	0.6548	0.0765					>7	0.6475	0.0719					>7	0.6157	0.0641
		C3(S)	0.1314	0.0674	0–500 m	0.7306	0.0492		C3(S)	0.1356	0.0683	0–500 m	0.7273	0.0496		C3(S)	0.1370	0.0650	0–500 m	0.7143	0.0464
					500–1000 m	0.1884	0.0127					500–1000 m	0.1896	0.0129					500–1000 m	0.1929	0.0125
					>1000 m	0.0810	0.0055					>1000 m	0.0831	0.0057					>1000 m	0.0928	0.0060
		C4(S)	0.1314	0.0674	0–500 m	0.7306	0.0492		C4(S)	0.1356	0.0683	0–500 m	0.7273	0.0496		C4(S)	0.1370	0.0650	0–500 m	0.7143	0.0464
					500–1000 m	0.1884	0.0127					500–1000 m	0.1896	0.0129					500–1000 m	0.1929	0.0125
					>1000 m	0.0810	0.0055					>1000 m	0.0831	0.0057					>1000 m	0.0928	0.0060
		C5(S)	0.0704	0.0361	0–500 m	0.7306	0.0264		C5(S)	0.0719	0.0362	0–500 m	0.7273	0.0263		C5(S)	0.0718	0.0341	0–500 m	0.7143	0.0243
					500–1000 m	0.1884	0.0068					500–1000 m	0.1896	0.0069					500–1000 m	0.1929	0.0066
					>1000 m	0.0810	0.0029					>1000 m	0.0831	0.0030					>1000 m	0.0928	0.0032
		C6(S)	0.0704	0.0361	0–500 m	0.7306	0.0264		C6(S)	0.0719	0.0362	0–500 m	0.7273	0.0263		C6(S)	0.0718	0.0341	0–500 m	0.7143	0.0243
					500–1000 m	0.1884	0.0068					500–1000 m	0.1896	0.0069					500–1000 m	0.1929	0.0066
					>1000 m	0.0810	0.0029					>1000 m	0.0831	0.0030					>1000 m	0.0928	0.0032
		C7(S)	0.0704	0.0361	0–500 m	0.7306	0.0264		C7(S)	0.0719	0.0362	0–500 m	0.7273	0.0263		C7(S)	0.0718	0.0341	0–500 m	0.7143	0.0243
					500–1000 m	0.1884	0.0068					500–1000 m	0.1896	0.0069					500–1000 m	0.1929	0.0066
					>1000 m	0.0810	0.0029					>1000 m	0.0831	0.0030					>1000 m	0.0928	0.0032
		C8(S)	0.0704	0.0361	0–500 m	0.7306	0.0264		C8(S)	0.0719	0.0362	0–500 m	0.7273	0.0263		C8(S)	0.0718	0.0341	0–500 m	0.7143	0.0243
					500–1000 m	0.1884	0.0068					500–1000 m	0.1896	0.0069					500–1000 m	0.1929	0.0066
					>1000 m	0.0810	0.0029					>1000 m	0.0831	0.0030					>1000 m	0.0928	0.0032
Climate	0.2615	C9(C)	0.4673	0.1222	<100 kWh/m^2	0.1047	0.0128	0.2666	C9(C)	0.4492	0.1198	<100 kWh/m^2	0.1073	0.0128	0.2784	C9(C)	0.4406	0.1227	<100 kWh/m^2	0.1186	0.0145
					100–125 kWh/m^2	0.2583	0.0316					100–125 kWh/m^2	0.2632	0.0315					100–125 kWh/m^2	0.2847	0.0349
					>125 kWh/m^2	0.6370	0.0778					>125 kWh/m^2	0.6295	0.0754					>125 kWh/m^2	0.5967	0.0732
		C10(C)	0.2772	0.0725	<300 h	0.1047	0.0076		C10(C)	0.2838	0.0757	<300 h	0.1073	0.0081		C10(C)	0.2750	0.0766	<300 h	0.1186	0.0091
					300–350 h	0.2583	0.0187					300–350 h	0.2632	0.0199					300–350 h	0.2847	0.0218
					>350 h	0.6370	0.0462					>350 h	0.6295	0.0476					>350 h	0.5967	0.0457
		C11(C)	0.1601	0.0419	S, SE, SW	0.6548	0.0274		C11(C)	0.1673	0.0446	S, SE, SW	0.6475	0.0289		C11(C)	0.1779	0.0495	S, SE, SW	0.6157	0.0305
					W, E	0.2499	0.0105					W, E	0.2550	0.0114					W, E	0.2776	0.0137
					N, NE, NW	0.0953	0.0040					N, NE, NW	0.0975	0.0043					N, NE, NW	0.1067	0.0053
		C12(C)	0.0954	0.0249	0–2.5°	0.6548	0.0163		C12(C)	0.0997	0.0266	0–2.5°	0.6475	0.0172		C12(C)	0.1064	0.0296	0–2.5°	0.6157	0.0182
					2.5–5.0°	0.2499	0.0062					2.5–5.0°	0.2550	0.0068					2.5–5.0°	0.2776	0.0082
					>5.0°	0.0953	0.0024					>5.0°	0.0975	0.0026					>5.0°	0.1067	0.0032
Legal	0.1290	C13(L)	1.0000	0.1290	area for development	0.6694	0.0864	0.1319	C13(L)	1.0000	0.1319	area for development	0.6623	0.0874	0.1413	C13(L)	1.0000	0.1413	area for development	0.6311	0.0892
					area for further development and addenda	0.2426	0.0313					area for further development and addenda	0.2479	0.0327					area for further development and addenda	0.2711	0.0383
					remaning	0.0879	0.0113					remaning	0.0898	0.0118					remaning	0.0978	0.0138
Technical	0.0634	C14(T)	1.0000	0.0634	0–250 m	0.6548	0.0415	0.0648	C14(T)	1.0000	0.0648	0–250 m	0.6475	0.0420	0.0698	C14(T)	1.0000	0.0698	0–250 m	0.6157	0.0043
					250–500 m	0.2499	0.0158					250–500 m	0.2550	0.0165					250–500 m	0.2776	0.0194
					>500 m	0.0953	0.0060					>500 m	0.0975	0.0063					>500 m	0.1067	0.0074
Environmental	0.0333	C15(E)	1.0000	0.0333	0–33%	0.6548	0.0218	0.0333	C15(E)	1.0000	0.0333	0–33%	0.6475	0.0216	0.0359	C15(E)	1.0000	0.0359	0–33%	0.6157	0.0221
					33–66%	0.2499	0.0083					33–66%	0.2550	0.0085					33–66%	0.2776	0.0100
					66–100%	0.0953	0.0032					66–100%	0.0975	0.0032					66–100%	0.1067	0.0038

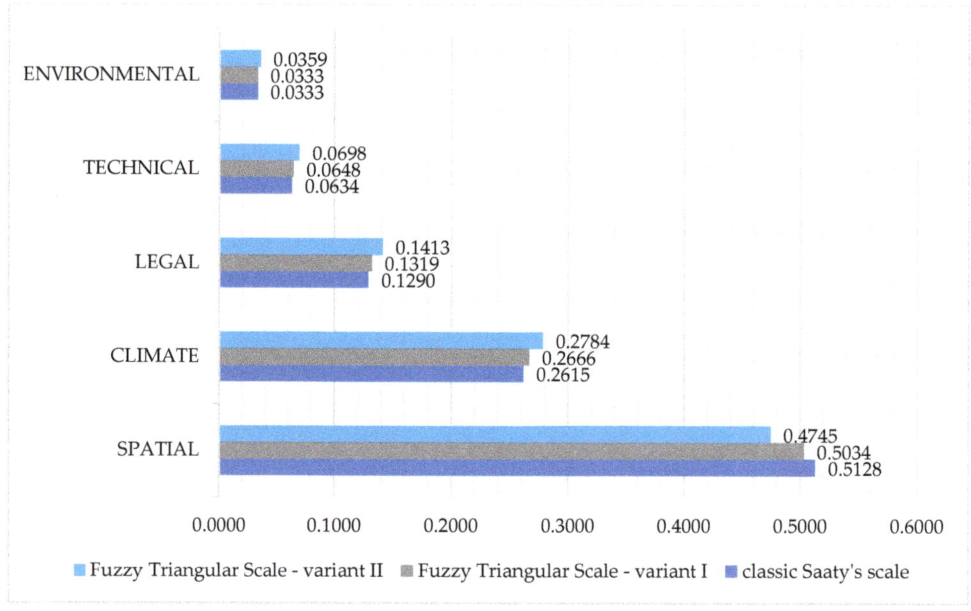

Figure 4. Group weights criteria calculated using three scales. Source: author's own work.

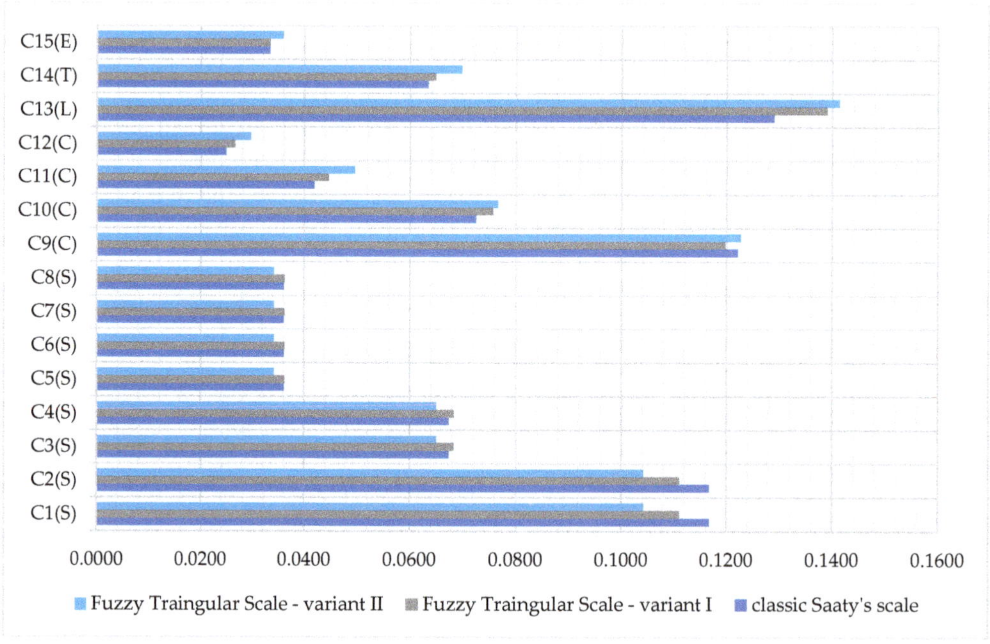

Figure 5. Global weight criteria calculated using three scales. Source: author's own work.

Based on the calculations, it can be stated, that in the case of groups of criteria, regardless of the scale, the group of spatial criteria obtained the highest weight, followed by the group of criteria pertaining to the selected climatic parameters. Relatively lower weights were obtained successively by legal, technical, and environmental criteria. The differences between the weights of groups of decision criteria obtained using the classic scale and the first variant of the fuzzy triangular scale were not significant. In the case of group of spatial criteria, this difference was 0.0094, for group of climatic criteria: 0.051, group of legal criteria: 0.029, and for group of technical criteria: 0.0014. The weights of the group of environmental criteria were the same in the both scale (classic scale and the first variant of triangular scale). Slightly higher differences were noted between the weights obtained using the second fuzzy scale, for example in the case of a group of spatial criteria this difference was 0.0383. Similar properties can be observed at the lower levels of the hierarchical structure.

4. Discussion

Based on the results obtained, it can be concluded that the differences between the criteria weights obtained using the classic Saaty's scale and the weights calculated using fuzzy scales, (especially its first variant) are not significant. First of all, despite the application of different scales, the general order of criteria did not change. The global weights of criteria groups influenced the weights of criteria and sub-criteria located at the lower levels of the structure. However, it should be emphasized here that the developed hierarchical structure, together with weights, is only an example aimed at assessing the impact of using different scales (existing in the literature on the subject) on the final set of decision-making factor weights at individual levels.

The FAHP method, just like its classic version, has its advantages and disadvantages. Among the advantages, we should undoubtedly indicate the simplified—from the point of view of the decision-maker—procedure of pairwise comparison, especially in the situation of uncertainty or incomplete information about a given decision-making challenge. On the other hand, it could be considered a certain limitation of FAHP that it operates under a more complicated algorithm when compared to the classical AHP method. The calculation algorithm presented in this paper, based on Buckley's assumptions, is not the only procedure of operation, see [65]. In addition, applying FAHP to more complex decision-making challenges and performing sensitivity analyses without the appropriate computer support significantly increases calculation times.

It should be emphasized that a comparative analysis of the AHP and FAHP methods has already been subject to research, for example during the analysis of the selected decision-making challenges in the food industry [66], resource analysis based on a selected example from the energy industry, [67], supplier selection [68], or a decision-making model that is closest to the subject matter of this paper, i.e., a model in the field of spatial planning [69]. The authors of the aforementioned studies evaluated the application potential of Saaty's method extensions in the context of selected decision-making challenges. In modern times, especially in complex economic conditions where many decisions are accompanied by a level of uncertainty, the MCDM/MCDA methods on fuzzy sets may serve as support tools [65]. According to the authors Kabir Hasin [67], the fuzzy AHP's approach allows for a more accurate description of the decision-making process, allowing one to grasp the vagueness of human thinking. The authors recommend using FAHP if information/evaluations are not certain. On the other hand—in the case of spatial planning and based on their research—the authors stated that: "if the planning aims to identify priority areas for development as a focal point, simpler MCDM methods such as AHP should be sufficient. In this situation, selecting more sophisticated techniques like Fuzzy AHP, which can only be seen as a black box by stakeholders, will not necessarily generate different outcomes" [69] (p. 64).

In summary, both AHP and FAHP have certain limitations. Therefore, the key aspect is correct method selection and adjusting the calculation techniques to the specificity of the decision-making challenge under consideration. In the case of such fields as architecture, urban planning or energy-efficient construction, it is often necessary to take into account the preferences of residents/users of facilities/space. For that reason, the AHP method—together with its extensions—can significantly support the design processes. The algorithm of pairwise comparison, using the traditional Saaty scale, as well as the triangular number scales, can be used when in questionnaire for collecting and assessing the preferences of decision-makers see e.g., [70]. Examples of decision-making problems that require research the preferences of residents/users include, e.g., delineation of development areas in the city, identifying the preferred urban revitalization scenarios or selection of the best modernization method. Importantly, there are already practical examples of using the MCDM/MCDA methods, for example for the research of residential preferences in Poznan [71].

5. Conclusions

Based on the subject-matter literature studies and the fuzzy multi-criteria analysis, the following conclusions can be made:

- MCDM/MCDA methods can serve as effective support tools at the stage when various decisions are being made by engineers. The overview of possible applications, which was elaborated within this paper, can constitute a set of good practices, facilitating the selection of an appropriate method or a set of methods tailored to a specific decision-making challenge;
- MCDM/MCDA methods are constantly evolving, with new methods appearing and numerous modifications to the existing ones coming into the picture; the development of fuzzy set theory in multi-criteria decision-making support is also an important research direction;
- The AHP and FAHP methods can be particularly useful when dealing with challenges that require taking into account the preferences of various entities; the popular examples of such challenges within the scope of the analyzed fields are e.g., revitalization models of both entire districts and individual facilities, delimitation and assessment of selected urban areas (developmental, neglected etc.,) or improving the energy efficiency of facilities;
- In a situation when complete information is available, it is recommended to use the classic AHP method that has the advantage of a simpler algorithm and wide variety of available computer programs that can perform the analysis; the example from the elaborated case study and the results of previous studies show that the differences between the weights obtained using AHP and FAHP are not significant;
- The AHP method and its extensions are some of the most popular methods that enable weighting of decision-making factors, but it is not the only one; therefore, among future research directions, one can indicate a comparative analysis of AHP modifications with other methods, e.g., SWARA method.

Funding: The research has been completed within the research project WZ/WBiIŚ/6/2019 and financed by public funds of Polish Ministry of Science and Higher Education.

Conflicts of Interest: The author declare no conflict of interest.

References

1. Bossel, H. *Indicators for Sustainable Development: Theory, Method, Applications*; A Report to the Balaton Group; International Institute for Sustainable Development: Winnipeg, MB, Canada, 1999.
2. International Institute for Sustainable Development. Available online: https://www.iisd.org/ (accessed on 14 October 2019).
3. Zavadskas, E.K.; Šaparauskas, J.; Antucheviciene, J. Sustainability in Construction Engineering. *Sustainability* **2018**, *10*, 2236. [CrossRef]
4. Williams, K.; Burton, E.; Jenks, M. *Achieving Sustainable Urban. Form: an introduction*; Spon Press: London, UK, 2004.
5. Ogrodnik, K. Contemporary concepts of sustainable urban development-selected examples in theory and practice. In *Spatial Management in Conditions of Sustainable Development*; Broniewicz, E., Ed.; Publishing House of the Bialystok University of Technology: Bialystok, Poland, 2017; pp. 67–84.
6. Harputlugil, T. Analytic Hierarchy Process (AHP) as an Assessment Approach for Architectural Design: Case Study of Architectural Design Studio. *Int. J. Archit. Plan.* **2018**, *6*, 217–245. [CrossRef]

7.	Della Spina, L.; Giorno, C.; Galati Casmiro, R. Bottom-Up Processes for Culture-Led Urban Regeneration Scenarios. In *Computational Science and Its Applications—ICCSA 2019*; Misra, S., Ed.; Lecture Notes in Computer Science; Springer: Berlin/Heidelberg, Germany, 2019; pp. 93–107.

8.	Jalilzadehazhari, E.; Vadiee, A.; Johansson, P. Achieving a Trade-Off Construction Solution Using BIM, an Optimization Algorithm, and a Multi-Criteria Decision-Making Method. *Buildings* **2019**, *9*, 81. [CrossRef]

9.	Behzadian, M.; Otaghsara, S.K.; Yazdani, M.; Ignatius, J. A state-of the-art survey of TOPSIS applications. *Expert Syst. Appl.* **2012**, *39*, 13051–13069. [CrossRef]

10.	Zavadskas, E.K.; Turskis, Z. Multiple Criteria Decision Making (MCDM) methods in economics: An overview. *Technol. Econ. Dev. Econ.* **2011**, *17*, 397–427. [CrossRef]

11.	Zavadskas, E.K.; Antucheviciene, J.; Kapliński, O. Multi-criteria decision making in civil engineering: Part I—A state-of-the-art survey. *Eng. Struct. Technol.* **2015**, *7*, 103–113. [CrossRef]

12.	Dytczak, M. *Selected Methods for Multi-Criteria Decision Analysis in Civil. Engineering*; Opole University of Technology: Opole, Poland, 2008.

13.	Kobryń, A. *Multi-Criteria Decision Support in Space Management*; Difin: Warsaw, Poland, 2015.

14.	Zavadskas, E.K.; Turskis, Z.; Kildienė, S. State of art surveys of overviews on MCDM/MADM methods. *Technol. Econ. Dev. Econ.* **2014**, *20*, 165–179. [CrossRef]

15.	Emrouznejad, A.; Marra, M. The state of the art development of AHP (1979–2017): A literature review with a social network analysis. *Int. J. Prod. Res.* **2017**, *55*, 6653–6675. [CrossRef]

16.	Jato-Espino, D.; Castillo-Lopez, E.; Rodriguez-Hernandez, J.; Canteras-Jordana, J.C. A review of application of multi-criteria decision making methods in construction. *Autom. Constr.* **2014**, *45*, 151–162. [CrossRef]

17.	Tamosaitiene, J.; Sipalis, J.; Banaitis, A.; Gaudutis, E. Complex model for the assessment of the location of high-rise buildings in the city urban structure. *Int. J. Strateg. Prop. Manag.* **2013**, *17*, 93–109. [CrossRef]

18.	Bielinskas, V.; Burinskienė, M.; Palevičius, V. Assessment of neglected areas in Vilnius city using MCDM and COPRAS methods. *Procedia Eng.* **2015**, *122*, 29–38. [CrossRef]

19.	De Toro, P.; Iodice, S. Evaluation in urban planning: A multi-criteria approach for the choice of alternative Operational Plans in Cava De' Tirreni. *Aestimum* **2016**, *69*, 93–112. [CrossRef]

20.	Ogrodnik, K. The application of the PROMETHEE method in evaluation of sustainable development of the selected cities in Poland. *Econ. Environ.* **2017**, *3*, 19–36.

21.	Zinatizadeh, S.; Azmi, A.; Monavari, S.M.; Sobhanardakani, S. Multi-criteria decision making for sustainability evaluation in urban areas: A case study for Kermanshah City, Iran. *Appl. Ecol. Environ. Res.* **2017**, *15*, 1083–1100. [CrossRef]

22.	Pujadas, P.; Pardo-Bosch, F.; Aguado-Renter, A.; Aguado, A. MIVES multi-criteria approach for the evaluation, prioritization, and selection of public investment projects. A case study in the city of Barcelona. *Land Use Policy* **2017**, *64*, 29–37. [CrossRef]

23.	Chen, C.S.; Chiu, Y.H.; Tsai, L. Evaluating the adaptive reuse of historic buildings through multicriteria decision-making. *Habitat Int.* **2018**, *81*, 12–23. [CrossRef]

24.	Tian, G.; Zhang, H.; Feng, Y.; Wang, D.; Peng, Y.; Jia, H. Green decoration materials selection under interior environment characteristics: A grey-correlation based hybrid MCDM method. *Renew. Sustain. Energy Rev.* **2018**, *81*, 682–692. [CrossRef]

25.	Masoumi, Z.; Genderen, J.V. Investigation of sustainable urban development direction using Geographic Information Systems (case study: Zanjan city). *Int. Arch. Photogramm. Remote Sens. Spat. Inf. Sci.* **2019**, *13*, 1313–1320. [CrossRef]

26.	Ribera, F.; Nesticò, A.; Cucco, P.; Maselli, G. A multicriteria approach to identify the Highest and Best Use for historical buildings. *J. Cult. Herit.* **2019**. [CrossRef]

27.	Pons, O.; Franquesa, J.; Amin Hosseini, S.M. Integrated Value Model to Assess the Sustainability of Active Learning Activities and Strategies in Architecture Lectures for Large Groups. *Sustainability* **2019**, *11*, 2917. [CrossRef]

28.	Yang, Y.; Li, B.; Yao, R. A method of identifying and weighting indicators of energy efficiency assessment in Chinese residential buildings. *Energy Policy* **2010**, *38*, 7687–7697. [CrossRef]

29.	Kuzman, M.K.; Grošelj, P.; Ayrilmis, N.; Zbašnik-Senegačnik, M. Comparison of passive house construction types using analytic hierarchy process. *Energy Build.* **2013**, *64*, 258–263. [CrossRef]

30.	Šiožinytė, E.; Antuchevičienė, J. Solving the problems of daylighting and tradition continuity in a reconstructed vernacular building. *J. Civ. Eng. Manag.* **2013**, *19*, 873–882. [CrossRef]

31. Ruzgys, A.; Volvačiovas, R.; Ignatavičius, Č.; Turskis, Z. Integrated evaluation of external wall insulation in residential buildings using SWARA-TODIM MCDM method. *J. Civ. Eng. Manag.* **2014**, *20*, 103–110. [CrossRef]

32. Šiožinytė, E.; Antuchevičienė, J.; Kutut, V. Upgrading the old vernacular building to contemporary norms: Multiple criteria approach. *J. Civ. Eng. Manag.* **2014**, *20*, 291–298. [CrossRef]

33. Medineckiene, M.; Dziugaite-Tumeniene, R. Energy simulation in buildings with the help of multi-criteria decision making method. In Proceedings of the 9th International Conference on Environmental Engineering (ICEE), Vilnius, Lithuania, 22–23 May 2014.

34. Chen, L.; Pan, W. A BIM-integrated Fuzzy Multi-criteria Decision Making Model for Selecting Low-Carbon Building Measures. *Procedia Eng.* **2015**, *118*, 606–613. [CrossRef]

35. Sedláková, A.; Vilčeková, S.; Krídlová Burdová, E. Analysis of material solutions for design of construction detailsof foundation, wall and floor for energy and environmental impacts. *Clean Technol. Environ. Policy* **2015**, *17*, 1323–1332. [CrossRef]

36. Motuzienė, V.; Rogoža, A.; Lapinskienė, V.; Vilutienė, T. Construction solutions for energy efficient single-family house based on its life cycle multi-criteria analysis: A case study. *J. Clean. Prod.* **2016**, *112*, 532–541. [CrossRef]

37. Vujosevic, M.L.; Popovic, M.J. The comparison of the energy performance of hotel buildings using PROMETHEE decision-making method. *Therm. Sci.* **2016**, *20*, 197–208. [CrossRef]

38. Tomczak, K.; Kinash, O. Assessment of the Validity of Investing in Energy-Efficient Single-Family Construction in Poland - Case Study. *Arch. Civ. Eng.* **2016**, *62*, 119–138. [CrossRef]

39. Seddiki, M.; Anouche, K.; Bennadji, A.; Boateng, P. A multi-criteria group decision-making method for the thermal renovation of masonry buildings: The case of Algeria. *Energy Build.* **2016**, *129*, 471–483. [CrossRef]

40. Zavadskas, E.K.; Antucheviciene, J.; Kalibatas, D.; Kalibatiene, D. Achieving Nearly Zero-Energy Buildings by applying multi-attribute assessment. *Energy Build.* **2017**, *143*, 162–172. [CrossRef]

41. Cortes, J.P.R.; Ponz-Tienda, J.L.; Delgado, J.M.; Gutierrez-Bucheli, L. Choosing by advantages; benefits analysis and implementation in a case study, Colombia. In Proceedings of the 26th Annual Conference of the International Group for Lean Construction (IGLC), Chennai, India, 16–22 July 2018.

42. Guzmán-Sánchez, S.; Jato-Espino, D.; Lombillo, I.; Diaz-Sarachaga, J.M. Assessment of the contributions of different flat roof types to achieving sustainable development. *Build. Environ.* **2018**, *141*, 182–192. [CrossRef]

43. Moghtadernejad, S.; Chouinard, L.E.; Mirza, M.S. Multi-criteria decision-making methods for preliminary design of sustainable facades. *J. Build. Eng.* **2018**, *19*, 181–190. [CrossRef]

44. Zolfani, S.H.; Pourhossein, M.; Yazdani, M.; Zavadskas, E.K. Evaluating construction projects of hotels based on environmental sustainability with MCDM framework. *Alex. Eng. J.* **2018**, *57*, 357–365. [CrossRef]

45. Arroyo, P.; Mourgues, C.; Flager, F.; Correa, M.G. A new method for applying choosing by advantages (CBA) multicriteria decision to a large number of design alternatives. *Energy Build.* **2018**, *167*, 30–37. [CrossRef]

46. Beltrán, R.D.; Martínez-Gómez, J. Analysis of phase change materials (PCM) for building wallboards based on the effect of environment. *J. Build. Eng.* **2019**, *24*, 100726. [CrossRef]

47. Velasquez, M.; Hester, P.T. An Analysis of Multi-Criteria Decision Making Methods. *Int. J. Oper. Res.* **2013**, *10*, 56–66.

48. Aruldoss, M.; Lakshmi, T.M.; Venkatesan, V.P. A Survey on Multi Criteria Decision Making Methods and Its Applications. *Am. J. Inf. Syst.* **2013**, *1*, 31–43. [CrossRef]

49. Ho, W. Integrated analytic hierarchy process and its applications – A literature review. *Eur. J. Oper. Res.* **2008**, *186*, 211–228. [CrossRef]

50. Vaidya, O.S.; Kumar, S. Analytic Hierarchy Process an Overview of Applications. *Eur. J. Oper. Res.* **2006**, *169*, 1–29. [CrossRef]

51. Forman, E.H. Facts and fictions about the analytic hierarchy process. *Math. Comput. Model.* **1993**, *17*, 19–26. [CrossRef]

52. Antucheviciene, J.; Kala, Z.; Marzouk, M.; Vaidogas, E.R. Solving Civil Engineering Problems by Means of Fuzzy and Stochastic MCDM Methods: Current State and Future Research. *Math. Probl. Eng.* **2015**, *2015*, 1–16. [CrossRef]

53. Mandic, K.; Delibasic, B.; Knezevic, S.; Benkovic, S. Analysis of the financial parameters of Serbian banks through the application of the fuzzy AHP and TOPSIS methods. *Econ. Model.* **2014**, *43*, 30–37. [CrossRef]

54. Ayhan, M.B. A fuzzy AHP approach for supplier selection problem: A case study in a Gearmotor company. *Int. J. Manag. Value Supply Chain.* **2013**, *4*, 11–23. [CrossRef]
55. Saaty, T.L. Decision making with the analytic hierarchy process. *Int. J. Serv. Sci.* **2008**, *1*, 83–98. [CrossRef]
56. Li, L.; Shi, Z.H.; Yin, W.; Zhu, D.; Ng, S.L.; Cai, C.F.; Lei, A.L. A fuzzy analytic hierarchy process (FAHP) approach to eco-environmental vulnerability assessment for the danjiangkou reservoir area, China. *Ecol. Model.* **2009**, *220*, 3439–3447. [CrossRef]
57. Łuczak, A.; Wysocki, F. Linear ordering of objects from application of fuzzy AHP and TOPSIS. *Stat. Rev.* **2011**, *1–2*, 1–23.
58. Stoltmann, A.; Bućko, P.; Jaskólski, M. Multi-criteria investment decision support model using fuzzy Analytic Hierarchy Process (F-AHP) method for power industry. *Sci. Pap. Fac. Electr. Control. Eng. Gdan. Univ. Technol.* **2015**, *47*, 179–182.
59. Online Output Softwares. Available online: http://www.onlineoutput.com/ (accessed on 18 November 2019).
60. Kolendo, Ł.; Ogrodnik, K. The selected criteria of location of solar housing development. In *Urban Planning in Spatial Economy*; Chmielewski, J.M., Ed.; Warsaw University of Technology: Warsaw, Poland, 2016; pp. 127–141.
61. Kolendo, Ł.; Ogrodnik, K. Multi-criteria analysis of the usability of the area of Bialystok for solar housing development. In *Urban Planning in Spatial Economy*; Chmielewski, J.M., Ed.; Warsaw University of Technology: Warsaw, Poland, 2016; pp. 142–154.
62. Hejmanowska, B.; Hnat, E. Multi-factoral evaluation of residential area locations: Case study of Podegrodzie local authority. *Arch. Photogramm. Cartogr. Remote Sens.* **2009**, *20*, 109–121.
63. Majerska-Pałubicka, B. The quality of built environment in aspects of quality of natural environment based on examples of eco–settlement. *Archit. Et Artibus* **2010**, *2*, 57–62.
64. Jaroszewicz, J.; Bielska, A.; Szafranek, A. Application of map algebra to determine the lands preferred for building development. *Arch. Photogramm. Cartogr. Remote Sens.* **2012**, *23*, 127–137.
65. Ahmed, F.; Kilic, K. Fuzzy Analytic Hierarchy Process: A performance analysis of various algorithms. *Fuzzy Sets Syst.* **2019**, *362*, 110–128. [CrossRef]
66. Özdağoğlu, A.; Özdağoğlu, G. Comparison of AHP and Fuzzy AHP for the multicriteria decision making processes with linguistic evaluations. *İstanbul Ticaret Üniversitesi Fen Bilimleri Derg. Yıl* **2007**, *1*, 65–85.
67. Kabir, G.; Hasin, M.A.A. Comparative analysis of AHP and Fuzzy AHP models for multicriteria inventory classification. *Int. J. Fuzzy Log. Syst.* **2011**, *1*, 1–16.
68. Ishizaka, A. Comparison of Fuzzy logic, AHP, FAHP and Hybrid Fuzzy AHP for new supplier selection and its performance analysis. *Int. J. Integr. Supply Manag.* **2014**, *9*, 1–22. [CrossRef]
69. Mosadeghi, R.; Warnken, J.; Tomlinson, R.; Mirfenderesk, H. Comparison of Fuzzy-AHP and AHP in a spatial multi-criteria decision making model for urban land-use planning. *Comput. Environ. Urban. Syst.* **2015**, *49*, 54–65. [CrossRef]
70. Prusak, A.; Stefanów, P.; Gardian, M. Graphic Form of Questionnaire in AHP/ANP Research. *Mod. Manag. Rev.* **2013**, *20*, 171–189.
71. Matusiak, M.; Palicki, S. *Multi-Criteria Analysis: Housing Needs and Preferences of Poznan Residents-Recommendations for Housing Policy*; Poznan University of Economics and Business: Poznan, Poland, 2015.

 *buildings*

Article

On the Search of Models for Early Cost Estimates of Bridges: An SVM-Based Approach

Michał Juszczyk

Faculty of Civil Engineering, Cracow University of Technology, 31-155 Kraków, Poland; mjuszczyk@izwbit.pk.edu.pl

Received: 14 November 2019; Accepted: 13 December 2019; Published: 19 December 2019

Abstract: The completion of a bridge construction project within budget is one of the project's key factors of success. This prerequisite is more likely to be achieved if the cost estimates, especially those provided in the early stage of a project, are realistic and close to the actual costs. The paper presents the research results on the development of a cost prediction model based on machine learning, namely the support vector machines (SVM) method, for which the input represents basic information and parameters of bridges, available in the early stage of projects. Several SVM-based regression models were investigated with the use of data collected for a number of bridge construction projects completed in Poland. Having finished the machine learning and testing processes, five of the models, of satisfying knowledge generalization ability and comparable performance, were preselected. The final selection of the best model was based on the comparison and analysis ability to predict bridge construction costs with accuracy appropriate for the early stage of projects. The general testing metrics of the finally selected model, named $BCCPM_{SVR}2$, were as follows: root mean square error: 1.111; correlation coefficient of real-life bridge construction costs and costs predicted by the model: 0.980; and mean absolute percentage error: 10.94%. The research resulted in the development and introduction of an original model capable of providing early estimates of bridge construction costs with satisfactory accuracy.

Keywords: cost estimates; construction costs; bridge construction projects; machine learning; support vector machines; regression

1. Introduction

Bridges, which are without a doubt of high significance for transportation networks, can be also seen as results or products of construction projects. The completion of a project within budget is one of the project's success key factors. It is more likely to achieve success if the cost estimates are realistic and close to the actual costs. Therefore, there is a need for cost estimates provided at the successive stages of a construction project. Early cost estimates rely on basic information and parameters of a project. Although their expected accuracy is relatively low (they can be considered as qualitative predictions rather than precise cost estimates), they are delivered when the crucial decisions are made and thus the impact on the final cost is great.

Along with the intensive development and modernization of transport infrastructure in Poland, bridge construction has also increased over the past few years. On the one hand, it is important to start the process of cost estimation for a bridge project as early as possible. On the other hand, some artificial intelligence and machine learning tools offer capabilities, such as learning from experience and knowledge generalization, which make them applicable for the early cost estimation models. Especially for bridge projects, the development of such models is supposed to provide early estimates or forecasts of the final cost.

The aim of this paper is to introduce a cost estimation model for bridge construction projects based on machine learning, namely the support vector machine (SVM) method. The goal of the research was to develop a model supporting fast cost estimates of total construction costs of bridges in the early stages of construction projects.

1.1. Literature Review

The problem of cost modeling for bridge projects is present in scientific publications. One can distinguish various approaches to this issue.

Part of the research is focused on the development of models for estimating the costs of either selected cost components or elements of bridge structures. In [1], the costs of doing preliminary engineering as cost components of the total costs of newly built bridges are addressed. The authors introduced statistical models that link variation in preliminary engineering costs with specific parameters. A conceptual model aiding cost estimates of bridge foundations is presented in [2]. A three-stage decision process including the foundation system selection, materials' quantities estimation, and foundation cost estimation is supported by the proposed model. In this study, stepwise regression analysis was applied. Another work [3] reports analysis which aimed to develop material quantity models of the abutment and caisson as components of a whole bridge structure, with prestressed concrete I-girder superstructure. The research and application of multiple regression analysis resulted in a number of equations proposed for estimates of concrete volume and reinforcing steel weight of abutment and caisson as components of a whole bridge structure. Another study [4] presents the problem of bridge superstructures cost estimates. The proposed method, based on linear regression and a bootstrap resampling, provides estimates in the early stages of road projects.

Another part of the research presents efforts on development models for estimating construction costs of specific kinds of bridges. The authors of [5] proposed a model for the cost estimation of timber bridges based on artificial neural networks. The performance of the proposed neural network-based model is reported to be better than the model based on linear regression. Another work [6] introduces a model for approximate cost estimation for prestressed concrete beam bridges based on the quantity of standard work. The proposed method supports cost estimates for a typical beam bridge structure using three parameters: length of span, total length of bridge, and width. Another paper [7] presents the methodology for estimates of railroad bridges. The proposed model combines case-based reasoning, genetic algorithms, and multiple regression as tools. Another work [8] introduced a computer-aided system providing cost estimates of prestressed concrete road bridges. The system, built upon the database including data collected from completed bridge projects, allows estimating the material quantities and costs of all bridge elements. The estimating models that constitute the core of the system were developed with the use of statistical analysis. The authors of [9] focus on the use of Bridge Information Modeling (BrIM) for detailed cost estimates. The authors discussed the issue of extraction of information from the bridge model and cost estimation process prepared on this basis. The methodology for generating cash flow and required payments are presented as well.

The problem of risk analysis in bridge construction is addressed in [10]. The research aimed to identify and analyze risks associated with bridge construction. Impacts of risks on cost and schedule in bridge projects are discussed.

Some publications refer to the issue of replacement, renovation, repair, and maintenance costs of bridges. Replacement cost prediction models, developed with the use of regression techniques, are introduced in [11], in which the authors investigated the applicability of nonlinear and log-linear models for the task. Another work [12] presents the development of a model for cost estimation of repair and maintenance of bridges using artificial neural networks. Another paper [13] presents the development, discussion, and performance assessment of a set of regression models for estimating the costs of rehabilitating bridges. One of the papers addresses specifically the issue of repair or replacement costs damaged by hurricane Katrina [14]. The authors analyzed and compared damage patterns to bridges and examples of repair measures. Relationships between storm surge elevation, damage level, and repair costs were developed. The issue of potential design considerations for bridges in vulnerable coastal regions is discussed. Some studies address the topic of life-cycle costs of bridges. In another report [15], the life-cycle cost-effectiveness of fiber-reinforced-polymer bridge decks is investigated and analyzed. The author used life-cycle cost method analysis, tailored for comparing new materials with conventional ones. Publications on cost optimization of concrete bridge components and systems are reviewed in [16] along with the presentation of the state-of-the-art in life-cycle cost analysis and design of concrete bridges.

SVM are machine learning systems with the ability to learn from experience (hidden in the data presented to the systems) and knowledge generalization. The theory of SVM, developed by Vapnik and co-workers, is based on the principles of statistical learning [17,18]. The methodology and theory of SVM are also broadly presented in the literature by other authors, e.g., [19–21]. SVM can be applied for either classification or regression problems. Some SVM implementations in construction management, introduced in works published in recent years, are the automated document classification for improving information flow in construction management systems [22], methodology of legal decision support aiming at mitigation of negative impacts of conflicts that occur in the course of construction projects [23], risk hedging prediction for construction material suppliers [24], modeling construction contractors default prediction [25], prediction of company failure in the construction industry [26], and dynamical prediction of construction project success [27].

In the field of cost analyses in construction, specifically supported by SVM, one can also find recent works. SVM-based modeling variations of construction prices with the use of construction cost index in Taiwan were introduced in [28]. The study established a hybrid intelligence system based on the fusion of SVM and Differential Evolution for estimation of construction cost index in construction. The system is reported to perform with a satisfying, high accuracy. In another work [29], the authors developed models supporting the prediction of construction project cost and schedule success, as the input early project planning status information was used. The alternative models, based on either ANN or SVM, were compared—the latter proven to perform better. In one of the works [30], SVM-based machine learning, along with interval estimation and differential evolution, is implemented for modeling the cost at completion of construction projects (one of the metrics known from the Earned Value Management method). The proposed model proved its capability of delivering reliable forecasts. The authors of [31] focused on conceptual cost estimates of school buildings. Models based on linear regression, ANN and SVM, were developed and compared. The study on the estimation of costs and durations of urban road construction supported by alternative artificial intelligence tools, that are ANN or SVM, is presented in [32]. The SVM-based model is reported to perform with significantly better accuracy in terms of costs; whereas, for duration prediction, the SVM-based model is just slightly better than the one based on ANN.

1.2. Research Objectives

The aim of this paper is to present the results of studies on the development of a machine learning-based regression model, using the support vector machine (SVM) method, to support early estimates of total construction costs of bridges. The paper content includes an introduction and review of the literature. The following section presents the synthesis of the SVM-based regression methodology and assumptions for the prediction of the total construction costs of bridges as a regression problem to be solved. These are followed by the introduction of the results of the SVM-based regression analysis and the discussion. The last section includes conclusions and recapitulation.

The main assumption for the model proposed in this paper is the use of the SVM method. The rationale for this assumption is the method's capability of dealing with great dimensional data, applicability to non-linear regression and the fact that the method allows finding a global solution for a given task. Moreover, SVM works well on small sets of training data. The following remarks that refer to the mentioned can be made. First: it is possible to take into account many variables that play the role of cost predictors in the problem of early cost estimation of bridges. Second: nonlinear relationships between the cost predictors and the total construction costs of bridges can be modeled with the use of the SVM machine learning-based regression model. Third: The SVM-based model can be built upon a moderate amount of training data that characterize bridges and their costs.

The novelty of the introduced model relies on the fact that it offers cost predictions of bridges as whole objects. Moreover, several types of bridge structures are considered. Earlier works [2–4] focused mostly on estimates of either the substructure or superstructure. On the other hand, some works are limited to specific types of bridges [5–8]. The application of the SVM-based regression method for the development of a cost estimation model allows overcoming some drawbacks of the models built on the basis of regression analysis [2–4] or ANN [5]. When compared to linear regression, the SVM method does not require a priori assumptions about the functional relationship for the developed model. When compared to ANN, SVM is not at risk of the so-called local minima problem.

2. Methodology and Concept of a Model

The development of a model capable of providing early cost estimates of bridges based on the SVM method is understood here by solving the regression problem with the use of machine learning. The dependent variable of the sought-for regression model was the total construction cost of a bridge, later denoted as y. On the other hand, independent variables such as vectors of cost predictors, later denoted as x, represent information such as the features, characteristics, and specificity of bridges. The sought-for model was intended to provide multidimensional mapping from the set of cost predictors to the set of values representing total construction costs. Formally, the implicit regression function f, which is supposed to provide the mapping $x \rightarrow y$ denoted as:

$$y = f(x), \tag{1}$$

is supposed to be found with the use of machine learning-based on the SVM method. This method is based on knowledge generalization and learning from examples (that represent some experiences) presented to a machine.

2.1. Support Vector Machines Method in Regression Analysis

The following fundamentals of the method were compiled and summarized after [17–21]. The SVM method allows approximating f as a linear hyperplane. The linear approximation is achieved specifically for nonlinear problems due to a transformation of independent variable space to a higher dimensional, linear feature space. If the set of training examples is given as χ such that: $\{ \chi = [x, y] \in R^m \times R \}$ and Φ is a nonlinear transformation used to determine a new feature space H for the inputs: $\Phi: R^m \rightarrow H, \Phi(x) \in H, y \in R$, then the function f can be given as follows:

$$f(x) = w^{\mathrm{T}} \Phi(x) + w_0 \tag{2}$$

The transformation $\Phi(x)$ is supposed to increase the expressive power of the representation, and the approximation function is computed in the higher dimensional, linear feature space H. Support vectors (sv) are the training data points that lie closest to the hyperplane and thus they affect its optimal location.

To measure the errors of the training process, Vapnik's ε-insensitive loss function is assumed:

$$l(f(x),y) = |y - f(x)|_\varepsilon, \tag{3}$$

where:

$$|y - f(x)|_\varepsilon = 0 \text{ for } |y - f(x)| \leq \varepsilon \text{ and } |y - f(x)|_\varepsilon = |y - f(x)| - \varepsilon \text{ for } |y - f(x)| > \varepsilon, \tag{4}$$

Here, ε defines a tube of insensitiveness used to fit the training examples around the true values y. In other words, the value of ε affects the number of support vectors.

Following this the, problem comes down to optimization by machine learning:

$$\frac{1}{2}\|w\|^2 + C\Sigma(\xi - \xi^*) \rightarrow \min, \tag{5}$$

subject to the constraints for the both sides of ε-tube:

$$w^T \Phi(x) + w_0 - y \leq \varepsilon + \xi \text{ and } y - (w^T \Phi(x) + w_0) \leq \varepsilon + \xi^* \text{ and } \xi, \xi^* \geq 0 \tag{6}$$

The use of loss function (3) results in toleration of deviations smaller than ε. The C represents the regularization parameter in the SVM method, and determines a compromise between decision function's margin against training accuracy. It determines the compromise between the complexity of a model and ξ, and ξ^* in (5) and (6) are slack variables that penalize predictions out of the ε-tube. The optimization of (5) is solved with the use of Lagrange multipliers:

$$f(x) = \Sigma_{nsv}(\alpha - \alpha^*)\Phi(x)^T\Phi(x') + w_0, \tag{7}$$

where nsv stands for the number of support vectors and α, α^* are the multipliers for the optimal solution such as:

$$0 \leq \alpha \leq C \text{ and } 0 \leq \alpha \leq C \tag{8}$$

The choice of appropriate transformation Φ and explicit calculation of $\Phi(x)^T\Phi(x')$ is difficult and computationally complex. To simplify the computations, the kernel functions $K(x, x')$ are introduced instead:

$$K(x, x') = \Phi(x)^T\Phi(x'), \tag{9}$$

The kernel functions which are mostly mentioned for the use in the SVM method are: polynomial (10), radial basis (11), and sigmoidal (12):

$$K(x, x') = \tanh(\gamma x \cdot x' + c), \tag{10}$$

$$K(x, x') = \exp(-\gamma\|x - x'\|^2), \tag{11}$$

$$K(x, x') = (\gamma x \cdot x' + c)^d, \tag{12}$$

Taking into account the above, the approximation function can be given finally as:

$$f(x) = \Sigma_{sv}(\alpha - \alpha^*)K(x, x') + w_0, \tag{13}$$

2.2. Variables of the Model and the Concept of Model Development

Before the start of actual regression analysis, data that reflected the values of model variables were collected and analyzed. The collected data included information about road bridges, rail bridges, and animal bridges (as wildlife crossings) built in Poland between 2005 and 2018. In terms of total construction costs, the real-life values were updated to be comparable—regardless of the date of project completion—with the use of price indices of construction assembly production published by the General Statistical Office in Poland. Later in the paper, the updated costs of bridges given in millions of PLN (e.g., PLN 10.53 m) are referred to as y. For better recognition, the costs are given in millions of EUR as well (e.g., EUR 2.45 m). The conversion was made on the basis of the Polish National Bank official exchange rate for the PLN/EUR pair of currencies published for 31.12.2018. The values of y varied between PLN 2.46 m (EUR 0.57 m) and PLN 23.48 m (EUR 5.46 m).

The cost predictors, as the independent variables, brought to the model information about the type of bridge, type of project, structural and material solutions, types of supports and their foundations, and load class. All the mentioned information was initially recorded as nominal data. Moreover, basic size measures, in terms of the decks' total length and width, as well as the number of spans, were taken into account. The independent variables of a model are presented in Table 1. In this table, one can see that finally the characteristics of bridges recorded initially as nominal data were coded as binary values (0 or 1). Information recorded as numerical data was scaled to the range <0; 1>. In the case of structural solution, type of intermediate supports and load class, the values for x_{14}, x_{22}, and x_{27} were introduced to represent more than one nominal value that were ARCHED/BOX, COLUMNS/PILES, and k/C/D/E, respectively (see also the footnotes under Table 1). This was done due to the fact that some nominal values were not numerous enough in the dataset to be represented alone by one binary variable. It is important to note that for each of the characteristics listed in Table 1, only one nominal value was allowed, so only one of the binary variables belonging to this characteristic could take value 1. For example, for the type of a structure of which the nominal value was VIADUCT, the values $x_1 - x_3$ equaled $x_1 = 0$, $x_2 = 1$, $x_3 = 0$.

Table 1. Input data for regression model—independent variables.

Characteristic	Nominal Values	Coding	Symbol
	BRIDGE	binary	x_1
Type of a structure	VIADUCT	binary	x_2
	WHARF	binary	x_3
	ROAD BRIDGE	binary	x_4
Type of a bridge	RAIL BRIDGE	binary	x_5
	ANIMAL BRIDGE	binary	x_6
Type of a project	BUILD	binary	x_7
	DESIGN&BUILD	binary	x_8
Total length	LENGTH [m]	numerical	x_9
Width of a structure	WIDTH [m]	numerical	x_{10}
Number of spans	SPANS	numerical	x_{11}
	BEAM	binary	x_{12}
Structural solution	FRAME	binary	x_{13}
	ARCHED/BOX	binary	x_{14}
	REINFORCED CONCRETE	binary	x_{15}
Material solution	PRESTRESSED CONCRETE	binary	x_{16}
	STEEL	binary	x_{17}
Bridgehead supports	SOLID-WALLED	binary	x_{18}
	COLUMNS	binary	x_{19}
	NONE	binary	x_{20}
Intermediate supports	SOLID-WALLED	binary	x_{21}
	COLUMNS/PILES	binary	x_{22}
Supports' foundations	SHALLOW	binary	x_{23}
	DEEP	binary	x_{24}
	A	binary	x_{25}
Load class *	B	binary	x_{26}
	k/C/D/E [1]	binary	x_{27}

[1] k for rail bridges or C, D, E for other bridges; * according to standards applied in Poland.

Table 2 presents a random sample of the coded variables x and y as used for model development, and p stands for pattern number.

The selection of the cost predictors was based on the availability of information in the early stages of the bridge construction projects. The characteristics and their values that became independent variables of the model (as presented in Table 1) can be easily identified in at beginning of the design process.

Overall, the number of patterns to be used for the process of machine learning and testing models equaled 167. The data was collected from the public clients responsible for bridge construction projects in Poland. The data was divided into two subsets—the first subset (later denoted as L) was used for the machine learning purposes, the second subset (later denoted as T) was used for the models' testing purposes. Both subsets were selected so as to be equivalent and to ensure their representativeness in terms of the features of the investigated bridges and the range of construction costs as well. The cardinality of subset L equaled 131, whereas the cardinality of subset T equaled 36. One can easily note that the number of patterns belonging to subset T accounted for more than 20% of the overall number of collected data patterns.

Table 2. Random sample of the model's variable values.

p	10	77	83	104	109	111	119	150	166
x_1	1	0	0	0	1	0	1	0	0
x_2	0	1	1	0	0	1	0	1	0
x_3	0	0	0	1	0	0	0	0	1
x_4	0	0	0	0	0	1	1	0	0
x_5	0	1	1	1	0	0	0	0	1
x_6	1	0	0	0	1	0	0	1	0
x_7	1	1	1	1	1	0	0	1	0
x_8	0	0	0	0	0	1	1	0	1
x_9	0.069	0.151	0.150	0.219	0.197	0.180	0.456	0.095	0.715
x_{10}	0.114	0.057	0.027	0.114	0.114	0.092	0.097	0.426	0.049
x_{11}	0.000	0.000	0.143	0.143	0.143	0.000	0.286	0.071	0.500
x_{12}	1	1	0	1	1	0	0	0	1
x_{13}	0	0	1	0	0	0	0	1	0
x_{14}	0	0	0	0	0	1	1	0	0
x_{15}	0	0	1	0	0	0	1	0	0
x_{16}	1	1	0	1	0	0	0	1	1
x_{17}	0	0	0	0	1	1	0	0	0
x_{18}	1	0	1	1	1	1	1	1	1
x_{19}	0	1	0	0	0	0	0	0	0
x_{20}	0	0	1	0	1	0	0	0	0
x_{21}	0	0	0	1	0	0	1	1	1
x_{22}	1	1	0	0	0	1	0	0	0
x_{23}	0	1	0	1	0	1	0	1	0
x_{24}	1	0	1	0	1	0	1	0	1
x_{25}	1	0	0	1	1	0	0	0	0
x_{26}	0	1	1	0	0	0	0	0	1
x_{27}	0	0	0	0	0	1	1	1	0
y [PLN]	3.02	5.49	6.11	9.84	10.82	12.54	14.15	6.83	19.85
y [EUR] [1]	0.70	1.28	1.42	2.29	2.52	2.92	3.29	1.59	4.62

[1] training and testing of the model was done with the use of costs given in millions of PLN.

The research included an investigation of the number of SVM-based regression models. A schematic diagram of the investigated models is presented in Figure 1.

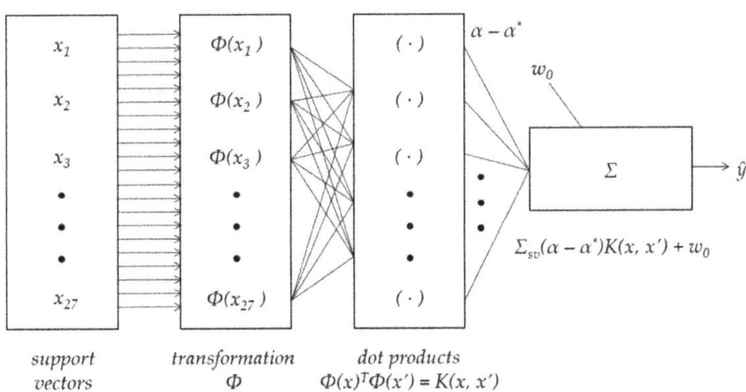

Figure 1. Schematic diagram of the investigated support vector machine (SVM)-based regression models.

The SVM-based models' performance rely on the assumed kernel function and its parameters as well as C and ε meta-parameters.

For the purposes of transformation Φ, the use of the three aforementioned kernel functions (10)–(12) were investigated, however the best results were obtained for radial basis function (11). Thus, in the two following sections, the author focused on a presentation and discussion of the models in which this particular type of function was applied.

The selected methods of the parameters C and ε can be summarized after [17,18,33–35] as follows:

- The choice is made on the basis of the a priori knowledge of the problem and/or users' expertise;
- Values are selected on the basis of the grid search;
- Determination of the parameters directly from the data;
- Assuming C equal to the range of output values;
- Tuning ε parameter to the training data noise density.

The choice of the two parameters for the models proposed herein compromised the above-mentioned approaches, namely determination of the parameters on the basis of the training data and grid search.

Each of the models was analyzed and its predictive performance was assessed in terms of correlation between the real-life values of the bridges' total construction costs y and the predicted values $\hat{y}$, the predictions' errors, and the residuals analysis. The following equations were used for computations of Pearson's correlation coefficient (R), root mean squared error ($RMSE$), mean absolute percentage error ($MAPE$), and absolute percentage error for p-th case (APE^p):

$$R = cov(y;\hat{y})/(\sigma_y\sigma_{\hat{y}}), \tag{14}$$

$$RMSE = (1/n\cdot\Sigma(y - \hat{y})^2)^{0.5}, \tag{15}$$

$$MAPE = 1/100\%\cdot\Sigma[(|y - \hat{y}|)/y], \tag{16}$$

$$APE^p = 100\%\cdot (|y^p - \hat{y}^p|)/y^p, \tag{17}$$

where $cov(y;\hat{y})$—covariance of real values of the bridges' total construction costs and values predicted by a model, σ_y and $\sigma_{\hat{y}}$ standard deviations of real values of the bridges total construction costs and values predicted by a model, respectively; n—cardinality of either L or T subset, $y - \hat{y}$—prediction errors, computed after completion of the machine learning process for either L or T subset; and p—pattern index. The SVM machine learning process was made with the use of STATISTICA^{TM} software suite.

According to the literature [36–38] and remarks about the expected accuracy of cost estimates provided at the early stages of construction projects (also called conceptual estimates), the error of estimates should fall into the ranges <−30%/−25% and +25%/+30%> when compared to the actual, final construction costs. If the proposed models' predictions and APE^p are considered, the above rule can be reformulated into the expectation about the desired range of APE^p between 0% and +25%/+30%. What is obvious is that the predictions of the bridges' total construction costs are still required to be provided by the models with errors as small as possible. However, the rule can be used for the purposes of the models' performance comparison and assessment.

3. Results

For the investigated SVM-based regression models, the parameter γ (for radial basis kernel function) was assumed as the inverse of the number of inputs, thus $\gamma = 1/27 = 0.037$. The γ value can be explained as the inverse of the radius of influence of samples selected in the course of machine learning to be support vectors.

Regularization meta-parameter C was initially assessed following the rule [35]:

$$C = \max\{|E(y) + 3\sigma_y|; |E(y) - 3\sigma_y|\}, \tag{18}$$

where $E(y) = 6.61$ and $\sigma_y = 4.22$ computed for y^p belonging to subset L resulted in $C = 19.27$. After this, it was assumed that 20 will constitute the upper boundary of C. Values of C were sought for with the use of grid search; the values of ε (threshold of the loss function) were also sought for with the use of grid search. The considered ranges of C and ε, as well as the grid search details, are given in Table 3.

Table 3. Considered ranges of length axis (C) and depth axes (ε) parameters.

Parameter	Lower Boundary	Step	Upper Boundary
C	5	1	20
ε	0.05	0.05	0.20

The machine learning process for each of the models was carried out with the use of 10-fold cross-validation. Having finished the process, the performances of the models were compared. *RMSE* values were computed for both L and T subsets for all of the obtained models. The *RMSE* values obtained for the subset that was used in the course of machine learning (subset L) are presented in Figure 2. Figure 3 depicts *RMSE* values computed for testing subset T. The values of errors (height axes in Figures 2 and 3) are presented as 3D surfaces with regard to C (length axes) and ε (depth axes). One can see that in the case of *RMSE*, the values computed for subset L are decreasing with the increase of C and decrease of ε. On the other hand, the tendency for errors computed for subset T is similar with regards to ε, however the opposite with regard to C.

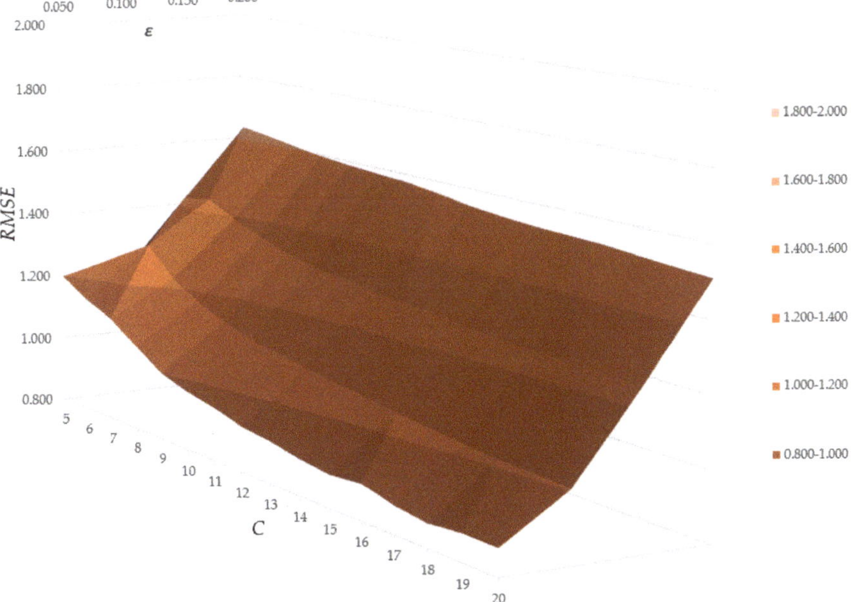

Figure 2. *RMSE* errors obtained for subset L.

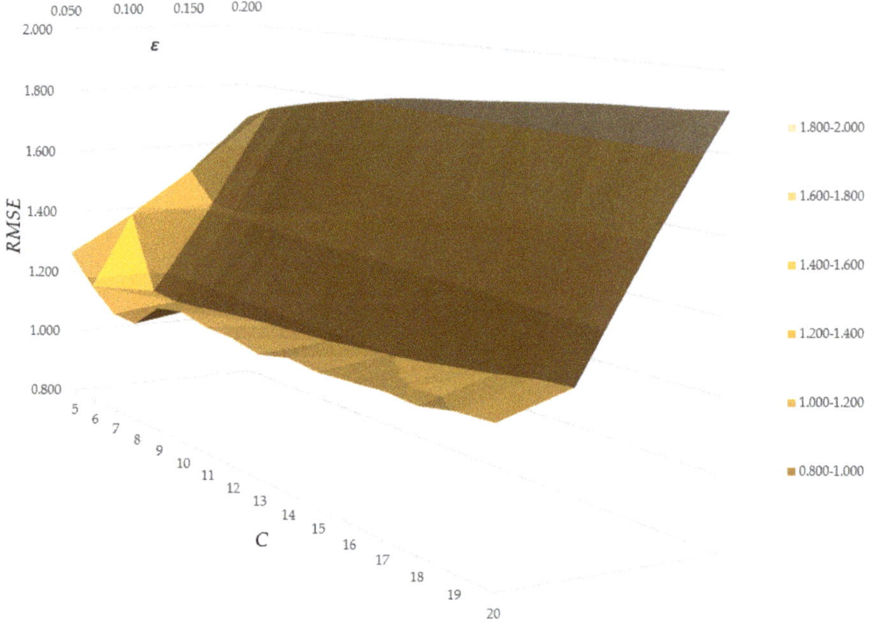

Figure 3. *RMSE* errors computed for subset *T*.

When considering the values of *RMSE* for both subsets *L* and *T* together, one can find the points in the grid representing errors of learning and testing computed for certain models, where the values of *RMSE* for testing reach minimums; moreover, the values of *RMSE* for machine learning are close.

The analysis of *RMSE* values allowed for the selection of five models that were further investigated. The five bridges' construction cost prediction models based on support vector regression (later referred to as BCCPM$_{SVR}$) are introduced in Table 4. Characteristics of the models include values of meta-parameters *C* and ε, number of support vectors (*sv*), and number of bounded support vectors and values of the constants w_0. The support vectors are the data patterns belonging to subset *L* that determine the position of the regression hyperplane for a certain model. Furthermore, errors of 10-fold cross-validation are also presented. General error and performance measures *RMSE*, *R*, and *MAPE* for the five BCCPM$_{SVR}$ models, computed for *L* and *T* subsets, are set together in Table 5.

Table 4. Five selected models and their characteristics.

Model	*C*	ε	*sv*	Bounded *sv*	w_0	Cross-Validation Error
BCCPM$_{SVR}$1	7	0.050	91	50	−0.108761	0.038
BCCPM$_{SVR}$2	8	0.050	85	47	−0.118497	0.037
BCCPM$_{SVR}$3	8	0.100	59	24	−0.132814	0.037
BCCPM$_{SVR}$4	9	0.100	58	23	−0.137849	0.036
BCCPM$_{SVR}$5	10	0.100	56	22	−0.130312	0.035

The values of *RMSE* and *R* (in Table 5), when comparing the five selected models, are relatively close. Thus, in light of the *RMSE* and *R* values analysis, the performance of the models can be assessed as comparable. In terms of *MAPE* values, the differences are slightly more evident. The final choice of the model, however, was based on the comparison of the distribution of *APE^p* errors and the rule, (presented in Section 2.2) that refers to the desired range of *APE^p* values for bridge construction early cost estimates.

Table 5. Measures of errors and performance obtained for the five selected models.

Model	$RMSE_L$	$RMSE_T$	R_L	R_T	$MAPE_L$	$MAPE_T$
BCCPM$_{SVR}$1	1.115	1.112	0.971	0.979	14.64%	11.33%
BCCPM$_{SVR}$2	1.058	1.111	0.974	0.980	13.85%	10.94%
BCCPM$_{SVR}$3	1.175	1.141	0.968	0.978	17.03%	11.44%
BCCPM$_{SVR}$4	1.139	1.152	0.970	0.978	16.69%	11.28%
BCCPM$_{SVR}$5	1.115	1.161	0.971	0.978	16.56%	11.30%

Table 6 presents the distributions of APE^p errors of predictions of total bridge construction costs both for L and T subsets under the conditions that $APE^p \leq 25\%$ or $APE^p \leq 30\%$. In light of the analysis of the values in Table 6, model BCCPM$_{SVR}$ 2 was proven to perform better than the others—the model reached the highest shares of $APE^p \leq 25\%$ for L and T subsets and the same shares of $APE^p \leq 30\%$ for L and T subsets as BCCPM$_{SVR}$1.

Table 6. Comparison of absolute percentage error for p-th case (APE^p) errors for the five selected models.

	Subset L		Subset T	
Model	$APE^p \leq 25\%$	$APE^p \leq 30\%$	$APE^p \leq 25\%$	$APE^p \leq 30\%$
BCCPM$_{SVR}$1	85.38%	92.31%	81.08%	91.89%
BCCPM$_{SVR}$2	86.92%	92.31%	83.78%	91.89%
BCCPM$_{SVR}$3	72.31%	80.00%	81.08%	89.19%
BCCPM$_{SVR}$4	73.08%	80.77%	81.08%	89.19%
BCCPM$_{SVR}$5	73.85%	82.31%	83.78%	89.19%

For the finally selected model of BCCPM$_{SVR}$2, the scatter plots of values of y (actual bridge construction costs, presented on the horizontal axes) and $\hat{y}$ (bridge construction cost predictions by model BCCPM$_{SVR}$ 2, presented on the vertical axes) are depicted in Figures 4 and 5. The former shows the scatter plot of y and $\hat{y}$ values for subset L, the latter for subset T. The charts include also the cones of errors ±25% and ±30%.

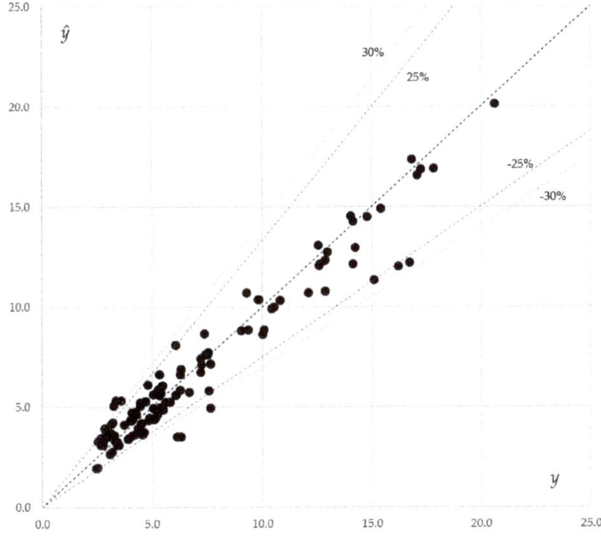

Figure 4. Scatter plot of y and $\hat{y}$ predicted by BCCPM$_{SVR}$2 for subset L.

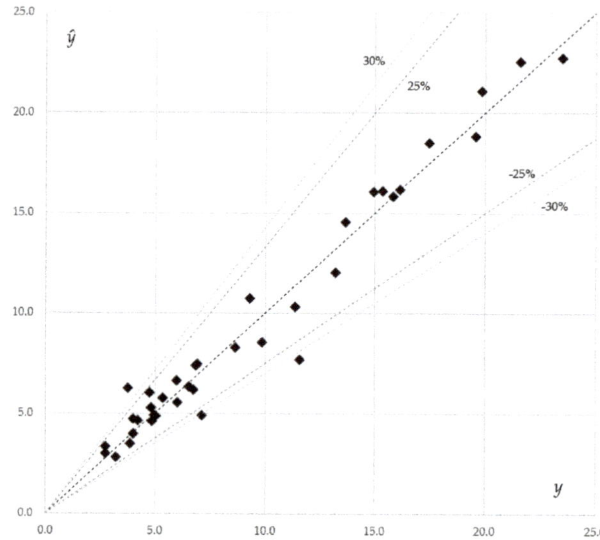

Figure 5. Scatter plot of y and $\hat{y}$ predicted by BCCPM$_{SVR}$2 for subset T.

Table 7 presents the percentage shares of APE^p errors of bridge construction cost predictions provided by the model BCCPM$_{SVR}$2 (both for L and T subsets) divided into intervals of a range equal to 5%. Additionally, distributions (cumulated shares) of APE^p errors are given in the Table.

Table 7. Shares and distribution of APE^p values for BCCPM$_{SVR}$2.

	Subset			Subset	
Share	**L**	**T**	**Distribution**	**L**	**T**
$APE^p \leq 5\%$	20.77%	27.03%	$APE^p \leq 5\%$	20.77%	27.03%
$5\% < APE^p \leq 10\%$	20.77%	37.84%	$APE^p \leq 10\%$	41.54%	64.86%
$10\% < APE^p \leq 15\%$	24.62%	10.81%	$APE^p \leq 15\%$	66.15%	75.68%
$15\% < APE^p \leq 20\%$	10.77%	5.41%	$APE^p \leq 20\%$	76.92%	81.08%
$20\% < APE^p \leq 25\%$	10.00%	2.70%	$APE^p \leq 25\%$	86.92%	83.78%
$25\% < APE^p \leq 30\%$	5.38%	8.11%	$APE^p \leq 30\%$	92.31%	91.89%
$APE^p > 30\%$	7.69%	8.11%	$APE^p > 30\%$	100.00%	100.00%

The distribution of points $(y^p; \hat{y}^p)$ in the scatter plots (in Figures 4 and 5) is even along the line of a perfect fit. Moreover, for both of the subsets L and T, the vast majority of bridge construction cost predictions are located within the ±25% cone of errors; almost all of the predictions are located within the ±30% cone of errors.

The values of the APE^p, (in Table 6), as complementary information, confirm that most of the bridge construction cost predictions made by BCCPM$_{SVR}$2 meet the condition of early cost estimates.

The general conclusion on the results presented above is that the proposed model provides the predictions of costs for bridge construction projects with satisfactory accuracy regarding the expectations for estimates at the early stages of projects.

4. Discussion

When compared to the models proposed by other authors, some significant differences of the model introduced herein can be indicated. The previous works that aimed at modeling costs of bridges in the early stages of projects were focused on cost estimates of either parts of bridge structures [2–4] or specific types of bridges [5–8]. The model introduced herein offers cost predictions of bridges as a whole object (the substructure and superstructure together). Moreover, the predictions are made for different types of bridges with regard to their structure, purpose, and structural and material solutions.

On the other hand, most of the previously proposed models are based either on regression analysis [2–4] or ANN [5]. The former requires a priori assumptions about the functional relationship binding bridge construction cost as a dependent variable with cost predictors as independent variables. The latter are at risk of the so-called local minima problem. Both of these drawbacks are overcome by the use of the SVM-based regression method for the development of the model for prediction costs of bridges.

The results of the research confirmed the assumptions made for the application of the SVM method for bridge construction cost prediction. Several SVM-based regression models were investigated with the use of data collected for a number of bridge construction projects completed in Poland. Having finished machine learning and testing processes, five of the models, of satisfactory knowledge generalization ability and comparable performance, were preselected. An important fact to be mentioned here is that in the case of repetitions of machine learning processes with given constraints, the results obtained for each of the investigated models were exactly the same every time. Application of the SVM method for early estimates of bridge construction costs eliminates the risks of local minima problem.

The final selection of the best model was based on the comparison and analysis ability to predict the bridge construction costs with accuracy appropriate for the early stage of the projects.

The general performance of the selected model, namely $BCCPM_{SVR}2$, and its measures are presented in Section 3. The predictions of the bridge construction costs provided by the model can also be analyzed in a way that focuses on selected characteristics and features of bridges as the model's input.

Tables 8–11 present relative percentage shares of APE^p, computed for the machine learning subset, belonging to certain intervals (compare Table 7) with regard to variables of a nominal type (coded as binary values for machine learning). The relative percentage shares of APE^p for variables of nominal type were computed as follows:

- For each of the variables x_j for $j = 1 - 8$ or $j = 12 - 27$, the number of predictions that fulfilled the condition of having corresponding APE^p that fell into the certain interval were counted and divided by the number of occurrences of $x_j = 1$.

Table 8. APE^p predictions' errors for machine learning with regard to the type of bridge its structure and type of a project.

	Relative Percentage Share of APE^p						
	0–5%	5–10%	10–15%	15–20%	20–25%	25–30%	>30%
BRIDGE (x_1)	35.71%	14.29%	14.29%	7.14%	7.14%	14.29%	7.14%
VIADUCT (x_2)	15.31%	21.43%	28.57%	12.24%	11.22%	3.06%	8.16%
WHARF (x_3)	60.00%	40.00%	0.00%	0.00%	0.00%	0.00%	0.00%
ROAD BRIDGE (x_4)	17.65%	11.76%	26.47%	8.82%	14.71%	5.88%	14.71%
RAIL BRIDGE (x_5)	18.39%	25.29%	25.29%	10.34%	9.20%	5.75%	5.75%
ANIMAL BRIDGE (x_6)	60.00%	10.00%	10.00%	20.00%	0.00%	0.00%	0.00%
BUILD (x_7)	15.65%	22.61%	26.96%	11.30%	10.43%	6.09%	6.96%
DESIGN&BUILD (x_8)	62.50%	6.25%	6.25%	6.25%	6.25%	0.00%	12.50%

Table 9. APE^p predictions' errors for machine learning with regard to the structural and material solutions.

	Relative Percentage Share of APE^p						
	0–5%	**5–10%**	**10–15%**	**15–20%**	**20–25%**	**25–30%**	**>30%**
BEAM (x_{12})	19.64%	20.54%	26.79%	9.82%	8.93%	5.36%	8.93%
FRAME (x_{13})	15.31%	21.43%	28.57%	12.24%	11.22%	3.06%	8.16%
ARCHED/BOX (x_{14})	54.55%	9.09%	9.09%	9.09%	9.09%	9.09%	0.00%
REINFORCED CONCRETE (x_{15})	18.33%	15.00%	26.67%	8.33%	15.00%	5.00%	11.67%
PRESTRESSED CONCRETE (x_{16})	20.83%	29.17%	25.00%	8.33%	6.25%	6.25%	4.17%
STEEL (x_{17})	30.43%	17.39%	17.39%	21.74%	4.35%	4.35%	4.35%

Table 10. APE^p predictions' errors for machine learning with regard to the types of bridgehead and intermediate supports and supports' foundations.

	Relative Percentage Share of APE^p						
	0–5%	**5–10%**	**10–15%**	**15–20%**	**20–25%**	**25–30%**	**>30%**
SOLLID-WALLED (x_{18})	21.77%	20.16%	23.39%	10.48%	10.48%	5.65%	8.06%
COLUMNS (x_{19})	14.29%	28.57%	42.86%	14.29%	0.00%	0.00%	0.00%
NONE (x_{20})	11.11%	14.29%	34.92%	12.70%	11.11%	6.35%	9.52%
SOLLID-WALLED (x_{21})	36.36%	18.18%	22.73%	9.09%	9.09%	0.00%	4.55%
COLUMNS/PILES (x_{22})	28.26%	30.43%	10.87%	8.70%	8.70%	6.52%	6.52%
SHALLOW (x_{23})	12.68%	22.54%	30.99%	7.04%	14.08%	5.63%	7.04%
DEEP (x_{24})	31.67%	18.33%	16.67%	15.00%	5.00%	5.00%	8.33%

Table 11. APE^p predictions' errors for machine learning with regard to the load class.

	Relative Percentage Share of APE^p						
	0–5%	**5–10%**	**10–15%**	**15–20%**	**20–25%**	**25–30%**	**>30%**
A (x_{25})	24.71%	22.35%	21.18%	11.76%	8.24%	5.88%	5.88%
B (x_{26})	0.00%	36.36%	45.45%	9.09%	9.09%	0.00%	0.00%
k/C/D/E [1] (x_{27})	20.00%	11.43%	25.71%	8.57%	14.29%	5.71%	14.29%

[1] (compare with Table 1).

Analyzing the Tables 8–11, one can see how the predictions accuracy depends relatively on the certain, chosen characteristics of the bridges described by the nominal values.

Tables 12–14 present the relative percentage shares of APE^p, computed for the machine learning subset, belonging to certain intervals (compare Table 6) with regard to variables of a numerical type.

The relative percentage shares of APE^p for these variables were computed as follows: for each of the variables x_j for $j = 9 - 11$:

- The values were divided regarding the ranges given in the Tables 12–14;
- Predictions for variables values that fulfilled the conditions of falling into certain range of values and having corresponding APE^p from a certain error's interval were counted and divided by the number of occurrences.

Analyzing the Tables 12–14 one can see how the predictions accuracy depends relatively on the certain, chosen characteristics of the bridges described by the structure's length, width or number of spans.

A limitation of the model that should be mentioned here is that the real-life bridge construction costs were updated for a certain moment in time for the data that was used both in the machine learning and in the testing processes. Thus, for now, dynamical predictions are not provided by the developed model. The reason for this limitation is the number of collected data patterns which does not currently allow for dynamical predictions that comply to the changes of costs in time.

Future research plans cover the issue of database expansion and further collection of training data, and development of models capable of dynamical predictions. One of the possible future research directions, which also rely on the database expansion, is the decomposition of the problem, development of separate models for certain types of bridges and combining the models in a so-called *committee machine*.

Table 12. APE^p predictions' errors for machine learning with regard to the total length of bridge (x_9).

LENGTH (x_9)	Relative Percentage Share of APE^p						
	0–5%	5–10%	10–15%	15–20%	20–25%	25–30%	>30%
up to 25 m	0.00%	16.67%	36.67%	6.67%	16.67%	6.67%	16.67%
25–50 m	14.63%	19.51%	26.83%	21.95%	9.76%	2.44%	4.88%
50–75 m	18.18%	22.73%	31.82%	4.55%	13.64%	9.09%	0.00%
75–100 m	36.36%	31.82%	13.64%	4.55%	4.55%	0.00%	9.09%
more than 100 m	45.45%	9.09%	0.00%	4.55%	0.00%	9.09%	4.55%

Table 13. APE^p predictions' errors for machine learning with regard to the width of bridge (x_{10}).

WIDTH (x_{10})	Relative Percentage Share of APE^p						
	0–5%	5–10%	10–15%	15–20%	20–25%	25–30%	>30%
up to 11 m	11.76%	29.41%	35.29%	5.88%	11.76%	0.00%	5.88%
11–14 m	15.87%	12.70%	26.98%	9.52%	11.11%	9.52%	14.29%
14–17 m	29.73%	29.73%	21.62%	13.51%	5.41%	0.00%	0.00%
17–20 m	5.41%	5.41%	2.70%	5.41%	2.70%	2.70%	0.00%
more than 20 m	8.11%	2.70%	0.00%	0.00%	2.70%	0.00%	0.00%

Table 14. APE^p predictions' errors for machine learning with regard to the of number of spans (x_{11}).

NUMBER OF SPANS (x_{11})	Relative Percentage Share of APE^p						
	0–5%	5–10%	10–15%	15–20%	20–25%	25–30%	>30%
1	9.09%	13.64%	33.33%	15.15%	13.64%	6.06%	9.09%
2	15.00%	40.00%	10.00%	10.00%	15.00%	5.00%	5.00%
3	34.48%	31.03%	20.69%	3.45%	3.45%	0.00%	6.90%
4	3.45%	0.00%	6.90%	3.45%	0.00%	6.90%	0.00%
5 and more	27.59%	3.45%	0.00%	0.00%	0.00%	0.00%	3.45%

5. Conclusions

As a result of the research, an original model capable of supporting early estimates of bridge construction costs, based on machine learning and SVM method, was developed and introduced. The input variables bring to the model information, available in the early stage of a bridge construction project, that represent the features of bridges.

According to the presented results and discussion, as well as the accuracy expectations applicable for conceptual estimates, the model offers good performance. Applied kernel functions are of the radial basis type, and the meta-parameters of the model are $C = 8$ and $\varepsilon = 0.050$. The values of the general measures of the model's performance, respectively for machine learning and testing, are:

- $RMSE$: 1.058 and 1.111;
- Pearson's correlation coefficient R of real-life bridge construction costs and costs predicted by the model: 0.974 and 0.980;
- $MAPE$: 13.85% and 10.94%.

The model provides cost predictions with satisfactory accuracy, within the range of errors appropriate for early estimates (conceptual estimates) that is ±25%/30%.

The proposed approach is prospective for early cost estimates (conceptual cost estimates) in bridge construction projects. The study contributes to the body of knowledge by the application of machine learning methods for cost analyses in construction.

Funding: This research was funded by statutory activities of Cracow University of Technology.

Acknowledgments: Computations for SVM machine learning were done with the use of STATISTICA™ software suite.

Conflicts of Interest: The author declares no conflicts of interest.

References

1. Hollar, D.A.; Rasdorf, W.; Liu, M.; Hummer, J.E.; Arocho, I.; Hsiang, S.M. Preliminary engineering cost estimation model for bridge projects. *J. Constr. Eng. Manag.* **2012**, *139*, 1259–1267. [CrossRef]
2. Fragkakis, N.; Lambropoulos, S.; Tsiambaos, G. Parametric model for conceptual cost estimation of concrete bridge foundations. *J. Infrastruct. Syst.* **2010**, *17*, 66–74. [CrossRef]
3. Alhusni, M.K.; Triwiyono, A.; Irawati, I.S. Material quantity estimation modelling of bridge sub-substructure using regression analysis. *MATEC Web Conf.* **2019**, *258*, 02008. [CrossRef]
4. Fragkakis, N.; Lambropoulos, S.; Pantouvakis, J.P. A cost estimate method for bridge superstructures using regression analysis and bootstrap. *Org. Technol. Manag. Constr. Int. J.* **2010**, *2*, 182–190.
5. Creese, R.C.; Li, L. Cost estimation of timber bridges using neural networks. *Cost Eng.* **1995**, *37*, 17–22.
6. Kim, K.J.; Kim, K.; Kang, C.S. Approximate cost estimating model for PSC Beam bridge based on quantity of standard work. *KSCE J. Civ. Eng.* **2009**, *13*, 377–388. [CrossRef]
7. Kim, B.S. The approximate cost estimating model for railway bridge project in the planning phase using CBR method. *KSCE J. Civ. Eng.* **2011**, *15*, 1149–1159. [CrossRef]
8. Fragkakis, N.; Lambropoulos, S.; Pantouvakis, J.P. A computer-aided conceptual cost estimating system for pre-stressed concrete road bridges. *Int. J. Inf. Technol. Proj. Manag.* **2014**, *5*, 1–13. [CrossRef]
9. Marzouk, M.; Hisham, M. Applications of building information modeling in cost estimation of infrastructure bridges. *Int. J. 3-D Inf. Model.* **2012**, *1*, 17–29. [CrossRef]
10. Choudhry, R.M.; Aslam, M.A.; Hinze, J.W.; Arain, F.M. Cost and schedule risk analysis of bridge construction in Pakistan: Establishing risk guidelines. *J. Constr. Eng. Manag.* **2014**, *140*, 04014020. [CrossRef]
11. Saito, M.; Sinha, K.C.; Anderson, V.L. Statistical models for the estimation of bridge replacement costs. *Transp. Res. Part A Gen.* **1991**, *25*, 339–350. [CrossRef]
12. Bouabaz, M.; Hamami, M. A cost estimation model for repair bridges based on artificial neural network. *Am. J. Appl. Sci.* **2008**, *5*, 334–339. [CrossRef]
13. Chengalur-Smith, I.N.; Ballou, D.P.; Pazer, H.L. Modeling the costs of bridge rehabilitation. *Transp. Res. Part A Policy Pract.* **1997**, *31*, 281–293. [CrossRef]
14. Padgett, J.; DesRoches, R.; Nielson, B.; Yashinsky, M.; Kwon, O.S.; Burdette, N.; Tavera, E. Bridge damage and repair costs from Hurricane Katrina. *J. Bridge Eng.* **2008**, *13*, 6–14. [CrossRef]
15. Ehlen, M.A. Life-cycle costs of fiber-reinforced-polymer bridge decks. *J. Mater. Civ. Eng.* **1999**, *11*, 224–230. [CrossRef]
16. Hassanain, M.A.; Loov, R.E. Cost optimization of concrete bridge infrastructure. *Can. J. Civ. Eng.* **2003**, *30*, 841–849. [CrossRef]
17. Vapnik, V. *Statistical Learning Theory*; John Wiley & Sons: New York, NY, USA, 1998.
18. Vapnik, V. *The Nature of Statistical Learning Theory*; Springer: New York, NY, USA, 2013.
19. Smola, A.J.; Schölkopf, B. A tutorial on support vector regression. *Stat. Comput.* **2004**, *14*, 199–222. [CrossRef]
20. Cristianini, N.; Shawe-Taylor, J. *An Introduction to Support Vector Machines (and Other Kernel-based Learning Methods)*; Cambridge University Press: Cambridge, UK, 2000.
21. Gunn, S.R. *Support Vector Machines for Classification and Regression*; Technical Report; University of Southampton: Southampton, UK, 1998.
22. Caldas, C.H.; Soibelman, L. Automating hierarchical document classification for construction management information systems. *Autom. Constr.* **2003**, *12*, 395–406. [CrossRef]

23. Mahfouz, T.; Kandil, A. Construction legal decision support using support vector machine (SVM). In Proceedings of the Construction Research Congress 2010: Innovation for Reshaping Construction Practice, Banff, AB, Canada, 8–10 May 2010; pp. 879–888. [CrossRef]
24. Chen, J.H.; Lin, J.Z. Developing an SVM based risk hedging prediction model for construction material suppliers. *Autom. Constr.* **2010**, *19*, 702–708. [CrossRef]
25. Tserng, H.P.; Lin, G.F.; Tsai, L.K.; Chen, P.C. An enforced support vector machine model for construction contractor default prediction. *Autom. Constr.* **2011**, *20*, 1242–1249. [CrossRef]
26. Horta, I.M.; Camanho, A.S. Company failure prediction in the construction industry. *Expert Syst. Appl.* **2013**, *40*, 6253–6257. [CrossRef]
27. Cheng, M.Y.; Wu, Y.W.; Wu, C.F. Project success prediction using an evolutionary support vector machine inference model. *Autom. Constr.* **2010**, *19*, 302–307. [CrossRef]
28. Cheng, M.Y.; Hoang, N.D.; Wu, Y.W. Hybrid intelligence approach based on LS-SVM and Differential Evolution for construction cost index estimation: A Taiwan case study. *Autom. Constr.* **2013**, *35*, 306–313. [CrossRef]
29. Wang, Y.R.; Yu, C.Y.; Chan, H.H. Predicting construction cost and schedule success using artificial neural networks ensemble and support vector machines classification models. *Int. J. Proj. Manag.* **2012**, *30*, 470–478. [CrossRef]
30. Cheng, M.Y.; Hoang, N.D. Interval estimation of construction cost at completion using least squares support vector machine. *J. Civ. Eng. Manag.* **2014**, *20*, 223–236. [CrossRef]
31. Kim, G.-H.; Shin, J.-M.; Kim, S.; Shin, Y. Comparison of School Building Construction Costs Estimation Methods Using Regression Analysis, Neural Network and Support Vector Machine. *J. Build. Constr. Plan. Res.* **2013**, *1*, 1–7. [CrossRef]
32. Peško, I.; Mučenski, V.; Šešlija, M.; Radović, N.; Vujkov, A.; Bibić, D.; Krklješ, M. Estimation of Costs and Durations of Construction of Urban Roads Using ANN and SVM. *Complexity* **2017**, *2017*, 2450370. [CrossRef]
33. Scholkopf, B.; Burges, J.; Smola, A. *Advances in Kernel Methods: Support Vector Learning*; MIT Press: Cambridge, MA, USA, 1998.
34. Cherkassky, V.; Mulier, F. *Learning from Data. Concepts, Theory, and Methods: Second Edition*; John Wiley & Sons: Hoboken, NJ, USA, 2006. [CrossRef]
35. Cherkassky, V.; Ma, Y. Practical selection of SVM parameters and noise estimation for SVM regression. *Neural Netw.* **2004**, *17*, 113–126. [CrossRef]
36. Brook, M. *Estimating and Tendering for Construction Work*; Routledge: Abingdon, UK, 2016. [CrossRef]
37. Kasprowicz, T. *Inżynieria Przedsięwzięć Budowlanych in KAPLIŃSKI O*; Metody i modele badań w inżynierii przedsięwzięć budowlanych; PAN KILIW: Warszawa, 2007; pp. 35–78.
38. Potts, K. *Construction Cost Management: Learning from Case Studie*; Taylor & Francis: Abingdon, UK, 2008.

Article

Research and Development Directions for Design Support Tools for Circular Building

Charlotte Cambier [1,*, Waldo Galle [1,2] and Niels De Temmerman [1]**

[1] Department of Architectural Engineering, Vrije Universiteit Brussel, 1050 Brussels, Belgium; Waldo.Galle@vub.be (W.G.); Niels.De.Temmerman@vub.be (N.D.T.)
[2] Transition Platform, Flemish Institute for Technical Research (VITO), Boeretang 200, 2400 Mol, Belgium
* Correspondence: Charlotte.Cambier@vub.be; Tel.: +32-(0)2-629-1872

Received: 30 June 2020; Accepted: 12 August 2020; Published: 18 August 2020

Abstract: To support the construction sector in its transition to a circular economy, many design instruments and decision support tools have been and are still being developed. This development is uncoordinated and raises confusion among building designers and advising engineers, slowing down the tools' adoption in practice. Moreover, it is unclear if the available design tools are able to fulfil the needs of design professionals at all. Therefore, this research identifies the knowledge challenges for the "supply and demand" of design tools for a circular construction practice. It focuses on Flanders, given the importance the topic receives in the region's policy programme and among practitioners. This study builds on a thorough literature review, and on inventorying and categorising instruments and ongoing developments. By comparing that review with the needs that were identified during interviews with a focus group, it was possible to pinpoint designers' needs for support tools and outline three urgent research tracks. More generally, it was found that the needs of our focus group are only partially reflected by the available design tools and the ongoing developments. This identified mismatch advocates for a more participatory and practice-oriented research approach when developing design support tools for circular building.

Keywords: design for circularity; design support tools; circular construction; circular economy

---

1. Introduction

In recent years, a growing academic, political and industrial interest has been arising in transitioning from a linear to a circular economy (CE) [1,2]. A CE can be defined as "an economic system that is based on business models which replace the 'end-of-life' concept with reducing, alternatively reusing, recycling and recovering materials [...] with the aim to accomplish sustainable development [...].", according to Kirchherr et al. [3], who identified 114 different CE definitions.

The construction sector has an important role in the transition to a CE, as it accounts for about 50% of all extracted material and for over 35% of the EU's total waste generation [4]. Therefore, the European Commission identified "construction and buildings" as one of the seven key product value chains in its Circular Economy Action Plan [4]. Also, subsequent Flemish Governments have set the transition of the construction sector to a circular economy as one of their priorities [5,6].

Since 2014, the policy programme "Material-Aware Construction through Circular Supply Chains—A Sustainable Materials Management Prevention Program for the Construction Sector 2014–2020" [7] aims to establish an economy of closed material loops through socio-technical innovations in the Flemish construction sector [8]. The programme is governed by the Flemish Government's Agency for Public Waste, Materials and Soil (Openbare Vlaamse Afvalstoffenmaatschappij OVAM) under the auspices of the regional government of the Belgian region Flanders. As stated by Silva et al. [9], "the Flemish Sustainable Material Management program initiated by OVAM, was the first larger scale waste to materials policy restructure in the world, […] and won a Circular award at the World Economic Forum for its dedication to shift towards a circular economy". Paredis [10] explained earlier:

> "All in all, the change in discourse from waste to sustainable materials management is undeniable. It is not only taken up in the Materials Decree and propagated by OVAM as main government actor, it also seems to find support with all actors involved in the waste/materials system: advisory councils, different sectors of the industry, knowledge actors, such as universities and VITO, and NGOs. Politically, the build-up of the discourse coalition benefited from the possibility to link it to ongoing developments at European level and to the innovation and green economy debate at Flemish level".

More recently, to accelerate the implementation of the Flemish policy programme and to increase its impact, sector-oriented initiatives were set-up to foster new demolition and design practices, for example, waste management organisations (e.g., Tracimat) and design instruments [11], or so-called green deals [12] and project calls [13].

When taking a broader look at the rather complex landscape of the circular economy, one can notice that multiple pathways and directions are taken in various regions [14]. Circular economy has become a political ambition in the European Union and other countries worldwide, such as China and Japan [2], where each country or region has its own focal points. Even with given direction from the EU [4] and international standards [15], variations among different continents and even among EU-member states can be noticed in various sectors, including in construction. Whereas, in Portugal, for example, the CE concept is mostly applied in the area of waste management, in countries like Belgium and The Netherlands they also emphasise the implementation of CE principles in the design stage.

This diversity and divergence was also identified by Bauwens et al. who posit "that a CE can be conceptualized in very different ways and that it is essential to better examine the trade-offs between these conceptual models and their societal consequences" [16]. This diversity is not uncommon for challenging sustainability transitions. Transition Management researchers Geels and Schot name these multidimension processes "co-evolution" or "co-construction" and argue that the conjuncture of multiple developments is important for any transition's success [17].

Given that Flanders can be considered as one of the forerunners in the transition towards a circular construction economy (with varying success), it makes the region and its ongoing initiatives a well-documented and instructive case for reflection and learning about the transition itself.

Furthermore, in addition to the different directions of CE, today, many innovative experiments are being performed and various collaborations are taking shape. Also, in Flanders, experimenting with CE principles and exchanging knowledge and experiences is encouraged by, amongst others, the Flemish transition hub Circular Flanders. As a result, applications and analytical studies are each taking their own approach, and different practitioners have to navigate the increasing "methodological noise" and try to make sense of the available information and means for their own working context.

Hence, the transition to a CE within the construction sector still faces major challenges. The transition implies radical changes at different levels and scales: from organisational changes within the sector to new building design methods [14,18,19]. Due to the numerous challenges and the rather complex landscape of the circular economy, and even though there is a demand from Belgian construction stakeholders to implement circular building concepts [20], the construction sector still struggles to effectively put circularity into practice. For example, building designers find it difficult to design circular construction products or buildings, when there is a lack of interest, knowledge, skills or incentives [21,22].

As an answer to the struggles of this specific stakeholder group (i.e., the building designers and advising engineers), the demand and supply of design support tools for circular building is rising [21–23]. Design support tools intend to facilitate the design process. They can be defined as instruments of any form or kind that address architects and/or advising engineers, include circular design principles and/or evaluation criteria, and aim to make better informed design choices. These tools can be an important enabler in the transition towards a circular building sector [24] through providing guidance on, for instance, waste generation, material selection, making reversible connections between building elements, and on the reuse and recycling potential [25]. However, when using existing design support tools for circular building, or before developing new ones, one should know which tools are available, what their effectiveness and limitations are, and which tools or features are still missing. Due to the lack of an overview of the available tools [20] and a comparative framework, it remains unclear for designers and advising engineers which tools fit their way of working and the context of their projects. Further, there is a lack of understanding of what designers need of (features in) design tools [20]. Addressing the needs of these stakeholders is crucial to understand the potential uptake of tools, and it lowers the risk of putting effort into developing new design support tools without answering any need [26].

The present study was set-up to classify available design support tools for circular building, to identify building designers' and advising engineers' needs and expectations from such tools, and to reveal which research tracks on design support tools for circular building are currently being developed. The tools as well as the needs were classified per building design aspect and by design stage. This way, the tools and the needs could be compared. Subsequently, conclusions could be drawn on the effectiveness and limitations of the available design support tools, on opportunities to improve available tools, and on prospects to develop new tools for circular building.

We assumed that guiding the practitioners through this experimental phase, with finding the tools and methods that support their specific situation best, can accelerate the learning and transition process, or at least make it as effective as possible.

2. Method

This study was done in five phases (Figure 1). The first phase entailed the selection and review of relevant design support tools for circular building. The design support tools were initially collected based on the authors' own knowledge of existence of such tools, enlisting other researchers and practitioners on their awareness and attending various events and presentations where design support tools were (partly) discussed. The relevant tools for this study were selected by using four criteria in line with the adopted definition mentioned earlier: instruments of any form or kind that address architects and/or advising engineers, include circular design principles and/or evaluation criteria and aim to make better informed design choices. The first selection criterium was "relevant for the Flemish building sector", related to the location context of this study, where tools were selected that are developed by Flemish or Dutch developers or by internationally known developers (e.g., The Ellen MacArthur Foundation, the European Commission and Pré (SimaPro)). The second selection criterium was "(claim to) support circular building". The tools were screened on the adoption of the circular design principles "closed-material loops" and "life cycle design". The third selection criterium was "available for use". Only tools which were ready for immediate use were selected. Tools that were still in the research phase or in the development were eliminated. The fourth and last criterium was that they should "address building designers and advising engineers". For example, written documents without a coupled action were considered redundant for this stakeholder group, and 38 tools remained after the selection process.

The process of categorising the tools was similar to defining themes in transcribed interviews [27]. A predetermined list of categories was set-up before defining the categories of the selected tools. This list contained general categories such as assessment tools, principle tools, economic tools and describing tools. This list was subsequently complemented during the review of their role in the design process. The six resulting categories were: Circular design strategies, Circularity score, Environmental impact, Product and material choice, Practical examples and Circular business models.

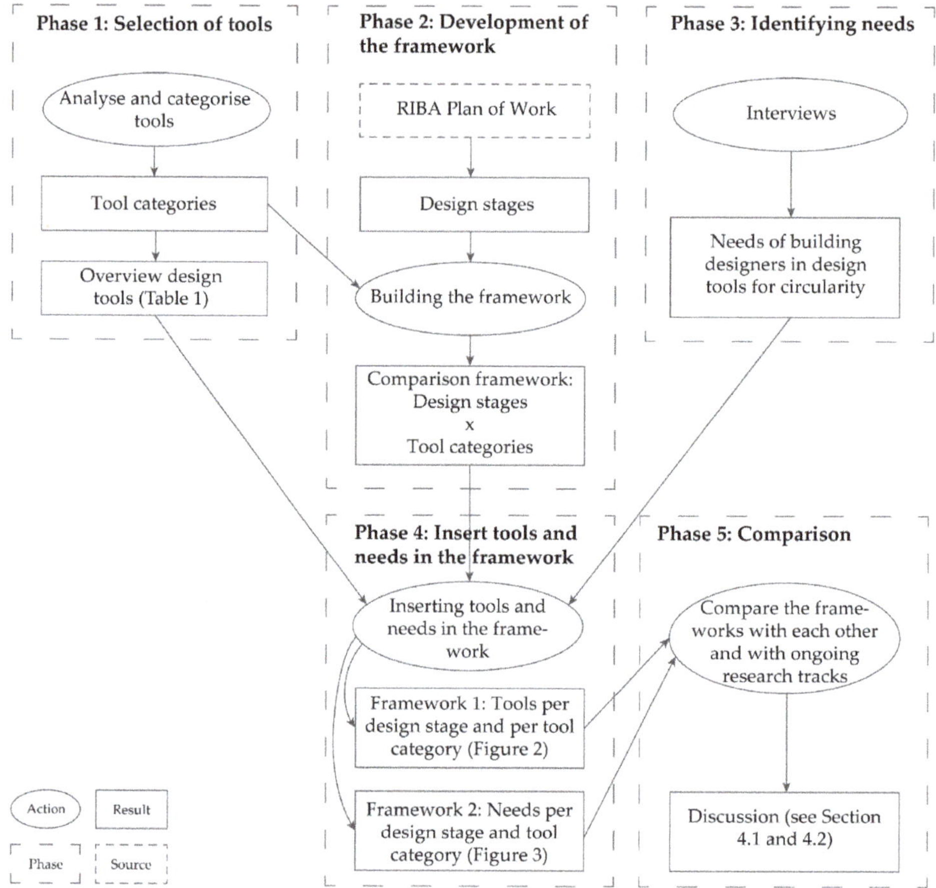

Figure 1. Diagram that shows the actions and results of the five research phases. RIBA is the Royal Institute of British Architects.

In the second phase, a framework that allows the consistent mapping of tools and needs was set-up. On the horizontal axis of the framework different building design stages were listed. Those stages were based on the Royal Institute of British Architects' (RIBA) Plan of Work [10] which is the definitive UK model for the building design and construction process and which was translated and verified for Flanders by Lespagnard [28]. On the vertical axis, the identified and abovementioned design tool categories were lined up.

During the third phase, stakeholders were interrogated through face-to-face, semi-structured interviews. The aim of the interviews was to obtain a preliminary but thorough idea of what the needs were of building designers concerning design support tools for circular building. Through in-depth individual interviews, valuable information could be provided on the personal thoughts and perspectives of the stakeholders on these needs. Seven interviews were conducted with different interviewees: a researcher on sustainable buildings (15 January 2019, Heerlen, The Netherlands), a facade contractor/designer (21 January 2019, Velp, The Netherlands), an architect (22 January 2019, Antwerp, Belgium), a sustainability engineer (23 January 2019, Louvain-la-Neuve, Belgium), an architect (03 September 2019, Antwerp, Belgium), an architect (04 November 2019, Brussels, Belgium) and an architect (06 November 2019, Brussels, Belgium). The interviewees were Flemish or Dutch forerunners active in Flanders and were selected by the researchers on the basis of their familiarity with circular design principles and their practical design experience. This was assessed by reviewing their portfolio and their explicit circularity ambitions in various media. Although not representative for the whole sector, working with forerunners, as advised by Geels et al. [29], was important to have an in-depth understanding of the needs in design support tools for the still uncommon but generally envisioned practice of circular building. Concretely, interviewees were asked about their perceived needs and motivations related to the abovementioned tool categories in order to be able to proceed in designing circular buildings. The interview guide is available in Appendix A.

The fourth phase consisted of filling out the developed framework twice: once with the identified needs and once with the reviewed tools. To determine to which design phase(s) the tools belonged, three questions of the RIBA Plan of Work were adopted:

1. For the feasibility design phase: does the tool assist in making design decisions on project objectives, sustainability aspirations, concept design or programme?
2. For the developed design phase: does the tool assist in making design decisions on the proposals for structural design, building services systems, outline specifications, cost information or project strategies?
3. For the detailed design phase: does the tool assist in making design decisions on the coordinated and updated architectural, structural and building services proposals? Is material and dimension-specific information needed to do the calculations?

During the fifth and last phase, the two filled-out frameworks were compared with each other. The similarities and mismatches among the tools and the needs were identified, reviewed and analysed. Additionally, the ongoing main research and development tracks on the development of design support tools for circular building design were outlined. These research paths were also compared with the needs of the building designers and advising engineers. Last, further research and development paths were identified and proposed for practitioners, tool developers and researchers, based on a synthesis of the previous phases.

3. Results

3.1. Existing Design Support Tools for Circularity

The 38 tools that met the four set criteria are listed in Table 1. They were inventoried during the period October 2019–April 2020. It should be noted that the list is not exhaustive, and other tools may exist, as tools and their features can change rapidly in this field.

Table 1. Non-exhaustive list of design support tools for circular building in alphabetical order.

Tool	Developer	Publisher	Year Published	Source
16 Design Qualities for a Circular Economy (Design principles ((DP))	Vrije Universiteit Brussel (VUB) Architectural Engineering	Le Bati Bruxellois Source de Nouveaux Materiaux (BBSM) Research Consortium	2019	[30]
24 Design Principles for Design for Change (DP)	VUB Architectural Engineering, Vlaamse Instelling voor Technologisch Onderzoek (VITO) and KULeuven	Openbare Vlaamse AfvalstoffenMaatschappij (OVAM)	2016	[31]
Business Model Innovation Grid	Nancy Bocken, Samuel Short, Padmakshi Rana and Steve Evans (University of Cambridge)	Circular Flanders	-	[32]
Bouwcatalogus Veranderingsgericht Bouwen (DP)	Vlaams Instituut voor Bio-ecologisch Bouwen en Wonen (VIBE)	OVAM	2019	[33]
Building Circularity Index	Alba Concepts, Verberne Jeroen (TU Eindhoven)	Alba Concepts	-	[34]
C-calc	Cenergie	Cenergie	2018	[35]
Circular Building Assessment Prototype	Building Research Establishment (BRE), VITO, University of Twente	Buildings As Material Banks (BAMB) Research Consortium	2018	[36]
Circular Design Guide	The Ellen McArthur Foundation and Ideo	The Ellen McArthur Foundation	2018	[37]
Circular Transition Indicators	World Business CouncilFor SustainableDevelopment	Circular IQ	-	[38]
Circularity Calculator	IDEAL and CO Explore BV	IDEAL and CO Explore BV	2017	[39]
Circulator	VITO, Circular Flanders, TU Delft, Rasboud University	EIT RawMaterials	-	[40]
Circulytics	The Ellen McArthur Foundation	The Ellen McArthur Foundation	-	[41]
Closing the Loop by Design	UTwente	Remeha BV	2018	[42]
Ecolizer Ontwerptool	OVAM, VITO	OVAM	2011	[43]
Green Deal Circulair Bouwen (Platform)	Circular Flanders, OVAM, Vlaamse Confederatie Bouw	Circular Flanders	2019	[12]
GaBi Circularity Toolkit (Life Cycle Assessment (LCA))	Sphera	Sphera	-	[44]
GRO	Het Facilitair Bedrijf	Het Facilitair Bedrijf	2020	[45]
Harvestmap/OogStkaart (Reused Materials (RM))	Superuse Studios	Superuse Studios	-	[46]
IMPACT (LCA)	BRE Group	BRE Group	-	[47]
Insert Marktplaats (RM)	Insert, Buro Boot	Insert	-	[48]
Kernmeetmethoe	Action team (36 participants)	Platform CB'23	2020	[49]
Level(s)	European Commission Joint Research Centre	EuropeanCommission	2020	[50]
Madaster platform (Materials Passport (MP))	Madaster Services	Madaster Services	-	[51]

Table 1. *Cont.*

Tool	Developer	Publisher	Year Published	Source
MarketplaceHUB (RM)	World Business Council for Sustainable Development	World Business Council for Sustainable Development	-	[52]
Material EIA for Single-Family Dwellings	Elke Meex et al.	UHasselt	2019	[53]
Milieuclassificaties Bouwproducten	Nederlands Instituut voor Bouwbiologie en Ecologie (NIBE)	NIBE	2019	[54]
Online Material Flow Analysis Tool (Material Flow Analysis (MFA))	Team Metabolism Of Cities	Metabolism Of Cities	2020	[55]
One Click LCA (LCA)	Bionova Ltd.	Bionova Ltd.	-	[56]
Opalis (RM)	Rotor vzw, Atelier 4|5	Rotor vzw	-	[57]
OpenLCA (LCA)	GreenDelta	GreenDelta	-	[58]
Platform CB'23 (Platform)	13 companies	Circulair Bouwen 2023 (CB'23)	2018–2023	[59]
Pixii (Platform)	Pixii	Pixii	-	[60]
ReCiPe method (LCA)	Rijksinstituut voor Volksgezondheid en Milieu (RIVM), Radboud University Nijmegen, Leiden University, PRé Sustainability	Dutch National Institute for Public Healthand the Environment	2018	[61]
Scenario based Life Cycle Costing (LCC)	Waldo Galle et al.	VUB Architectural Engineering	2016	[62]
SimaPro (LCA)	PRé Sustainability	PRé Consultants BV	-	[63]
Stan (MFA)	TU Wien, Institute for Water Quality, Resource and Waste Management	TU Wien, Institute for Water Quality, Resource and Waste Management	2012	[64]
Totem	VITO/EnergyVille, KU Leuven, Wetenschappelijk en Technisch Centrum Voor Het Bouwbedrijf (WTCB)	OVAM, Brussels Environment, Wallonie Service Public	2020	[65]
Werflink (RM)	Floow2	Werflink	-	[66]

When the available design support tools from Table 1 are situated in the developed framework and sorted per design phase and per design tool category (Figure 2), it becomes clear that some tools serve the same purpose and can be clustered. As Figure 2 shows, this results in eight subcategories of tools (darker coloured bars), in addition to 17 more unique tools (light coloured bars). Although the exact impact and role of each tool might vary from project to project, depending on the adoption by its users, the similarity among the majority of tools raises questions about the tools' effective complementarity in terms of goal and scope.

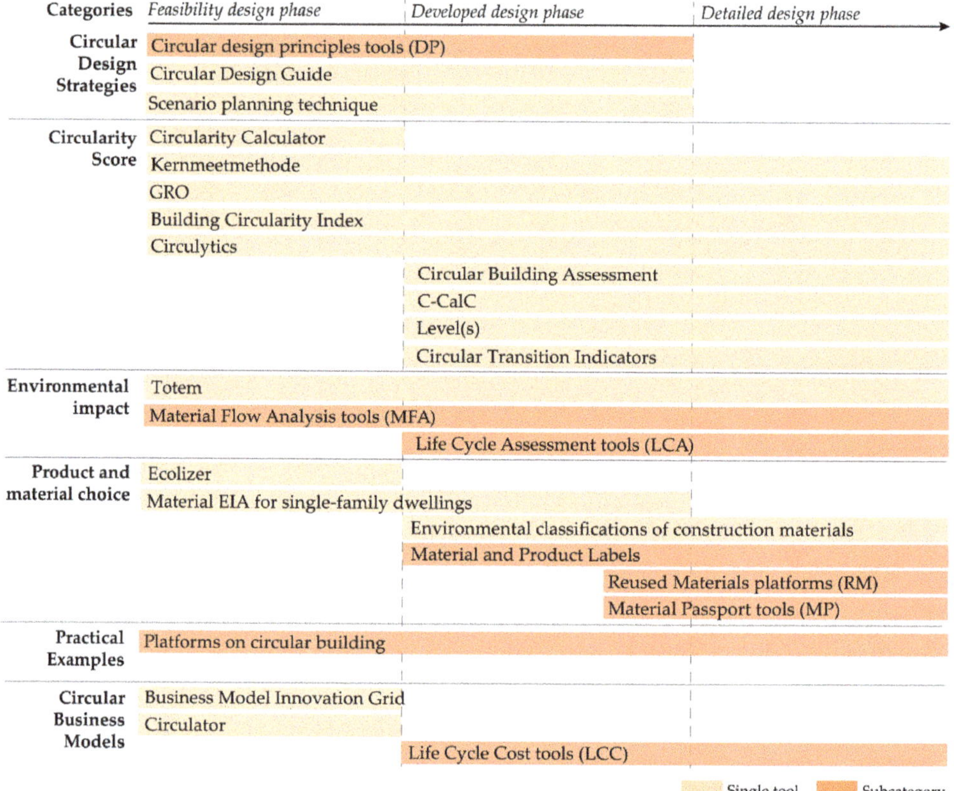

Figure 2. Framework 1: Categorisation of the design support tools for circular building per design phase and per design tool category. Tools that serve the same purpose and have the same target audience are further clustered in subcategories (darker bars).

From the review and the filled-out framework, a series of findings becomes apparent. They can be discussed per design tool category:

- Circular Design Strategies: A circular design strategy tool, such as the "Circular Design Guide" of The Ellen McArthur Foundation, aims to offer guidance in considering alternative design decisions through design strategies. One subcategory was identified: the Circular Design principle tools (DP), containing tools that have a number of principles, strategies or qualities that can be set as ambitions but which are not linked to a scoring system. For example, the "Design Qualities" (VUB Architectural Engineering) presents 16 circular design considerations linked to different design strategies and circularity principles, and "Closing the Loop by Design" (Remeha B.V.) presents 36 guidelines in four different categories (architecture, component, connection, and material).

- Circularity score: Circularity scoring tools aim to objectify the circularity performance of a building or a building element through a scoring or assessment system. A scoring system often forms the basis for comparative analysis [24]. There are various attempts to develop a tool that calculates or measures the level of circularity of a building. The way circularity is evaluated and scored still differs largely from one tool to another. For example, C-Calc (Cenergie) evaluates on three aspects which are equally weighted in all projects: material use, adaptability and information flow, while GRO (Het Facilitair Bedrijf) takes into account 29 design aspects and also considers the level of ambitions and project-specific information. This issue is further tackled in the discussion section below.

- Environmental impact: Life cycle assessment (LCA) and material flow analysis (MFA) are well-known methods for assessing the environmental impact on the built environment, and they are suitable for analysing the environmental performance of circular systems and designs [67–69]. For each of these methods, several tools were developed (Table 1) and which are translated into subcategories LCA and MFA (Figure 2). The Totem tool focusses on the Belgian construction sector and aims to measure the environmental effect of building elements through 17 environmental impact indicators. The scores of each indicator are expressed per kg CO_2 eq. for global warming [70]. The developer's intend to further develop the tool with regard to the circular design aspect.

- Product and material choice: This tool category includes design support tools that focus on the building product and material level. For example, NIBE developed the "Environmental classification of building materials and products". Furthermore, the subcategory "Material and Products Labels" can be seen as a scoring system at the material and product level. Before the label is obtained, some (environmental and perhaps circular) criteria must be met which is thoroughly checked by an independent inspection body [71].

- Practical examples: Practical examples, such as technical details and case studies, can be of great value when working out a circular construction project. There are several platforms that try to gather knowledge and such examples in one central place, gathered in the subcategory "Platforms on Circular Building". The collected knowledge and information are usually shared partly online and partly through events and workshops. Moreover, it is frequently the intention to link partners concerning circular building and do matchmaking. In Flanders, the most common platform concerning circular building is the Green Deal Circular Building platform.

- Circular business models: The linear model of "take, make and dispose" has reached its limits. Also, the fragmented supply chain in the construction sector is a key challenge in the transition towards a circular economy [21]. A number of tools have been developed to help realise this transition in terms of the business model such as the Business model innovation grid (Circular Flanders) and Circulator (EIT RawMaterials). The subcategory "Life Cycle Costing Tools" (LCC) includes tools that study long-term costs and revenues and can calculate the financial consequences of applying circular business models or circular principles [62].

3.2. Building Stakeholders' Needs for Design Support Tools

Having categorised the design tools, the question is whether or not they answer the needs of the building designers and advising engineers who are trying to integrate circular principles in their practice. Therefore, the needs that are expressed by our focus group are placed in the developed framework, where they are categorised per design stage and per circular design tool category (Figure 3).

For a consistent categorisation of the expressed needs, the specific role of architects and advising engineers in Flanders was taken into account. That role is broader than "designing plans" and, thus, different from some other regions. Architects and advising engineers in Flanders are a pivotal figure between the client and contractors, they are obliged to supervise the construction works and remain responsible 10 years after completion of the building. Moreover, regulations require architects to implement increasingly more diverse knowledge on specific themes such as energy and water use or building safety and health. They are also expected to have insight into the society, whereby they have the ability to respond to future needs of building users and to be informed of the latest (innovative) building trends and solutions. The emergence of the circular economy in construction is consequently considered as "yet another constraint" by multiple designers [72].

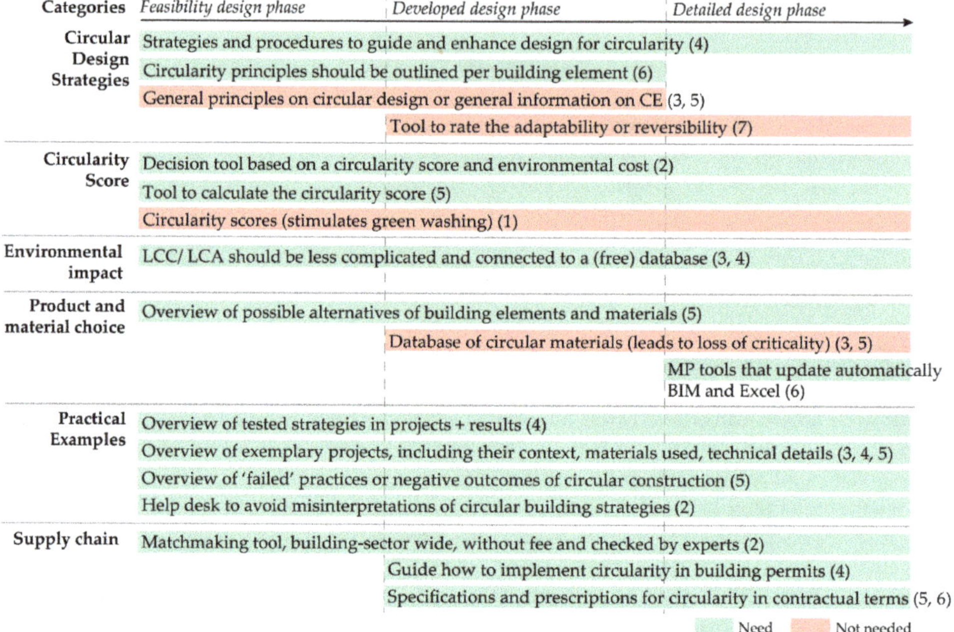

Figure 3. Framework 2: Categorisation of the building designers' and advising engineers' needs for tools for circular building per design phase and per design tool category (green bars). Related interviews are indicated as: (1) researcher sustainable buildings; (2) facade contractor/designer; (3) architect (engineering office); (4) sustainability engineer; (5) architects; (6) architect; (7) architect.

From the interviews and the filled-out framework, a series of findings becomes apparent. They can be discussed per design tool category:

- Circular design strategies: Multiple interviewees expressed that general guidelines or principles on circular building are sufficiently available during information sessions and in literature. However, it has emerged from Interview 6 that principles or guidelines that focus on specific building elements would be useful too but are lacking, for example, a guide that explains how to apply general principles to the facade.

- Circularity score: The need for a circularity score, expressed by the interviewees, shows a demand for some sort of measurability and straightforward evaluation of the level of circularity of a building project, resembling sustainability assessment tools, such as the Building Research Establishment Environmental Assessment Method (BREEAM) and Leadership in Energy and Environmental Design (LEED), which are widely used in Flanders. Some interviewees expressed that a scoring tool is not useful in the context of circular building. These observations are further discussed in Section 4.1. below.

- Environmental impact: Life cycle assessment and costing are considered time-consuming and complex to base design choices on. Moreover, some interviewees complained that (much of) the necessary data are not free of charge in order to perform a proper LCA. Hence, building actors demand less complex and more affordable methods to perform an environmental or financial impact analysis.

- Product and material choice: There is no unified opinion among the interviewees on whether certain building products or materials should be classified in a kind of "circular materials/product" database. Such a database's advantage could be a clear overview that allows to work faster and more informed. Conversely, the disadvantages include the risk that actors lose their criticality on materials and the difficulty to keep the overviews up to date in this transitioning sector.

- Practical examples: There is a much louder call for overviews of projects where circular principles are already applied—this resonates with the need for more specific design strategies mentioned above. In such overviews, the outcomes and lessons could be structured and shared in a coherent way, according to the interviewees. This finding is consistent with the conclusion of Thelen et al. [73], where the lack of demonstration projects is mentioned as a barrier to the transition towards a circular economy. Moreover, the interviewees emphasize that failed practices must be shared too. While successful practices cannot seem to be shared fast enough, less successful experiments are seldomly shared [74]. As a result, today, there is still a lot of insecurity about the positive and negative outcomes of circular building strategies [75]. Therefore, showing the result of "failed practices" would be useful as learning opportunities towards more circular buildings, both economically and technically [76,77]. In the context of, for example, learning networks, not sharing the "failed practices" holds the risk that stakeholders will repeat the made mistakes rather than learning from them. The social conditions should be that people feel safe and joint responsibility is taken for failure, rather than attempting to find a scapegoat, as we tend to remember failure more and longer than success.
- Supply chain: Finally, the need was identified for clear management and monitoring tools to transfer information between partners in an efficient way, taking into consideration the growing complexity of the design process [20]. Linked to that, also matchmaking tools, to connect stakeholders with each other, are demanded. One interviewee stated that in a matchmaking tool, different than in platforms, the professionality of the participants should be verified [76]. Apart from the technical support from tools, there is a need for legal support too: some interviewees mentioned, for example, the need for guidance preparing building permit applications and contracts in line with circular design strategies (e.g., reuse of reclaimed components) or circular business models (e.g., as-a-service contracts).

4. Discussion

Comparing the available tools (see Section 3.1) and the building actors' needs (see Section 3.2) yields three possible outcomes: (1) there is a need but a corresponding tool is not (yet) available; (2) there is a need and a tool is available; and (3) there is no need, however, a tool is available. Each of these three situations is discussed hereunder and can be related to one or more tool (sub)categories and their role in the design process. Those situations are subsequently put in contrast with ongoing research tracks and developments.

4.1. Comparing the Available Tools and the Building Actors' Needs

First, from the interviews it can be concluded that there is an important need for practical examples and best practices. The current platforms and learning networks do not seem sufficient to bring those examples and best practices together and share them further. Although platforms enable information exchange and direct interactions [26], each has its flaws with regard to a scale-up of circular building. Some for example, accept only a small "expert" group, require a financial contribution, or lack a coherent structure [76]. Further, interviewees expressed an absence of practical examples with sufficient technical information [78], learned lessons, etc. Others repeat their demand for cases that did tackle and report on juridical challenges of circular building in today's context [75,76,79]. Moreover, platforms and learning networks do not, or only in one direction, facilitate matchmaking between professional partners interested in circular building. The coverage, usefulness and effect of platforms as they are today must therefore be questioned.

Second, during the interviews, needs were identified while tools are already available. Possible explanations for tools not being used are: marketing strategies are lacking, they are not satisfying the users' expectations or they are too complex, time-consuming, and expensive [80]. The ease of use, next to the associated costs, is one of the most important criteria when selecting design supporting tools [20]. User-friendliness of tools could include, for instance, compatibility with other software and compliance with standards and regulations [20]. This results in the dismissal of some tools, the remaining need for such tools and the oversupply of certain other tools. For example, this research identifies a need from practitioners for tools to calculate a circularity score, while there are already tools available that could (partly) do that. However, not all interviewees agree on the necessity and relevance of circularity scoring. Various possible disadvantages were mentioned: first, the term "circularity" could be greenwashed, i.e., a scoring tool that does not lead to more circular outcomes and falsely promotes its efforts (strategies, method and goals) as environmentally friendly or circular [81]. This means that a scoring tool can, for instance, focus on achieving certain circularity goals, such as using Building Information Modeling (BIM) to manage your project or designing the building elements in such a way that they could be reused after the building's lifespan, but does not necessarily serve the general goal of lowering the environmental footprint of the constructing sector. For this reason, vigilance and a critical eye are important to ensure a relevant implementation. That the terms "circularity" and "circular economy" are ambiguous and the broader socio-economic implications of a CE are often side-stepped [82] is extensively described in the growing literature that is analysing the CE and its problematic relationship with perceptions, policy and consumption such as by Hobson and Lynch [82], Gregson et al. [83], Torelli et al. [84] and Testa et al. [85]. Second, each project is different which makes it difficult to quantitatively measure indicators and to compare them as such. Until today, there is no agreed framework for circular building benchmarking. This means that the results generated by circularity score tools might differ significantly from each other, which could lead to confusion or misinterpretation. Developers of circularity score tools should thus be careful which indicators they measure and how they can avoid greenwashing and stimulate circular building. The different opinions on circularity scoring tools show that the needs are dependent per user and their expertise. However, in order to make any further conclusions, more research is needed on this topic.

Third, there is an oversupply of certain tools that are developed with the same purpose, for the same design phase, for the same design aspect and usually also for the same target group. On the one hand, this oversupply is represented by the eight subcategories of tools in Figure 3 and, on the other, by the red "Not Needed" bars in Figure 3. For example, the subcategory "Design Principle Tools" contains four tools available in Flanders. Two interviewees expressed that there was no need for more such tools that elaborate on theoretical principles of circular design. When developing a new tool concerning circular design principles, the added value of that tool should therefore be questioned and doing a focused survey on the remaining needs for information and guidance in design principles of circular construction is thus recommended.

4.2. Ongoing Research Tracks and Developments on Design Support Tools for Circular Building

Next to the available tools, there are ongoing research tracks on the development of design support tools for circular building design. These trends show that the circular economy still has a significant traction in academia and in the building practice. Most developments are focusing on (1) the integration of BIM with LCA; (2) the integration of BIM with circular design strategies; and (3) developing tools to measure circularity in a building. In this study, BIM was not seen as a design support tool as such but as a concept and method to enhance creating and managing information of a construction project by its linked database of geometric and nongeometric data attached to building elements. The main reason to integrate BIM with design support tools is to deliver accurate and adequate information of the building design, to decrease the (calculation) effort and to speed up the project process [86,87]. Additionally, in Belgium, 29% of the architect officesalready used BIM in their projects and 67% are aware of its functionalities in design practice [88]. This awareness could lead to a large implementation percentage of tools linked to BIM.

The first research track, the integration of BIM in LCA, has been progressively published in scientific literature in the last five years and new tools have been developed [86,89,90]. Building information modelling tools can facilitate quantitative assessments of design options, with automated inventories of material flows and waste [24,91]. This trend answers to the need of making LCA user friendlier (Figure 3). This is an important aspect, as ease of use is considered essential (85%), even more than the cost of a tool (63%) according to a survey done with 224 Flemish architects [20].

Similarly, the second research track concerns circular design strategies tools that become increasingly BIM compliant. For example, the disassembly network analysis method uses BIM and network analysis to analyse the interdependency between building elements to define which elements are recovered and lost during the disassembly of a building and to calculate how long the disassembly takes [91]. Other examples are the BIM-Based Deconstructability Assessment Score, which determines the extent to which a building could be deconstructed [24], and the disassembly planning method of Sanchez et al. [92]. Integrating BIM in design tools for reversible buildings is a result of the support that BIM can offer in this respect: efficiently developing three-dimensional representations of a reversible building, identifying spatial conflicts in the design and collaboratively resolving them with clash detection software [93].

A third development path contains tools and methods that aspire to quantify circularity in buildings. They do so in two different ways. First, the research on circularity indicators in construction is growing [94–96]. For example, Verberne et al. [95] developed a tool that determines the degree of circularity. However, also in this case, the comment was made that the assessment model is meant to provide guidance in concretising the ambitions and should not be seen as an absolute outcome. In the same vein, Flanders developed the GRO tool which is based on sustainability criteria and performance levels [45]. Second, adaptive capacity quantification tools arise as well [97]. For example, the Spatial Assessment of Generality and Adaptability (SAGA) method uses weighted graphs to quantify a building's capacity to support changes [98], the FLEX 4.0 uses a point-based system to assess the adaptive capacity of buildings [99], and the AdaptSTAR model is based on a weighted checklist scoring system to evaluate future adaptation potential in newly designed buildings [100].

Next to those three main research tracks, there are also several tools that do not fit a well-developed research track. There are some initiatives to develop collaboration tools for circular economy in the building sector that is published in the scientific literature such as by Leising et al. [101] and Simons et al. [102]. The conceptual tool "Circular Building Components"-generator aims to support designers in creating and reviewing circular design options [103]. Furthermore, when looking at a broader perspective than the construction sector, more tools on circular design are being researched and developed. For example, the tool of Gehin et al. [104] on implementing sustainable end-of-life strategies in the product development phase, the value mapping tool for sustainable business modelling by Bocken et al. [105], the Circular Material Library of Virtanen et al. [106] and the assessment tool for end of life product recovery strategies by Alamerew et al. [107] are studies aiming to develop design support tools for all kinds of circular products.

4.3. The Relevance of Design Tools

In the busy and increasingly complex practice of building designers, the question is to what extent tools are picked up and used to steer design choices. In general, design practice is largely based on experience and less on tools. This statement is confirmed by the survey of Weytjens et al. [20]. They concluded that design decisions are mainly based on experience, the client's demands and on regulations. Intuition and reference projects are taken into account by over 35% of the respondents when making design decisions, whereas only 21% use design decision tools as a deciding factor [20]. This does, however, not mean that tools do not have an unconscious impact on the considered alternatives and course of the design process.

Similarly, the development and use of design support tools cannot only guarantee the creation of circular buildings. Several other enablers and barriers that have an influence on the transition to a circular economy in the construction sector have been outlined too [21,22,108,109]. Furthermore, users and developers must be aware about every project's specific context [2,14] and the broad scope every design process entails.

5. Conclusions

The present study was set-up to classify available design support tools for circular building, to identify building actors' needs and expectations from such tools and to reveal which research tracks on design support tools for circular building are currently being developed. This section divides the conclusions drawn and the actions to be taken for three types of actors: practitioners (building designers and advising engineers), tool developers and researchers.

First, from a practitioner's perspective, the present study categorises available design support tools for circular building and identifies designers' needs for such tools. This way, it offers an overview and reference to building designers and advising engineers which tools fit their particular needs, their way of working and the context of their projects(?), accelerating the tools' purposeful adoption in practice. Concretely, this study resulted in eight subcategories of tools each having its specific added value during different design stages: Design principles tools, Material flow analysis tools, Life cycle assessment tools, Material and product labels, Reused material platforms, Material passport tools, Life cycle cost tools and Knowledge sharing platforms.

Second, from a developer's perspective, this study reveals opportunities to work on new design support tools, to improve the already available tools and further accelerate their adoption. By comparing the identified needs through the set-up framework with ongoing developments on design support tools for circular building, it became clear that the needs of designers are only partially reflected by the available design support tools and the ongoing developments, and it is now better understood which features are overrepresented and which needs have received little attention so far. When developers create additional design support tools, they should investigate the tools already available, the support base of potential users and the added value of their tool. The resulting research agenda therefore includes these recommendations:

1. There is an oversupply of tools that illustrate the basic principles of circular building. In contrast, there is a loud call for a structured and detailed overview of practical examples and best practices on circular building. The current platforms and learning networks do not seem sufficient to bring those practical insights effectively together and share them further;

2. There is a need for clear workflow management and monitoring tools to transfer the information between partners in a more efficient way. Beyond sharing information, providing new insights and allowing better informed design choices by individual stakeholders, interviewees have put clear importance on collaboration from the start of a project;

3. Research tracks on design tools for circular building are mainly focused on the integration of BIM in LCA, on the integration of BIM in circular design strategies and on quantifying circularity in buildings. On the one hand, this is not in line with the practices and needs of designers to use this kind of integration as a design support tool. On the other hand, in Belgium, 29% of the architecture practices have already used BIM in their projects, and 67% are aware of its functionalities in the design practice [88]. This awareness could lead to a large implementation percentage of tools linked to BIM;

4. Not only development but also guidance in the use of design support tools are of high importance. The framework developed in this research could guide practitioners towards appropriate tools for applying circular building in their projects. However, in order to provide proper guidance, more research and development is needed with attention on user friendliness, prior knowledge and integral accessibility.

Third, from a researcher's perspective, the frameworks in Figures 2 and 3 are a first attempt to compare systemically the "supply and demand" in this field. This framework allows to further monitor the lack and oversupply of certain design support tools for circular building and guide researchers and developers further in their endeavours. Nevertheless, it also needs further refinement and validation. Concretely, possible research paths are:

1. A more elaborate study should be done to clarify how specific contextual aspects, such as the user profile and their expertise, affect the demands and aspirations on guidance in this transition to a circular building sector. Several stated needs in this study were not shared among all the interviewees such as the need for a circularity score or the classification of circular building materials and products;
2. It is recommended to develop the identified missing features and functionalities, albeit with feedback from the envisioned users. Following from the identified mismatches, this research advocates for a more participatory and practice-oriented approach when studying and developing design support tools for circular building;
3. Considering the mismatch between both, it might be interesting to investigate how the current research tracks on design support tools could become more in line with the needs of the building designers and advising engineers, although fundamental exploration must be possible too in particular about the impact of design tools on the design process in general. How can conceptual but scientifically based tools reach practice and influence the design process?
4. In this study, life cycle management and corresponding tools were not identified. Considering the close relation between design, circular business strategies (like PSS) and the life cycle management of assets [110]; therefore, the role of the product service system (PSS) approach and related tools should be further investigated. This might build on the changing role of the architectural designer in the transition to a circular economy, including a change from short-term involvement of the designer to a long-term engagement with the building [72].

Some of the shortfalls of this study include the focus on the region Flanders, which is a small region that has a specific construction sector culture and political landscape, and the small interview sample. However, given that Flanders can be considered as one of the forerunners in the transition towards a circular construction economy, makes the region and its ongoing initiatives a well-documented and instructive case for reflection and learning about the transition itself. The main limitation of the interview sample and the in-depth interviews is that generalizations about the results cannot be made because a small sample was chosen and random sampling methods were not used. These in-depth interviews, however, provide valuable information for a preliminary idea of what the needs are concerning design support tools for circular building and to be able to compare these needs with the available tools and research directions. Furthermore, this study focused on design support tools as a possible aid to provide guidance for the practitioners in the design process. Other tools and institutions can also be of significance as guidance, where the expertise of policy makers, researchers and consultants can be employed.

Through these insights and recommendations, this research wants to contribute to a much needed debate within the building sector supply chain to better understand and prioritise the key issues concerning supply and demand of design support tools for circular building practice and, consequently, supporting the construction sector in its transition to a circular, sustainable economy.

Author Contributions: The conceptualization, the outlining of the methodology, the original draft preparation and the making of the figures was performed by C.C.; For writing—review and editing, C.C. and W.G. were involved. The study was supervised by N.D.T. and W.G. All authors have read and agreed to the published version of the manuscript.

Funding: This research was funded by Fonds Wetenschappelijk Onderzoek (FWO), grant number 1S55518N.

Conflicts of Interest: The authors declare no conflict of interest.

Appendix A. Interview Guide

Face-to-face, semi-structured interviews were conducted to identify the specific needs of building designers and advising engineers who are trying to integrate circular principles in their practice and designs for design support tools (and their features) for circular building. The following questions served as an interview guide:

- First, the interviewer asked some general questions on the implementation of circularity in their design and on their use of tools: What does the term circularity mean to you? In what way have you tried to apply circularity in practice? Where did you gain knowledge about circularity? Do you know tools to design circular buildings? Do you use them? Why do you use them? Would you use them? What are the barriers to applying circular principles? What is the biggest difficulty you encounter when implementing circular principles in the projects?
- Second, per tool category, the interviewer asked more in-depth questions: Do you know such tools? Does it seem useful to you? Would you use it? Why or why not?

For example, for the category "Circularity Score", the following questions were asked: According to you, is a tool that measures the degree of circularity in a building useful? What criteria should be used to assess whether a product, component, element or building as a whole is circular or not?

References

1. Geissdoerfer, M.; Savaget, P.; Bocken, N.M.P.; Hultink, E.J. The Circular Economy—A new sustainability paradigm? *J. Clean. Prod.* **2017**, *143*, 757–768. [CrossRef]
2. Ghisellini, P.; Ripa, M.; Ulgiati, S. Exploring environmental and economic costs and benefits of a circular economy approach to the construction and demolition sector. A literature review. *J. Clean. Prod.* **2018**, *178*, 618–643. [CrossRef]
3. Kirchherr, J.; Reike, D.; Hekkert, M. Conceptualizing the circular economy: An analysis of 114 definitions. *Resour. Conserv. Recycl.* **2017**, *127*, 221–232. [CrossRef]
4. European Commission. *EU Circular Economy Action Plan: A new Circular Economy Action Plan for a Cleaner and More Competitive Europe*; European Union: Brussels, Belgium, 2020.
5. OVAM. *Startverklaring Vlaanderen Circulair*; OVAM: Mechelen, Belgium, 2016.
6. Demir, Z. Beleidsnota 2019–2024. *Omgeving*; Vlaams Parlement: Brussels, Belgium, 2019.
7. OVAM. *Materiaalbewust Bouwen in Kringlopen: Preventieprogramma Duurzaam Materialenbeheer in de Bouwsector 2014–2020*; OVAM: Mechelen, Belgium, 2013.
8. Ellen MacArthur Foundation. Case studies: Flanders Materials Programme. Available online: https://www.ellenmacarthurfoundation.org/case-studies/belgium-flanders-materials-programme (accessed on 3 August 2020).
9. Silva, A.; Rosano, M.; Stocker, L.; Gorissen, L. From waste to sustainable materials management: Three case studies of the transition journey. *Waste Manag.* **2017**, *61*, 547–557. [CrossRef]
10. Paredis, E. A Winding Road: Transition Management, Policy Change and the Search for Sustainable Development. Ph.D. Thesis, Ghent University, Ghent, Belgium, 2012.
11. TOTEM: Environmental Profile of Building Elements. Available online: https://www.totem-building.be/ (accessed on 14 October 2019).
12. Vlaanderen Circulair Green Deal Circulair Bouwen. Available online: https://vlaanderen-circulair.be/nl/onze-projecten/detail/green-deal-circulair-bouwen (accessed on 25 June 2020).
13. Open Call: Innovatieve Circulaire Economieprojecten. Available online: https://vlaanderen-circulair.be/nl/aan-de-slag/open-call (accessed on 27 July 2020).
14. Hossain, M.U.; Ng, S.T. Critical consideration of buildings' environmental impact assessment towards adoption of circular economy: An analytical review. *J. Clean. Prod.* **2018**, *205*, 763–780. [CrossRef]
15. International Organization for Standardization (ISO). *Sustainability in Buildings and Civil Engineering Works—Design for Disassembly and Adaptability—Principles, Requirements and Guidance*; ISO 20887:2020; ISO: London, UK, 2020.
16. Bauwens, T.; Hekkert, M.; Kirchherr, J. Circular futures: What Will They Look Like? *Ecol. Econ.* **2020**, *175*, 106703. [CrossRef]
17. Geels, F.W.; Schot, J. Typology of sociotechnical transition pathways. *Res. Policy* **2007**, *36*, 399–417. [CrossRef]

18. Debacker, W.; Manshoven, S.; Apelman, L.; Beurskens, P.; Biberkic, F.; Denis, F.; Durmisevic, E.; Dzubur, A.; Hansen, K.; Henrotay, C.; et al. *Synthesis of the State of the Art: Key Barriers and Opportunities for Materials Passports and Reversible Building Design in the Current System*; Buildings as Materials Banks: Brussels, Belgium, 2016; p. 103.

19. Vandenbroucke, M.; De Temmerman, N.; Paduart, A.; Debacker, W. Opportunities and obstacles of implementing transformable architecture. In Proceedings of the Int. Conf. on Sustainable Building, University of Minho, Guimarães, Portugal, 20 February 2013.

20. Weytjens, L.; Verdonck, E.; Verbeeck, G. Classification and Use of Design Tools: The Roles of Tools in the Architectural Design process. *Des. Princ. Pract. Int. J.* **2009**, *3*, 289–303. [CrossRef]

21. Adams, K.; Osmani, M.; Thorpe, T.; Thornback, J. Circular economy in construction: Current awareness, challenges and enablers. *Waste Resour. Manag.* **2017**, *170*, 15–24. [CrossRef]

22. Hart, J.; Adams, K.; Giesekam, J.; Tingley, D.D.; Pomponi, F. Barriers and drivers in a circular economy: The case of the built environment. *Procedia CIRP* **2019**, *80*, 619–624. [CrossRef]

23. Bocken, N.M.P.; de Pauw, I.; Bakker, C.; Grinten, B. van der Product design and business model strategies for a circular economy. *J. Ind. Prod. Eng.* **2016**, *33*, 308–320. [CrossRef]

24. Akinade, O.O.; Oyedele, L.O.; Bilal, M.; Ajayi, S.O.; Owolabi, H.A.; Alaka, H.A.; Bello, S.A. Waste minimisation through deconstruction: A BIM based Deconstructability Assessment Score (BIM-DAS). *Resour. Conserv. Recycl.* **2015**, *105*, 167–176. [CrossRef]

25. Osmani, M.; Glass, J.; Price, A.D.F. Architects' perspectives on construction waste reduction by design. *Waste Manag.* **2008**, *28*, 1147–1158. [CrossRef]

26. Taranic, I.; Behrens, A.; Topi, C. *Understanding the Circular Economy in Europe, from Resource Efficiency to Sharing Platforms: The CEPS Framework*; Social Science Research Network: Rochester, NY, USA, 2016.

27. *Interviewing Experts*; Bogner, A.; Littig, B.; Menz, W. (Eds.) Palgrave Macmillan: London, UK, 2009; ISBN 978-0-230-20680-9.

28. Lespagnard, M. Renovation of Residential Post-War Highrise: Understanding the dEsign Process and Impact of Reversible Design Tools and Strategies. Master's Thesis, Vrije Universiteit Brussel, Brussels, Belgium, 2019.

29. Geels, F.; Grin, J.; Loorbach, D.; Rotmans, J.; Schot, J.; Schot, J.; Grin, J. *Transitions to Sustainable Development: New Directions in the Study of Long Term Transformative Change*; Routledge: New York, NY, USA, 2011; ISBN 978-0-415-87675-9.

30. VUB Architectural Engineering. *Building a Circular Economy. Design Qualities to Guide and Inspire Building Designers and Clients*; Vrije Universiteit Brussel: Brussels, Belgium, 2019.

31. OVAM. *24 Ontwerprichtlijnen Veranderingsgericht Bouwen*; OVAM: Mechelen, Belgium, 2016.

32. Circular Flanders. The Business Model Innovation Grid. Available online: https://www.vlaanderen-circulair.be/bmix/ (accessed on 22 October 2019).

33. Vandenbroucke, M. *Bouwcatalogus Veranderingsgericht Bouwen*; OVAM: Mechelen, Belgium, 2019.

34. Alba Concepts Building Circularity Index. Available online: https://albaconcepts.nl/building-circularity-index/ (accessed on 16 October 2019).

35. Cenergie C-CalC. Available online: https://www.cenergie.be/nl/diensten/advies/c-calc (accessed on 16 October 2019).

36. BAMB. Circular Building Assessment Prototype. Available online: https://www.bamb2020.eu/post/cba-prototype/ (accessed on 14 October 2019).

37. Ellen MacArthur Foundation. The Circular Design Guide. Available online: https://www.circulardesignguide.com/ (accessed on 20 March 2019).

38. Circular IQ Circular Transition Indicators. Available online: https://ctitool.com/ (accessed on 10 February 2020).

39. IDEAL&CO. Explore BV Circularity Calculator. Available online: http://www.circularitycalculator.com/ (accessed on 14 October 2019).

40. EIT. RawMaterials Find a Circular Business Model that Fits. Available online: http://www.circulator.eu/ (accessed on 10 December 2019).

41. Ellen MacArthur Foundation. Circulytics—Measuring Circularity. Available online: https://www.ellenmacarthurfoundation.org/resources/apply/circulytics-measuring-circularity (accessed on 14 October 2019).

42. Remeha, B.V. Toolbox. Available online: https://home.et.utwente.nl/designtool/5d33e8a3cfb35737ec484429954f0ea98fed69d8/toolbox.html (accessed on 14 October 2019).

43. OVAM. Bepaal de Milieu-Impact van uw Product en Maak het Verschil. Available online: http://www.ecolizer.be/ (accessed on 13 October 2019).

44. Sphera GaBi Circularity Toolkit—Circular Economy Software. Available online: http://www.gabi-software.com/international/software/gabi-software/gabi-circularity-toolkit/ (accessed on 11 February 2020).

45. Het Facilitair Bedrijf. *GRO Gebruikershandleiding*; Het Facilitair Bedrijf: Brussels, Belgium, 2019.

46. Superuse Studios NL Oogstkaart. Available online: https://www.oogstkaart.nl/ (accessed on 29 October 2019).

47. BRE Group IMPACT. Available online: https://www.bregroup.com/impact/ (accessed on 11 February 2020).

48. Insert. Available online: https://www.insert.nl/ (accessed on 29 October 2019).

49. Platform CB'23. *Meten van Circulariteit*; Platform CB'23: Delft, The Netherlands, 2020.

50. European Commission. *Level(s): Building Sustainability Performance*; European Commission: Brussels, Belgium, 2020.

51. Madaster Services Madaster Platform. Available online: https://www.madaster.com/nl/our-offer-2/Madaster-Platform (accessed on 11 October 2019).

52. World Business Council for Sustainable Development MarketplaceHUB. Available online: https://marketplacehub.org/ (accessed on 29 October 2019).

53. Meex, E. Early Design Support for Material Related Environmental Impact Assessment of Dwellings. Ph.D. Thesis, UHasselt: Hasselt, Belgium, 2018.

54. NIBE Milieuclassificaties. Available online: https://www.nibe.info/nl (accessed on 12 October 2019).

55. Metabolism of Cities Online Material Flow Analysis Tool (OMAT). Available online: https://metabolismofcities.org/ (accessed on 18 February 2020).

56. Bionova Ltd OneCLick LCA. Available online: https://www.oneclicklca.com/ (accessed on 11 February 2020).

57. Rotor vzw Bouwen en Renoveren met Hergebruik. Available online: https://opalis.eu/nl (accessed on 29 October 2019).

58. openLCA. Available online: http://www.openlca.org/ (accessed on 13 November 2019).

59. Platform CB'23. Available online: https://platformcb23.nl/ (accessed on 14 November 2019).

60. Pixii Kennisplatform Energieneutraal Bouwen. Available online: https://pixii.be/ (accessed on 11 October 2019).

61. RIVM. LCIA: The ReCiPe Model. Available online: https://www.rivm.nl/en/life-cycle-assessment-lca/recipe (accessed on 11 February 2020).

62. Galle, W. Scenario Based Life Cycle Costing: An Enhanced Method for Evaluating the Financial Feasibility of Transformable Building. Ph.D. Thesis, Vrije Universiteit Brussel, Brussels, Belgium, 2016.

63. Pré Consultants, B.V. LCA Software for Fact-Based Sustainability. Available online: https://simapro.com/ (accessed on 10 February 2020).

64. TU Wien STAN. Available online: http://www.stan2web.net/ (accessed on 11 February 2020).

65. Totem. Available online: https://www.totem-building.be/ (accessed on 10 April 2020).

66. Werflink Werflink. Available online: https://www.werflink.com/nl-werflink.html (accessed on 29 October 2019).

67. Pomponi, F.; Moncaster, A. Circular economy for the built environment: A research framework. *J. Clean. Prod.* **2017**, *143*, 710–718. [CrossRef]

68. Haupt, M.; Zschokke, M. How can LCA support the circular economy? *Int. J. Life Cycle Assess.* **2017**, *22*, 832–837. [CrossRef]

69. Scheepens, A.E.; Vogtländer, J.G.; Brezet, J.C. Two life cycle assessment (LCA) based methods to analyse and design complex (regional) circular economy systems. Case: Making water tourism more sustainable. *J. Clean. Prod.* **2016**, *114*, 257–268. [CrossRef]

70. FAQ—Gebruik van de TOTEM-Tool. Available online: https://www.totem-building.be/pages/faq.xhtml (accessed on 10 April 2020).

71. How does Labelinfo Work? Available online: https://www.labelinfo.be/nl/hoe-werkt-labelinfo (accessed on 14 May 2020).

72. Galle, W.; Herthogs, P.; Vandervaeren, C.; De Temmerman, N. *The Architect's Role in a Change-Oriented Construction Sector: A Belgian Perspective*; Proceedings of Open Building for Resilient Cities: Los Angeles, CA, USA, 2018; pp. 69–75.

73. Thelen, D.; van Acoleyen, M.; Huurman, W.; Thomaes, T.; van Brunschot, C.; Edgerton, B.; Kubbinga, B. *Scaling the Circular Built Environment: Pathways for Businesses and Government*; World Business Council for Sustainable Development: Geneva, Switzerland, 2018.

74. Laksov, K.B.; McGrath, C. Failure as a catalyst for learning: Towards deliberate reflection in academic development work. *Int. J. Acad. Dev.* **2020**, *25*, 1–4. [CrossRef]

75. Architects, interview by authors, Antwerp, Belgium. 3 September 2019.

76. Sustainability Engineer, interview by authors, Louvain-la-Neuve. 23 January 2019.

77. Jungic, V.; Creelman, D.; Bigelow, A.; Côté, E.; Harris, S.; Joordens, S.; Ostafichuk, P.; Riddell, J.; Toulouse, P.; Yoon, J.-S. Experiencing failure in the classroom and across the university. *Int. J. Acad. Dev.* **2020**, *25*, 31–42. [CrossRef]

78. Architect/Advisory Office, interview by authors, Antwerp, Belgium. 22 January 2019.

79. Architect, interview by authors, Brussels, Belgium. 4 November 2019.

80. Weytjens, L.; Verbeeck, G. Towards "architect-friendly" energy evaluation tools. In Proceedings of the 2010 Spring Simulation Multiconference, Society for Computer Simulation International, Orlando, FL, USA, 11–15 April 2010; pp. 1–8.

81. Becker-Olsen, K.; Potucek, S. Greenwashing. In *Encyclopedia of Corporate Social Responsibility*; Idowu, S.O., Capaldi, N., Zu, L., Gupta, A.D., Eds.; Springer: Berlin/Heidelberg, 2013; pp. 1318–1323, ISBN 978-3-642-28036-8.

82. Hobson, K.; Lynch, N. Diversifying and de-growing the circular economy: Radical social transformation in a resource-scarce world. *Futures* **2016**, *82*, 15–25. [CrossRef]

83. Gregson, N.; Crang, M.; Fuller, S.; Holmes, H. Interrogating the circular economy: The moral economy of resource recovery in the EU. *Econ. Soc.* **2015**, *44*, 218–243. [CrossRef]

84. Torelli, R.; Balluchi, F.; Lazzini, A. Greenwashing and environmental communication: Effects on stakeholders' perceptions. *Bus. Strategy Environ.* **2020**, *29*, 407–421. [CrossRef]

85. Testa, F.; Iovino, R.; Iraldo, F. The circular economy and consumer behaviour: The mediating role of information seeking in buying circular packaging. *Bus. Strategy Environ.* **2020**. [CrossRef]

86. Hollberg, A.; Genova, G.; Habert, G. Evaluation of BIM-based LCA results for building design. *Autom. Constr.* **2020**, *109*, 102972. [CrossRef]

87. Akinade, O.O.; Oyedele, L.O.; Omoteso, K.; Ajayi, S.O.; Bilal, M.; Owolabi, H.A.; Alaka, H.A.; Ayris, L.; Henry Looney, J. BIM-based deconstruction tool: Towards essential functionalities. *Int. J. Sustain. Built Environ.* **2017**, *6*, 260–271. [CrossRef]

88. Mirza & Nacey Research. *The architectural profession in Europe*; Architects' Council of Europe: West Sussek, UK, 2019.

89. Bueno, C.; Fabricio, M.M. Comparative analysis between a complete LCA study and results from a BIM-LCA plug-in. *Autom. Constr.* **2018**, *90*, 188–200. [CrossRef]

90. Santos, R.; Costa, A.A.; Silvestre, J.D.; Vandenbergh, T.; Pyl, L. BIM-based life cycle assessment and life cycle costing of an office building in Western Europe. *Build. Environ.* **2020**, *169*, 106568. [CrossRef]

91. Denis, F.; Vandervaeren, C.; De Temmerman, N. Using Network Analysis and BIM to Quantify the Impact of Design for Disassembly. *Buildings* **2018**, *8*, 113. [CrossRef]

92. Sanchez, B.; Rausch, C.; Haas, C.; Saari, R. A selective disassembly multi-objective optimization approach for adaptive reuse of building components. *Resour. Conserv. Recycl.* **2020**, *154*, 104605. [CrossRef]

93. van den Berg, M.C.; Durmisevic, E. BIM uses for reversible building design. In Proceedings of the Vital Cities and Reversible Buildings: Conference Proceedings; Sarajevo Green Design Foundation, Mostar, Bosnia and Herzegovina, 4–7 October 2017.

94. Heisel, F.; Rau-Oberhuber, S. Calculation and evaluation of circularity indicators for the built environment using the case studies of UMAR and Madaster. *J. Clean. Prod.* **2020**, *243*, 118482. [CrossRef]

95. Verberne, J. Building Circularity Indicators: An Approach for Measuring Circularity of a Building. Master's Thesis, Eindhoven University of Technology, Eindhoven, The Netherlands, 2016.

96. Nuñez-Cacho, P.; Górecki, J.; Molina-Moreno, V.; Corpas-Iglesias, F.A. What Gets Measured, Gets Done: Development of a Circular Economy Measurement Scale for Building Industry. *Sustainability* **2018**, *10*, 2340. [CrossRef]

97. Rockow, Z.R.; Ross, B.; Black, A.K. Review of methods for evaluating adaptability of buildings. *Int. J. Build. Pathol. Adapt.* **2019**, *37*, 273–287. [CrossRef]

98. Hertogs, P.; Debacker, W.; Tunçer, B.; De Temmerman, N.; Yves, D.W. Quantifying the Generality and Adaptability of Building Layouts using Weighted Graphs: The SAGA Method. *Buildings* **2019**, *9*, 20. [CrossRef]

99. Geraedts, R. FLEX 4.0, A Practical Instrument to Assess the Adaptive Capacity of Buildings. *Energy Procedia* **2016**, *96*, 568–579. [CrossRef]

100. Conejos, S.; Langston, C.; Smith, J. AdaptSTAR model: A climate-friendly strategy to promote built environment sustainability. *Habitat Int.* **2013**, *37*, 95–103. [CrossRef]

101. Leising, E.; Quist, J.; Bocken, N. Circular Economy in the building sector: Three cases and a collaboration tool. *J. Clean. Prod.* **2018**, *176*, 976–989. [CrossRef]

102. Simons, C.; Galle, W.; De Temmerman, N. Proposal for a Guiding Toolbox for Circular Design in Renovation Projects. Master's Thesis, Vrije Universiteit Brussel, Brussel, Belgium, 2019.

103. van Stijn, A.; Gruis, V. Towards a circular built environment: An integral design tool for circular building components. *Smart Sustain. Built Environ.* **2019**. [CrossRef]

104. Gehin, A.; Zwolinski, P.; Brissaud, D. A tool to implement sustainable end-of-life strategies in the product development phase. *J. Clean. Prod.* **2008**, *16*, 566–576. [CrossRef]

105. Bocken, N.; Short, S.; Rana, P.; Evans, S. A value mapping tool for sustainable business modelling. *Corp. Gov.* **2013**, *13*, 482–497. [CrossRef]

106. Virtanen, M.; Manskinen, K.; Eerola, S. Circular Material Library. An Innovative Tool to Design Circular Economy. *Des. J.* **2017**, *20*, S1611–S1619. [CrossRef]

107. Alamerew, Y.A.; Brissaud, D. Circular economy assessment tool for end of life product recovery strategies. *J. Remanuf.* **2019**, *9*, 169–185. [CrossRef]

108. Geldermans, R.J. Design for Change and Circularity—Accommodating Circular Material & Product Flows in Construction. *Energy Procedia* **2016**, *96*, 301–311. [CrossRef]

109. Mahpour, A. Prioritizing barriers to adopt circular economy in construction and demolition waste management. *Resour. Conserv. Recycl.* **2018**, *134*, 216–227. [CrossRef]

110. Fargnoli, M.; Lleshaj, A.; Lombardi, M.; Sciarretta, N.; Di Gravio, G. A BIM-based PSS Approach for the Management of Maintenance Operations of Building Equipment. *Buildings* **2019**, *9*, 139. [CrossRef]

 *buildings*

Article

Value of Technical Wear and Costs of Restoring Performance Characteristics to Residential Buildings

Beata Nowogońska * and Jacek Korentz

Faculty of Civil Engineering, Architecture and Environmental Engineering, University of Zielona Góra, Szafrana 1, 65-516 Zielona Góra, Poland; j.korentz@ib.uz.zgora.pl
* Correspondence: b.nowogonska@ib.uz.zgora.pl; Tel.: +48-68-3282-290

Received: 25 November 2019; Accepted: 6 January 2020; Published: 8 January 2020

Abstract: Each building, over the course of subsequent years of use, undergoes wear, with a deterioration of its technical condition. As a result of this, the performance characteristics of a building decrease with the passing of time, with their complete or partial restoration requiring repair and renovation works to be carried out. It is the task of real estate managers or owners to maintain the building in a non-deteriorating technical and functional condition. In order to preserve the technical and functional condition of a building at an adequate level, methodological support of decision-making processes pertaining to the conducting of rational maintenance management is necessary. The present article presents a proposal of a model allowing for the accurate assessment of the costs of renovation and repair works on a building at a given stage of its use, and their relationship with the value of the technical wear of the building in the same time period. Residential buildings constructed using traditional technology were subjected to analysis. In the carried out analysis, temporal methods applied for calculating the level of technical wear were applied, with the PRRD (prediction of reliability according to Rayleigh distribution) model of changes in the performance characteristics applied to determine the costs of renovation works necessary for restoring performance characteristics to the building.

Keywords: technical condition; performance characteristics; prediction; degree of wear

1. Introduction

The technical wear of a building changes as a result of the aging process. Along with the passing of time, a decrease in the performance characteristics of a building [1,2] takes place, and their total or partial restoration is only possible as a result of repair works [3–7].

The date of complete renovation in the case of a non-renovated building ought to be planned in accordance with the guidelines of rational management [8–12]. This date should result from the principle of the lowest costs of achieving the aim, which is maintaining the technical condition of the building at an adequate level. Earlier renovation restoring the technical condition of a building leads to limiting further wear, which results in lower total costs of repairs. Repair works pushed back to further years of use lead to significant deterioration of the technical conditions, and, at the same time, increased costs of repair.

Making decisions connected with the choice of the scope, type and date of repairs on buildings is very problematic for managers. The problem of the current assessment of the technical conditions is a subject of many studies. However, in addition to assessing the technical conditions, algorithms supporting decision-making regarding repair works are also necessary. The paper [13] discusses a technical and economic analysis that compares three proactive strategies (preventive, predictive, and improvement) for two types of coatings typically used in maintenance interventions of facades of social housing units in Lisbon: Emulsion, and elastomeric coatings. Bucoń and Sobotka [1] propose a decision model of the choice of the scope of repairs based on three assessments of a building: The technical, energy, and functional condition. The synthetic indicator of the use value of a building obtained from the assessments is the basis for selecting a renovation solution which allows the highest increase in use value in relation to the amount of financial resources engaged to be obtained. The optimal selection of proposed interventions across the broad spectrum of assets is also problematic, and currently, it is performed in a subjective manner. The paper [14] identifies a number of prioritization techniques that can be used to compare and rank repair and renewal projects.

The subject matter connected with developing methods of analyzing the costs of the life cycle of technical objects (life cycle cost analysis) is becoming increasingly popular [15,16]. More and more frequently, attempts are made to develop tools aiding the indication of economically validated directions of actions pertaining to the management of the life cycle of building structures [10–12]. Reference [17] presents the BdMS (building management systems) model of managing a building, based on predicting the durability of building components on the basis of deterministic and stochastic methods. In reference [18], the optimization method is described, which is based on minimizing the expected total cost of the life cycle of a building while maintaining acceptable reliability of its structure over the entire course of its use. Analyses connected with the relationship of the cost of the life cycle and time also occur in many other models pertaining to various building structures [19–21]. The paper [22] discusses a set of 17 criteria to help the maintenance choice for building facades, from three viewpoints: Physical performance, risk, and costs.

Risk and uncertainty are typical for the cost calculation of a life cycle and require an effective system of managing risk to constitute an integral part of the life cycle cost analysis [23,24]. Reference [25] contains a model of estimating the costs of the life cycle and the entire cost of the life of a building, which makes it possible to quantify the increase in costs resulting from the carried and assessed risk. The model makes use of Mamdani fuzzy inference.

In references [26,27], a stochastic and multi-task system of aiding decision-making was proposed in an effort to optimize management of roof maintenance, both at the level of planned renovation works, as well as at the design stage. The discrete Markov chain was used to predict the effectiveness of the roof system over the time of its use.

Another method of budgeting the costs of maintenance and renovation is Schroeder's method [28]. This method is based on determining the technical state of various building components in the function of time.

In reference [29] a new stochastic dynamic programming model is proposed where optimality conditions are derived through the Hamilton–Jacobi–Bellman equations. The model defined the joint production and repair, major maintenance switching strategies minimizing the total cost over an infinite planning horizon. Two probabilistic approaches are described in [30], one approach using a mathematical function (Weibull) to describe the performance of a component over time and one approach using discrete Markov chains.

In reference [31] a method combines the use of failure mode and effect analysis was proposed to permit identifying likely failure modes from which maintenance actions could be planned and the limit states method to assess the durability of the given retrofit action. The purpose of the research reported in reference [32] was to develop a model that allows for the identification of the owner's needs in all phases of the building life cycle. A six-level classification system for the information required in the project and a two-dimensional model that maps the life cycle were corroborated and improved by applying the Delphi technique to a panel of ten experts in two rounds. Reference [33] concluded that the planning and the application of the condition-based maintenance strategy have significant characteristics and make reference to the resulting prediction model.

The authors of this article suggest another approach to the cost analysis of renovation works necessary to maintain a building at a given level of performance characteristics. The proposed model allows for estimating the costs of renovation and repair works on a building at a given stage of its use with the value of the technical wear of the building in the same time period. It would seem that the consumption values and the cost of renovation works are at a similar level. However, it turns out that the cost of restoring the building's performance differs from the value of the building's consumption. The proposed method takes into account the change of performance over time. With this action plan repair, it is possible at any given moment of time.

Estimating the cost of renovation work is a very important problem. The existing methods are based on working ad hoc. The proposed method takes into account changes in performance and degree of technical wear over time.

2. Main Objectives of the Proposed Method

It is the task of real estate managers or owners to maintain the building in a non-deteriorating technical and functional condition. In order to preserve the technical and functional condition of a building at an adequate level, methodological support of decision-making processes pertaining to carrying out rational maintenance management is necessary. The purpose of the research is a proposal of a model allowing for the estimation assessment of the costs of renovation and repair works on a building at a given stage of its use, and their relationship with the value of technical wear of a building in the same time period. Residential buildings constructed in traditional technology were subjected to analysis.

In the carried out analysis, temporal methods applied for calculating the level of technical wear were applied, with the PRRD [2] (prediction of reliability according to Rayleigh distribution) model of changes in the performance characteristics was applied to determine the costs of renovation works necessary for returning performance characteristics to the building.

The research consisted in determining the value of the technical wear of the building in good, satisfactory, average, and poor technical condition. The value of technical wear of each of the examined buildings resulting from the degree of wear was compared with the cost of restoring the functional properties of the building to their original state.

3. Functions Describing the Aging Process of a Building

In accordance with the recommendations specified by standards [34–36], determining the changes in the performance characteristics of a building requires the application of PSLDC (predicted service life of components) risk curves as tools aiding the planning of renovation deadlines resulting from these recommendations. Models of predicting changes in the performance characteristics of building components, as well as the entire building, were proposed [2,37]. Methods of non-linear aging processes of the entire building accounting for the role and weight of individual building components were developed. The PRRD (prediction of reliability according to Rayleigh distribution) [2,37,38] allows for the prospective assessment of the technical conditions of a building. The developed models of predicting the aging process of residential buildings constructed using traditional technology, as well as its components, account for Weibull distribution as a distribution of the reliability over the time a building structure is in service.

The aging process of a building can be described by the following functions of the PRRD model: Function of changes in the performance characteristics of a building $R(t)$, function of changes failure of a building $F(t)$ as well as the function of building wear $S_Z(t)$.

A building comprises many components. For an individual i-th element, the functions of the aging process are described by three relationships:

$$R_i(t) = \exp\left(-\left(\frac{t}{T_i}\right)^2\right), \tag{1}$$

$$F_i(t) = 1 - R_i(t) \tag{2}$$

$$S_{Zi}(t) = \frac{t^2}{T_i^2} \tag{3}$$

where:

t—service life of a component,
T—life span of a component.

Technical wear is defined as the loss of service, functional or technological capabilities of an object or its components [6,7]. Wear is described by the function of time as well as the effect of external factors. A numerical parameter describing the technical condition of a building or its components is the level of wear expressed in percentages. Among methods of describing the level of wear are temporal, visual, and weighted average methods.

In the proposed method, the level of wear, was described in a different manner, on the basis of the intensity of damage. However, eventually, after appropriate transformation formulas, we obtain the same pattern as in the method of the time.

$$S_{Zi}(t) = \int_0^t \lambda(t)dt \tag{4}$$

where the intensity of damage $\lambda(t)$ is the rate of deterioration of performance characteristics:

$$\lambda(t) = \frac{dF(t)}{dt}\frac{1}{R(t)} \tag{5}$$

Each building is made up of many components. These elements serve various functions, are created from dissimilar construction materials, each is characterized by different properties and different service lives. Functions of the aging process for the entire building, accounting for weights Ai of components are described by the relationships:

$$R(t) = \sum A_i\, R_i(t) \tag{6}$$

$$F(t) = 1 - R(t) \tag{7}$$

$$S_Z(t) = \sum A_i \frac{t^2}{T_i^2} \tag{8}$$

where:

t—service life of a component,
T—life span of a component.

The proposed PRRD method of predicting the aging process of buildings allows for a description of the changes in the technical wear, assessing the rate at which a building ages, a description of the aging process of a building, predicting changes in the performance characteristics of a renovated building for any given cycle between repairs and an analysis of the consequences of the lack of repairs in a building.

4. Building Constructed Using Traditional Technology—Case Study

The method was applied for a building with chosen material and construction solutions. The building subjected to analysis was built using traditional technology.

The research included 592 residential buildings located within the municipality of Zielona Góra. All buildings are two-story, founded on a rectangular plan, almost all frontages of houses overlook the streets. The buildings were built between 1915–2015. The administrator of all the analyzed buildings is the Department of Municipal and Housing Administration in Zielona Góra.

All the buildings of the collected research material have similar material and structural solutions. The walls of the analyzed objects are made of full brick on cement-lime mortar, the ceilings—wooden beam, stairs and roof truss—wooden, roof cover—ceramic tiles. The differences are only in the spans of floors, numbers of floors and roof truss constructions (usually purlin-tick, sometimes collar beam). For all the buildings, a periodic assessment of technical condition was carried out by experts. During the evaluation, the consumption rates of individual elements of buildings were established.

Each building was divided into 25 components. Components serving a structural function have the most significant influence on the service life. Other supporting elements influence the performance characteristics of the building to a lesser degree. When indicating the performance characteristics for the entire building, which is a collection of components, the intensity of the influence of performance characteristics of components was accounted for in the form of a scale of weights of A components. In determining functions (6), (7), and (8): Changes in the performance characteristics of the entire building RA(t), functions of changes in building unreliability F(t), as well as functions of changes in the technical wear of a building S(t), relationships (1), (2), and (3) determining changes in 25 building components were used. The course of functions describing the aging process of the entire building has been illustrated in Figure 1.

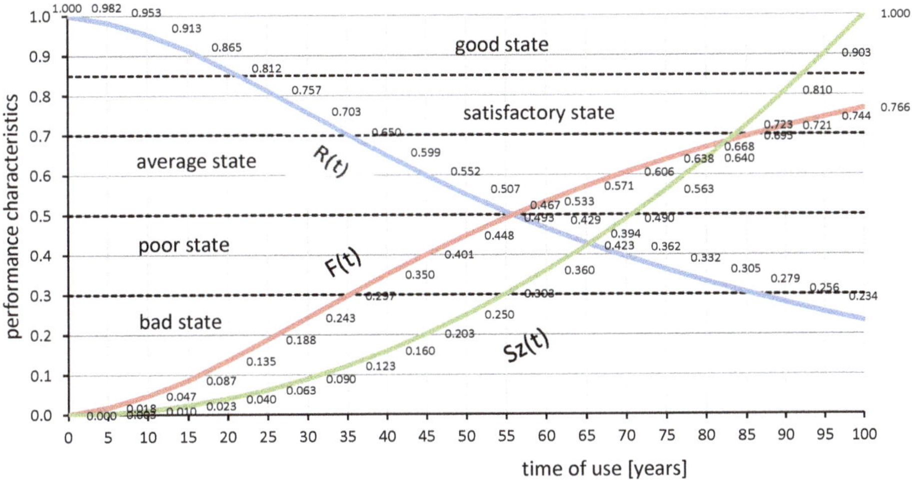

Figure 1. Functions describing the aging process of a building.

5. Value of Technical Wear and the Decrease in Performance Characteristics

The level of wear is a measure of the technical condition of a building. With the passing of time, the level of wear in continuously increasing in non-renovated buildings. Value of technical wear $W_Z(t)$ signifies the cost k_P recreating the level of wear:

$$W_Z(t) = k_P S_Z(t) \tag{9}$$

The performance characteristics R(t) of a building decrease over the course of its use. Restoring the performance characteristics to their maximal value 1.0 may be possible only as a result of carrying out renovation.

The value of loss of performance characteristics $U_R(t)$ signifies the cost k_U of restoring performance characteristics. Changes in the performance characteristics of a building over the course of its use expressed by cost are the loss of use values.

$$U_R(t) = k_U(1 - R(t)) \tag{10}$$

Figure 2 presents the values of the technical wear (red field) as well as values of loss in performance characteristics (blue field) for subsequent years that a building with a normative service life of T = 100 years is in service.

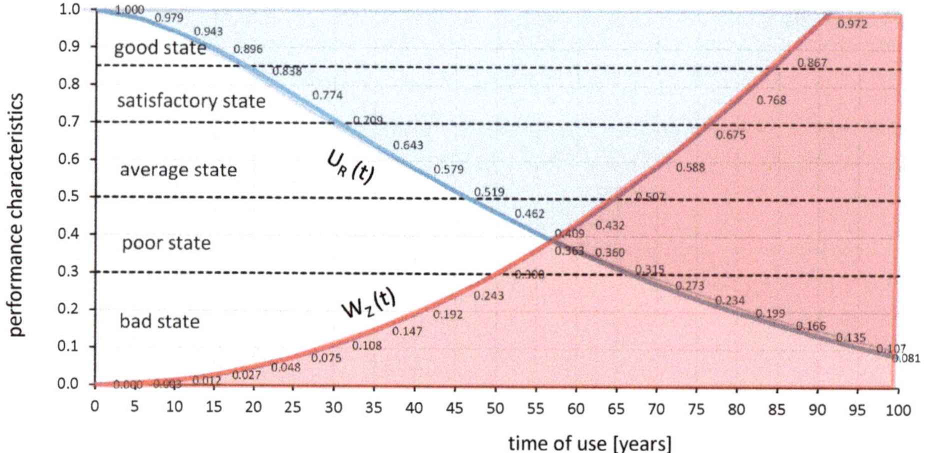

Figure 2. Decrease in use value and wear value of a building over the course of its use.

The difference between the values of the technical wear $W_Z(t)$ and the value of the decrease in the performance characteristics $U_R(t)$ determines how much higher the costs of renovation are from the value of building wear:

$$K(t) = U_R(t) - W_Z(t) \tag{11}$$

The optimization task comes down to seeking the aim function, which is the extreme of the function that is the efficiency of the renovation date $E_R(t)$:

$$E_R(t) = \frac{dK(t)}{dt} = 0 \tag{12}$$

Figure 3 shows a graph of the function of the efficiency of the date of renovation. The date at which function $K(t)$ reaches the maximum value is the most cost-effective for renovation. The derivative of the function is equal to zero. The date t_K is the period when it is most worth carrying out complete renovation.

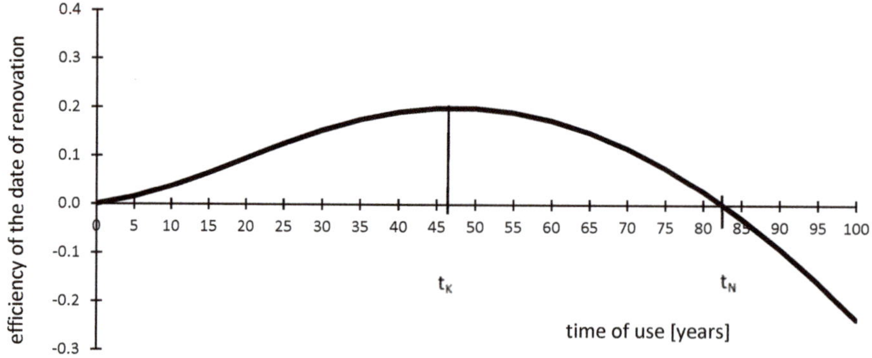

Figure 3. Function of the efficiency of renovation time $K(t)$.

After the period t_K, both the costs of restoring performance characteristics as well as the values of technical wear of a building are high. With the lack of renovation works, the building undergoes degradation, with the costs of renovation works continuously increasing. The deadline for carrying out renovation works is the period t_N, after which renovation is not economically efficient.

For the examined example, a non-renovated building made in traditional technology, it turned out that after 47 years, both the costs of restoring performance and the value of technical wear are high. Renovation is economically unprofitable if the building has not been renovated for 82 years.

Negative values of the K(t) function indicate the fact that long-term lack of renovation works is the reason behind high wear and, at the same time, low performance characteristics. Negative values of the K(t) signify the necessity of demolishing the building due to the unprofitability of renovation.

6. Conclusions

The basic task over the course of maintaining a building is the prospective planning of renovation works and predicting the resources necessary for their execution. The dates of renovation works ought to be planned in accordance with the guidelines of rational management. The proposed method may be helpful when planning renovation works on buildings.

The proposed model allows for the proper estimation of the costs of renovation and repair works and their relationship with the value of the technical wear of a building at each period of its use. Residential buildings constructed using traditional technologies were subjected to analysis. In the carried out analysis, the PRRD (prediction of reliability according to Rayleigh distribution) model of changes in the performance characteristics was applied.

The conclusions resulting from the conducted analysis are the basis for further research on the costs of restoring performance. The value of technical wear signifies the cost recreating the level of wear. Changes in the performance characteristics of a building over the course of its use expressed by cost are the loss of performance characteristics. It might seem that the consumption values and the cost of renovation works are at a similar level, but in practice, it turns out that the cost of restoring the functional properties of the building is different from the value of the building's consumption. And this is the essence of the problem presented.

Anticipating the aging process is essential when maintaining a residential building. However, the problem remains unrecognized in many respects. No data indicate the need for research and analysis related to this topic. Building degradation curves are just the beginning of the research. Forecasting damage to a building can be developed and predicting changes in operating costs over the entire life cycle of a building requires many simplifications. Further research will cover the subject of LCA, the impact of renovation works on the environment. It will also analyze changes in prices of building materials and renovation works over time in Poland and in other countries.

Author Contributions: Conceptualization, B.N., J.K.; methodology, B.N., J.K.; formal analysis, B.N., J.K.; investigation, B.N., J.K.; resources, B.N., J.K.; data curation, B.N., J.K.; writing—original draft preparation, B.N., J.K.; writing—review and editing, B.N., J.K. All authors have read and agreed to the published version of the manuscript.

Funding: This research received no external funding.

Conflicts of Interest: The authors declare no conflict of interest.

References

1. Bucoń, R.; Sobotka, A. Decision-making model for choosing residential building repair variants. *J. Civ. Eng. Manag.* **2015**, *21*, 893–901. [CrossRef]
2. Nowogońska, B. Diagnoses in the Aging Process of Residential Buildings Constructed Using Traditional Technology. *Buildings* **2019**, *9*, 126. [CrossRef]
3. Prieto, A.J.; Vásquez, V.; Silva, A.; Horn, A.; Alejandre, F.J.; Macías-Bernal, J.M. Protection value and functional service life of heritage timber buildings. *Build. Res. Inf.* **2019**, *47*, 567–584. [CrossRef]

4. Lacasse, M.A. Advances in service life prediction—An overview of durability and methods of service life prediction for non-structural building components. In Proceedings of the Annual Australasian Corrosion Association Conference, Wellington, New Zealand, 16–19 November 2008; pp. 1–13.

5. Silva, A.; de Brito, J.; Gaspar, P.L. *Methodologies for Service Life Prediction of Buildings*; Springer International Publishing: New York, NY, USA, 2016; Volume VII, p. 432. [CrossRef]

6. Bucoń, R. Model supporting decisions on renovation and modernization of public utility buildings. *Open Eng.* **2019**, *1*, 178–185. [CrossRef]

7. Zavadskas, E.; Antuchevičienė, J.; Kapliński, O. Multi-criteria decision making in civil engineering: Part I—A state-of-the-art survey. *Eng. Struct. Technol.* **2016**, *7*, 103–113. [CrossRef]

8. Moretti, N.; Re Cecconi, F. A Cross-Domain Decision Support System to Optimize Building Maintenance. *Buildings* **2019**, *9*, 161. [CrossRef]

9. Ortega, L.; Serrano, B.; Fran, J.M. Proposed method of estimating the service life of building envelopes. *Revis. Constr.* **2015**, *14*, 60–68. [CrossRef]

10. Prieto, A.J.; Silva, A. Service life prediction and environmental exposure conditions of timber claddings in South Chile. *Build. Res. Inf.* **2019**, *48*, 191–206. [CrossRef]

11. Paulo, P.; Branco, F.; Brito, J.; Silva, A. Buildings Life—The use of genetic algorithms for maintenance plan optimization. *J. Clean. Prod.* **2016**, *121*, 84–98. [CrossRef]

12. Matulionis, R.C.; Freitag, J.C. *Preventive Maintenance of Buildings*; Van Nostrand Reinhold: New York, NY, USA, 1990.

13. Colen, I.F.; de Brito, J. Discussion of proactive maintenance strategies in facade's coatings of social hoisin. *J. Build. Apprais.* **2010**, *5*, 223–240. [CrossRef]

14. Vanier, D.; Tesfamariam, S.; Sadiq, R.; Lounis, Z. Decision models to prioritize maintenance and renewal alternatives. In Proceedings of the Joint International Conference on Computing and Decision Making in Civil and Building Engineering, Montréal, QC, Canada, 14–16 June 2006; pp. 2594–2604.

15. Biolek, V.; Hanák, T. LCC Estimation Model: A Construction Material Perspective. *Buildings* **2019**, *9*, 182. [CrossRef]

16. Yi-Kai, J.; Nai-Pin, H. BIM-Based Approach to Simulate Building Adaptive Performance and Life Cycle Costs for an Open Building Design. *Appl. Sci.* **2017**, *7*, 837. [CrossRef]

17. Paulo, P.; Branco, F.; Brito, J. Buildings Life: A building management system. *Struct. Infrastruct. Eng.* **2014**, *10*, 388–397. [CrossRef]

18. Frangopol, D.M.; Lin, K.-Y.; Estes, A.C. Life-cycle cost design of deteriorating structures. *J. Struct. Eng. ASCE* **1997**, *123*, 1390–1401. [CrossRef]

19. Anysz, H.; Krzemiński, M. Cost approach to the flow-shop construction scheduling. In *the E3S Web of Conferences, Proceedings of International Science Conference SPbWOSCE-2018, Petersburg, Russia, 10–12 December 2018*; EDP Sciences: Les Ulis, France, 2018; p. 02048. [CrossRef]

20. Drozd, W.; Leśniak, A. Ecological Wall Systems as an Element of Sustainable Development—Cost Issues. *Sustainability* **2018**, *10*, 2234. [CrossRef]

21. Leśniak, A.; Zima, K. Cost Calculation of Construction Projects Including Sustainability Factors Using the Case Based Reasoning (CBR) Method. *Sustainability* **2018**, *10*, 1608. [CrossRef]

22. Flores-Colen, I.; de Brito, J.; Freitas, V. Discussion of Criteria for Prioritization of Predictive Maintenance of Building Façades: Survey of 30 Experts. *J. Perform. Constr. Facil.* **2009**, *24*, 337–344. [CrossRef]

23. Flanagan, R.; Kendell, A.; Norman, G.; Robinson, G.D. Life cycle costing and risk management. *Constr. Manag. Econ.* **1987**, *5*, S53–S71. [CrossRef]

24. Macedo, M.; de Brito, J.; Silva, A.; Oliveira Cruz, C. Design of an Insurance Policy Model Applied to Natural Stone Facade Claddings. *Buildings* **2019**, *9*, 111. [CrossRef]

25. Wieczorek, D.; Plebankiewicz, E.; Zima, K. Model estimation of the whole life cost of a building with respect to risk factors. *Technol. Econ. Dev. Econ.* **2019**, *25*, 20–38. [CrossRef]

26. Vanier, D.J.; Lacasse, M.A. BELCAM project: Service life, durability, and asset management research. In Proceedings of the 7th International Conference on Durability of Building Materials and Components, Stockholm, Sweden, 19–23 May 1996; pp. 848–856.

27. Lounis, Z.; Vanier, D.J.; Lacasse, M.A. A discrete stochastic model for performance prediction of roofing systems. In Proceedings of the CIB World Congress, Gävle, Sweden, 7–12 June 1998; pp. 305–313.

28. Christen, M.; Schroeder, J.; Wallbaum, H. Evaluation of strategic building maintenance and refurbishment budgeting method Schroeder. *Int. J. Strateg. Prop. Manag.* **2014**, *18*, 393–406. [CrossRef]

29. Rivera-Gómez, H.; Oscar Montaño-Arango, O.; Corona-Armenta, J.R.; Garnica-González, J.; Hernández-Gress, E.S.; Barragán-Vite, I. Production and Maintenance Planning for a Deteriorating System with Operation-Dependent Defectives. *Appl. Sci.* **2018**, *8*, 165. [CrossRef]

30. Rudbeck, K. *Methods for Designing Building Envelope Components Prepared for Repair and Maintenance*; Department of Buildings and Energy, Technical University of Denmark: Lyngby, Denmark, 1999; p. R-035.

31. Morelli, M.; Lacasse, M.A. A systematic methodology for design of retrofit actions with longevity. *J. Build. Phys.* **2019**, *42*, 4. [CrossRef]

32. Alshubbak, A.; Pellicer, E.; Catala, J.; Teixeira, J. A Model for identifying owner's needs in the building life cycle. *J. Civ. Eng. Manag.* **2015**, *21*, 1046–1060. [CrossRef]

33. Chen, C.J.; Juan, Y.K.; Hsu, Y.H. Developing a systematic approach to evaluate and predict building service life. *J. Civ. Eng. Manag.* **2017**, *23*, 890–901. [CrossRef]

34. ISO 7162:1992 Performance Standards in Building—Contents and Format of Standards for Evaluation of Performance. Available online: https://www.iso.org/standard/13758.html (accessed on 8 January 2020).

35. ISO 19208:2016 Framework for Specifying Performance in Buildings. Available online: https://www.iso.org/standard/63999.html (accessed on 8 January 2020).

36. ISO 15686-2:2012 Buildings and Constructed Assets—Service Life Planning—Part 2: Service Life Prediction Procedures. Available online: https://www.iso.org/standard/51826.html (accessed on 8 January 2020).

37. Nowogońska, B. The Method of Predicting the Extent of Changes in the Performance Characteristics of Residential Buildings. *Arch. Civ. Eng.* **2019**, *65*, 81–89. [CrossRef]

38. Korentz, J.; Nowogońska, B. Assessment of the life cycle of masonry walls in residential buildings. In Proceedings of the Environmental Challenges in Civil Engineering ECCE Opole, MATEC Web of Conferences, Opole, Poland, 23–25 April 2018; Volume 174. [CrossRef]

Article

Suitability of Eye Tracking in Assessing the Visual Perception of Architecture—A Case Study Concerning Selected Projects Located in Cologne

Małgorzata Lisińska-Kuśnierz [1] and Michał Krupa [2,*]

[1] College of Management and Quality Science Cracow, University of Economics, Kraków 31-510, Poland; liskusm@uek.krakow.pl
[2] Faculty of Architecture, Cracow University of Technology, Kraków 30-084, Poland
* Correspondence: michal.krupa@pk.edu.pl; Tel.: +48-12-6282430

Received: 23 December 2019; Accepted: 22 January 2020; Published: 26 January 2020

Abstract: This article discusses the visual perception of selected buildings located in the historic centre of Cologne, Germany, that have been designed by outstanding architects. It presents eye-tracking research, both from a theoretical perspective and that of its application potential in, among other fields, psychology, management, architecture and urban planning. It also presents an experiment which was performed to evaluate the suitability of eye tracking in the assessment of the visual perception of architecture and its surroundings, utilising the case study method and members of Generation Z as the subject population. Analysis of the experiment's results enabled the authors to formulate commentary on findings concerning typically observed attractors and distractors in the perception of architecture and its surroundings depending on context-specific conditions. The study provided evidence of the suitability of eye tracking in the assessment of the visual perception of works of architecture and indicated the possibility of continuing research concerning the assessment and shaping of the state of awareness and knowledge of architecture and urban planning, which can significantly affect public participation in urban governance.

Keywords: eye tracking; visual perception; the architecture of Cologne; case study; application in architecture and management

---

1. Introduction

The problem of objectively studying the scope of the perception of works of architecture by persons who look at them is important from the point of view of the proper governance of space and the zones located within it [1–4]. Knowledge about various types of perception and the evaluation of structures and spaces by professionals or designers and persons with no architectural education can also be helpful in educating new architectural design personnel, as well as—in a broader sense—being a part of the management of knowledge about architecture and urban planning [5,6].

The aim of this work is to present the utility of video-oculographic studies in assessing the visual perception of architecture, dependent on the type of space and external stimuli.

An eye-tracking-based experiment was performed on two selected buildings located in the historical centre of Cologne that were designed by world-famous architects, which made it possible to analyse perception and visual cognition. This experiment has been presented in the form of a case study and is an element of studying and analysing the contemporary architecture of Cologne.

Video-oculographic studies currently find practical application in a wide range of marketing, market and utility studies [7–9]. Concerning the fact that vision and the cognitive processes associated with it occur almost always and everywhere, eye-tracking studies have increasingly become a part of studies of many areas of life. They are recommended for use in areas like category management, traditional and online advertisement, information technology (IT) system ergonomics, human–machine interaction, information management systems, medicine and psychology, education, sports, entertainment and the military [10–12]. Studies of the perception of art and architecture have also recently started to make use of them [13,14].

The use of eye tracking in this new field of use should allow a relatively objective determination of the form of perceiving various types of information that reach us as humans and that affect how we build assessments and the structure of our knowledge and awareness [3].

2. The Essence of Eye-Tracking Studies

The subject of the video-oculographic method has been discussed in works presenting research assumptions concerning the experiment and the considerations indicating its application potential. The most important items of the literature on the subject include publications by D. Richardson [15], G.D.M. Underwood [16], A. Duchowski [17], as well as Z. Hoolmovist, M. Nystrom, R. Anderson et al. [18], which together form a compendium of knowledge on the method. Other notable publications include those by A. Bojko [19]; as well as those edited by M. Horsley, M. Eliot, B. Knight and R. Reilly [20]; and by J. Nielsen and K. Pernice [21].

The subject matter of the possible application of the method in studies of architecture and spaces has recently been discussed by the following teams: Ch. Lebrun, A. Sussman, W. Crolius, G. van der Linde from the Institute for Human Centered Design in Boston [22]; D. Junker and Ch. Nollen from the University of Applied Science in Osnabruck [23]; Z. Zou and S. Ergan; as well as A. Radwan from New York University [24,25]; L. Dupont, K. Ooms, A. Duchowski and V. Van Eetvelde from Ghent University [26,27], R. Noland, M. Weiner, D. Goo, M. Cook and A. Nelessen from State University of New Jersey [28], J. Hollander from Tufts University [5] and by M. Rusnak, W. Fikus and J. Szewczyk from the Faculty of Architecture of the Wrocław University of Technology [1,29], in addition to the authors of this article [3,4,30].

Previous studies focusing on the suitability of eye tracking in architecture, urban planning and landscape architecture performed by other research teams focused on three main aspects, namely: the use of eye tracking itself, coupled with a stationary or mobile eyetracker, and the use of eye tracking in combination with tools from other research methods. Those that have been found to be the most numerous in the literature are studies presenting results based only on the use of eye tracking with the implementation of stationary devices to investigate the visual perception of works of architecture and landscape architecture.

These, among others, include an experiment performed by a team of researchers focusing on architecture, interior architecture and cognitive science from the Institute of Human-Centred Design in Boston, who used a stationary device and focused on the visual perception of architecture and its surroundings in Boston. The experiment was performed on volunteers from different professions and of various ages, while the object of research were photographs of buildings and their interiors, either individually, or with persons present. The results pointed to a varied visual perception of buildings, often independent of their type, depending on the presence of human figures, their faces and other elements of the landscape. According to the publication's authors, eye tracking is a very good method in allowing scholars to understand the visual aspects of experiencing architecture by persons who are not architects. Furthermore, it pointed to the justification of obtaining knowledge about how an architectural design communicates with the public, the client and professionals, particularly when viewers see buildings with people present around said buildings or the buildings alone. The authors pointed to the utility of this knowledge in teaching architectural theory, the history of architecture and architectural design [22].

Studies conducted at the Faculty of Architecture of the Wrocław University of Technology focused on determining the utility of this method in investigating the visual perception of historical structures. According to the authors of this publication, knowledge based on video-oculographic study findings could aid in facilitating the objectivisation of historical zone evaluation, and would make it easier to manage them [1]. Another study, which was performed using a stationary eyetracker, also pertained to the visual perception of the interior of a Gothic church depending on its height and depth. The experiment focused on depth perception in modelled interiors and changes in interest in reading depth relative to increases in the cathedral's height. The results made it possible to conclude that the change in a layout's length causes more complex consequences in matters of depth perception than merely changes in nave height. The study pointed out that more in-depth research using other methods is justified [29].

Studies performed at another Polish research facility, namely the Faculty of Architecture of the Cracow University of Technology, performed using a stationary device, concerned the perception of selected historical buildings and the space of the Rabka-Zdrój health resort in Poland. The study was aimed at determining the scope and manner of perception of buildings of high cultural significance that suffer from decay. Its findings have confirmed the effect of perceptual competition between the details of the buildings and historical and contemporary spaces. According to the study's author, the focus on and perception of these details instead of entire buildings can be the reason for a lack of valuation of the perceived surroundings [3]. Further studies have enabled the detection of the strong impact of various types of advertisements and information boards on the disruption of a building's perception. The authors demonstrated the utility of video-oculographic studies in formulating guidelines and planning measures associated with the protection of heritage sites and conducting education efforts [30].

Another type of study conducted using stationary devices were those of the impact of the level of urbanisation on the landscape presented on photographs on the visual exploration of images by viewers. The experiments, conducted at Ghent University, concerned the assessment of the visual perception of various landscapes, ranging from rural to urban ones, as seen on photographs. More extensive and scattered exploration was observed in more urbanised landscapes. In poorly urbanised landscapes, fixations were more focused. Meanwhile, when no buildings were visible on the photograph, unexpectedly broad exploration was observed. The results of this study provided evidence for the conclusion that the level of urbanisation is positively correlated with visual complexity, as indicated by its potential impact on the viewer's behaviour [26]. Furthermore, studies concerning the use of significance maps, which are theoretical prognoses of the pattern of human vision, with the aim of comparing the visibility of various designs of simulated constructs placed on photographs of original landscapes, have been performed. The results of the experiment, in the form of a high correlation of significance maps with human focus maps, made it possible to formulate conclusions as to the suitability of eye tracking and the significance maps themselves in planning structures within the landscape, as it was concluded that visual impact is lower when the visual perception of a structure decreases and an optimal integration of a structure with the existing landscape can be achieved [27].

The second aspect of previous studies was the sole use of eye tracking through mobile devices. Studies of this type were performed at, among other places, Tufts University and New York University, and focused on the impact of urban environments on the mental states of people present within them. Experiments performed using a mobile eye tracker made it possible to identify urban environments associated with more positive reactions, suggesting a feeling of relaxation and the desire to spend time there. The authors of the publication also pointed to the significance of the study and the detection of such environments as a part of formulating new principles of urban design [5,24,25].

Studies performed at the University of Applied Science in Osnabrück, which similarly utilised a mobile device, involved long-term experiments in real-world open urban environments ("Grosser Garten" in Hanover and "Stourhead" in Wiltshire). The experiments made it possible to prove the suitability of this method in studies of open space with the purpose of obtaining knowledge about how people behave and react and how they enter said reactions. This knowledge is particularly noteworthy and can be useful in designing the best possible spaces, consequently enabling quality of life improvements. The research team behind the study pointed to the purposefulness of conducting holistic research, e.g., by using a combination of eye tracking with other methods. According to the authors, a combination of eye tracking with electroencephalography (EEG) along with mobile measurements intended to record interactions in detail and analyse the reactions of the human body, could prove particularly useful. Knowledge obtained through such studies should form a repository of data concerning user requirements and provide utility in the planning and management of green areas and public buildings, while taking subjective feelings of safety and contemporary aesthetic tastes into account [23].

The third aspect of eye tracking studies is using them in conjunction with or by using other research tools. One example of such studies are experiments performed at the State University of New Jersey. As a part of the said experiment, qualitative survey studies concerning visual preferences were performed and were followed by eye-tracking experiments. The survey study focused on the perception and assessment of various urban objects placed in green surroundings, amidst pedestrian traffic and in proximity of public transport. Eye-tracking studies with the use of mobile devices have made it possible to expand the knowledge declared by the subjects with an objective image of their perception of this environment. According to the publication's authors, studies following this methodology provide knowledge that is necessary to urban and transport planners and municipal governing bodies, allowing them to improve the functioning of the city, e.g., by increasing pedestrian activity and introducing vehicular traffic constraints in cities [28].

Analysis of the state of the art in terms of the application of eye tracking in studies of architecture revealed significant factual discrepancies between individual studies. The experiments that have thus far been performed concern the utility of this method in solving specific scientific problems that are the focus of the given research teams. Therefore, it can be said the studies are selective in nature. Regardless of the context of the method's application, all completed experiments point to its significant application potential in architecture, urban planning and landscape architecture. The authors wish to highlight the significance of the findings concerning the visual perception of various works of architecture, both historical and contemporary ones, located in various spaces, landscapes and surrounded by people and various forms of technical infrastructure. The findings are based on analysis of so-called descriptive statistic parameters. Despite all of the publication's authors declaring an awareness of the need for further studies in broader, more comprehensive perspectives, they are of the opinion that their findings provide research material that can prove highly useful in formulating assumptions and measures concerning the planning, design and construction of contemporary architecture, as well as protecting and managing heritage sites. Furthermore, in the opinion of the authors, the presented findings should be used in educating future architectural staff and social education efforts.

In the literature, video-oculography is presented as a set of research and study techniques used to measure, record and analyse data concerning the position and motion of the eye. It supplies quantitative measurement data without referring to subjective, verbal reactions of the subject, instead referring to psychophysical and neuropsychological processes that accompany the collection and processing of visual information and oculomotor reactions to stimuli received from the environment. Eye tracking and visual perception are fundamentally interlinked.

In cognitive psychology (U. Neisser), perception is understood as a process of rationality and abstraction. Abstraction activities are present at two levels of perception: first taking place after the sensory reception phase. The sensory reception that precedes it is equated with the initial process of passive information gathering, during which input is detected in receptors ("burned" in the photoreceptors of the retina). Reception inaugurates the process of information processing. In the case of visual perception, during this stage it is already possible to indicate so-called property detectors, i.e., neurons that selectively react to lines with a specific spatial orientation, or the general outline of the human face. It is the second stage of experiencing that is considered rational, and which involves the active process that follows reception. It is based on interpreting sensory data using contextual suggestions, attitude and previously gained knowledge. Perception involves activities such as discrimination, recognition, orientation or perceptual categorisation. Rationality is present during this stage of higher-order cognition, which is equated with the capacity to think, distinguish certain common traits in objects at the cost of ignoring others (either indistinct or non-general traits), which are then used to form generalisations in creating cognitive representations. It can be said that in U. Neisser's cognitive psychology, which is derived from experimental psychology, references are made to the notion of cognition as construing and creating knowledge. Meanwhile, in the concept of "visual thinking" or "thinking with images" (R. Arnheim), it is assumed that perception processes feature rational principles that govern the seeing and imaging of an object. Here, this notion is seen as an equivalent to the image and treated as a word. The language of images is less arbitrary and wealthy than the language of words, featuring more analogies and non-isomorphic relations between the sign and the object. In this concept, the geometric shape is seen as one of the most stable notions of the language of images. In Gestalt theory, it is assumed that sensory experience and cognition are not based on the passive reception of individual stimuli, but on the creative perception of a certain whole—the image of an object as a whole that is not reduced to the sum of its parts. The object of cognition is also its construct—a creation dependent on such factors like one's memory, experience, knowledge, attitude or desire. Principles of extracting the whole or the figure from the background (spatial proximity, similarity, good figure, symmetry), the principles of simplicity and the illusory character of perception (illusion is an essential component of images and gives them continuity) function here. In summary, it can be concluded that there are many different concepts of visual perception. This model of perception is referenced by some theories from the field of philosophy and psychology, namely: Gestalt psychology and other psychologies of aesthetic perception, such as psycho-aesthetics, neuro-aesthetics or visual psychology, and philosophical epistemology in part. The aforementioned disciplines are a part of broader theoretical and experimental perception studies [31–36].

The following definition of visual perception was adopted for the purposes of the experiment.

Visual perception is a complex cognitive process based on the interpretation of objects, phenomena and processes in the environment based on specific stimuli picked up by the visual system. Receiving a stimulus begins the process of perception, which makes it possible for one to understand what has been seen and to implement the information one has received within one's system of knowledge and values and to memorise it. Perception is conditioned by the aesthetic and artistic elements of an exposition and by the individual characteristics of the person performing the observation [18,19,29,30].

The use of eye tracking began to see wider methodological application in the second half of the twentieth century, along with the development of academic disciplines and specialisations such as psychology, cognitive science or human-computer interactions. The period towards the end of the twentieth century saw the technological development of tools enabling the manufacture of small, mobile devices, along with the development of applications for computing data, obtaining specific results and their presentation and interpretation. At present, video-oculographic studies are typically based on using a system of video cameras placed near the subject's eyes or at a close proximity to their face. During tests, the cameras track the movement of the subject's eyes and their video feed is recorded by a computer and analysed using specialist software. We can extract a lot of useful information from this data, including which elements attracted the attention of the subject and after what length of time; which element attracted the subject's attention the longest and which elements were observed repeatedly, what is the direction of the sequence of scanning space or whether the subject was confused or showed interest [15,18,23,37].

A typical eye-tracking measurement is based on recording two types of information:

- Fixations, which are points that the subject looked at. Visual information is collected during fixation. Knowing where the subject looks (the locations they fixate their eyes on) helps in identifying what they noticed. These locations are recorded in the form of dots of varying size, with the size of each dot denoting the duration of the fixation.
- Saccades, which are eye movements from one fixation to another. During a saccade (20–40 milliseconds), the brain records no visual information. Saccades are recorded in the form of lines that connect dots (fixations) [17,19].

The most often used forms of the graphical presentation of data obtained during tests include: heat maps, gaze plots and area of interest analyses.

The heat map (or hotspot map) makes it possible to determine which element attracted the subject's attention. In the case of each of the materials presented on-screen, it is possible to display the points that the subject fixated their eyes on, presenting summary attention focus results for each subject group. A longer fixation time is marked by a more intense warm colour, while cool colours denote a shorter focus time. Places without colour denote fragments that were completely ignored by the subjects. A particular case of heat map is its inverted version, called the focus map, which shows only those areas the subjects fixated their eyes on, with the remaining areas blacked out [18,20,32].

The second form of graphical presentation is the gaze plot, which indicates the sequence of fixations on individual areas during the observation of the presented image. Circles are used to mark each gaze point (fixation). The longer the subject looked at a given point, the larger the diameter of the corresponding circle. The number presented inside the circle shows the sequence in which it was observed, while lines symbolise saccades, presenting the path that the subject's gaze travelled between fixations.

The third form of graphical presentation is the area of interest (AOI). Here, it is possible to separate a large number of gazes that concern distinguishable areas presented on-screen. AOIs can themselves be individually designed by the person designing the study or generated automatically, with a recorded attention distribution percentage. The advantage of using areas of interest over heat maps is the possibility of obtaining specific numerical values that enable a more precise quantitative analysis of fixations and the use of parametric metrics. The so called statistics in use here are different from study to study and depend on their objective [17].

When analysing areas of interest, the following measurements are typically taken, among others:

- time to first fixation; this makes it possible to determine how much time subjects require to find a given area that is significant for the study.
- the number of fixations within a given area of interest during observation for one subject and for all subjects; it is assumed that a larger number of fixations indicates greater interest in an area;
- total time of all fixations on a given area for a single subject and for all subjects;
- the number of persons who made at least 1 fixation relative to the number of study participants; it is assumed that the greater the percentage of these persons is, the more attractive the area is to the subjects;
- the number of revisits to a given area of interest during observation for one subject and for all subjects; it is assumed that the greater the number of visits to an area, the more interesting it appears to the viewer (it can represent interest in novelty or content that is difficult, hence the revisit to help understand the information) [38,39].

After outlining areas of interest, every area or image used in the study typically shows areas that are unclassified, or not an AOI. These areas, although uninteresting from the point of view of the degree of perception of objects that make up AOIs, can also be included in the analysis from the point of view of the presence of other attractors (elements that attract attention) or distractors (elements that distract one's attention) [18].

In conclusion, the main advantage of video-oculographic studies is that they allow one to study the perception activity of test subjects objectively. Eye tracking makes it possible to pinpoint those elements of the image of an analysed object that the observer actually looks at. Therefore, results are based on facts, instead of declarations or conjecture [3].

3. Research Material, Methods Used in the Experiment, and the Course of the Experiment

The primary objective of the study was to determine the scope of utility of eye-tracking research in relatively objectively assessing the visual perception of space and the works of architecture within it. Mindful of the results of experiments concerning the use of this method, performed by various research teams and the authors themselves for the purposes of studying architecture, urban layouts and landscape architecture, the authors assumed that one can describe and then attempt to define mechanisms of perception that are dependent on external factors through appropriately planned experiments. Such measures are necessary to create repositories of data concerning the perception of various objects and buildings and to describe and spread good practices so as to modify the process of building design from the point of view of improving the quality of life of their users. The authors also assumed that case studies can prove to be the appropriate research method for such experiments. This method is widely used in studies in the social sciences (psychology, management) and can be considered useful in this experiment as, based on an analysis of an individual case with its detailed description, it enables one to draw conclusions as to the causes and results of its course and concerning a certain model of behaviour and any observed technical, cultural and social determinants. Furthermore, in methodological terms, it can include various quantitative and qualitative research techniques [40,41].

The case study included the following stages:

- preparation of the experiment;
- selection of research material for analysis;
- selection of the target population;
- performance of eye-tracking tests with the participation of test subjects;
- calculating the sequence and scope of the analysis of generated quantitative and qualitative data;
- research findings presentation and commentary.

The preparation of the experiment included a pilot study intended to perform an initial validation of research assumptions. Due to the experiment comprising a portion of studies and analyses of contemporary Cologne, 10 photographs of the city's architecture were taken, featuring various views of buildings within space. The photographs, with a vertical and horizontal orientation, presented architecture from up close and from afar, with various technical infrastructure and surrounded by greenery, as well as with persons present in the view and without the surrounding landscape. These views of architecture were selected and based on an analysis of studies presented in publications by Ch. Lebrun et al. [22], R. Noland et al. [28], L. Dupont et al. [26,27], L. Garcia Moruno et al. [42] and the authors of [3,30]. In terms of selecting the target population of the experiment, a random selection of volunteers declaring no visual impairments was chosen for the pilot study. According to recommendations concerning the number of subjects presented by A. Duchowski [17], Z. Hoolmovist et al. [18], J. Nielsen and K. Pernice [21], and Tobii [39], the company that supplied the stationary equipment, as well as taking into consideration experiments performed previously by the aforementioned research teams, 30 subjects were chosen. It was assumed that the time of exposure to the photographs on a computer screen would be between 10 to 15 s. It was assumed that results generated in the form of heat maps, focus maps and gaze plots of around half a dozen subjects for each photograph should be sufficient material for analysis during the pilot study.

The pilot study was carried out according to plan and the analysis of its findings made it possible to make appropriate decisions concerning the preparation of the main experiment. It was assumed that, in light of the use of case studies, there was no need to present a large number of photographs to the subjects. Three photographs of works of architecture located in the historical centre of Cologne were taken: one with a purely random view and two with a pre-determined selection of buildings and their surroundings. The first photograph was not intended to be used as research material, as it was meant to stimulate the focus of the experiment's subjects and lower the level of emotion associated with their initial impressions. The two photographs depicting works of architecture built as designed by award-winning and world-famous architects were the essential research material. One photograph showed a vertically-oriented frame, while the other showed a horizontal one, with the buildings being shown either closer or farther away from the camera and were thus seen from different perspectives. There were no people visible in the two photographs, with exposed elements including greenery and public space with its accompanying technical infrastructure, cars and public transport vehicles. The elimination of the presence of people, who are considered to be perceptual distractors, was assumed to allow for an analysis of other distractors within the surroundings of historical buildings.

The presence of greenery could allow a confrontation of the eye-tracking experiment's findings with those of other studies concerning landscape architecture [43]. Concerning the target population of the experiment, it was decided to increase considerably the number of subjects taking part in the experiment relative to the number recommended in the literature and to the number of persons who participated in the pilot study. After deciding on a population count of 100, the authors decided that the population type should be more homogenous in terms of age and to exclude persons with an architectural education. It was assumed that persons belonging to generation Z could be a good population choice in light of the experiment's objective. The recruitment of volunteers was performed by telephone, with two assumptions regarding age and educational background. The time of exposure to each photograph displayed on-screen was 10 seconds. Furthermore, the authors assumed that data would be generated by the eye-tracking system for pre-determined areas of interest.

The research material prepared in accordance with guidelines that had been formulated after the pilot study encompassed two visualisations of works of architecture in the space of the historical centre of Cologne, Germany.

The first building to be studied was the "Kolumba" museum, designed by the well-known Swiss architect and Pritzker prize [44] laureate Peter Zumthor [45].

The museum is located in the historic centre of the city and occupies a part of the urban block delineated by the following streets: Brückenstraße, Kolumbastraße, Ludwigstraße and Minoritenstraße. Its vicinity is primarily composed of multi-storey housing and office buildings with commercial spaces on the ground floors. It is of varied cultural value as only some of the buildings near the museum are actual historic townhouses. The townhouse at 17 Brückenstraße is one such building. The ground level of its street-facing facade is composed of large windows that are highlighted and separated by pilasters that are slightly extruded from the facade's surface. These pilasters are crowned by reliefs depicting human figures. The aforementioned windows fulfil the role of storefronts. The first-floor level, similarly to the ground floor, has a notable rhythm of window openings (in a shape more or less resembling elongated rectangles), which form windows or porte-fenêtres providing access to balconies. The upper storeys are much less architecturally varied. They feature narrow windows that are rhythmically placed along the entire width of the facade.

The part of the city under analysis does not have many green spaces. In this context, the green square near the museum gains a particular significance. It is located at the intersection of Brückenstraße and Ludwigstraße.

The museum building was built in 2007 and incorporated the adaptive reuse of the ruins of the medieval church of St. Kolumba that was destroyed during a bombing raid of the city in 1943, along with a bold contemporary design above them [46].

The main idea devised by the architect was to incorporate the museum into the extant context and to continue the accruing of historical layers. The building's massing is massive, has three storeys and is enclosed in cuboid forms. Notably, the walls of the museum not only feature the relics of the Gothic temple and its predecessors (a Carolingian and a Romanesque church), but also those of a villa from the period of the end of the second and the beginning of the third century CE, which remembers the ancient Roman beginnings of the city.

The facades of the museum were built from bright, elongated, beige-coloured brick, which contrasts with the stone relics of the church of St. Kolumba, which are visible on southern and western facades. Other important elements of the massing include the remains of the Gothic sacristy, which were left in place in accordance with the concept of the "permanent ruin", as an original archaeological reference in the south-western corner of the museum.

The facades were designed to feature few windows, which causes the observer to see them as almost complete solids, interspersed in places with openwork strips. This produces the impression of an austere and minimalist work and does not compete with the well-preserved Gothic relics in the form of openings with pointed arches.

The main entrance to the museum is located in the southern facade. It is formed by a simple, recessed opening [39–41]. A view of a fragment of the museum's facade and that of the historical townhouse and square from the side of Bruckenstrasse and Ludwigstrasse is presented in Figure 1.

Figure 1. View of a fragment of the "Kolumba" museum's facade and of the historical townhouse at 17 Brückenstraβe and the square with greenery near the intersection of Brückenstraβe and Ludwigstraβe in Cologne, Germany (image 1).

The second of the investigated buildings is the "Weltstadthaus" department store designed by the outstanding Italian architect and likewise Prtizker prize [42] laurieae Renzo Piano [43].

The building that houses the facility, similarly to the "Kolumba" museum, is located in the historic city centre. It occupies the rather narrow space between Cäcilienstraße, Antonsgasse and Schildergasse. The building is adjacent to contemporary commercial and office buildings. Most of them were built in the 1970s. Another notable piece of information is that the late Gothic church of St. Anthony is located nearby, to the south-east of Renzo Piano's building.

The building's most important facade faces southwards. It is highlighted thanks to the broad circulation arterial of Cäcilienstraße, which forms an appropriately wide open space in front of it, facilitating observation.

The building, erected in 2005, despite the use of modern materials such as glass and glued-laminated timber which enabled its flowing, organic form, harmoniously blends into the extant cultural context of Cologne's city centre. The "Weltstadthaus" is another "blob" architecture project (a part of so-called "blobism"), which, as a current of postmodernism, is characterised by curved and rounded building forms or "droplet architecture".

The building is composed of two parts. From the west, from the side of Antonsgasse, it is composed of a facade from natural stone with a classical, cubic shape. The northern facade (from the side of Schildergasse) and the western facade were designed in contrast to this simple form. They are composed of organic, rounded forms made from glass curtain walls with high thermal insulation and optical parameters. The form of the building necessitated that the glass panels and profiles had to be manufactured using high-precision CNC (computer or computerized numerical control) technology.

Thanks to the abovementioned measures and the use of then-cutting-edge construction materials and technologies, the "Weltstadthaus" has remained one of the most recognisable buildings of the centre of Cologne [47–52]. A view of a fragment of the "Weltstadthaus" and the church of St. Anthony from the south is presented in Figure 2.

Figure 2. View of a fragment of the "Weltstadthaus" and the church of St. Anthony in Cologne, from the side of the Cäcilienstraße arterial (image 2).

The subject population for the experiment was selected in a deliberate manner. The subjects were selected from among young people belonging to generation Z (persons born after 1995, who were mostly still university students). The distinct characteristics of this generation were the reason behind this decision. Members of generation Z are generally reported to have a realistic outlook on life, while also being creative and ambitious. They have been brought up under the influence of talent shows and iEverything—iPhones and iPods. They also stand out because of their approach to knowledge—they learn by themselves using online sources. They are also characterised by high mobility, their knowledge of foreign languages, their frequent travels and having acquaintances from all across the world. Members of generation Z display a natural approach to online and information and communication technology (ICT) studies. Generation Z is sometimes labelled as Generation C—which stands for connected [53]. It was assumed that young people from Generation Z who studied in Krakow can constitute a test subject population for this experiment. The subject population encompassed 100 individuals within the Generation Z age group, with additional selection criteria such as: the character of their education being unrelated to architecture and a lack of visual impairment. The size of the population was deemed sufficient from the point of view of the study's methodological assumptions, in addition to being much higher than population samples taking part in similar studies from other fields [8,17,20–22,38]. The population's structure in terms of gender was as follows: 66% of the population was composed of women and 34% was composed of men.

Prior to the start of the test, the participants were informed about its course. However, no information about the objective of the study and the manner of analysis of the gaze plot of the subject was divulged. Withholding this information was intentional so as not to suggest areas of interest and attention focus. The study was performed using a stationary Tobii X2-30 Eye Tracker with specialist equipment [39]. Prior to displaying photographs of the selected buildings on a computer screen, the eye-tracking device was calibrated for each subject to adapt it to their eyes. Afterwards, each of the photographs was displayed on the screen for 10 seconds.

The areas that the subjects fixated their eyes on were recorded by the computer and processed using specialist software and could be generated in the form of various types of plots. Due to the objective and goal of the study (two buildings held in high regard by professionals, along with their surrounding space and architectural context), the following sequence and scope of analysis were adopted for each image:

- determining areas of interest and the "unclassified" area;
- the generation of so-called descriptive statistic metrics, i.e., the quantitative data for the selected areas;
- the presentation of cumulative quantitative data in graphical form, as a heat map;
- the generation of quantitative data for a single gaze plot, including the outlined areas, followed by their qualitative graphical presentation.

Three areas of interest were outlined for image 1 with the "Kolumba" museum building (1AOI - formed by the museum building; 1AOI2-a fragment of the historical residential and office townhouse with store windows at 17 Bruckenstrasse 17; 1AOI3-formed by the greenery of the square adjacent to the museum at the intersection of Bruckenstrasse and Ludwigstrasse).

Area of interest 1AOI3 was included because of the possibility of greenery acting as an attractor—drawing in the gaze of the observers. Furthermore, an unclassified area (Not on 1AOI) was included because of the possibility of it featuring distractors, i.e., elements that could distract the subjects, in the form of parked cars and road infrastructure.

For image 2, which features the "Weltstadthaus" commercial building, four areas of interest were defined (2AOI1—the "Weltstadthaus" building itself; 2AOI2—the church of St. Anthony; 2AOI3—the frontage of commercial and office buildings; 2AOI4—the visible greenery). It was assumed that the area of greenery located near the buildings could play the role of an attractor. An unclassified area (Not on 2AOI) was also included, in the form of the Cäcilienstraße circulation arterial, along with vehicles and road infrastructure, which could distract the subjects.

The data generated for the outlined areas—i.e., parameters—included: average time to first fixation, the number of fixations during observation, the average number of fixations per subject, the total duration of all fixations, the number of persons who performed at least 1 fixation relative to the total number of participants, the number of visits during observation and the average number of visits per person. The numerical data that were generated was grouped by subject gender. Apart from heat maps, inverted heat maps were also plotted to be used in distractor analysis.

The data generated for each gaze plot, including the outlined areas, concerned the following parameters: time to first fixation, the number of fixations and their percentage in each area, the total duration of all fixations during observation, the number of visits and the minimum and maximum fixation time. A graphical presentation of eye movement paths was generated based on this data for a representative individual from among the subject population, whose parameters were the closest to the average parameter values for areas associated with buildings of the highest value as rated by professionals.

4. Research Findings and Analysis

Image 1 is oriented vertically. The main building (the "Kolumba" museum), in the context of the analysis, is set back relative to the observer. In the first plane, to the left, we can see the facade of the historical building, while to the right we can see trees. A significant portion of the image's surface area below the view of the museum is occupied by the intersection of Bruckenstrasse and Ludwigstrasse.

The areas of interest and the unclassified area taken into consideration in the assessment of the visual perception of architecture and its surroundings by subjects who observed image 1 has been presented in Figure 3. The parameters describing the process of perceiving these areas have been presented in Table 1.

Figure 3. Areas of interest (AOI) and the unclassified area outlined for image 1, legend: area 1AOI1—pink, area 1AOI2—red, area 1AOI3—green, area Not on 1AOI—blue.

Table 1. Parameters describing the process of the observation of areas outlined on image 1.

Item No.	Parameter	Subject Population	1AOI1	1AOI2	1AOI3	Not on 2AOI
1.	Average time to first fixation (s)	Women	1.84	0.59	4.59	0.45
		Men	1.90	0.72	4.61	0.51
		Total	1.84	0.63	4.60	0.47
2.	Number of fixations during observation	Women	315.00	521.00	106.00	715
		Men	161.00	284.00	68.00	409
		Total	476.00	805.00	174.00	1124
3.	Average number of fixations per person	Women	4 (4.77)	7 (7.89)	1 (1.61)	10 (10.83)
		Men	4 (4.74)	8 (8.35)	2 (2.00)	12 (12.03)
		Total	4 (4.73)	8 (8.54)	2 (2.05)	11 (11.24)
4.	Length of time of all fixations (s)	Women	119.73	193.80	38.34	270.88
		Men	61.29	97.96	23.45	144.83
		Total	181.02	291.76	61.79	415.71
5.	Number of persons who have performed at least 1 fixation relative to the entire population (attractiveness estimator) (%)	Women	98.49	98.49	68.18	100
		Men	100	100	73.53	100
		Total	99	99	69	100
6.	Number of visits during observation	Women	229.00	275.00	81.00	344
		Men	117.00	142.00	52.00	185
		Total	346.00	417.00	133.00	529
7.	Average number of visits (in proportion to number of research participants)	Women	3 (3.47)	4 (4.17)	1 (1.23)	5 (5.21)
		Men	3 (3.44)	4 (4.18)	1 (1.53)	5 (5.44)
		Total	3 (3.54)	4 (4.43)	1 (1.59)	5 (5.29)

Analysis of the generated numerical data concerning average time to first fixation, the number of all fixations during observation and their duration have provided evidence that the unclassified area (Not on 1AOI) was not only noticed the quickest, but was also the most interesting to the observers. The share of the number of fixations on this area amounted to over 43% of all fixations. The area's greatest attractiveness was confirmed not only by the fact that it was noticed by all observers, but that it was also the most explored, as proven by the number of observer revisits to this area. The share of revisits to this area relative to all revisits amounted to 37%. The area representing the space of the streets behind the Kolumba museum from the south-west, up to the urban block outlined by Bruckenstraße and Ludwigstraße along with its infrastructure and parked cars, appeared to be the most interesting to the observers. This area, constituting the surroundings of works of architecture, occupies a significant portion of the photograph. It can be assumed that, similar to the case of studies performed by Ch. Lebrun et al. [22], this space has a significant impact on the perception of works of architecture. Numerical data indicate interest in this area, which features various elements of street infrastructure and standing cars. These elements, as indicated in studies by R. B. Noland et al. [28] and the authors of [30], can dominate the interest of viewers.

The second area shown to attract the greatest attention and focus was area of interest 1AOI2. The fixations on this area constituted 31% of all fixations performed on the entire image. Only one person out of the entire population was uninterested in this area and did not look at it. The share of revisits to this area amount to almost 29% of all revisits made by the viewers to each area. The area outlined by the view of a fragment of the facade of the historical townhouse at 17 Bruckenstraße along with its ground floor commercial section was also interesting to the subjects, although less so. The initially greater interest in this building as an area relative to the building of greater cultural value can be a result of its elements, which are important in cognition theory, such as: its perspective view, spatial character, shape, lines, colour and texture [34,35].

The third area in terms of interest shown by those who viewed the image was area 1AOI1, which is a view of a part of the Kolumba museum's facade from its south. It is the area where we can see the building with the greatest cultural value and acknowledged architectural design. The average time to first fixation on this area was four times as long as the value for the unclassified area. The number of fixations on this area was significantly lower than the values for the previous two areas and constituted only 18% of all fixations. Despite a similar degree of interest to other areas (almost every subject performed at least one fixation), the degree of exploration was lower, at a level of slightly over 25% of all revisits to all areas. The above can be considered proof that elements that form perceptual notions that are significant in perception theory (according to R. Arnheim), such as shape, light, colour and texture did not have as significant of an impact as might be expected [35]. Therefore, it can be stated that the architectural design incorporating the medieval church of St. Kolumba which made it a part of a modern museum building held in high regard by professionals and that has won numerous prizes, along with its visually perceived informative content, was not a particularly strong stimulus to the viewers.

The area that the observers noticed the slowest and the one that attracted the least amount of interest from observers was area 1AOI3. It was formed by the greenery of the small green square adjacent to the Kolumba museum. The average time to first fixation was 10 times as long as in the case of the area spotted as the first (Not on 1AOI). As many as 31% of the viewers did not perform a single fixation on this area, while in the case of those that did, the area was visited only once in most cases. Fixations on this area constituted only around 7% of all fixations performed by those who observed the image, providing evidence of its lack of attractors and distractors relative to the areas featuring works of architecture. The late spotting of greenery, as indicated by average time to first fixation, is theoretically understandable, as studies conducted by, among others, D. Junker and Ch. Nollen and L. Dupont et al. on the perception of greenery among buildings indicate its perception as varied [23,26]. The number of fixations and the low level of exploration indicate a lack of interest in greenery. Therefore, no conclusive evidence of the role of greenery as an estimator of "mystery" to the viewers was found [43].

The statistical analysis of the data generated by the computer system for the entire studied population concerning the degree of their variability (standard deviation and coefficient of variation) showed high variability irrespective of the subject's gender. No correlation was found with the findings of M. Schissel et al. [30] concerning significant gender-dependent differences in perceiving photographs, particularly concerning the much stronger attention focus reportedly displayed by men relative to women. However, the statement that the elements conditioning visual perception were mostly individual character traits, such as aesthetic and artistic sensitivity or one's awareness level [18,19,29,30] was confirmed. Such views are presented by U. Neisser in the concept of cognitive psychology, R. Arnheim in the concept of "visual thinking" and in Gestalt psychology [33,35,36]. This was also the motivation behind analysing average values for quantitative data for such a large population group, whose size significantly exceeded recommendations featured in the literature [8,17,20–22,39].

Heat map no. 4 is a graphical presentation of the places and elements that attracted the attention of all subjects and held it the longest (Figure 4).

Figure 4. Visualisation of results in the form of a heat map for all subjects for image 1.

The heat map in question shows three major hotspots. The first, larger hotspot and the second, smaller hotspot are formed by the surface of the image of cars parked along Bruckenstraße near the Kolumba museum. They attracted and held the gaze of viewers significantly longer than other areas. These red, hot areas extended through yellow areas to extensive green ones. These mutually adjacent areas were fragments of the unclassified area (not on 1AOI). Another clear, yet smaller, focus of viewer attention was also shown near the traffic light infrastructure and on the street surface, forming yellow and green areas. The third major hotspot on the map is a fragment of the facade of the historical building (area 1AOI2), in its storefront-featuring ground-floor section. An area of intense red coloration formed by numerous long fixations on the area of a round advertisement and the lighting above the storefront awning expanded through yellows to fading greens. A number of smaller green-coloured areas can be seen in the commercial section of this townhouse.

No coloured signs of interest were observed for original facade elements such as sculptures above the ground-floor section of the building, which constitute its cultural value, despite greater interest in this building as an area than the museum. Heat maps did not point to the perception of the entirety of the building in accordance with cognition theory's "visual thinking" and Gestalt psychology [35,36].

The area outlined by the facade of the museum (area 1AOI1) showed various clusters of eye attractors, with yellow and green colours. They were present both in the historical section, i.e., the stone remains of the Gothic temple, and in the openwork section of the facade and the part made from bright, contrasting brick. The varied facade, which appeared to attract the attention of viewers, inspired viewer interest in the entirety of the building and not just a few of its elements. The museum's design included numerous attractors for viewers—particularly those of high aesthetic awareness and sensitivity, hence the yellow and green colours. To perform a detailed analysis of distractors, inversed heat maps were prepared (according to a procedure used by Bric Visual Solution) [38]. The inversed heat map confirmed the presence of distractors in the form of an area (a sequence of cars), an advertisement and the lighting above the shop window's awning.

The sequence and length of the observation of the aforementioned areas and objects that are essential in the aspect of the process of visual perception for a selected representative of the studied population has been presented using generated statistical metrics in Table 2, as well as in graphical form, as shown in Figure 5.

Table 2. Parameters characterising the gaze plot for one person relative to the areas outlined for analysis on image 1.

Item No.	Parameter		1AOI1	1AOI2	1AOI3	Not on 1AOI
1.	Time to first fixation (s)		1.95	0.77	2.85	0.12
2.	data	n	5	6	3	10
		%	20.83	25.00	12.50	41.67
3.	Fixation time (s)		1.97	3.47	0.96	2.91
4.	Number of visits in an area		3	4	2	4
5.	Fixation duration range (s)	min	0.267	0.273	0.280	0.225
		max	0.778	0.981	0.327	0.811

Numerical data from Table 2, presented in a corresponding view with Table 1 and supplemented by the minimum and maximum fixation times, enable the analysis of the perception of buildings by a specific person. In this case the population representative was male.

The analysis of statistical metrics of the descriptive heat map for the entire population, as well as the gaze plot for the selected person, provide evidence in favour of stating that the processes of perceiving buildings and spaces were practically analogous. There were some differences that had probably been caused by the individual traits of the person performing the observation, e.g., their state of socio-cultural awareness. The difference between the averaged perception characteristic and that of the selected person was largely rooted in the much quicker fixation on and interest of the selected person in the area outlined by the green square near the museum (1AOI3). The duration of the fixations on this area was close to the average value, but the number of fixations and revisits was greater. Furthermore, no clear difference in the number of fixations on areas with works of architecture (1AOI1 and 1AOI2), as in the average for all subjects, was observed here. Instead, the difference pertained to the duration of all fixations and a greater focus of attention on the historical commercial townhouse.

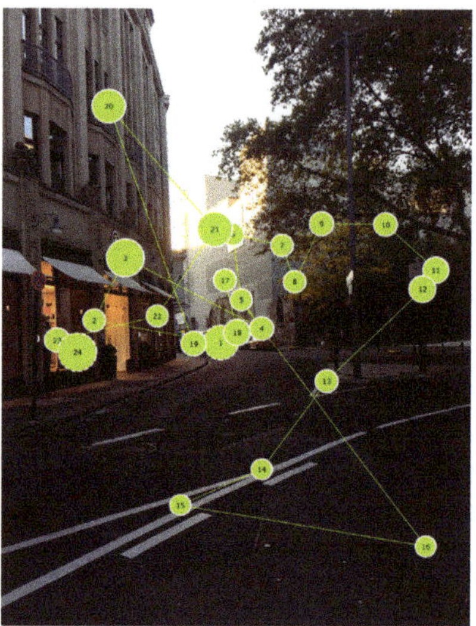

Figure 5. Gaze plot of the observation of image 1 by a single person.

The first element of the image to be noticed by the person whose gaze plot is discussed were the cars along the street near the museum (area Not in 1AOI1). The first fixation took place after an observation time close to the average value for the entire population, whilst its duration (811 ms) was relatively long when compared to the average and the longest among all of the fixations of this person in this area. The two subsequent fixations were within area 1AOI2, while the elements that were observed were associated with the facade of the ground-floor section of the townhouse. The third fixation has been presented as a circle placed near to the round advertisement board and the lighting, and has the largest diameter of all the circles comprising the gaze plot, as it lasted for 981 ms.

The first sign of interest in the facade of the museum is illustrated by fixations 5 and 6. The duration of each was relatively short, at around 300 ms. They focused on an element (a Gothic window) in the historical section and the bright section of the contemporary facade of the building. Subsequently, the gaze plot shows how greenery, traffic light infrastructure and the road surface and its information lines were viewed, in the order listed. Of note is the return of the viewer's focus to the openwork section of the museum's facade (fixation 17) and the bright section above it (fixation 21). The previous gaze plot path and fixations in area 1AOI1 could be considered evidence of the perception of the museum as an entire object by the person. Therefore, it can be stated that perception was based on it shape, light, colour and texture. It can also be presented as R. Arnheim's "visual thinking" [35]. Only fixation 20 pointed to renewed interest in the facade of the historical townhouse—specifically, the balcony on one of the upper floors. This observation lasted a substantial amount of time, with its fixation time amounting to around 720 ms. The last three fixations were also focused on the townhouse, but on its ground-floor section featuring storefronts. The occurrence of the last and longest fixation (778 ms) here can point to the subject becoming bored with the image, instead of merely showing heightened interest.

The individual subject's gaze plot analysed above, analysed against the heat map for all subjects, shows how the different elements and objects within the space surrounding the professionally acknowledged and prize-winning architectural projects can shape the process of their visual perception by viewers through the sequence of observations and interest in them, followed by their study.

The numerical data obtained in the study and their analysis can be used to theoretically present the process of visual perception of presented photographs in accordance with the chosen concept of cognition theory, e.g., Gestalt psychology [34,36]. However, this subject is a separate research problem that exceeds the factual scope of this experiment.

The second photograph used in the experiment, labelled as image 2, has a horizontal orientation. The central placement of the main building (the "Weltstadthaus") was deliberate in relation to the photograph's analysis. To the left we can see the commercial and office building adjacent to the "Weltstadthaus", while to the right is the greenery surrounded by buildings. On a further plane there is a fragment of the late Gothic church of St. Andrew. A large portion of the photograph below the "Weltstadthaus" is occupied by a fragment of the Cäcilienstraße arterial, with its typical road infrastructure, cars and tram.

The areas of interest and the unclassified area outlined for the purposes of the analysis of this photograph have been presented in Figure 6. The descriptive statistical metrics—the parameters—are shown in Table 3.

Table 3. Parameters describing the process of the observation of areas outlined on image 2.

Item No.	Parameter	Subject Population	2AOI1	2AOI2	2AOI3	2AOI4	Not on 2AOI
1.	Average time to first fixation (s)	Women	0.23	6.60	1.16	3.48	1.89
		Men	0.29	4.72	1.81	4.25	1.60
		Total	0.25	5.66	1.38	3.82	1.79
2.	Number of fixations during observation	Women	584.00	9.00	282.00	28.00	582
		Men	332.00	8.00	135.00	23.00	327
		Total	916.00	17.00	417.00	51.00	909
3.	Average number of fixations per person	Women	8 (8.85)	0 (0.14)	4 (4.27)	0 (0.42)	8 (8.82)
		Men	9 (9.76)	0 (0.24)	3 (3.97)	0 (0.68)	9 (9.62)
		Total	9 (9.16)	0 (0.17)	4 (4.17)	0 (0.51)	9 (9.09)
4.	Length of time of all fixations (s)	Women	241.19	3.16	125.82	8.45	216.91
		Men	123.87	2.56	56.41	7.45	128.83
		Total	365.06	5.72	182.23	15.90	345.74
5.	Number of persons who have performed at least 1 fixation relative to the entire population (attractiveness estimator) (%)	Women	100	10.61	98.49	28.78	98.49
		Men	100	20.59	100	44.12	100
		Total	100	14	99	34	99
6.	Number of visits during observation	Women	269.00	8.00	195.00	26.00	225
		Men	144.00	8.00	85.00	20.00	124
		Total	413.00	16.00	280.0	46.00	349
7.	Average number of visits (in proportion to number of research participants)	Women	4 (4.08)	0 (0.12)	2 (2.95)	0 (0.39)	3 (3.41)
		Men	4 (4.24)	0 (0.24)	2 (2.50)	0 (0.59)	3 (3.65)
		Total	4 (4.13)	0 (0.16)	2 (2.80)	0 (0.46)	3 (3.49)

After performing a holistic analysis of the parameter values that describe the process of the visual perception of architecture and its surroundings, it can be concluded that the perception of the building selected for analysis—a prize-winning work of architecture widely acknowledged by professionals—was different in this case. Area 2AOI1 (the facade of the "Weltstadthaus") and the unclassified area (Not on 2AOI) were determined to be the most readily noticeable and most interesting to those who viewed the image. The share of the fixations on each of these areas among all fixations amounted to around 40% each. The facade of the "Weltstadthaus", a building with a curved and rounded shape ("droplet architecture"), was the first to be looked at and was explored numerous times (high number of revisits) by all of the viewers when compared to the unclassified area, which featured numerous elements that, by attracting the viewers' attention, could act as distractors and hinder the perception of the value of the design of Renzo Piano's project. It should be added that area 2AOI1 was the only one to achieve a 100% notice rate among all subjects. Apart from shape, material type and colour, essential perceptual elements most probably included light and shadow, and the repeating shape of the reflected building. Light and shadow can significantly affect the visual expression of a building's massing and composition and the perception of the aesthetic of an architectural form [35].

Figure 6. Areas of interest and the unclassified area outlined for image 2, legend: area 2AOI1—orange, area 2AOI2—blue, area 2AOI3—pink, area 2AOI4—green, area Not on 2AOI—purple.

The second building, located to the left of the "Weltstadthaus" (area 2AOI3) was noticed by almost all of the viewers, who did so in a relatively short time (time to first fixation) relative to the remaining areas of interest. However, the share of the number of fixations on the facade of this building amounted only to 18% of all fixations performed on the photograph by the subjects. Its degree of exploration was at a similar level to that of the unclassified area, which had a much higher number of fixations. One can assume that the elements of the facade of this building inspired interest and a desire to inspect them, hence the number of visits. These areas dominated the attention and visual perception of the photograph's viewers.

The remaining areas of interest (2AOI2 and 2AOI4) were not only noticed to a small degree (attractiveness estimated at 14% and 34%, respectively), but their share of fixations relative to all fixations was below 1% and slightly above 2%, respectively. It should be concluded that this view of the historical church of St. Andrew was practically invisible to the viewers. The area with greenery was more noticeable (almost a half of all men fixated their eyes on it at least once) and neutral from a cognitive point of view.

Similarly as in the case of data concerning image 1, no significant gender-based differences were found in the perception of the elements of the photograph among the experiment's subjects. This provides evidence that an analysis performed on a large population pool can ignore gender as a factor.

The graphical presentation of the places and elements that attracted and focused the attention of all subjects in the form of a heat map has been presented on Figure 7.

The heat map has only one major hotspot, which attracted and held the attention of the subjects. The area, with a red colour and a yellow outline that fades into horizontal, blurry green elements, is located in the upper portion of the commercial office building's facade, below a horizontal advertisement featuring a white inscription. It was the major attractor within area 2AOI3, which attracted the most attention. The second attractor on the facade was another advertisement with an analogous black and white colour scheme, located at the height of the first floor, near the corner of two outer walls. A yellow and green, circular-shaped area covered almost the entirety of the advertisement. In area 2AOI1, the facade of the "Weltstadthaus", there are four yellow spots that blend into a green surface that stretches vertically in two directions. They are present in the "droplet-shaped" part of the facade. A blurry green colour is also visible along the entire height of the object. The building's form, which proved attractive to viewers, resulted in interest in the entire building and its key perceptual elements. The heat map for the unclassified area, which attracted around 40% of all fixations, also displayed yellow and green areas that covered elements which attracted the attention of the viewers. Analogous to the area Not on 1AOI, in this case they were also mostly associated with cars and the surface of the road, in addition to a public transport vehicle, near its section with an inscription denoting its route. This was also visible on an inverse heat map, similarly as in the case of the "Kolumba" museum.

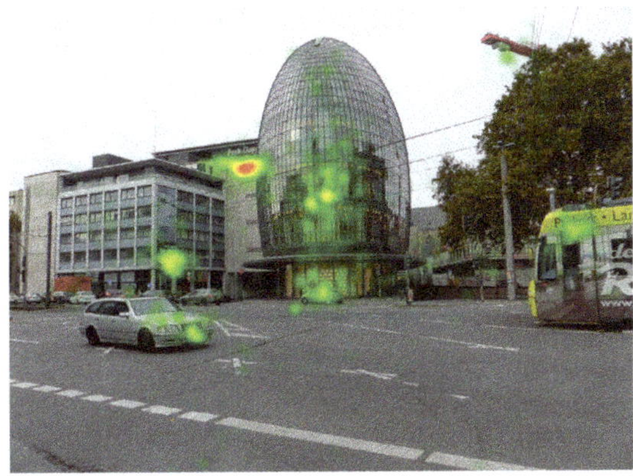

Figure 7. Visualisation of results in the form of a heat map for all subjects for image 2.

The gaze path and the duration of fixations on the abovementioned areas and elements have been presented on data generated for one of the subjects, meant to serve as an example, and presented in Table 4. The gaze plot for this person, who was also male, analogous to the case of photograph 1, has been presented in Figure 8.

Table 4. Parameters characterising the gaze plot for one person relative to the areas outlined for analysis on image 2.

Item No.	Parameter		2AOI1	2AOI2	2AOI3	Not on 2AOI
1.	Time to first fixation (s)		0.18	5.83	1.44	4.23
2.	data	n	9	1	6	8
		%	37.50	4.17	25.00	33.33
3.	Fixation time (s)		3.03	0.49	3.38	2.45
4.	Number of visits in an area		2	1	3	4
5.	Fixation duration range (s)	min	0.185	-	0.221	0.173
		max	0.657	0.488	1.638	0.467

Analysis of quantitative data describing the heat map for all subjects and the gaze plot for the selected person provides evidence of an above-average interest of that person in buildings and their architectural elements relative to the entire population. This is corroborated by a higher number of fixations on area 2AOI3 and their greater duration, making one fixation in area 2AOI2, which was noticed only by 14% of all subjects, while maintaining an analogous interest in area 2AOI1. Furthermore, the aforementioned statement is also supported by the time to first fixation in the unclassified area being twice as long relative to the average for the entire population. This male subject completely ignored area 2AOI4 (the greenery), despite a higher attraction estimator value for this area indicated for the remainder of the population in reference to area 2AOI2.

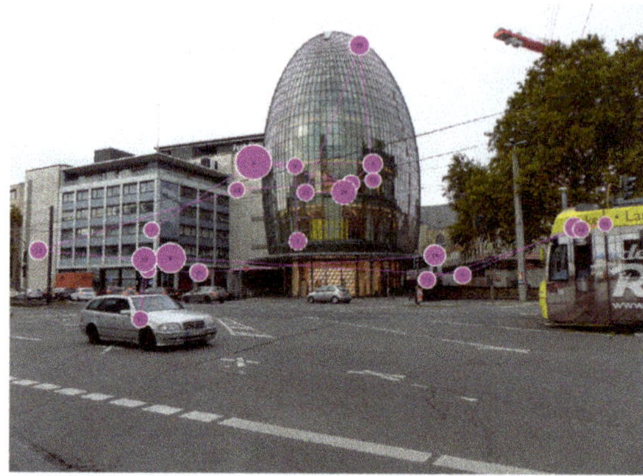

Figure 8. Gaze plot of the observation of image 2 by a single person.

The gaze plot for the subject selected for analysis, similarly to the one pertaining to photograph 1, was composed of 24 circles of varying diameter, linked by saccades.

The first five fixations were on the "Weltstadthaus" building's facade, the first landing on the base of the glass wall comprising the rounded form, with the following landing on its subsequent floor, whose interior is visible from outside. The duration of each fixation was around average, and in the case of fixation 3 was even the shortest of all the fixations on this area. Fixation 6, which landed in the area formed by the facade of the nearby commercial and office building, had the longest duration of all of the fixations on this photograph (1638 ms). This fixation, indicated by the circle with the largest diameter, which almost completely covers the advertisement, corresponded with the heat map's major hotspot. The two subsequent fixations also landed on the facade of this building, with fixation 9, which ended this exploratory visit, also having a long duration (914 ms). The five subsequent fixations landed in the unclassified area, on the opposite side of the photograph. The first focused on an element of a building that was under construction, while the three successive ones were aimed at the tram car in its upper section featuring an inscription detailing its route. Fixation 14 was the only one to target the church of St. Anthony, located to the south of the "droplet" building and visible on the far plane of the frame, from the side of the Cäcilienstraße arterial. This object held the attention of the viewer for around 500 ms. The following fixations amounted to 3 visits to the unclassified area and 3 to the area outlined by the facade of the commercial and office building. Within the area Not on 2AOI, the elements that were noticed and that attracted attention included the traffic light infrastructure, overland power cables and a car visible on the first plane of the photograph. It was a distractor, although it became active relatively late (fixation 19). The final four fixations, including fixation 24, which had the longest duration in this area (657 ms), focused on the building inspiring the greatest interest, namely the "Weltstadthaus", a project by Renzo Piano and one of the most recognisable buildings of Cologne's city centre. The gaze plot, along with its fixations and the saccades linking them, constitutes valuable analytical material along with the statistical metrics that describe it, allowing a better understanding of the process of visual perception, presenting the sequence and type of elements that were noticed, the implementation of the collected information within the system of the knowledge and values of the person observing a photograph. It is possible to determine attractors and distractors for specific persons with a defined aesthetic and artistic sensitivity and awareness concerning values within the space that surrounds them.

5. Conclusions

To summarise the experiment discussed in the article, it can be stated that it enabled the achievement of its initial research goal—namely, determining the utility of eye tracking in architecture. The use of the case study method, along with the selection of research material in the form of the visualisations of two works of architecture by acclaimed designers, which were present on the photographs and framed from different perspectives as surrounded by competing buildings and elements, was justified from the point of view of visual perception analysis. Furthermore, it was proved that population selection in terms of its size and determining its socio-demographic characteristics along with the adopted analysis scenario featuring descriptive statistic metrics enables the description of typical observations concerning attractors and distractors and their determinants based on an analysis of a real-world case.

The presented findings collected during the authors' experiment and the analysis of the state of the art concerning the application potential of eye tracking in studies of architecture provide evidence of said potential in urban design, architecture and landscape architecture. Also notable is the significance of knowledge about the visual perception of various works of architecture, whether contemporary or historical, which are located in different spaces, landscapes, are surrounded by people and various forms of infrastructure—knowledge that could be obtained through this method. This knowledge can prove useful in formulating design assumptions for contemporary architecture and heritage preservation, as well as in the education of future architects and society.

Depending on specific research goals, eye tracking—utilising either stationary or mobile devices, as well as the use of this method in combination with research tools from other methods—can be considered to be highly useful. Studies of visual perception can be comprehensively conducted using tools associated with research methods from the social sciences sociology, cognitive psychology, management) and medicine (electroencephalography—EEG). Such studies can allow us to investigate not only the visual perception of works of architecture and their surroundings in terms of objectivised numerical data, but also to learn the expectations, preferences and state of awareness concerning the knowledge and attitudes of subjects, e.g., representatives of various social groups.

Knowledge of the perception and evaluation of contemporary and historical architecture by city residents is important and should be the focus of studies within the framework of integrated city governance. It requires public participation in all types of action, including the implementation of the "Creating and ensuring high-quality public spaces" strategy adopted in the "LEIPZIG CHARTER on Sustainable European Cities" [54]. The implementation of this strategy through good construction standards ("Baukultur") is particularly important in the preservation of architectural heritage. It allows the multi-aspect use of eye tracking in architecture and governance.

Author Contributions: Both authors are responsible of the whole article and participated equally in the research process. All authors have read and agreed to the published version of the manuscript.

Funding: This research was funded by Cracow University of Economics and Cracow University of Technology.

Acknowledgments: This publication has received funding from a subsidy given to the University of Economics in Krakow.

Conflicts of Interest: The authors declare no conflict of interest.

References

1. Rusnak, M.; Szewczyk, J. Eye tracker as innovative conservation tool. Ideas for expanding range of research related to architectural and urban heritage. *Wiadomości Konserwatorskie J. Herit. Conserv.* **2018**, *54*, 25–35.
2. Uttley, J.; Simpson, J.; Qasem, H. Eye-tracking in the real world: Insights about the urban environment. In *Handbook of Research on Perception-Driven Approaches to Urban Assessment and Design*; IGI Global: Hershey, PA, USA, 2018; pp. 368–396.
3. Krupa, M. *Rabka-Zdrój: Aspekty Urbanistyczno-Architektoniczne Dziedzictwa Kulturowego*; DWE: Wroclaw/Krakow, Poland, 2018.
4. Kabaja, B.; Krupa, M. Possibilities of using the eye tracking method for research on the historic architectonic space in the context of its perception by users (on the example of Rabka-Zdrój). Part 1. Preliminary remarks. *Wiadomości konserwatorskie J. Herit. Conserv.* **2017**, *52*, 74–85.

5. Hollander, J.; Purdy, A.; Wiley, A.; Foster, V.; Jacob, R.; Taylor, A.; Brunye, T. Seeing the city: Using eye-tracking technology to explore cognitive responses to the built environment. *J. Urban. Int. Res. Placemaking Urban Sustain.* **2019**, *12*, 156–171. [CrossRef]
6. Junker, D.; Nollen, C. Mobile eye tracking in landscape architecture: Discovering a new application for research on site. In *Landscape Architecture—The Sense of Places, Models and Applications*; Almusaed, A., Ed.; IntechOpen: London, UK, 2018; pp. 45–66.
7. Wedel, M.; Pieters, R. *Eye Tracking for Visual Marketing*; Now Publishers Inc.: Hanover, Germany, 2008.
8. Nielsen, J.; Pernice, K. How to Conduct Eyetracking Studies. Available online: https://media.nngroup.com/media/reports/free/How_to_Conduct_Eyetracking_Studies.pdf (accessed on 22 November 2019).
9. Vila, J.; Gomez, Y. Extracting business information from graphs: An eye tracking experiment. *J. Bus. Res.* **2016**, *69*, 1741–1746. [CrossRef]
10. Bergstrom, J.; Schall, A. (Eds.) *Eye Tracking in User Experience Design*; Elsevier Inc.: Waltham, MA, USA, 2014.
11. Walla, P.; Brenner, G.; Koller, M. Objective measures of emotion related to brand attitude: A new way to quantify emotion-related aspects relevant to marketing. *PLoS ONE* **2011**, *6*, 1–7. [CrossRef] [PubMed]
12. Wedel, M.; Pieters, R. A review of eye-tracking research in marketing. In *Review of Marketing Research*; Naresh, K.M., Ed.; Emerald Publishing Limited: Bingley, UK, 2008; Volume 4, pp. 123–147.
13. Massaro, D.; Savazzi, F.; Di Dio, C.; Freedberg, D.; Gallese, V.; Gilli, G.; Marchetti, A. When art moves the eyes: A behavioral and eye-tracking study. *PLoS ONE* **2012**, *7*, e37285. [CrossRef] [PubMed]
14. Santini, T.; Brinkmann, H.; Reitstätter, L.; Leder, H.; Rosenberg, R.; Rosenstiel, W.; Kasneci, E. The art of pervasive eye tracking: Unconstrained eye tracking in the Austrian Gallery Belvedere. In Proceedings of the 7th Workshop on Pervasive Eye Tracking and Mobile Eye-Based Interaction, Warsaw, Poland, 15–16 June 2018.
15. Richardson, D. Eye-tracking: Characteristics and methods. In *Encyclopedia of Biomaterials and Biomedical Engineering*; CRC Press: Boca Raton, FL, USA, 2004; Volume 3, pp. 1028–1042.
16. Underwood, G.D.M. *Cognitive Processes in Eye Guidance*; Oxford University Press: New York, NY, USA, 2005.
17. Duchowski, A. *Eye Tracking Methodology. Theory and Practice*; Springer: London, UK, 2007.
18. Holmqvist, K.; Nystrom, M.; Andersson, R.; Dewhurst, R.; Jarodzka, H.; Weijer, J. *Eye Tracking: A Comprehensive Guide to Methods and Measures*; Oxford University Press: New York, NY, USA, 2011.
19. Bojko, A. *Eye Tracking the User Experience. A Practical Guide for Research*; Rosenfeld: New York, NY, USA, 2013.
20. Horsley, M.; Eliot, M.; Knight, B.; Reilly, R. (Eds.) *Current Trends in Eye Tracking Research*; Springer International Publishing: Cham, Switzerland, 2013.
21. Nielsen, J.; Pernice, K. *Eyetracking Web Usability*; New Riders: Berkeley, CA, USA, 2009.
22. Lebrun, C.; Sussman, A.; Crolins, W.; Van der Linde, G. *Eye Tracking Architecture: A Pilot Study of Building in Boston*; Institute for Human Centered Design & Ecole de Design Nantes Atlantique: Boston, MA, USA, 2016.
23. Kiefer, P.; Giannopoulus, I.; Kremer, D.; Schlieder, C.; Rauball, M. Starting to get bored. An outdoor eye tracking study of tourists exploring a city panorama. In Proceedings of the Symposium on Eye Tracking Research and Applications, (ETRA '14), Safety Harbor, FL, USA, 26–28 March 2014; pp. 315–318.
24. Zou, Z.; Ergan, S. Where do we look? An eye-tracking study of architectural features in building design. In Proceedings of the 35th CIB W78 2018 Conference: IT in Design, Construction, and Management, Chicago, IL, USA, 1–3 October 2018.
25. Radwan, A.; Ergan, S. Quantifying human experience in interior architectural spaces. In Proceedings of the ASCE International Workshop on Computing in Civil Engineering 2017, Seattle, WA, USA, 25–27 June 2017.
26. Dupont, L.; Ooms, K.; Duchowski, A.; Antrop, M.; Eetvelde, V. Investigating the visual exploration of the rural-urban gradient using eye-tracking. *J. Spat. Cogn. Comput. Interdiscip. J.* **2017**, *17*, 65–88. [CrossRef]
27. Dupont, L.; Ooms, K.; Antrop, M.; Eetvelde, V. Comparing saliency maps and eye-tracking forms maps: The potencial use in visual impact assessment base on landscape photographs. *Landsc. Urban Plan.* **2016**, *148*, 17–26.
28. Noland, R.B.; Weiner, M.D.; Gao, D.; Cook, M.P.; Nelessen, A.; Blonstein, E. Eye-tracking technology, visual preference surveys, and urban design: Preliminary evidence of an effective methodology. *J. Urban. Int. Res. Placemaking Urban Sustain.* **2017**, *10*, 98–110. [CrossRef]
29. Rusnak, M.; Fikus, W.; Szewczyk, J. How do observers perceive the depth of a Gothic cathedral interior along with the change of its proportions? Eye tracking survey. *Architectus* **2018**, *1*, 77–88.

30. Lisińska-Kuśnierz, M.; Krupa, M. Eye tracking in research on perception of objects and spaces. *Czasopismo Techniczne Tech. Trans.* **2018**, *12*, 5–22.

31. Zuyagina, N.; Talceva, A.; Kuznetsova, D. Physiological merkers of visual environment comfort in the North. *IOP Conf. Ser. Earth Environ. Sci.* **2019**, *263*, 1–7.

32. Schiessel, M.; Duda, S.; Thölke, A.; Fischer, R. Eye tracking and its application in usability and media research. *J. Sonderheft Blickbewegung MMI-interaktiv J.* **2003**, *6*, 41–50.

33. Wade, N.; Swanston, M.S. *Visual Perception. An Introduction*; Psychology Press: London, UK, 2013.

34. Thomas, H. *Advances in Visual Perception Research*; Nova Science Publishers Inc.: Hauppauge, NY, USA, 2015.

35. Arnheim, R. *Dynamis of Architectural Form*, 30th ed.; University of California Press: Berkeley, CA, USA, 2009.

36. Jules, F.A. *A Comparison of the Application to Architecture of the Ecological and Gestalt Approaches to Visual Perception*; University of Wisconsin-Milwaukee: Milwaukee, WI, USA, 1984.

37. Poole, A.; Ball, L.J. Eye tracking in HCI and usability research. In *Encyclopaedia of Human Computer Interaction*; Idea Group: Hershey, PA, USA, 2006; pp. 211–219.

38. Brick. Eye on Architecture Vizualization Brick Visual. Available online: https://brickvisual.com/eye-architecture-visualization/ (accessed on 10 December 2019).

39. Tobii. *User Manual—Tobii Studio*; Version 3.2. Rev A; Tobii Technology AB: Stockholm, Sweden, 2012.

40. Creswell, J. *Research Design: Qualitative, Quantitative, and Mixed Methods Approaches*, 4th ed.; Sage Publications: Thousand Oaks, CA, USA, 2014.

41. Yin, R. *Case Study Research Design and Method*, 3rd ed.; Sage Publications: Thousand Oaks, CA, USA, 2003.

42. Moruno, L.G.; Montero, M.J.; Hernandez, J.; Lopez-Casares, S. Analysis of lines and forms in building to rural landscape integration. *Span. J. Agric. Res.* **2010**, *8*, 833–847. [CrossRef]

43. Ikemi, M. The effects of mystery on preference for residential fasades. *J. Environ. Psychol.* **2005**, *25*, 167–173. [CrossRef]

44. The Hyatt Foundation. The Pritzker Architecture Prize. 2009. Available online: https://www.pritzkerprize.com/laureates/2009 (accessed on 15 November 2019).

45. Stec, B. Trzy rozmowy z Peterem Zumthorem. *Architektura Biznes* **2003**, *2*, 20–38.

46. Skolimowska, A. Muzeum „Kolumba" w Kolonii. Available online: https://architektura.nimoz.pl/2013/12/04/muzeum-kolumba-w-kolonii/ (accessed on 18 November 2019).

47. Plotzek, J.M. *Kolumba: Ein Architekturwettbewerb in Köln 1997*; Walther König: Cologne, Germany, 1997.

48. Backes, E. *Kolumba. Die Evolution eines Museums*; Kühlen Verlag: Mönchengladbach, Germany, 2015.

49. Węcławowicz-Gyurkovich, E. Daring modern realisations in historic surroundings. *Wiadomości Konserwatorskie J. Herit. Conserv.* **2010**, *28*, 70–77.

50. The Hyatt Foundation. The Pritzker Architecture Prize. 1998. Available online: https://www.pritzkerprize.com/laureates/1998 (accessed on 18 November 2019).

51. Januszkiewicz, K. Performative architecture in Cologne. *Archivolta* **2012**, *2*, 32–45.

52. AZoBuild. Renzo Piano builds the "Weltstadthaus" in Cologne. *AZoBuild*. 27 November 2005. Available online: https://www.azobuild.com/news.aspx?newsID=1772 (accessed on 22 November 2019).

53. The Everything Guide to Generation Z by Visioncritical with Research by Maru/VCR&C. Available online: https://cdn2.hubspot.net/hubfs/4976390/E-books/English%20e-books/The%20everything%20guide%20to%20gen%20z/the-everything-guide-to-gen-z.pdf (accessed on 4 December 2019).

54. Leipzig Charter on Sustainable European Cities, Final Draft (2 May 2007). Available online: https://ec.europa.eu (accessed on 15 December 2019).

 *buildings*

Article

Testing Joints between Walls Made of AAC Masonry Units

Radosław Jasiński [1] and Iwona Galman [2,*]

[1] Department of Building Structures, Silesian University of Technology, Akademicka 5, 44-100 Gliwice, Poland; radoslaw.jasinski@polsl.pl

[2] Department of Structural Engineering, Silesian University of Technology, Akademicka 5, 44-100 Gliwice, Poland

* Correspondence: iwona.galman@polsl.pl; Tel.: +48-32-237-22-88

Received: 15 December 2019; Accepted: 30 March 2020; Published: 2 April 2020

Abstract: Joints between walls are very important for structural analysis of each masonry building at the global and local level. This issue has often been neglected in the case of traditional joints and relatively squat walls. At present, the issue of wall joints is becoming particularly important due to the continuous drive for simplifying structures, introducing new technologies and materials. Eurocode 6 and other standards (American, Canadian, Chinese, and Japanese) recommend inspecting joints between walls, but no detailed procedures have been specified. This paper presents our own tests on joints between walls made of autoclaved aerated concrete (AAC) masonry units. Tests included reference models composed of two wall panels joined perpendicularly with a standard masonry bond (six models), with classic steel and modified connectors (twelve models). The shape and size of test models and the structure of a test stand were determined on the basis of the analysis of the current knowledge, pilot studies and numerical FEM (Finite Element Method) - based analyses. The analyses referred to the morphology and failure mechanism of models. Load-displacement relationships for different types of joints were compared and obtained results were related to results for reference models. The mechanisms of cracking and failure was found to vary, and clear differences in the behaviour and load capacity of each type of joint were observed. The individual working phases of joints were determined and defined, and an empirical approach was proposed for the determination of forces and displacement of wall joints.

Keywords: masonry structures; stiffening walls; wall joints; connectors; bed joint reinforcement

1. Introduction

The relationship between the type of bond between intersecting walls and the load-bearing capacity of test models was investigated by Castro et al. [1]. The models with no bond and with traditional masonry bond were tested. In the models without a bond, failure was caused by the loss of stability of a shorter wall component, while in the models with full bond, shearing along the whole height of the joint appeared. In addition to the joints constructed with masonry units, tests were also performed on reinforced joints. Paganoni and D'Ayala [2] investigated the effectiveness of steel anchors at the connections of intersecting walls. Similar tests were conducted by Maddaloni et al. [3,4]. However, in that case, investigations covered the effectiveness of innovative clamp anchors (rods made of carbon fibres wrapped longitudinally and spirally with a stainless steel mat). Unfortunately, there are only results from tests on connectors. Therefore, it is difficult to interpret their effectiveness because of the lack of any reference to the load-bearing capacity of joints made with a traditional masonry bond.

It is also worth mentioning tests performed by the authors in [5], in which pilot tests on masonry wall joints were presented. They were the first tests of that kind performed in Poland, and were among the few that has been performed in Europe. Within this testing programme, three types of wall joints were compared: traditional masonry bonds, bonds with the use of steel L-shaped profiles and two-arm steel punched flat profiles. Traditional bonds exhibited almost five times higher load-bearing capacity than joints with steel L-profiles; the capacity of joints with flat profiles was almost twice higher.

The obtained test results encourage further investigations and continued work on the detailed description of joints and on the use of new methods for the construction of joints using other types of connectors, a higher number of connectors and the optimization of their shape. The performed pilot tests also demonstrated imperfections in test models and the testing method. Asymmetric failure images of two identical joints made it impossible to understand the work of a single joint. Despite the application of point forces close to the contact plane, cracks also occurred in the lower part of the web wall, which indicates bending of this part of the model and, consequently, complicates analyses. Another worrying phenomenon observed during the tests was the variation in deformations of steel connectors depending on the location of joints in relation to the loaded edge of the web wall, meaning the non-uniform work of the joints. Therefore, in further tests the authors decided to change the shape of test models and the method of load application.

The review of tests on joints described in [6] showed the lack of comprehensive studies on the behaviour of wall joints. That did not only refer to walls made of autoclaved aerated concrete (AAC) masonry units, but also made of other masonry units. A poor insight into the issue of joints and the mutual action of walls resulted in the neglect of calculations for such structures. Design standards lack guidelines for determining internal forces and stresses acting on wall intersections, and for determining conditions to verify the Ultimate Limit State (ULS) and the Serviceability Limit State (SLS). Those few tests are insufficient to describe the mechanism of joints work, much less to develop guidelines for their design and construction. Moreover, there is the need to design a connector in a new shape to satisfy the demands of the market, which is aimed at optimizing existing solutions. A new connector should meet requirements of ultimate states and simultaneously should have a simple construction, easy assembly and much higher performance reliability in the phase after reaching the greatest loading. Therefore, the overall aims of our own tests were specified and they included:

- Determination of the cracking and failure mechanisms of joints between AAC walls;
- Comparison of load capacity of wall joints using traditional masonry bonds and steel connectors;
- Optimization of the shape of a steel connector.

Moreover, the authors made an attempt to build simplified models representing the behaviour of reinforced and unreinforced joints, a process which was described in this paper and in [7]. Tests and analyses presented in this paper were completed with a new series of tests.

2. Programme of Our Own Tests

Three series (12 test models in total) with the same shape and dimensions were prepared and tested. T-shaped models were monosymmetric, with a web and flange length of ~89 cm. A vertical joint, whose structure varied intentionally, was formed between loaded and unloaded walls. A series of test models marked with **P** had traditional masonry joints between the web and the flange (Figure 1a). Those elements were regarded as reference models, whose mechanical parameters and behaviour at loading and failure were compared with results from other tests. In two other series, joints between webs and flanges were made with steel connectors (wall geometry acc. to Figure 1b). They were single punched flat profiles in series **B10** (Figure 1c), and modified flat profiles with a widened central part in series **BP10** (Figure 1d). This solution was proposed on the basis of our own tests [7] on perforated connectors. The widening of the central part was intended to increase the flexural capacity of the connector and its stiffness. The proposed shape is copyrighted via an application to the Polish Patent Office [8]. Joints made of galvanized perforated steel with a thickness of 1 mm were used in both series.

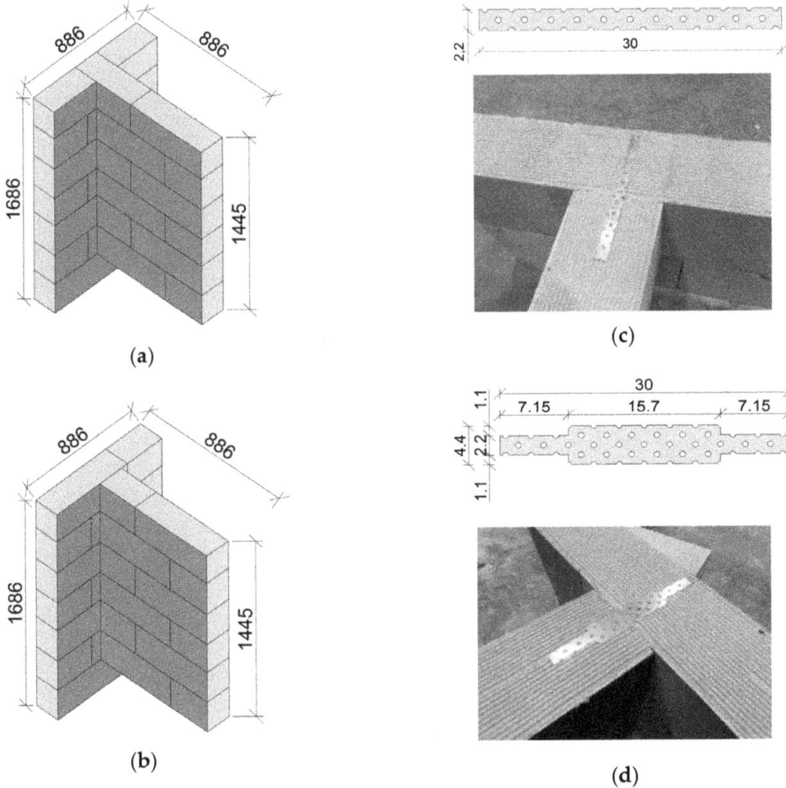

Figure 1. Geometry and details of test models and: (**a**) traditional masonry bond, P series, (**b**) walls with steel joints (B10 and BP10 series), (**c**) joining method with a punched flat bar, (**d**) joining method with a punched widened flat profile (mm).

Tests were conducted in a test stand specially designed for that purpose—see Figure 2. Models 1a and 1b with confining elements 3 and elements taking load 2 were put on the strong floor (panel 1b) and placed on a dynamometer 6, which with a resistor 4 acted as a fixed articulated support. Models were placed below a steel frame 8, to which a hydraulic actuator was fixed (with an operating capacity of 1000 kN), generating shearing at a constant displacement gain equal to 1 mm/min. The structure response was registered using an inductive force transducer with an operating capacity of 250 kN and reading accuracy of ±2.5 kN. Prestress of 0.1 MPa was exerted using reinforced concrete elements 3 and steel strands 7 to model the considerable length of a joined wall in panel *1b*. Models were loaded in one cycle until failure. Vertical load generating shear was transmitted linearly along the whole height of the wall through elements 2. As a result, shear stresses on joints were distributed uniformly. The loading and displacement of a loaded wall against the unloaded one were continually registered during tests. Two independent types of software were used to register data. One side of the test model was monitored using ARAMIS—an optical sensor of displacements. Another side was monitored with inductive transducers of displacement of type PJX-10, with an operating capacity of 10 mm and an accuracy of ±0.002 mm.

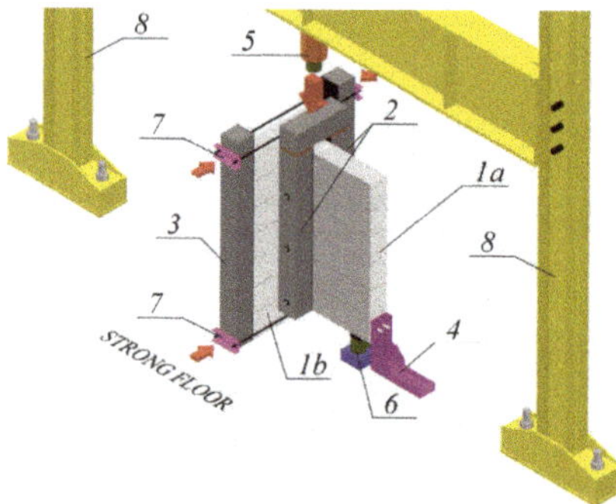
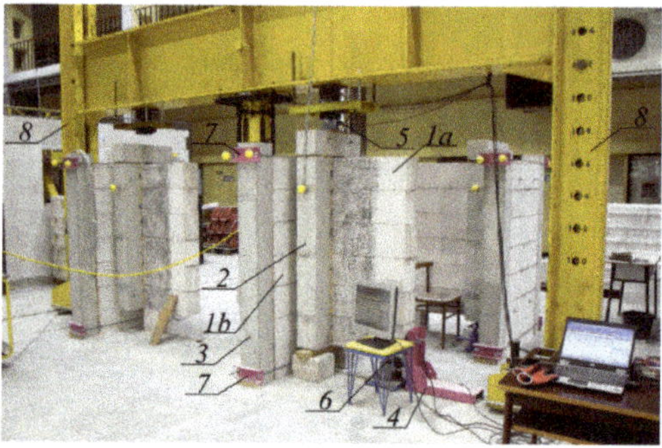

Figure 2. A scheme and photo of the test stand (longitudinal wall (*1a*), transverse wall (*1b*), reinforced concrete column transferring shear load (*2*), reinforced concrete pillars limiting horizontal deformation (*3*), horizontal support (*4*), system of the hydraulic cylinder and the force gauge used to induce shear stress (*5*), force gauge, vertical reaction (*6*), horizontal tie (*7*), steel frame (*8*)).

Tests were performed on models made of AAC masonry units with system mortar M5 class for the thin joints and the unfilled head joint. The compressive strength of the masonry specified in the code PN-EN 1052-1:2000 [9] and presented in [10] was f_c = 2.97 N/mm^2, and the modulus of elasticity was E_m = 2040 N/mm^2. The initial shear value determined according to the code PN-EN 1052-3:2004 [11] and presented in [12] was f_{vo} = 0.31 N/mm^2.

The mean friction coefficient in joints without mortar was μ = 0.92 [13]. The shear modulus determined according to the code ASTM E519-81 [14] and presented in [15] was G = 329 N/mm^2. Mortar for thin joints was used in the tested elements for the AAC blocks. This mortar is dedicated to the erection of AAC masonry walls Additional tests on steel connectors—see Figure 3—were conducted according to the standard [16]. Three elements were chosen randomly from each series of connectors and placed in the jaws of a testing machine. The basic mechanical parameters of the connectors were determined by controlling the displacement gain. The measurement of strains was non-contact with a video extensometer MEVIX 200. Strain was measured using a base with the length L_e = 53.5 mm in standard connectors and L_e = 75.0 mm in thickened connectors. Figure 3 illustrates (stress σ –strain ε) relationships. The stress–strain relationship of tested connectors was found to have no clear yield point. Therefore, results were approximated with a bi-linear relationship. A theoretical yield point f_y was determined at the intersection of straight lines. The slope of the tangent straight line presented within the range of 0–f_y was assumed as the mean initial modulus of elasticity E_s. Tensile strength was determined at failure of specimens, and the tangent of the straight-line slope within the range of f_y–f_t was determined as the mean secant modulus of elasticity E_t. Test results for connectors and the research programme are compared in Table 1.

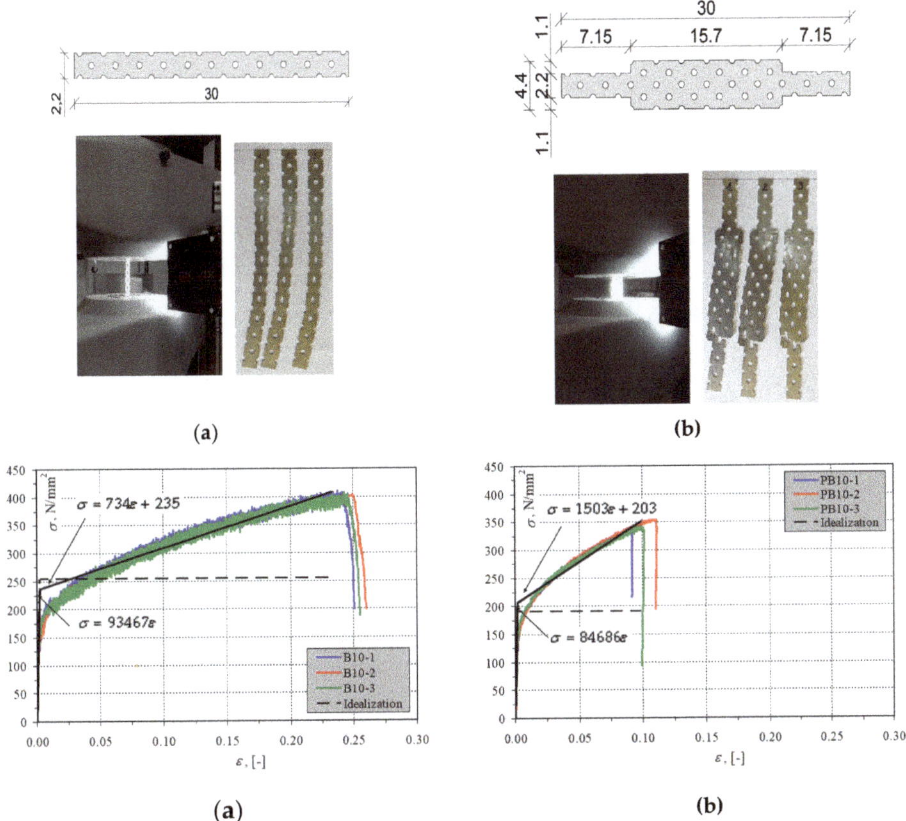

Figure 3. Test tested connectors: (**a**) connectors B10, (**b**) connectors BP10.

Table 1. Programme of tests and basic characteristics of connectors.

Name of Series	Type of Joint	A mm^2	I mm^4	f_y/f_t N/mm^2	E_s/E_t N/mm^2	Number of Models
P	Traditional masonry bond	–	–	–	–	6
B10	Punched steel flat profile $B \times t = 22 \times 1$ mm	22	1.83	236/408	93467/743	3
BP10	Punched widened flat profile $B \times t = 44 \times 1$ mm	44	3.67	207/345	84686/1503	3

Area of gross section (A), moment of inertia of gross section (I), mean yield stress (f_y), tensile strength (f_t), mean initial modulus of elasticity (E_s), mean secant modulus of elasticity (E_t).

3. Test results and Analysis

3.1. Unreinforced Models

The behaviour of all unreinforced models was similar. No cracking noise and no visible splitting on the lateral surfaces of elements were noticed in the initial phase of loading. Non-dilatation strain in some parts of the wall was observed. That phase lasted until the appearance of the first diagonal cracks in the adjacent vicinity of wall joints—see Figure 4a,b. Load increments caused the distinct development of cracks present at the location of joints and propagation towards the reinforced concrete column which transferred loading (Figure 4c). The greatest force was registered in that phase. Continued loading led to the distinct growth of mutual displacements and the rotation of joined walls. The joint was removed after failure—see Figure 4d. Almost vertical shearing of elements forming the bond was found. No clear damage was reported in the case of other elements.

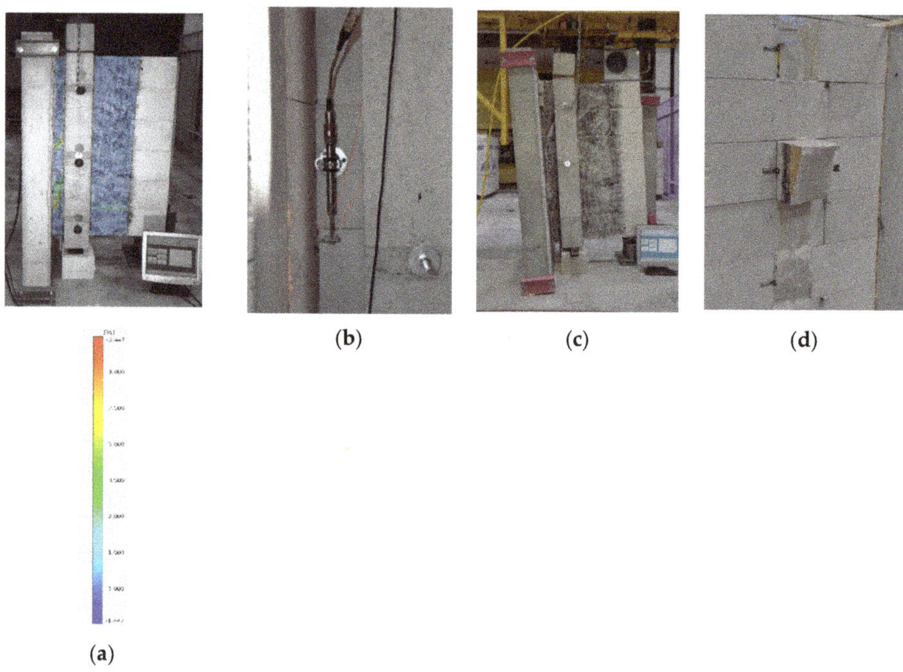

(a)   **(b)**   **(c)**   **(d)**

Figure 4. Destruction of models of series P (**a**) a first crack on the reference model P_2, (**b**) a first crack on the reference model P_6, (**c**) joint after failure P_5, (**d**) joint after failure P_3.

The cracking mechanism for elements is also visible on diagrams illustrating the relationship between the load N and relative (mutual) displacement u of bonded walls—see Figure 5. Until cracking of the contact surface observed under the load $N_{cr} = 27.3$–54.1 kN, increments in relative displacements u were almost directly proportional, and thus the working phase of the joint was called the elastic phase. After cracking in the post-elastic phase, stiffness was reduced. However, joints still had the capacity to take the load.

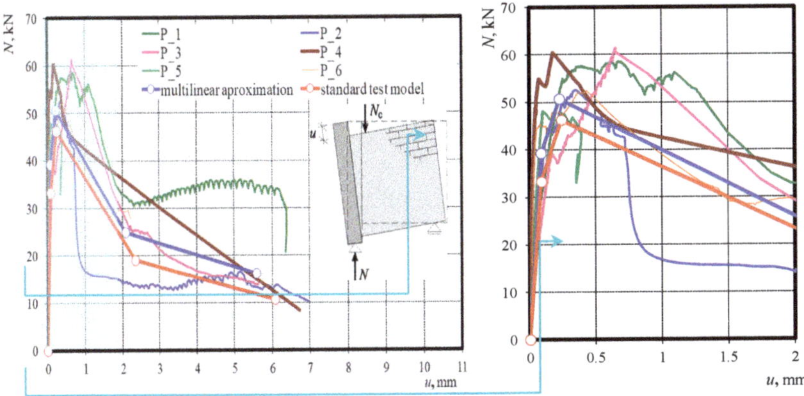

Figure 5. Relationship between the total force and mean displacement for test results and calculations.

This phase was completed at maximum force values within the range $N_u = 38.6$–59.8 kN. Continued attempts at loading in the failure phase resulted in a clear drop in the values of forces registered by a dynamometer, and an increment in relative displacements. Force was close to zero, and the joint had the capacity to take some load. In this phase, forces were called aggregate interlocking forces with values of $N_{ag} = 14.1$–31.1 kN. Further increment in joint displacements caused a minor load increase and hardening. The last registered forces, called residual forces, preceded the failure that resulted in the total splitting of bonded elements and their mutual rotation. Their value ranges were $N_r = 8.4$–42.9 kN. Forces and corresponding displacements are presented in Tables 2 and 3, and the linear approximation of results is shown in Figure 6. Joint stiffness was determined in each phase according to Equations (1)–(3) and they are presented in Table 4:

Table 2. Test results for joints between unreinforced walls.

Model	Force at the Time of Cracking		Maximum Force		Aggregate Interlocking Force		Residual Force	
	$N_{cr,i}$ kN	$N_{cr,mv}$ kN	$N_{u,i}$ kN	$N_{u,mv}$ kN	N_{ag} kN	$N_{ag,mv,i}$ kN	$N_{r,i}$ kN	$N_{r,mv}$ kN
P_1	27.3		56.3		31.1		20.7	
P_2	42.6		50.0		14.7		10.2	
P_3	31.2	39.2	38.6	50.7	25.5	24.9	13.8	16.2
P_4	54.1		59.8		–		8.36	
P_5	35.1		48.1		–		–	
P_6	45.1		51.6		28.264		27.9	

Table 3. Test results for joints between unreinforced walls (displacements).

Model	Displacement at the Time of Cracking		Displacement Right before Failure		Displacement at Aggregate Interlocking Force		Residual Displacement	
	$u_{cr,i}$ mm	$u_{cr,mv}$ mm	$u_{u,i}$ mm	$u_{u,mv}$ mm	$u_{ag,i}$ mm	$u_{ag,mv}$ mm	$u_{r,i}$ mm	$u_{r,mv}$ mm
P_1	0.07		0.31		2.43		6.36	
P_2	0.12		0.25		1.95		6.97	
P_3	0.12	0.09	0.16	0.23	2.22	2.08	5.64	5.58
P_4	0.07		0.17		–		6.72	
P_5	0.06		0.10		–		–	
P_6	0.08		0.36		1.71		2.22	

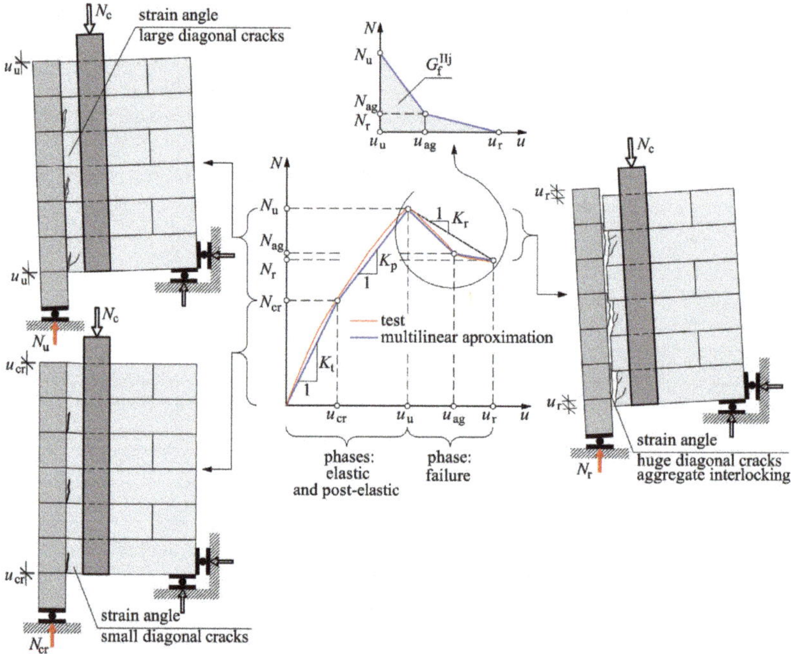

Figure 6. Approximation of work of unreinforced joints between masonry walls.

Table 4. Test results for joints between unreinforced walls (joint stiffness).

Model	Elastic Joint Stiffness		Post-Elastic Joint Stiffness		Residual Joint Stiffness	
	$K_{t,i}$ MN/m	$K_{t,mv}$ MN/m	$K_{p,i}$ MN/m	$K_{p,mv}$ MN/m	$K_{r,i}$ MN/m	$K_{r,mv}$ MN/m
P_1	413		119		5.89	
P_2	341		60		5.93	
P_3	268	496	163	123	4.51	7.39
P_4	804		52.8		7.86	
P_5	562		322		–	
P_6	590		23		12.75	

Joint stiffness in the elastic phase,

$$K_t = \frac{N_{cr}}{u_{cr}},\tag{1}$$

joint stiffness in the post-elastic phase,

$$K_p = \frac{N_u - N_{cr}}{u_u - u_{cr}},\tag{2}$$

joint stiffness in the failure phase,

$$K_r = \frac{|N_r - N_u|}{u_r - u_u}.\tag{3}$$

Validation of the Model with Unreinforced Wall Joints

Performed tests were used to generalize the obtained results by proposing the so-called standard model [17]. The following assumptions were made:

a) A non-linear relationship N–u determined from tests could be replaced with a multi-linear relationship expressing all observed phases:

    i. The elastic phase observed in the load range 0–N_{cr};
    ii. The post-elastic phase observed in the load range N_{cr}–N_u;
    iii. The failure phase observed in the load range N_u–N_{ag}–N_r.

b) It was suggested that all material parameters used in the model should be specified using standard and normalised methods;
c) The model would be subjected to statistical validation on the basis of performed tests.

The following empirical relationships were recommended to determine forces and displacements in particular phases:

Forces and displacements in the elastic phase:

$$N_{cr} = \alpha_1 \tau_{cr,RL} A,\tag{4}$$

$$u_{cr} = N_{cr}/K_t = N_{cr}/\alpha K_{RL}.\tag{5}$$

Forces and displacements in the post-elastic phase:

$$N_u = \beta_1 \tau_{u,RL} A,\tag{6}$$

$$u_u = u_{cr} + (N_u - N_{cr})/K_p = u_{cr} + (N_u - N_{cr})/\beta K_t,\tag{7}$$

where $A = 0.26$ m^2 is the joint area and α, α_1, β and β_1 are empirical coefficients.

Shear parameters determined during tests on diagonal compression performed in compliance with ASTM E519-81 were $\tau_{cr,RL} = 0.192$ MPa, $\tau_{u,RL} = 0.196$ N/mm^2 and stiffness $K_{RL} = 117.1$ MN/m were used as reference values in above equations. At the beginning of the failure phase, residual and aggregate interlocking forces were determined from the following equations:

$$N_r = \gamma \tau_{u,RL} A, \tag{8}$$

$$N_{ag} = \gamma_1 \tau_{u,RL} A, \tag{9}$$

where γ and γ_1 are empirical coefficients.

Displacements corresponding to the aggregate interlocking force were determined from the following empirical relationship:

$$u_{ag} = \omega \tau_{u,RL} / K_{RL}, \tag{10}$$

where ω is the empirical coefficient.

Values of empirical coefficients were calculated using the results from material tests and tests on individual elements. Furthermore, boundary values of mean coefficients α, α_1, β, β_1, γ, γ_1 and ω were determined at the significance level $\alpha = 0.8$ to create the reference model [18]. As the sample size was small $n < 30$, the following relationship was used:

$$P\left(\bar{x} - t_{1-\alpha/2}\frac{S}{\sqrt{n}} < m < \bar{x} + t_{1-\alpha/2}\frac{S}{\sqrt{n}}\right) = 1 - \alpha, \tag{11}$$

where: $\bar{x} = \sqrt{\sum(x-x)/n}$ is the mean value of the random sample, $S = \sqrt{\sum(x-x)^2/(n-1)}$ is the standard deviation of the sample, $t_{1-\alpha/2}$ are statistics with Student's t-distribution and n-1 degrees of freedom.

Lower and upper values from the confidence interval of mean coefficients are presented in Table 5.

Table 5. Validation of empirical coefficient of the model with unreinforced wall joints.

Model	x_i						
	$\alpha_i = \frac{K_{t,i}}{K_{RL}}$	$\beta_i = \frac{K_{p,i}}{K_{t,i}}$	$\alpha_{1,i} = \frac{N_{cr,i}}{\tau_{cr,RL}A}$	$\beta_{1,i} = \frac{N_{u,i}}{\tau_{u,RL}A}$	$\gamma_i = \frac{N_{r,i}}{\tau_{u,RL}A}$	$\gamma_{1,i} = \frac{N_{ag,i}}{\tau_{u,RL}A}$	$\omega_i = \frac{u_{ag,i}K_{RL}}{\tau_{u,RL}A}$
P_1	3.51	0.29	0.55	1.10	0.4	0.6	5.61
P_2	2.90	0.17	0.85	0.98	0.2	0.3	4.49
P_3	2.27	0.61	0.62	0.76	0.3	0.5	5.13
P_4	6.83	0.07	1.08	1.17	0.2	–	–
P_5	4.78	0.57	0.70	0.94	–	–	–
P_6	5.01	0.04	0.90	1.01	0.5	0.6	3.94
n	6	6	6	6	5	4	4
$\bar{x}$	4.22	0.29	0.79	0.99	0.32	0.49	4.80
S	1.7	0.2	0.2	0.1	0.2	0.1	0.7
$t_{1-\alpha/2}$	1.48	1.48	1.48	1.48	1.53	1.64	1.64
$\bar{x} - t_{1-\alpha/2}\frac{S}{\sqrt{n}}$	3.22	0.14	0.67	0.91	0.21	0.37	4.20
$\bar{x} + t_{1-\alpha/2}\frac{S}{\sqrt{n}}$	5.22	0.44	0.91	1.08	0.43	0.60	5.39

In the failure phase, during which dry shear fracture of separating walls was observed, the joint behaviour was mapped on the basis of standard behaviours specified in PN-EN 1052-3:2004. Those tests included measurements of relative displacements of two masonry units joined with mortar and determination of fracture energy of the joint $G_f^{II} = 2.37 \times 10^{-4}$ MN/m [13], which could be used to describe the behaviour of the brittle material in the failure phase in accordance with the continuum fracture mechanism. The failure phase was described on the basis of observations using two sections with forces varying from N_u to N_{ag}, and then from N_{ag} to N_r at corresponding displacements u_u, u_{ag} and u_r. Assuming that fracture energy per joint area G_f^{IIj} (expressed as the area below the diagram shown in Figure 6) was equal to fracture energy G_f^{II}, obtained from standard tests, the displacement corresponding to the residual force u_r was determined from the relationship:

$$AG_f^{II} = AG_f^{IIj} = \tfrac{1}{2}(N_u - N_{ad})(u_{ad} - u_u) + (N_{ad} - N_r)(u_{ad} - u_u)$$

$$+\tfrac{1}{2}(N_{ad} - N_r)(u_r - u_{ad}) \tag{12}$$

$$\Rightarrow u_r = \frac{2G_f^{II}A - N_u(u_{ag} - u_u) + N_{ag}u_u + N_r(u_{ag} - 2u_u)}{(N_{ag} - N_r)}.$$

Following that procedure, two values defining the lower and upper limits of yjr confidence intervals matched each of the seven coefficients (Table 5). Maximum and minimum values of displacement expressed by the relationship in (12) depended on the previously used values and could be considered as independent variables. Thus, there were $\binom{7}{2}$ different combinations (without any repetitions) for coefficients. The minimum value of the mean square error calculated separately for forces and displacements was applied as a selection criterion. Optimal values of those coefficients were calculated from 21 combinations. A Mean Percentage of Error (MPE) was calculated [19] $MPE = \frac{1}{N}\sum_{N=1}^{5}\frac{x_{obs} - x_{cal}}{x_{obs}}$ The minimum MPEs for calculated forces and displacements with respect to the coefficients listed in shaded cells in Table 5 were 16% (for force) and −6% (for displacement).

As a result, empirical relationships based on the results from model and standard tests, describing the work of joints in particular phases, are presented in Table 6, and calculated values and empirically obtained values are compared in Table 7 and Figure 5.

Table 6. Relationships describing the work of unreinforced joints between walls.

Joint Phase	Force	Stiffness	Displacement
Elastic phase	$N_{cr} = 0.67\tau_{u,RL}A$	$K_t = 3.22K_{RL}$	$u_{cr} = N_{cr}/3.22K_{RL}$
Post-elastic phase	$N_u = 0.91\tau_{u,RL}A$	$K_p = 0.14K_{RL}$	$u_u = (N_u - N_{cr})/0.14K_{RL}$
Failure phase	$N_{ag} = 0.37\tau_{u,RL}A$	$(N_u - N_{ag})/(u_u - u_{ag})$	$u_{ag} = 5.39\tau_{u,RL}A/K_{RL}$
	$N_r = 0.21\tau_{u,RL}A$	$K_r = (N_u - N_r)/(u_r - u_u)$	$u_r = \dfrac{2G_f^{II}A - N_u(u_{ag} - u_u) + N_{ag}u_u + N_r(u_{ag} - 2u_u)}{(N_{ag} - N_r)}$

Table 7. Compared test results and our own calculations for the standard model.

Test Results for				Calculated Results for			
Forces				Forces			
$N_{cr,mv}$ kN	$N_{u,mv}$ kN	$N_{ag,mv}$ kN	$N_{r,mv}$ kN	$N_{cr,cal}$ kN	$N_{u,cal}$ kN	$N_{ag,cal}$ kN	$N_{r,cal}$ kN
39.2	50.7	24.9	16.2	33.3	46.3	19.0	10.7
Displacements				Displacements			
$u_{cr,mv}$ mm	$u_{u,mv}$ mm	$u_{ag,mv}$ mm	$u_{r,mv}$ mm	$u_{cr,cal}$ mm	$u_{u,cal}$ mm	$u_{u,cal}$ mm	$u_{r,cal}$ mm
0.09	0.243	2.08	5.58	0.09	0.24	2.34	6.53

Following the assumptions, calculated forces determining coordinates for particular phases of joint work were smaller than those obtained from tests, which was consistent with the assumptions. Considering the force causing cracks, the difference was 15%, and for the failure force, it was −9%. The biggest differences were found in the failure phase. Then, the calculated values N_{ag} and N_r were lower by 36% and 44% than mean empirical values. For relative displacement in the elastic phase, the calculated displacement differed from the average empirical value by only 3%, and by 7% in case of the greatest force. In the failure phase, displacements corresponding to forces N_{ag} and N_r differed by 12% and 17%, respectively. The delivered results were sufficient to predict forces with satisfactory accuracy and thus to verify properly the SLS conditions for joints. Greater differences were found for displacements, which are crucial for verifying SLS conditions. The biggest discrepancy was obtained for the maximum load.

3.2. Reinforced Models

In the models of series **B10** and **BP10**, reinforced with steel connectors, no cracks on walls typical for unreinforced models were observed for the whole range of loading. Displacements of interconnected wall panels were unnoticeable in the initial phase of loading. At a given moment, a rapid increase in displacements was clearly visible to the naked eye. However, it was still possible to continue the loading of the models until the moment of failure. failure was rapid and caused shearing of the joint and a distinct vertical displacement (by ca. 17 mm) of the wall web—see Figure 7b. The wall settled on the wooden protection. Models at the point of failure are shown in Figure 7a. The failure of the models of series **B10** and **BP10** was caused by the yielding and bending of steel flat profiles in the vicinity of the contact surface (Figure 7c,d). Spalling of masonry units beneath each connector was observed at the wall edge (see arrows in Figure 7c,d). The measured length of spalling areas was ca. 15 mm. However, no shear fracture of the connector was observed in the mortar laid in bed joints due to the holes in the flat profile. Mortar penetrating through the holes was not subjected to shearing. It acted as a dowel and prevented displacement. For **B10** models, an increase in displacements was observed at lower values of the loading force when compared to **BP10** models.

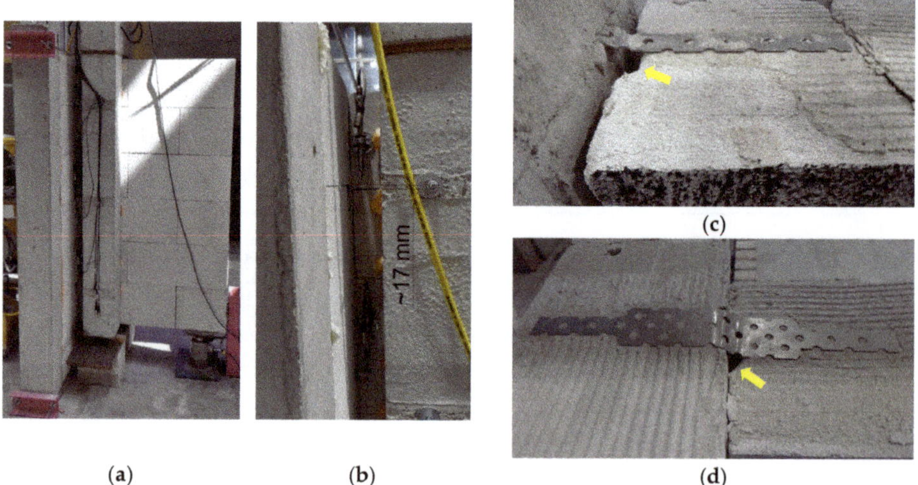

(a)   **(b)**   **(c)**   **(d)**

Figure 7. Failure of reinforced models: (**a**) damaged model (B10_1), (**b**) damaged model with dimensioned displacement between bed joints (B10_2), (**c**) typical bending of punched flat profile near the contact surface (B10_1), (**d**) typical bending of punched flat profile near the contact surface (BP10_3).

Bent connectors were removed from damaged models, inspected and their permanent deformation was evaluated—see Figure 8. The shape of connectors was originally flat in areas of anchoring in joints. However, permanent deformation occurred in the central area where connectors crossed the wall joints. A permanent displacement u_u perpendicular to thflate connector axis was observed in the section marked e_u. Additionally, the representative total extension of each connector δ_u was calculated. A permanent displacement e_u in models B10 ranged from 20 mm to 27 mm, and the mean was 23 mm (23t) at the mean displacement u_u between 8 mm and 17 mm and a mean of 11 mm (11t). A permanent displacement e_u in models BP10 ranged from 20 mm to 29 mm, and the mean was 23 mm (23t). The vertical displacement u_u was between 8 mm and 17 mm, and the mean was 12 mm (12t). Deformation seemed to be identical despite the shape of the connectors. The only reported difference was the position of the deformed area regarding the mid-length of the connector. Displacements observed for some connectors were of the order of ±20 mm with respect to the mid-length of the connector. As no regularity caused by, e.g., their position in joints was found, the above was assumed to be the effect of precisely made joints. The measured geometry of the connectors in each model and the mean values are presented in Table 8.

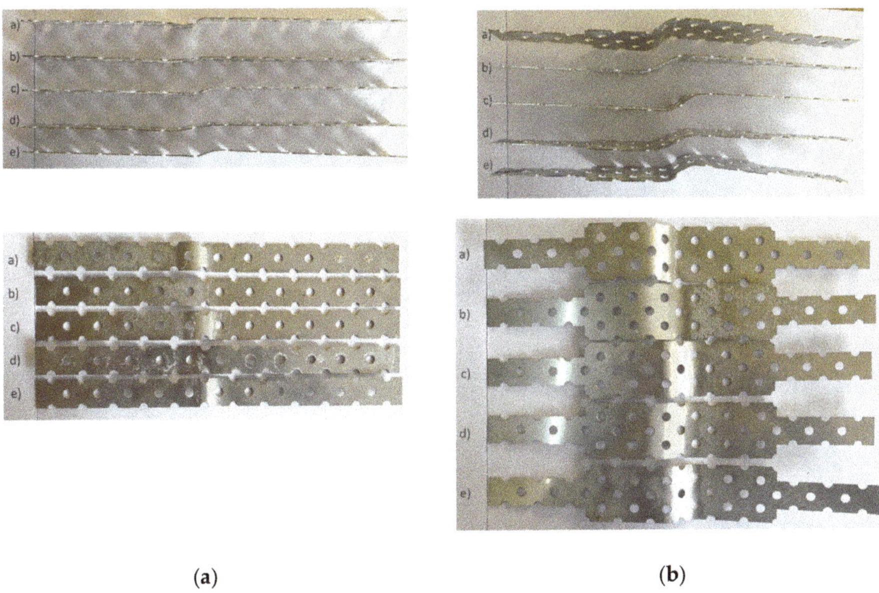

(**a**)   (**b**)

Figure 8. Deformed connectors removed from damaged test models after tests: (**a**) punched flat profiles in the wall B10_2, (**b**) punched widened flat profiles in the wall BP10_2.

Table 8. Measured geometry of deformed connectors.

Model	Layer of Connectors	Distance between Points of Contraflexure		Relative Displacement of Connector Ends		Connector Extension	
		$e_{u,i}$	$e_{u,mv}$	$u_{u,i}$	$u_{u,mv}$	$\delta_{u,i}=\sqrt{e_u^2+u_u^2}-e_u$	$\delta_{u,mv}$
		mm	mm	mm	mm	mm	mm
1	2	3	4	5	6	7	8
B10_1	a	23		9		1.70	
	b	21		8		1.47	
	c	20	22	9	9	1.93	1.8
	d	24		9		1.63	
	d	21		10		2.26	
B10_2	a	26		11		2.23	
	b	26		10		1.86	
	c	25	24	11	10	2.31	2.2
	d	22		10		2.17	
	d	20		10		2.36	
B10_3	a	21		11		2.71	
	b	23		13		3.42	
	c	23	23	15	14	4.46	3.7
	d	23		12		2.94	
	d	27		17		4.91	
			23 (23t)	–	**11 (11t)**	–	**2.57 (2.57t)**
BP10_1	a	19		12		3.47	
	b	27		14		3.41	
	c	24	24	13	13	3.29	3.33
	d	29		11		2.02	
	d	23		15		4.46	
BP10_2	a	23		17		5.60	
	b	22		14		4.08	
	c	27	23	14	15	3.41	4.22
	d	22		13		3.55	
	d	23		15		4.46	

Table 8. *Cont.*

1	2	3	4	5	6	7	8
BP10_3	a	22		12		3.06	
	b	23		10		2.08	
	c	19	23	9	10	2.02	2.01
	d	26		8		1.20	
	d	23		9		1.70	
			23 **(23t)**	–	**13** **(13t)**	–	**3.28** **(2.57t)**

As the in case of unreinforced joints, phases of reinforced joints can be presented in diagrams illustrating the relationship between the load N and relative (mutual) displacement u of joined walls—see Figure 9.

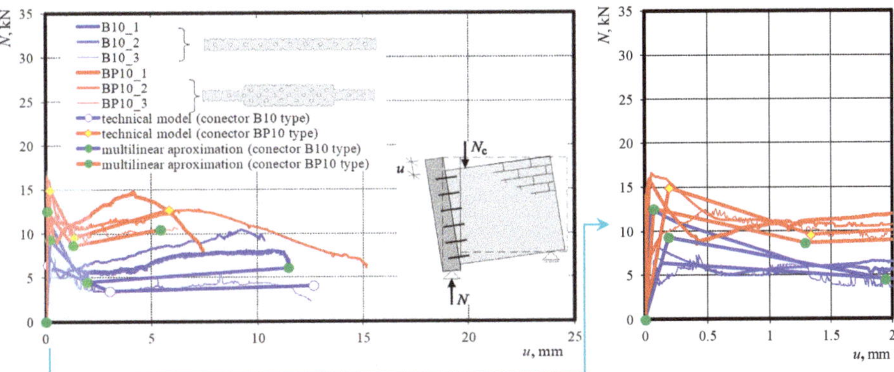

Figure 9. Relationship between the total force and mean displacement of joints.

Until the crack on the contact area that appeared under the maximum load $N_{cr} = N_u = 7.3–12.3$ kN for models **B10** and 12.5–16.5 kN for models **BP10**, an increment in displacement was nearly proportional and that phase was defined as the elastic phase. A clear increase in displacements and a drop in force to $N_d = 3.4–5.0$ kN in models **B10** and 8.9–10.5 kN in models **BP10** was observed after cracking in the failure phase. When the force N_d was reached in the failure phase, the joint demonstrated the capacity to take load, and a small hardening was noticed. The failure of the models caused by excessive displacements was observed under the maximum load $N_{cr} = N_u = 2.3–9.2$ kN for models **B10** and 10.6–14.8 kN for models **BP10**. Thus, a drop in the residual force of the maximum force was ca. 35% for models **B10** and only 15% for models **BP10**. Connectors **B10** produced lower values of the force in individual phases. Loading at the time of cracking was lower by 76%, and the maximum loading was lower by as much as 82%. Moreover, the residual force was lower by 63% when compared to the force determined for unreinforced models. Displacements in the reinforced models at the greatest force were lower only by 18% compared to the unreinforced joint. Displacements in reinforced joints greater than 100% were found under the residual force at the end of the failure phase. When compared to the unreinforced models, the cracking force acting on the models with connectors **BP10** with a widened central part was lower by 62% than in the model with the traditional joint. The maximum cracking force acting on the reinforced models was lower by 71% than in the case of the unreinforced models. Furthermore, the residual force was greater by more than 63%. Displacements in the reinforced models at the greatest force were lower by 15% than in the unreinforced joint. Moreover, displacements slightly greater by 4% than in unreinforced models were observed under the residual force at the end of the failure phase. A twofold widening of the connector in the models **BP10** resulted in ca. 60% increase in forces N_u and over 100% increase in forces N_d and N_r when compared to results obtained for models **B10**. Displacements in the models with wider connectors were as expected—almost identical in the elastic phase and lower by 30%–50% in the failure phase. The observed phases were the basis of a multi-linear diagram illustrating the N–u relationship for joints in AAC walls—see Figure 10. The elastic phase was defined within the loading range 0–$N_{cr} = N_u$, and the failure phase within the range N_u–N_d–N_r.

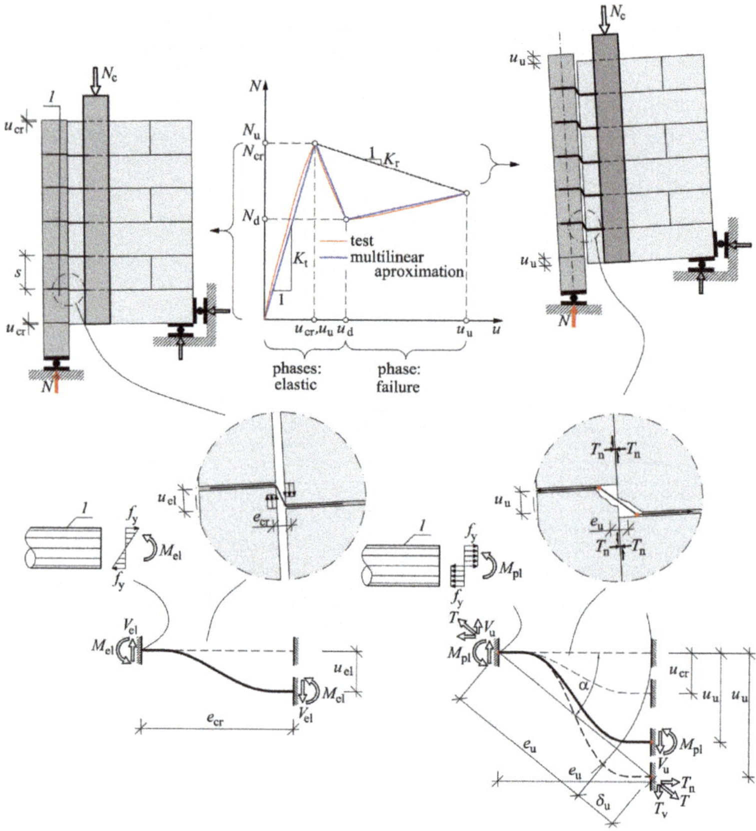

Figure 10. Approximation of work of reinforced joint between masonry walls (connector (*1*)).

Values of forces and corresponding displacements are presented in Tables 9 and 10. Joint stiffnesses were determined in each phase according to Equations (1)–(3) and they are presented in Table 11. A linear approximation of results is shown in Figure 10.

Table 9. Test results for reinforced joints (forces).

Model	Cracking Force		Force at Failure		Dowel Force		Residual Force	
	$N_{cr,i}$ kN	$N_{cr,mv}$ kN	$N_{u,i}$ kN	$N_{u,mv}$ kN	N_d kN	$N_{d,mv,i}$ kN	$N_{r,i}$ kN	$N_{r,mv}$ kN
B10_1	12.3		12.3		5.01		6.68	
B10_2	8.41	9.3	8.41	9.3	5.02	4.5	9.20	6.1
B10_3	7.27		7.27		3.39		2.32	
BP10_1	15.9		15.9		8.86		14.8	
BP10_2	16.5	14.9	16.5	14.9	10.5	9.6	12.6	12.7
BP10_3	12.4		12.4		9.36		10.6	

Table 10. Test results for reinforced joints (displacements).

Model	Displacement at the Time of Cracking		Displacement Right before Failure		Displacement at Dowel Force		Residual Displacement	
	$u_{cr,i}$ mm	$u_{cr,mv}$ mm	$u_{u,i}$ mm	$u_{u,mv}$ mm	$u_{d,i}$ mm	$u_{d,mv}$ mm	$u_{r,i}$ mm	$u_{r,mv}$ mm
B10_1	0.07		0.07		1.83		11.50	
B10_2	0.08	0.19	0.08	0.19	0.68	1.94	10.33	11.45
B10_3	0.41		0.41		3.32		12.52	
BP10_1	0.04		0.04		0.45		4.15	
BP10_2	0.05	0.19	0.05	0.19	1.61	1.33	7.04	5.80
BP10_3	0.49		0.49		1.94		6.22	

Table 11. Test results for reinforced joints (stiffness).

Model	Elastic Joint Stiffness		Residual Joint Stiffness	
	$K_{t,i}$ MN/m	$K_{t,mv}$ MN/m	$K_{r,i}$ MN/m	$K_{r,mv}$ MN/m
B10_1	180		0.496	
B10_2	102	100	0.077	0.327
B10_3	17.8		0.409	
BP10_1	432		0.269	
BP10_2	319	259	0.553	0.378
BP10_3	25.6		0.312	

Validation of the Model Representing Reinforced Joints in Walls

Like for unreinforced joints, the obtained results were generalised. The following assumptions were made:

d) A non-linear relationship N–u determined from tests was replaced with a multi-linear relationship expressing all observed phases:

    i.   the elastic phase observed in the load range $0 - N_{cr} = N_u$;

    ii.   the failure phase observed in the load range N_u–N_d–N_r.

e) It was suggested that all material parameters used in the model should be specified using standard and normalised methods;

f)   An elastic and perfectly plastic model of the connector was used;

g)   The model would be subjected to statistical validation on the basis of performed tests.

The behaviour of the joint was described with simplified solutions found in the literature [20,21]. According to the cited papers, connectors were working as bars fixed on both sides, and the value of the force causing the displacement u could be expressed as follows:

$$V = \frac{12E_sI}{e^3}u,\tag{13}$$

the corresponding bending moment in the connector is equal to:

$$M = \frac{6E_sI}{e^2}u,\tag{14}$$

where EI is the flexural stiffness of the connector, u is the relative displacement of the connector ends and e is the representative length of the connector (distance between points of contraflexure).

Stress values of extreme fibres in the connector fixed in bed joints increased proportionally to the displacement u. For some displacements u_{el}, stress at extreme fibres reached the yield point, and the bending moment and the shearing force were expressed via the following equations:

$$M_{el} = f_yW_{el} = \frac{6E_sI}{e^2_{el}}u_{el},\ V_{el} = \frac{2M_{el}}{e_{el}}.\tag{15}$$

where W_{el} is the elastic indicator of the transverse bending of the connector section, f_y is the representative yield point of steel in the connector and e_{el} is the connector length in the elastic phase.

An increase in the relative displacements of the ends of connectors was observed with the yielding of the total section of the connector, resulting in the highest bending moment and the greatest shearing force equal to:

$$M_{pl} = f_yW_{pl} = \frac{6E_sI}{e^2_{pl}}u_{pl},\ V_u = \frac{2M_{pl}}{e_{pl}},\tag{16}$$

where W_{pl} is the plastic index of the transverse bending of the connector section, f_y is the representative yield point of steel in the connector and e_{el} is the connector length in the plastic phase.

An increase in relative displacements could cause the spalling of the wall beneath the connector and an increase in the length of connectors. This, in turn, could produce a noticeable drop in the force in the joints. As in previous phases, bending moments in connectors and shearing forces were determined from the following relationship:

$$M_d = f_yW_{pl} = \frac{6E_sI}{e^2_d}u_d,\ V_d = \frac{2M_d}{e_d},\tag{17}$$

Tests demonstrated that a further increase in relative displacements could cause an increase in the forces in joints. In that phase, displacements were so considerable that connectors could work in a flexible and also a tendon mode. Consequently, friction force was generated between the joined walls. The bending moment and shearing forces in the joint can be expressed as:

$$M_u = f_yW_{pl} = \frac{6E_sI}{e^2_u}u_u,\ V_u = \frac{2M_u}{e_u},\tag{18}$$

And the axial force in the joint induced by tendon work was:

$$T = E_sA\frac{\delta_u}{e_u},\tag{19}$$

where δ_u is the extension of the connector, determined from the following equation:

$$\delta_u = \sqrt{e_u^2 + u_u^2} - e_u. \tag{20}$$

The horizontal and vertical components of force, being the effects of the tendon work (at $\alpha \approx 0$), were equal to:

$$T_n = T \cos \alpha \approx T,$$
$$T_v = T \sin \alpha \approx 0. \tag{21}$$

Taking into account the tendon work of connectors, the load capacity of reinforced joints in walls can be expressed as:

$$V_u = \frac{2f_y W_{pl}}{e_u} n_c + \alpha n_c E_s A \frac{\delta_u}{e_u} \mu, \tag{22}$$

where μ is the friction coefficient α is the empirical coefficient, e_u is the average length of the connector (distance between points of contraflexure acc. to Table 8) and $n_c = 5$ is the number of connectors.

The corresponding displacement is expressed by the following relationship:

$$u_u = \frac{f_y W_{pl} e_u^2}{6E_s I} \beta, \tag{23}$$

where μ is the friction coefficient and β is the empirical coefficient.

Forces in the failure phase can be determined similarly.

$$V_d = \frac{2f_y W_{pl}}{e_u} n_c + \alpha_1 n_c E_s A \frac{\delta_u}{e_u} \mu, \tag{24}$$

$$u_d = \frac{f_y W_{pl} e_u^2}{6E_s I} \beta_1, \tag{25}$$

$$V_r = \frac{2f_y W_{pl}}{e_u} n_c + \alpha_2 n_c E_s A \frac{\delta_u}{e_u} \mu, \tag{26}$$

$$u_r = \frac{f_y W_{pl} e_u^2}{6E_s I} \beta_2. \tag{27}$$

Above equations included not only the mechanical parameters of the connectors (E, f_y) but also the measured length of connectors e_u—the distance between points of contraflexure. However, this approach is not unconditional. The length of connectors measured in the tests was ca. 23t. The authors in [20] determined experimentally that the length of connectors from flat profiles was (1.6–2.5)t, provided that masonry units below the connector were not crushed as observed in the models made of AAC. Like for unreinforced models, the values of empirical coefficients were calculated using results from material tests and tests on individual elements. Boundary values of mean coefficients α, α_1, α_2, β, β_1, β_1 were determined at the significance level $\alpha = 0.8$. As the sample size was small, the relationship expressed by the relationship in Equation (11) was used. Lower and upper values from the confidence interval of mean coefficients are presented in Table 12.

Table 12. Validation of empirical coefficients of the model with reinforced wall joints.

Model	x_i					
	$\alpha_i = \dfrac{V_{u,i} - \frac{2f_y W_{pl}}{e_u} n_c}{n_c E_s A \frac{\delta u}{e_u} \mu}$	$\beta_i = \dfrac{6E_s I u_{u,i}}{f_y W_{pl} e_u^2}$	$\alpha_{1i} = \dfrac{V_{d,i} - \frac{2f_y W_{pl}}{e_u} n_c}{n_c E_s A \frac{\delta u}{e_u} \mu}$	$\beta_{1i} = \dfrac{6E_s I u_{d,i}}{f_y W_{pl} e_u^2}$	$\alpha_{2i} = \dfrac{V_{r,i} - \frac{2f_y W_{pl}}{e_u} n_c}{n_c E_s A \frac{\delta u}{e_u} \mu}$	$\beta_{2i} = \dfrac{6E_s I u_{r,i}}{f_y W_{pl} e_u^2}$
B10_1	0.01117	0.10	0.00421	2.73	0.00580	17.2
B10_2	0.00744	0.12	0.00422	1.02	0.00819	15.4
B10_3	0.00636	–	0.00268	–	–	18.7
n	3	2	3	2	2	3
$\bar{x}$	0.00832	0.11	0.003702	1.88	0.00699	17.11
S	0.00253	0.0146	0.0008880	1.21	0.00169	1.63
$t_{1-\alpha/2}$	1.89	3.08	1.89	3.08	3.08	1.89
$\bar{x} - t_{1-\alpha/2}\frac{S}{\sqrt{n}}$	0.00557	0.081	0.00274	−0.75	0.0033	15.33
$\bar{x} + t_{1-\alpha/2}\frac{S}{\sqrt{n}}$	0.01107	0.145	0.00467	4.50	0.0107	18.89
BP10_1	0.00619	0.06	0.00328		0.00573	
BP10_2	0.00645	0.08	0.00397	2.43	0.00484	10.6
BP10_3	0.00476		0.00348	2.91	0.00401	9.4
n	3	2	3	2	3	2
$\bar{x}$	0.00580	0.07	0.00358	2.67	0.00486	9.98
S	0.000911	0.0	0.000356	0.3	0.000859	0.877
$t_{1-\alpha/2}$	1.89	3.08	1.89	3.08	1.89	3.08
$\bar{x} - t_{1-\alpha/2}\frac{S}{\sqrt{n}}$	0.0048	0.0319	0.00319	1.93	0.0039	8.1
$\bar{x} + t_{1-\alpha/2}\frac{S}{\sqrt{n}}$	0.0068	0.1014	0.0040	3.41	0.0058	11.9

Following the procedure conducted for unreinforced joints, two values defining lower and upper limits of confidence intervals matched each of the six coefficients (Table 12). Thus, there were $\binom{6}{2}$ different combinations (without any repetitions) for coefficients. Similarly, as for unreinforced joints, the minimum value of the mean percentage error (MPE) [19] was applied as a selection criterion separately for forces and displacements. Optimal values of those coefficients were calculated from 15 combinations. For the values of coefficients in the shaded cells in Table 12, the minimum MPE for forces and displacements in connectors B10 was equal to 22%. For connectors BP10, the MPE for forces and displacements was 11%. Using results from the model and standard tests, empirical relationships describing the work of joints in particular phases are presented in Table 13, and calculated values and empirically obtained values are compared in Table 14 and Figure 10.

Table 13. Relationships expressing the work of reinforced joints in walls.

Joint Phase	Force	Stiffness	Displacement
	Connector B10		
Elastic phase	$V_u = \frac{2f_y W_{pl}}{e_u} n_c + 0.0056_c E_s A \frac{\delta_u}{e_u}\mu$	$K_t = V_u/u_u$	$u_u = 0.145\frac{f_y W_{pl} e_u^2}{6E_s I}$
Failure phase	$V_d = \frac{2f_y W_{pl}}{e_u} n_c + 0.0027 n_c E_s A \frac{\delta_u}{e_u}\mu$	$K_r = (V_u - V_r)/(u_r - u_u)$	$u_d = 4.50\frac{f_y W_{pl} e_u^2}{6E_s I}$
	$V_r = \frac{2f_y W_{pl}}{e_u} n_c + 0.0023 n_c E_s A \frac{\delta_u}{e_u}\mu$		$u_r = 18.9\frac{f_y W_{pl} e_u^2}{6E_s I}$
	Connector BP10		
Elastic phase	$V_u = \frac{2f_y W_{pl}}{e_u} n_c + 0.0048 n_c E_s A \frac{\delta_u}{e_u}\mu$	$K_t = V_u/u_u$	$u_u = 0.10\frac{f_y W_{pl} e_u^2}{6E_s I}$
Failure phase	$V_d = \frac{2f_y W_{pl}}{e_u} n_c + 0.0032 n_c E_s A \frac{\delta_u}{e_u}\mu$	$K_r = (V_u - V_r)/(u_r - u_u)$	$u_d = 1.93\frac{f_y W_{pl} e_u^2}{6E_s I}$
	$V_r = \frac{2f_y W_{pl}}{e_u} n_c + 0.0039 n_c E_s A \frac{\delta_u}{e_u}\mu$		$u_r = 8.1\frac{f_y W_{pl} e_u^2}{6E_s I}$

Table 14. Compared tests results and own calculations for the standard model.

Test results for connector B10			Calculations for connector B10		
forces			forces		
$N_{cr,mv} = N_{u,mv}$ kN	$N_{d,mv}$ kN	$N_{r,mv}$ kN	$N_{cr,cal} = N_{u,cal}$ kN	$N_{d,cal}$ kN	$N_{r,cal}$ kN
9.34	4.47	6.07	6.44	3.45	4.06
Displacements of connector B10			Calculated displacements of connector B10		
$u_{cr,mv} = u_{u,mv}$ mm	$u_{ag,mv}$ mm	$u_{r,mv}$ mm	$u_{cr,cal} = u_{u,cal}$ mm	$u_{u,cal}$ mm	$u_{r,cal}$ mm
0.19	1.94	11.45	0.10	3.01	12.6
Test results for connector BP10			Calculations for connector BP10		
force			force		
$N_{cr,mv} = N_{u,mv}$ kN	$N_{d,mv}$ kN	$N_{r,mv}$ kN	$N_{cr,cal} = N_{u,cal}$ kN	$N_{d,cal}$ kN	$N_{r,cal}$ kN
14.94	9.59	12.69	12.6	8.65	10.43
Displacements of connector BP10			Calculated displacements of connector BP10		
$u_{cr,mv} = u_{u,mv}$ mm	$u_{ag,mv}$ mm	$u_{r,mv}$ mm	$u_{cr,cal} = u_{u,cal}$ mm	$u_{u,cal}$ mm	$u_{r,cal}$ mm
0.19	1.33	5.80	0.07	1.29	5.40

For standard connectors **B10** without widening, calculated forces determining coordinates of particular phases were lower than those obtained during tests. The difference for the maximum force was equal to 31%, and for the aggregate interlocking force −23%. The value of the force Nr in the failure phase was lower by 33% than the empirical value. Similar results were obtained for connectors **BP10**. Determined force values were lower than experimental ones. The maximum force Nu was lower by 16%, and forces V_d and Vr in the failure phase were lower by 10% and 18%, respectively, when compared to forces determined experimentally. Calculated displacements of joints with connectors **B10** varied significantly. The calculated displacement at failure was lower by 48% than the experimentally determined values. Moreover, displacements in the failure phase corresponding to the force V_d were greater by over 55% than experimental values, and calculated displacements were greater only by 10%. For connectors **BP10**, displacements at the maximum force were underestimated at a level of over 65%, and overestimated by only 3% under the force V_d. Differences in calculated and measured displacements at failure were equal to just 7%.

The obtained results, particularly for forces, can be used to estimate, with safe margins, the forces in joints and to verify SLS conditions where no guidelines can be applied. As for unreinforced joints, the greatest differences were observed for displacements. The recommended relationships can cause a significant underestimation of displacement at failure, even at the level of ca. 50%.

4. Conclusions

Tests described in this paper are a part of a piece of complex research work conducted at the Silesian University of Technology. This paper presents results from testing three types of wall joints: a traditional mortar bonding (URM), joints with punched steel flat profiles (**B10**) and with connectors of genuine shape (**BP10**) protected by the patent.

The failure process and crack development on the walls bonded with mortar were mild and included three phases. Distinct wall cracks near the joint were observed prior to failure. Failure and cracking of models with steel elements, apart from lower load capacity, were completely different. No cracks preceding the wall destruction were observed, but there were rapid displacements and a drop in loading. For perforated flat profiles used as steel connectors, significantly lower values were obtained when compared to the models with mortar bonding. Forces at the time of cracking were lower by 62% (**BP10**) and 76% (**B10**), and the difference at the maximum force was 82% (**BP10**) and 71% (**B10**). Reinforced models were less deformed in the elastic phase. Differences at the maximum force were 18% (**B10**) and 15%(**BP10**). Greater differences were observed for displacements prior to the failure. Displacements in the models with reinforced joints **B10** were greater by over 100% than in unreinforced models. Generally, the same displacements were reported for the models with connectors **BP10**. A twofold widening of the connector in models **BP10** resulted in a ca. 60% increase in maximum forces when compared to results obtained for models **B10**. Displacements in the models with wider connectors were as expected and almost identical in the elastic phase and lower by 30%–50% in the failure phase.

Particular phases of joint work were determined and defined, and an empirical approach was proposed to determine the forces and displacement of wall joints using the results from less complicated standard tests. Values of cracking and failure forces were estimated with a safety margin for unreinforced joints. Moreover, they differed by 15% and 9% in comparison to the test results. On the basis of relationships described in the literature [20,21], a technical solution was proposed, which included the determination of forces producing cracks on the contact area and maximum forces in joints between walls reinforced with punched flat profiles. Due to the small number of elements per series, differences in the safe estimation of forces were of the order of 31% for maximum forces in connector **B10**, and 26% in connector **BP10**.

Work should be continued and additional test models should be constructed to define the statistically empirical parameters of models. Then, the results of validation can be expected to provide lower differences in extreme values. Moreover, FEM (Finite Element Method)-based analyses seem to be necessary to determine the real work of joints, particularly to determine their real length (e). The target model should also give consideration to the phase of joint weakening and to the estimation of forces N_{cr}, N_d and N_u and corresponding displacements with satisfactory accuracy.

Author Contributions: Conceptualization, R.J. and I.G.; methodology, R.J. and I.G.; validation, R.J.; formal analysis, I.G.; investigation, I.G.; writing—original draft preparation, R.J.; writing—review and editing, I.G.; visualization, R.J. and I.G.; supervision, R.J. and I.G. All authors have read and agreed to the published version of the manuscript.

Funding: The research was financed from the own funds of the Department of Building Structures and Department of Structural Engineering Silesian University of Technology and project: NB-323/RB-2/2017 Experimental tests of joints in masonry walls made of autoclaved aerated concrete, financed by Solbet Company.

Acknowledgments: The authors would like to express particular thanks to Solbet and NOVA companies for valuable suggestions and the delivery of masonry units, mortar, and connectors which were used to prepare test models and perform tests.

Conflicts of Interest: The authors declare no conflict of interest.

References

1. Castro, L.O.; Alvarenga, R.D.C.S.; Silva, R.M.; Ribeiro, J.C.L. Experimental evaluation of the interaction between strength concrete block walls under vertical loads. *Revista Ibracon de Estruturas e Materiais* **2016**, *9*, 643–681. [CrossRef]
2. Paganoni, S.; D'Ayala, D. Testing and design procedure for corner connections of masonry heritage buildings strengthened by metallic grouted anchors. *Eng. Struct.* **2014**, *70*, 278–293. [CrossRef]
3. Maddaloni, G.; Balsamo, A.; Di Ludovico, M.; Prota, A. Out of Plane Experimental Behavior of T-Shaped Full Scale Masonry Orthogonal Walls Strengthened with Innovative Composite Systems. In Proceedings of the Fourth International Conference on Sustainable Construction Materials and Technologies, Las Vegas, NV, USA, 7–11 August 2016.
4. Maddaloni, G.; Di Ludovico, M.; Balsamo, A.; Prota, A. Out-of-plane experimental behaviour of T-shaped full scale masonry wall strengthened with composite connections. *Compos. Part B Eng.* **2016**, *93*, 328–343. [CrossRef]
5. Galman, I.; Jasiński, R.; Hahn, T.; Konopka, K. Study of joints masonry walls. *Materiały Budowlane* **2017**, *10*, 94–96. (In Polish) [CrossRef]
6. Galman, I.; Jasiński, R. Joints in masonry walls. *Ce/papers* **2018**, *2*, 339–346. [CrossRef]
7. Galman, I.; Jasiński, R. Tests of joints in AAC masonry walls. *Arch. Civ. Eng. Environ.* **2018**, *11*, 79–92. [CrossRef]
8. Polish Patent Office. Wall Joint Connector. Niepodległości 188/192, 00-950 Warsaw (Poland). Patent No. W.128,153, 1 April 2019.
9. PN-EN 1052-1:2000 Methods of Tests for Masonry. *Part 1: Determination of Compression Strength*; Polish Committee for Standardization (PKN): Warsaw, Poland, 2000. (In Polish)
10. Jasiński, R.; Drobiec, Ł. Comparison Research of Bed Joints Construction and Bed Joints Reinforcement on Shear Parameters of AAC Masonry Walls. *J. Civ. Eng. Arch.* **2016**, *10*, 1329. [CrossRef]
11. PN-EN 1052-3:2004 Methods of Tests for Masonry. *Part 3: Determination of Initial Shear Strength*; Polish Committee for Standardization (PKN): Warsaw, Poland, 2004. (In Polish)
12. Drobiec, Ł.; Jasiński, R. Influence of the kind of mortar on mechanical parameters of AAC masonry subjected to shear – the basic strength parameters. *Materiały Budowlane* **2015**, *5*, 106–109. (In Polish) [CrossRef]
13. Jasiński, R. Research and Modelling of Masonry Shear. Walls. Thesis, Silesian University of Technology, Gliwice, Poland, 2017.
14. ASTM E519-81. *Standard Test Method for Diagonal Tension (Shear) of Masonry Assemblages*; American Society for Testing and Materials: West Conshohocken, PA, USA, 2000.
15. Drobiec, Ł.; Jasiński, R. Influence of the kind of mortar on mechanical parameters of AAC masonry subjected to shear—dilatational deformability. *Materiały Budowlane* **2015**, *7*, 116–119. (In Polish) [CrossRef]
16. PN-EN 10002-1:2004 Metallic Materials—Tensile Testing. *Part 1. Method of Test at Ambient Temperature*; Polish Committee for Standardization (PKN): Warsaw, Poland, 2004. (In Polish)
17. Galman, I.; Jasiński, R. Attempt to Describe the Mechanism of Work of Masonry Joints. *IOP Conf. Series Mater. Sci. Eng.* **2019**, *471*, 052054. [CrossRef]
18. Volk, W. *Applied Statistics for Engineers*; Literary Licensing, LLC: Whitefish, MT, USA, 2013.
19. David, F.; Robert, P.; Roger, P. *Statistics*, 4th ed.; W.W. Norton & Company: New York, NY, USA, 2007.

20. Simudic, G.; Page, A.W. Australian Developments in the Use of Walls of Geometric Section. In Proceedings of the 7th North American Masonry Conference, University of Notre Dame-South Bend, South Bend, IN, USA, 2–5 June 1996; Volume 2, pp. 1007–1018.
21. Phipps, M.E.; Montague, T.I. The Behaviour and Design of Steel Shear Connectors in Plain and Prestressed Masonry. In Proceedings of the 7th North American Masonry Conference, University of Notre Dame-South Bend, South Bend, IN, USA, 2–5 June 1996; Volume 2, pp. 789–798.

Article

Holistic Analysis of Waste Copper Slag Based Concrete by Means of EIPI Method

Wojciech Kubissa [1,*], Roman Jaskulski [1], Damian Gil [2] and Iwona Wilińska [1]

[1] Faculty of Civil Engineering, Mechanics and Petrochemistry, Warsaw University of Technology, Płock 09-400, Poland; roman.jaskulski@pw.edu.pl (R.J.); iwona.wilinska@pw.edu.pl (I.W.)

[2] Faculty of Civil Engineering and Architecture, Lublin University of Technology, Lublin 20-618, Poland; gildamian13@gmail.com

* Correspondence: wojciech.kubissa@pw.edu.pl; Tel.: +48-24-3672185

Received: 5 November 2019; Accepted: 16 December 2019; Published: 19 December 2019

Abstract: The aim of the research is a comprehensive evaluation of concrete using the EIPI method. In the evaluation the compressive strength of concrete and its durability properties represented by sorptivity and air permeability are taken into account. Since waste copper slag with increased natural radioactivity is used in the assessed concrete, additional evaluation is carried out taking into account the influence of natural radioactivity within the performance index. Additionally, the reference concrete, which is made without the use of waste copper slag, is evaluated for comparative purposes. In order to make the evaluation as comprehensive as possible, the concrete made with the use of three types of cement is subjected to CEM I, CEM II and CEM III assessments. If natural radioactivity is not taken into account in the evaluation, the best result of the most favourable value of Gross Ecological and Performance Indicator (GEPI) is obtained by the concrete made with waste copper slag, and if radioactivity is considered, the most favourable value of GEPI is obtained with concrete without addition of the waste. The results show that in both approaches the best result is achieved by concrete with CEM III cement. It follows from the above that although natural radioactivity has a significant impact on the EIPI evaluation result, the decisive factor is still the type of cement.

Keywords: concrete performance; concrete durability; EIPI method; waste copper slag; natural radioactivity

---

1. Introduction

The currently dominant model of goods production in the economy is linear. This assumes the acquisition of raw materials, the production of specific goods associated with the simultaneous production of waste, and then the goods produced after their consumption also become waste. This linear, unidirectional model begins to reach its limits due to the limited amount of natural resources. Another disadvantage is the production of large amounts of waste, which are deposited in landfills. Such landfills not only occupy a place, but can also be a source of emissions of harmful substances or radiation.

In order to be able to develop further in a harmonious manner we must follow the example of nature, which continually performs recycling processes [1,2]. Thanks to decay processes, which are an important part of its internal cycle, nature is an ideal example of a zero-waste economy. Trying to get at least a little closer to this model, it is worth making attempts to reuse post-production waste, treating it not as waste, but as raw materials of a new era. This is the basic premise of a circular economy, which is currently gaining more and more interest.

The cement and building materials industries offer great opportunities for using different mineral by-products. Materials, such as fly ash, silica fume and blast furnace slag, are commonly used as supplementary cementitious materials (SCMs) [3], the introduction of which into cement composites gives the possibility to reduce the amount of cement used and, consequently, a reduction of the adverse impacts of cement production on the environment. On the other hand, reduction of the amount of landfilled waste is possible. However, the introduction of SCMs into the concrete changes its chemical composition and rheological properties. In effect, the properties of the final composite are modified depending on the kind of SCM used, its quantity, and physicochemical properties. Therefore, obtaining hardened material with the required properties requires investigation and analysis of the physicochemical processes occurring over time in the system. In some cases, the starting material may require an additional treatment and modification procedure (e.g., chemical or physical activation) [4–7], and the composition of the mix should be optimized. It is also important that the final material does not adversely affect its user, so it is necessary to study, e.g., its natural radioactivity.

One such raw material, currently not often used in cement composite contrary to the SCMs mentioned above, is copper slag, which is a by-product from the process of copper extraction by smelting. The residues from the copper smelting process in the form of hot liquid are taken to landfills where they are cooled and then ground. The copper slag thus obtained contains a significant amount of SiO_2 and if it is cooled down quickly enough, this compound takes an amorphous form and exhibits a pozzolanic activity (the ability to react with $Ca(OH)_2$ in the presence of water to produce hydrated silicate and aluminate phases similar to those that are formed during Portland cement hydration). Additionally, its physical properties are similar to natural sand [8]. Copper slag obtained directly from smelters is a valued abrasive material used in surface blast-cleaning processes. Due to the morphology of the grains, it is more effective than sand.

Although the ground slag is, in large part (in Poland practically entirely), used as an abradant, after such use some of the material is treated and reused, but most of it is considered to be a waste, which is in major part disposed in landfills or stockpiles. It contains a small amount of corrosion products and corrosion protection coatings [9] and after the blast cleaning process its granulation is smoother. The fraction content of 0–0.125 mm and 0.125–0.25 mm is much higher than in the initial material. To distinguish between copper slag and the waste material from the blast cleaning procedure, the latter is referred to in the article as waste copper slag.

However, it can be utilised again, and its potential applications are described, amongst others, in [10,11]. Due to its composition and physical form, copper slag can be used in the production of concrete as a partial or total substitute for sand [12–15] even in lightweight concrete [16]. In contrast to e.g., fine fractions of recycled concrete aggregate, the material is also suitable for the production of high-quality concrete, without compromising its quality, and some properties even improve in comparison with concrete manufactured with sand [17,18]. Copper slag used instead of sand significantly improves the consistency of the mixture without changing the amount of mixing water which results in an increase in the compressive strength [13,17,19]. It is also possible to reduce the water content by about 20% while maintaining the same consistency, thus increasing the compression strength by up to 20%. The material used in the cleaning process does not have these particular advantages, as it deteriorates the consistency of the concrete due to its finer grain size, but it is still very useful in concrete technology. In [20] the use of blast-cleaning waste as a substitute for sand in concrete with a cement dosage of 300 kg/m^3 and w/c = 0.6 was tested and described. Shrinkage testing of concrete with copper slag as a substitute for sand has shown that such replacement does not have the negative consequences of increased shrinkage [12].

An important aspect of using waste materials in the production of concrete is their potential harmful impact on the natural environment. In [21] the authors suggested, that the copper slag is non-toxic and poses no environmental hazard. The slag can be safely considered for use in Portland cement and concrete manufacturing. It should be noted, however, that this material is one of the most intense sources of ionizing radiation among the materials used in construction due to its high content of natural radionuclides [22–25]. Of these, particular attention is paid to the content of radium isotopes ^{226}Ra. As a result of its decomposition radon ^{222}Rn is produced, which is a radioactive gas and can be absorbed into the human organism by breathing. There, it undergoes further radioactive decay, resulting in radioactive isotopes of lead and bismuth, which, as solids, accumulate in the body and act as mutagens on its cells [26]. The use of such a material as a concrete aggregate requires carrying out tests of the natural radioactivity of the concrete produced from it.

Studies on the radioactivity of building materials and waste used in their production are becoming more and more common [27–30]. So far, there is not a great deal of data about radon exhalation rate in building materials containing NORM residues [30]. For example, in [31] there are only 1100 pieces of data from 14 European countries on radon emanation/exhalation rate. The COST Action TU1301 project is being run: "NORM for Building materials (NORM4BUILDING)" with a view to promoting research into the reuse of waste containing increased concentrations of natural radionuclides (NORM) in customised building materials in the construction sector, while taking into account the impact on both external exposure of building users to gamma radiation and indoor air quality. Models are being developed to better simulate the behaviour of NORM residues in different types of building materials.

In this paper the use of waste copper slag obtained from blast-cleaning as a substitute for part of the sand in concrete with 360 kg/m^3 of 42.5 class cements, and w/c = 0.45 was tested and described. Some researchers pay attention to the large impact of the packing density on many concrete properties [32–35], therefore, the concrete mixtures were prepared in two variants which differed from each other in consistency and workability. For each cement type two mixtures with waste copper slag were made. In one, the same dosage of superplasticizer as in the reference series was used. In the second, the amount of superplasticizer was experimentally determined in order to obtain consistency similar to the reference series. It was 420 ± 30 mm in table flow test (near the limit between F2 and F3 class).

According to the requirements of the Polish law [36] the tests of natural radioactivity of waste copper slag and the concrete were performed. From the results the coefficients f_1 and f_2 were calculated and compared to the limit values which can be found in the relevant regulations. Leachability of hazardous elements (mainly heavy metals) was also assessed.

Optimization of the manufacturing process, the purpose of which is to obtain a material with required properties, needs consideration of many variables, including knowledge of the physicochemical processes occurring during the production process, as well as the impact of raw and final materials on the natural environment and on the user. In this work, the main emphasis was placed on evaluation of the composition of the concrete, taking into account its potential natural radioactivity. To evaluate the concrete studied, the method of multi-criteria EIPI assessment presented in [37] was applied, in which as the criteria were used: compressive strength, air permeability and sorptivity as parameters determining the durability of concrete, as well as radioactive activity indices f_1 and f_2 used for the evaluation of building materials. Concrete made of traditional fine aggregate (quartz sand) and concrete, in which waste copper slag characterized by higher values of indices f_1 and f_2, used as fine aggregate, were evaluated. Due to the co-existence of both positive (improvement of durability and mechanical properties of concrete) and negative (increase in the intensity of ionizing radiation of the material) effects of the use of waste copper slag, the valuation of the applied material solution encounters objective difficulties. The EIPI method allows this judgement to be reduced to a comparison of the value of one indicator, which significantly simplifies the evaluation.

2. Materials and Methods

2.1. Materials

Portland cement CEM I 42.5R, blast-furnace cement CEM III/A 42.5N from the Górażdże Cement Plant located in Poland and Portland-composite cement CEM II/B-V 42.5N from the Lafarge Cement Plant located in Poland, as per PN-EN 197, were used. Basic physical and chemical properties presented by the cement manufacturer are shown in Table 1.

Table 1. Basic physical and chemical properties of the cement.

Cement Type	Setting Time		Compr. Strength	Specific Surface Area	Specific Gravity	SO$_3$	Cl	Na$_2$O$_{eq}$
	Start	End		(Blaine)				
	(min)	(min)	(MPa)	(cm^2/g)	(g/cm^3)	(%)	(%)	(%)
CEM I 42.5R	176	231	57.9	3538	3.10	2.52	0.063	0.60
CEM II/B-V 42.5N	203	294	50.6	4888	2.82	2.66	0.063	1.12
CEM III/A 42.5N-LH/HSR/NA	201	306	58.3	4165	2.91	2.30	0.055	0.70

All concrete mixes contained 360 kg/m^3 of cement by a 0.45 w/c ratio. Fractions of river sand 0–2 mm and granite from the Strzegom stone mine fractions of 2–8 mm and 8–16 mm were used. Aggregates were at laboratory air-dry condition. Waste copper slag from blast cleaning was used as a partial replacement of sand. Average chemical composition of the slag is as follows: SiO$_2$ 30–45%, CaO 10–30%, Fe$_2$O$_3$ <25%, Al$_2$O$_3$ 7–15%, MgO 2–8% and the granulation was much finer than in the case of typical river sand. Waste copper slag is characterized by median diameter d$_m$ = 0.347 and the used sand by d$_m$ = 0.536. Grading of the mixes of the aggregates differed mainly in the amount of finest fractions 0–0.125 mm. The ratio of substitution was 66% of sand amount by volume. If only sand and granite were used, the portion of the finest fraction was about 0.3% while after replacing 66% of the sand with waste copper slag it increased to about 3.9%. The replacement rate allowed for the aggregate grading curves both in the reference concrete mixture and in the concrete mixture containing waste, fit between the boundary curves. Superplasticizer Chryso Optima 100 according to PN-EN 934-2 was used. Regular tap water was used as the mixing water.

Nine concrete mixtures were prepared. Mix IDs and proportions are presented in Table 2. The consistency of fresh concrete was measured by a slump test, in accordance with PN-EN 12350-2.

Table 2. Proportions of concrete mixtures(kg/m^3).

Material \ Mixture ID	CI0	CI66	CI66F	CII0	CII66	CII66F	CIII0	CIII66	CIII66F
CEM I 42.5R	360	360	360	0	0	0	0	0	0
CEM II/B-V 42.5N	0	0	0	360	360	360	0	0	0
CEM III/A 42.5N	0	0	0	0	0	0	360	360	360
natural sand 0–2 mm	598	199	198	587	196	195	591	197	196
granite aggregate 2–8 mm	621	621	618	610	610	608	614	614	612
granite aggregate 8–16 mm	659	659	655	659	647	645	651	651	649
waste copper slag	0	449	447	0	441	440	0	444	443
water	162	162	162	162	162	162	162	162	162
SP Optima Fluid 100% m.c.	0.65	0.65	1.65	0.70	0.70	1.30	0.80	0.80	1.50
W/C	0.45	0.45	0.45	0.45	0.45	0.45	0.45	0.45	0.45
W + Sp/C	0.457	0.457	0.467	0.457	0.457	0.463	0.458	0.458	0.465

Specimens were prepared and cured as per PN-EN 12390-2. They were cast in plastic moulds and compacted by double vibration (half and full) on a vibrating table. After one day they were stripped and then water-cured in the laboratory for 28 days.

2.2. Performed Tests

The compressive strength test was conducted on 100 mm cube specimens on the 28 day of hardening. The test were carried out in accordance with PN-EN 12390-3. The strength tests were performed by using a ToniTechnik instrument of 3000 kN compression force capacity. The rate of loading was maintained at 0.5 MPa/s.

A sorptivity test was conducted on the halves of cubic specimens of 100 mm edge length by means of the mass method described in [38]. Prior to the sorptivity test, the specimens were oven-dried to a stable mass at a temperature of 105 °C. The measurements were conducted at the temperature of approximately 20 °C. The specimens were weighed and arranged in a water containing vessel. Then they were immersed up to the height of 3 mm.

Air permeability testing of concrete was performed by means of the Torrent method with use of Proceq equipment. The test was conducted on two 150 mm cube specimens, which were cured in water for 28 days and then were stored in air-dry laboratory conditions (temperature t = 20 ± 2 °C and RH of air equal 55 ± 10%) until they reached age of 90 days. Moisture content was measured, before conducting the air permeability test, using Tramex CMEX II, which is recommended by Swiss Standard SIA 262/1 Annex E and by [39]. The testing procedure is described in [40].

Tests for the content of hazardous substances released from waste copper slag (i.e., leaching tests) were carried out in accordance with the applicable standards and regulations by the Laboratory of Solid Waste Analysis at the Central Environmental Monitoring Department of the Mining Institute in Katowice in accordance with Annex 3 to the Ordinance of the Minister of Economy of 16 July 2015 on the approval of waste for storage at landfills (Journal of Laws of 2015, item 1277).

The PI-MAZAR01 meter was used to perform tests of natural radioactivity. It is designed to determine the concentration of natural radioactive elements, such as radium, potassium or thorium. The measuring part is located in a lead shielded cabin, which includes a type SSU-70-2scintillation probe with a NaI (Tl) (thallium-doped sodium iodide) crystal, a preamplifier and a high voltage power supply, as well as a calibration isotope source Cs 137 used to stabilize the measuring path. In the reading part there is a microprocessor controller. The analyser is adapted to work with a PC, so that it is possible to visualize the spectrometric spectrum and save the measurement results on a hard disk.

The natural radioactivity measurement procedure begins with the calibration of the analyser according to the instrument manual and the recommendations of the instructions of Building Research Institute (ITB, Poland) [41] which recommends periodical calibration at least once a year and control measurements with the use of standards once a month or as a result of a change in conditions after 24 h (e.g., change in temperature at the place of measurement). Samples (so-called qualification samples) were prepared for testing, ground to a maximum grain size of 2 mm, then dried to a constant mass at 105 °C and left to cool under laboratory conditions to reach an air-dry state. The prepared material was placed in the Marinelli type containers with a volume of 1700 cm^3. The container and sample were then weighed, secured with adhesive tape and marked accordingly. The weight of the material of each sample was calculated on the basis of the performed weights. Afterwards, the samples were seasoned in containers for seven days at a significant distance from the measuring house (over 2 m). Before starting the measurements, the background of the samples was calculated on an aluminium mass standard and then the containers with samples were placed in the measuring chamber of the shielding house. During the study, the meter collected the measurement spectrum and then analysed the number of impulses recorded in potassium, radium and thorium windows, which were the basis for calculating concentrations of radioactive elements and qualification coefficients f_1 and f_2.

3. Results

3.1. Mechanical and Durability Properties

The results of compressive strength, sorptivity and air permeability tests are presented and discussed in detail in [40]. Table 3 presents the average values of those of all the obtained results, which were used for calculations in the EIPI analysis.

The results presented above show that compressive strength of CEM I and CEM II cement concretes containing waste copper slag increase both after the 28th and 90th days of hydration compared to the reference (CI0 or CII0 respectively). Only in the case of CEM III cement concrete, introduction of waste copper slag reduces the compressive strength. On the other hand, the presence of the sand replacement results in an improvement of the tightness of all investigated concrete compositions. The possible cause of sealing of the concrete structure is the pozzolanic reaction. The greatest share in the composition of waste copper slag is constituted by SiO_2 in amorphous form, which shows pozzolanic activity. As it is commonly known, the use of pozzolanic materials in the production of concrete improves, among other things, its tightness. An additional factor is the granulation of waste copper slag—a larger share of fine fractions. In summary, the results obtained indicate a predominance of benefits from the use of waste copper slag in concrete.

Table 3. Test results employed in EIPI calculations [40].

Test \ ID of Mixture	CI0	CI66	CI66F	CII0	CII66	CII66F	CIII0	CIII66	CIII66F
Flow (mm)	395	315R	410	410	310R	415	410	330R	440
Compressive strength 28d (MPa)	55.03	53.30	60.16	54.56	57.42	60.38	66.44	61.34	62.45
Compressive strength 90d (MPa)	60.78	61.98	68.18	63.00	67.32	70.50	73.67	68.60	71.96
Sorptivity $(cm^3/(cm^2 \cdot h^{0.5}))$	0.091	0.076	0.067	0.088	0.085	0.089	0.061	0.063	0.047
RH Tramex dry	0.40	0.54	0.68	0.58	0.89	0.81	1.42	1.39	1.33
Air permeability k_T $(\times 10^{-16}\ m^2)$	2.903	1.922	1.022	1.214	0.563	0.377	0.081	0.245	0.066

Flow: R- collapse of the specimen after lifting the cone.

3.2. Leaching and Natural Radioactivity Tests

Table 4 presents the results of a test of the leaching of hazardous substances from waste copper slag in comparison with the requirements of Polish legal regulations (The Ordinance of the Council of Ministers of 18 November 2014 on the conditions to be met when introducing sewage into water or soil and on the substances particularly harmful to the aquatic environment). The tests showed that the content of hazardous substances identified in the water extract does not exceed the permissible concentrations of these components specified in the applicable regulations.

Table 4. Hazardous substances released to water extract from waste copper slag.

Identified Ingredient or Parameter	Content in the Water Extract (mg/L)	Allowable Concentration (mg/L)
Cd	<0.001	0.2
Cr	<0.005	0.5
Cr(VI)	<0.01	0.1
Cu	0.052	0.5
Ni	<0.005	0.5
Pb	0.009	0.5
Zn	<0.05	2.0
Ba	<0.03	2.0
Sb	<0.005	0.3
As	0.026	0.1
Mo	0.011	1.0
Hg	<0.001	0.1
Se	<0.01	1.0
Chlorides	<5	1000
Fluorides	<0.1	25.0
Sulphates	3.6	500
DOC *	1.9	30
Soluble matter	33.2	—
pH of water extract	9.9	—

* Dissolved organic carbon.

The allowable content of natural radioactive isotopes in raw materials, building materials and waste used in construction is regulated by the Ordinance of the Council of Ministers of 2 January 2007 on requirements concerning the content of natural radioactive isotopes of potassium K-40, radium Ra-226 and thorium Th-228 in raw materials and materials used in buildings intended for human habitation and livestock, as well as in industrial waste used in construction, and control of the content of these isotopes. This ordinance also applies to waste used for the production of cement and concrete (such as fly ash, slag including copper slag used as an abrasive). Raw materials and building materials are qualified on the basis of two activity indicators f_1 and f_2.

The first of the above-mentioned indicators, f_1, identifies the exposure to radiation emitted by natural radionuclides (i.e., the nuclei of radioactive atoms): potassium (K), radium (Ra) and thorium (Th). This indicator takes into account the different activities of individual radioisotopes and is calculated using the Equation (1):

$$f_1 = \frac{C_K}{3000\ Bq/kg} + \frac{C_{Ra}}{300\ Bq/kg} + \frac{C_{Th}}{200\ Bq/kg} \tag{1}$$

where C_K, C_{Ra} and C_{Th} are concentration values of potassium ^{40}K, radium ^{226}Ra and thorium ^{228}Th in Bq/kg.

The f_2 indicator, calculated according to Equation (2), indicates the radium (Ra) content and indirectly the α radiation intensity emitted by radon (Rn) and products of its radioactive decay present in building materials:

$$f_2 = C_{Ra} \tag{2}$$

The results of tests of natural radioactivity of waste copper slag and coarse aggregate, i.e., granite, carried out using the method described above, are presented in Tables 5 and 6.

Table 5. Results of natural radioactivity tests of waste copper slag.

Radionuclide	Radioactivity (Bq/kg)
^{226}Ra	400 ± 12
^{228}Th	40.1 ± 3.1
^{40}K	749 ± 51

which translates into indicator values f_1 and f_2: $f_1 = 1.78 \pm 0.05$; $f_2 = 400 \pm 12$.

Table 6. Results of natural radioactivity tests of granite.

Radionuclide	Radioactivity (Bq/kg)
^{226}Ra	35.4 ± 6.1
^{228}Th	43.6 ± 4.4
^{40}K	1019 ± 69

which translates into indicator values f_1 and f_2: $f_1 = 0.67 \pm 0.05$; $f_2 = 35.4 \pm 6.1$.

According to the abovementioned ordinance, the activity rates f_1 and f_2 must not exceed by more than 20% the limit values of $f_1 = 2$ and $f_2 = 400$ Bq/kg for industrial waste used in the construction of ground structures built on built-up areas or intended to be built on in a local zoning plan and for the levelling of such areas. This means that the tested waste may be used in the production of concrete for the above-mentioned applications. Apart from testing the natural radioactivity of selected concrete components, samples of the concrete itself were also tested. The results of these tests in the case of concrete without and with waste copper slag are presented in Table 7.

The results presented in Table 7 allow to conclude that despite a relatively high level of values of indicators f_1 and f_2 obtained in the case of waste copper slag, concrete made with this material has a moderate level of radioactivity, although it is significantly higher than in the case of concrete made without the use of waste copper slag. Another important conclusion is the noticeably higher level of radioactivity of concrete, in which CEM II/B-V cement was used, compared to the series made with other cements and the same type of aggregate. The increased radioactivity of these concrete series should be linked to the presence of fly ash in the cement, which is a material with an increased radioactivity level [27,28,42,43]. Relative and absolute differences in the values of indicators f_1 and

f_2 in the case of CEM II/B-V cement concrete is significantly smaller when waste copper slag is used, which indicates the dominant influence of this component on the radioactivity of the obtained concrete. However, the impact of cement is not negligible and should be taken into account when designing the composition of concrete mix.

Table 7. Results of natural radioactivity tests of concrete.

Concrete ID	Radionuclide Activity (Bq/kg)			Indicator Value	
	^{226}Ra	^{228}Th	^{40}K	f_1	f_2
CI0	16.1 ± 5.1	34.5 ± 4.0	594 ± 48	0.42 ± 0.04	16.1 ± 5.1
CII0	60.8 ± 3.7	47.8 ± 3.5	758 ± 60	0.68 ± 0.03	60.8 ± 3.7
CIII0	16.0 ± 5.4	37.0 ± 4.2	612 ± 50	0.44 ± 0.04	16.0 ± 5.4
CI66F	101 ± 10	44.1 ± 2.9	759 ± 57	0.81 ± 0.05	101 ± 10
CII66F	127 ± 10	53.6 ± 3.5	855 ± 57	0.98 ± 0.05	127 ± 10
CIII66F	115 ± 10	45.3 ± 3.1	781 ± 58	0.87 ± 0.05	115 ± 10

4. Discussion

4.1. Assumptions and Calculation Method

Optimization of the composition of the concrete mix requires taking into account not only the properties of the final composite, but also the need to limit its broadly understood impact on the environment.

In the calculations using the EIPI method, emissions, consumption of raw materials and rarity of their occurrence were assumed according to the data presented in the article [37]. The value of PI is evaluated on the basis of the sum of normalized values of selected concrete properties. The compressive strength and sorptivity tested after 28 days were used for calculations. The reference values were adopted at the same level as in [37], i.e., f_{cm} = 60 MPa i S = 0.120 cm/h$^{0.5}$. As another concrete property, the air permeability k_T, measured with a Torrent apparatus on specimens dried at 65 °C, was included in the evaluation. As a reference value, the limit used for exposure classes XC4, XD1, XD2a, XF1 and, XF2 in Swiss Standard SIA 262 (SIA 262/1 Annex E) [39], i.e., 2.0×10^{-16} m^2, was used.

Equation (3), which contains the abovementioned concrete parameters, was used to calculate PI without taking into account radioactivity. The relevant quotients from normalization are multiplied by the respective weighting coefficients, whose values were taken as: w_{fcm} = 0.4, w_{kT} = 0.3 and w_S = 0.3 in the present study. The sum of the weighting coefficients should be equal to unity so that a concrete mix with reference values of selected properties will give a PI value of 1:

$$PI = \frac{f_{cm}}{60\ MPa} \times w_{fcm} + \frac{0.120\ cm/h^{0.5}}{S} \times w_S + \frac{2.0 \times 10^{-16}m^2}{kT} \times w_{kT} \tag{3}$$

In the further concrete assessment, the values of indicators f_1 and f_2 were taken into account in the PI calculations. In their case, the reference values were adopted according to the Ordinance mentioned above, i.e., f_1 = 2 and f_2 = 400 Bq/kg. To calculate so extended PI values Equation (4) was used:

$$PI = \frac{f_{cm}}{60\ MPa} \times w_{fcm} + \frac{0.120\ cm/h^{0.5}}{S} \times w_S + \frac{2.0 \times 10^{-16}m^2}{kT} \times w_{kT} + \frac{2}{f_1} \times w_{f1} + \frac{400}{f_2} \times w_{f2} \tag{4}$$

The higher values of PI the analysed concrete achieves, the more desirable engineering properties it possesses. The weighting coefficients in Equation (4), were assumed in a few variants which are presented and described in the next subsection.

The value of EI is calculated according to Equation (5) as the square root of the sum of the normalized total emission of CO_2 and the normalized total raw materials usage both multiplied by weights that sum to one:

$$EI = \sqrt{\frac{EM}{490 \; kg/m^3} \times w_{EM} + \frac{RM}{2000 \; kg/m^3} \times w_{RM}} \qquad (5)$$

To normalize the values of total emission of CO_2 (EM) and usage of raw materials (RM), which have to be calculated first, they are divided by the reference values. The reference values in this study were assumed as in [37] and equal approximately 490 kg of CO_2 emission and 2000 kg/m^3 of raw materials usage per cubic metre of concrete. The weighting coefficients were assumed as: $w_{EM} = 0.5$ and $w_{RM} = 0.5$.

A lower EI value means that analysed concrete is more environmentally friendly. Results of the calculations the EI for analysed concrete mixtures are presented in Table 8 and repeated in Table 9.

A comprehensive evaluation of concrete, taking into account both its ecological impact (EI) and engineering performance (PI), is expressed by Gross Ecological and Performance Indicator (GEPI), which is calculated using Equation (6):

$$GEPI = \sqrt{EI^2 + \frac{1}{PI^2}} \qquad (6)$$

Table 8. EI, PI and GEPI values without taking into account natural radiation.

	Concrete ID								
	CI0	**CI66**	**CI66F**	**CII0**	**CII66**	**CII66F**	**CIII0**	**CIII66**	**CIII66F**
EI	0.908	0.879	0.878	0.858	0.826	0.825	0.764	0.731	0.737
PI	0.856	1.024	1.592	0.898	1.114	1.156	1.262	1.463	1.956
GEPI	1.480	1.314	1.080	1.405	1.220	1.195	1.101	1.001	0.897

Table 9. Variants of weight values.

Weighting Coefficient	Weight Values in Variant:				
	0	**S1**	**S2**	**B1**	**B2**
w_{fcm}	0.40	0.28	0.28	0.12	0.12
w_S	0.30	0.21	0.21	0.09	0.09
w_{kT}	0.30	0.21	0.21	0.09	0.09
w_{f1}	0.00	0.15	0.10	0.35	0.24
w_{f2}	0.00	0.15	0.20	0.35	0.46

When designing a concrete mix in practice, a low GEPI is aimed for concrete with favourable concurrent EI and PI, while a high GEPI should be avoided.

It should be stressed very clearly here that the comparison of different variants of the designed concrete mixtures using the EIPI method in engineering practice will be only reasonable, if all the technical parameters of the concrete obtained from the designed concrete mixtures, taken into account in the PI calculations, meet the specified limit requirements defined by the construction designer or the relevant regulations or standards.

4.2. Results Analysis and Discution

The results of calculations conducted without taking into account the influence of radioactive nuclide content on the PI value are presented in Figure 1. The PI and EI values calculated under this assumption are presented in Table 8 together with the GEPI values calculated on their basis. Series with CEM III cement are characterized by the most favourable EI value due to lower clinker content than in other cements, resulting in a lower consumption of natural resources and a lower carbon dioxide emission. The highest PI values were achieved by the CI66F and CIII66F series. This is mainly due to higher tightness than in other series, which consists of the lowest values of sorptivity and one of the lowest values of air permeability. The overall assessment based on GEPI values indicates as the best series CIII66F (GEPI = 0.897) and CIII66 (GEPI = 1.001). The CI0 series (GEPI = 1.480) and CI66 series (GEPI = 1.314) were the least favourable from the point of view of the complete score.

In the next stage of the assessment, the impact of the radioactive nuclides contained in the concrete was also taken into account. This was done by using Equation (4) in the calculations of PI values. Four variants differing in the values of weights for the components of the formula taking into account indicators f_1 and f_2 were used in the calculations. Their influence on PI value was differentiated by assigning to them in the calculations a sum of weights equal to 0.3 (variants S) or 0.7 (variants B). Additional differentiation was based on taking equal weight values (variants S1 and B1) and assigning about twice as much weight to the f_2 indicator in relation to the f_1 indicator (variants S2 and B2). The list of adopted values of weights is presented in Table 9 and the obtained GEPI results are presented in Table 10.

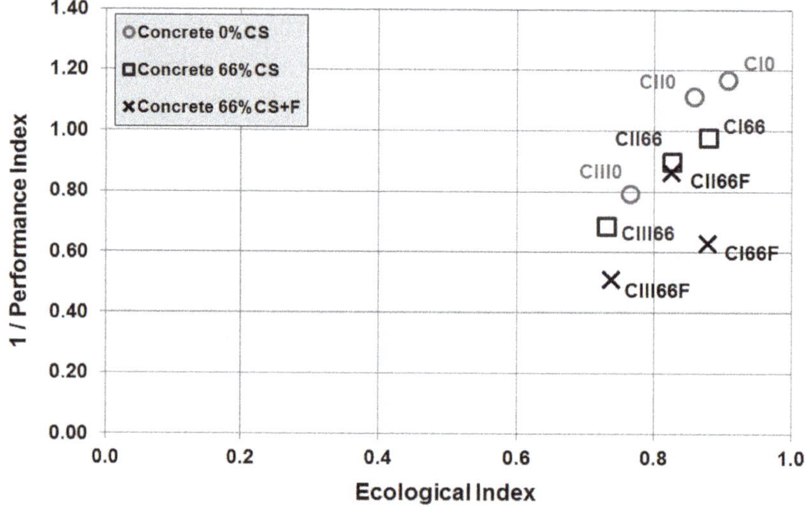

Figure 1. Ecological Index plotted against reciprocal of Performance Index in variant 0.

Table 10. Results of GEPI calculations.

Concrete ID	GEPI Values in Variant:					Max.	Min.
	0	S1	S2	B1	B2		
CI0	1.480	0.929	0.923	0.913	0.911	0.929	0.911
CII0	1.405	0.986	0.967	0.901	0.893	0.986	0.893
CIII0	1.101	0.787	0.781	0.770	0.768	0.787	0.768
CI66	1.314	1.062	1.048	0.962	0.953	1.062	0.953
CII66	1.220	1.046	1.033	0.948	0.936	1.046	0.936
CIII66	1.001	0.902	0.893	0.836	0.826	0.902	0.826
CI66F	1.080	1.001	0.993	0.952	0.944	1.001	0.944
CII66F	1.195	1.038	1.025	0.946	0.934	1.038	0.934
CIII66F	0.897	0.862	0.856	0.830	0.822	0.862	0.822

The analysis of the obtained results showed a clear but small variation in the calculated PI values obtained in the individual variants. Regardless of the adopted variant, the mutual proportions of GEPI values obtained in the case of individual series remained very close to each other. Therefore, it was found pointless to present in detail the results of EI and PI calculations of all variants and to visualize them in the figures. Only the results of calculations obtained in variant B2 were selected, in which the influence of natural radioactivity of concrete on the result of PI calculations was the greatest. The results obtained in this variant are presented in Table 11 and Figure 2.

Table 11. EI, PI and GEPI values taking into account the natural radiation-variant B2.

	Concrete ID								
	CI0	**CI66**	**CI66F**	**CII0**	**CII66**	**CII66F**	**CIII0**	**CIII66**	**CIII66F**
EI	0.908	0.879	0.878	0.858	0.826	0.825	0.764	0.731	0.737
PI	12.828	2.722	2.892	4.002	2.273	2.286	12.969	2.591	2.738
GEPI	0.911	0.953	0.944	0.893	0.936	0.934	0.768	0.826	0.822

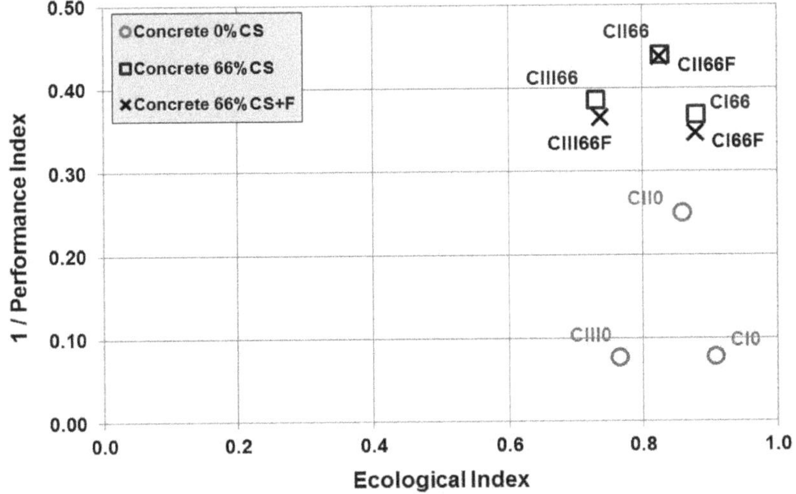

Figure 2. Ecological Index plotted against the reciprocal of the performance index in variant B2.

As can be seen, the weight variation in the adopted variants had the greatest impact on the GEPI values for CEM II cement concrete, regardless of the type of fine aggregate and plasticiser used, and for waste copper slag, regardless of the type of cement. However, this variation, understood as the difference between the highest and the lowest GEPI value, reaches a maximum of less than 12%.

The significantly lower natural radioactivity of concrete with CEM I and CEM III without the use of waste copper slag caused PI in these two series to be high, several times higher than in series with the same type of cement and waste. It is also about three times higher than that of CII0 series, in which cement with increased natural radioactivity due to fly ash content is used. CIII0 concrete (GEPI = 0.768) proved to be the best with such established assessment criteria. Despite increased natural radioactivity, mixtures with the waste and blast furnace slag cement were ranked in the next two places (GEPI = 0.822 and GEPI = 0.826). The worst results were obtained in the case of the series with CEM I cement and waste copper slag (GEPI = 0.953 and GEPI = 0.944). This allows us to state that the use of waste copper slag improves the performance of concrete so much that it reduces the negative impact of increased radioactivity in the assessment performed by the EIPI method.

It should be taken into account that when PI is calculated on the basis of other parameters (selected properties, reference values, weights), it is not possible to directly compare PI and GEPI results obtained in the calculation of the different variants. The comparisons make sense between the different concrete mixes assessed on the basis of the criteria adopted for the specific variant and adapted to the requirements of the specific conditions of concrete exploitation and, for example, the limitations related to natural radioactivity. The variant calculations of the impact of natural radioactivity of concrete on the PI value presented in the paper were aimed at analysing various variants of the differentiation of the weights and their impact on the final assessment of the concrete.

Despite favourable results of the calculations of GEPI values due to the relatively high natural radioactivity of waste copper slag, however, within acceptable limits, the authors do not recommend the use of concrete with this material for the construction of buildings intended for permanent human presence. This type of concrete materials can be used, e.g., for erecting farm buildings or road pavements (bottom layer) and structures (bridges, overpasses, etc.).

5. Conclusions

The results of the performed research allowed the formulation of the following conclusions:

- Replacing in the concrete mixture a part of the sand with waste copper slag does not aggravate any of the tested properties of concrete. The use of a plasticiser also allows obtaining the same consistency as in the reference series made with sand only.
- Concrete with the addition of waste copper slag is tighter than the reference concrete. This effect is particularly noticeable in the case of concrete of the same consistency as the reference concrete.
- Despite the high natural radioactivity of waste copper slag, it is possible to obtain concrete with radioactivity indices much lower than the maximum permitted values. The f_1 values of CI66F and CIII66F series of concrete are higher than those obtained with CII0 concrete without waste copper slag by 19% and 28% respectively.
- Excluding in the assessment the natural radioactivity of concrete, the highest GEPI rating was obtained by the series CEM III66F with waste copper slag.
- In applications where the natural radioactivity of concrete is of greater importance, the series with CEM III0 without waste copper slag obtained the most favourable result.
- The EIPI method allows for a comprehensive assessment of concrete properties, including among others natural radioactivity. Such an extended assessment may be useful in applications where increased natural radioactivity is not recommended, e.g., indoor areas for permanent human habitation.
- The EIPI evaluation showed that CEM III cement concrete is the best variant, among all those taken into account, regardless of whether natural radioactivity is considered or not. Omitting it in the evaluation leads to the conclusion that the best concrete is the one with the use of waste copper slag. However, taking into account natural radioactivity, the concrete without the addition of waste copper slag is moved to the leading edge of the CEM III series of concrete. This means that although the type of cement is the dominant factor in the EIPI evaluation, the level of natural radioactivity is also important.
- Taking into account mechanical properties of the composite, parameters relating to tightness of hardened structure as well as environmental impact of cement concrete (including CO_2 emission, radioactivity and consumption of natural resources), concrete made of CEM III cement is beneficial. Not only from the point of view of environmental friendliness, which would be quite obvious, but also regarding non-ecological reasons.

Author Contributions: W.K. planned and organized the experimental study, made the selection of materials and the mix design, wrote part of the article text. R.J. conducted an analysis of the results using the EIPI method, wrote part of the article text, edited the text. D.G. performed tests of natural radioactivity and developed the results of the tests. I.W. co-edited the text, participated in analysing the results and formulating conclusions. All authors have read and agreed to the published version of the manuscript.

Funding: This research was supported by the City of Płock through the Mayor's Research Grants Programme "Collaboration with Universities", grant number [502250200003]. The APC was funded by the City of Płock through the Mayor's Research Grants Programme "Collaboration with Universities", grant number [502250200003].

Acknowledgments: The work has been supported by the City of Płock through the Mayor's Research Grants Programme "Collaboration with Universities".

Conflicts of Interest: The authors declare no conflict of interest.

References

1. Murray, A.; Skene, K.; Haynes, K. The Circular Economy: An Interdisciplinary Exploration of the Concept and Application in a Global Context. *J. Bus. Ethics* **2017**, *140*, 369–380. [CrossRef]
2. Zaman, A. A Strategic Framework for Working toward Zero Waste Societies Based on Perceptions Surveys. *Recycling* **2017**, *2*, 1. [CrossRef]
3. Lothenbach, B.; Scrivener, K.L.; Hooton, R.D. Supplementary cementitious materials. *Cem. Concr. Res.* **2011**, *41*, 1244–1256. [CrossRef]
4. Hela, R.; Orsakova, D. The mechanical activation of fly ash. *Procedia Eng.* **2013**, *65*, 87–93. [CrossRef]
5. Garcia-Lodeiro, I.; Donatello, S.; Fernández-Jiménez, A.; Palomo, Á. Hydration of hybrid alkaline cement containing a very large proportion of fly ash: A descriptive model. *Materials* **2016**, *9*, 605. [CrossRef]
6. Wilińska, I.; Pacewska, B. Influence of selected activating methods on hydration processes of mixtures containing high and very high amount of fly ash: A review. *J. Therm. Anal. Calorim.* **2018**, *133*, 823–843. [CrossRef]
7. Wilińska, I.; Pacewska, B.; Ostrowski, A. Investigation of different ways of activation of fly ash–cement mixtures: Part 1. Chemical activation by Na_2SO_4 and $Ca(OH)_2$. *J. Therm. Anal. Calorim.* **2019**, *138*, 4203–4213. [CrossRef]
8. Meenakshi Sudarvizhi, S.; Ilangovan, R. Performance of Copper slag and ferrous slag as partial replacement of sand in Concrete. *Int. J. Civ. Struct. Eng.* **2011**, *1*, 918–927.
9. Rzechuła, J. Gospodarcze wykorzystanie odpadowego ścierniwa z żużla pomiedziowego. In *Fizykochemiczne Problemy Mineralurgii, Zeszyt 28*; Łuszczkiewicz, A., Ed.; Politechnika Wrocławska: Wrocław, Poland, 1994; pp. 207–218.
10. Gorai, B.; Jana, R.K. Premchand Characteristics and utilisation of copper slag-A review. *Resour. Conserv. Recycl.* **2003**, *39*, 299–313. [CrossRef]
11. Dhir, R.K.; de Brito, J.; Mangabhai, R.; Lye, C.Q. *Sustainable Construction Materials-Copper Slag*; Woodhead Publishing: Cambridge, UK, 2017.
12. Ayano, T.; Sakata, K. Durability of concrete with copper slag fine aggregate. *Spec. Publ.* **2000**, *192*, 141–158.
13. Al-Jabri, K.S.; Al-Saidy, A.H.; Taha, R.A. Effect of copper slag as a fine aggregate on the properties of cement mortars and concrete. *Constr. Build. Mater.* **2011**, *25*, 933–938. [CrossRef]
14. Alp, İ.; Deveci, H.; Süngün, H. Utilization of flotation wastes of copper slag as raw material in cement production. *J. Hazard. Mater.* **2008**, *159*, 390–395. [CrossRef] [PubMed]
15. Kubissa, W.; Jaskulski, R.; Ng, P.-L.; Chen, J. Utilisation of Copper Slag Waste and Heavy-weight Aggregates for Production of Pre-cast shielding Concrete Elements. *J. Sustain. Archit. Civ. Eng.* **2018**, *22*, 39–47. [CrossRef]
16. Jaskulski, R.; Kubissa, W. Lightweight concrete with copper slag waste as sand substitution. In Proceedings of the MATEC Web of Conferences, Lille, France, 8–10 October 2018; Volume 163.
17. Al-Jabri, K.S.; Hisada, M.; Al-Saidy, A.H.; Al-Oraimi, S.K.A. Performance of high strength concrete made with copper slag as a fine aggregate. *Constr. Build. Mater.* **2009**, *23*, 2132–2140. [CrossRef]
18. Ambily, P.S.; Umarani, C.; Ravisankar, K.; Ranjan, P.; Bharatkumar, B.H.; Iyer, N.R. Studies on ultra high performance concrete incorporating copper slag as fine aggregate. *Constr. Build. Mater.* **2015**, *77*, 233–240. [CrossRef]
19. Wu, W.; Zhang, W.; Ma, G. Optimum content of copper slag as a fine aggregate in high strength concrete. *Mater. Des.* **2010**, *31*, 2878–2883. [CrossRef]
20. Kubissa, W.; Jaskulski, R.; Simon, T. Surface blast-cleaning waste as a replacement of fine aggregate in concrete. *Archit. Civ. Eng. Environ.* **2017**, *3*, 89–94. [CrossRef]
21. Shanmuganathan, P.; Lakshmipathiraj, P.; Srikanth, S.; Nachiappan, A.L.; Sumathy, A. Toxicity characterization and long-term stability studies on copper slag from the ISASMELT process. *Resour. Conserv. Recycl.* **2008**, *52*, 601–611. [CrossRef]
22. Zapotoczna-Sytek, G.; Mamont-Cieśla, K.; Rybarczyk, T. Naturalna promieniotwórczość wyrobów budowlanych, w tym autoklawizowanego betonu komórkowego (ABK). *Prz. Bud.* **2012**, *83*, 39–42.
23. Keller, G.; Hoffmann, B.; Feigenspan, T. Radon permeability and radon exhalation of building materials. *Sci. Total Environ.* **2001**, *272*, 85–89. [CrossRef]
24. Skowronek, J.; Michalik, B.; Dulewski, J. NORM in Polish industry. In Proceedings of the an International Conference Held in Szczyrk, Szczyrk, Poland, 17–21 May 2004; pp. 1–9.

25. Cooper, M.B. *Naturally Occurring Radioactive Materials (NORM) in Australian Industries-Review of Current Inventories and Future Generation*; Envirorad Serv. Pty. Ltd.: Melbourne, Australia, 2005.
26. Clavensjö, B.; Åkerblom, G. The Radon Book. *J. Radiol. Prot.* **1996**, *16*, 129.
27. Kovler, K. Radiological constraints of using building materials and industrial by-products in construction. *Constr. Build. Mater.* **2009**, *23*, 246–253. [CrossRef]
28. Nuccetelli, C.; Pontikes, Y.; Leonardi, F.; Trevisi, R. New perspectives and issues arising from the introduction of (NORM) residues in building materials: A critical assessment on the radiological behaviour. *Constr. Build. Mater.* **2015**, *82*, 323–331. [CrossRef]
29. Campos, M.P.; Costa, L.J.P.; Nisti, M.B.; Mazzilli, B.P. Phosphogypsum recycling in the building materials industry: Assessment of the radon exhalation rate. *J. Environ. Radioact.* **2017**, *172*, 232–236. [CrossRef]
30. Leonardi, F.; Bonczyk, M.; Nuccetelli, C.; Wysocka, M.; Michalik, B.; Ampollini, M.; Tonnarini, S.; Rubin, J.; Niedbalska, K.; Trevisi, R. A study on natural radioactivity and radon exhalation rate in building materials containing norm residues: Preliminary results. *Constr. Build. Mater.* **2018**, *173*, 172–179. [CrossRef]
31. Trevisi, R.; Leonardi, F.; Risica, S.; Nuccetelli, C. Updated database on natural radioactivity in building materials in Europe. *J. Environ. Radioact.* **2018**, *187*, 90–105. [CrossRef]
32. Wong, H.H.C.; Kwan, A.K.H. Packing density of cementitious materials: Part 1-measurement using a wet packing method. *Mater. Struct.* **2008**, *41*, 689–701. [CrossRef]
33. Kwan, A.K.H.; Wong, H.H.C. Packing density of cementitious materials: Part 2-packing and flow of OPC + PFA + CSF. *Mater. Struct.* **2008**, *41*, 773–784. [CrossRef]
34. Li, L.G.; Kwan, A.K.H. Packing density of concrete mix under dry and wet conditions. *Powder Technol.* **2014**, *253*, 514–521. [CrossRef]
35. Li, L.G.; Lin, C.J.; Chen, G.M.; Kwan, A.K.H.; Jiang, T. Effects of packing on compressive behaviour of recycled aggregate concrete. *Constr. Build. Mater.* **2017**, *157*, 757–777. [CrossRef]
36. Council of Ministers of Republic of Poland Ordinance of 2 January 2007 on Requirements Concerning the Content of Natural Radioactive Isotopes of Potassium K-40, Radium Ra-226 and Thorium Th-228 in Raw Materials and Materials Used in Buildings Intended for Human and Live. Available online: http://prawo.sejm.gov.pl/isap.nsf/DocDetails.xsp?id=WDU20070040029 (accessed on 14 May 2019).
37. Kubissa, W.; Jaskulski, R.; Chen, J.; Ng, P.-L.; Godlewska, V.; Reiterman, P. Evaluation of ecological concrete using multi-criteria ecological index and performance index approach. *Archit. Civ. Eng. Environ.* **2019**, *12*, 97–107. [CrossRef]
38. Kubissa, W.; Jaskulski, R. Measuring and time variability of the sorptivity of concrete. *Procedia Eng.* **2013**, *57*, 634–641. [CrossRef]
39. Jacobs, F.; Leemann, A.; Teruzzi, T.; Torrent, R.; Denarie, E. Specification and site control of the permeability of the cover concrete: The Swiss approach Dedicated to Professor Dr. Bernhard Elsener on the occasion of his 60th birthday. *Mater. Corros* **2012**, *63*, 1127–1133.
40. Kubissa, W.; Jaskulski, R. Improving of concrete tightness by using surface blast-cleaning waste as a partial replacement of fine aggregate. *Period. Polytech. Civ. Eng.* **2019**. [CrossRef]
41. Brunarski, L.; Dohojda, M. *Badania Promieniotwórczości Naturalnej Wyrobów Budowlanych. Poradnik [Natural Radioactivity Testing of Construction Products. Guidebook]*; ITB: Warsaw, Poland, 2010.
42. Hasani, F.; Shala, F.; Xhixha, G.; Xhixha, M.K.; Hodolli, G.; Kadiri, S.; Bylyku, E.; Cfarku, F. Naturally occurring radioactive materials (NORMs) generated from lignite-fired power plants in Kosovo. *J. Environ. Radioact.* **2014**, *138*, 156–161. [CrossRef] [PubMed]
43. European Commission. *Radiological Protection Principles Concerning the Natural Radioactivity of Building Materials (Radiation Protection Report RP-112)*; European Commission: Luxembourg, 1999.

 *buildings*

Article

Computer Aided Assembly of Buildings

Roman Marcinkowski and Maciej Banach *

Faculty of Civil Engineering, Mechanics and Petrochemistry, Warsaw University of Technology,
00-661 Warszawa, Poland; roman.marcinkowski@pw.edu.pl
* Correspondence: maciej.banach@pw.edu.pl; +48-512-472-771

Received: 30 December 2019; Accepted: 11 February 2020; Published: 13 February 2020

Abstract: This article presents an interactive method of computer-aided assembly planning. It is estimated that such planning will be more and more desirable due to the increasing use of prefabrication in construction. Prefabrication meets the trends of sustainable development and digitization as it enables the application of intelligent control systems at the stage of highly specialized production, assembly and facility maintenance. The presented planning method is based on the Monte Carlo simulations and logical algorithms for assembly work planning. It was determined on the basis of the literature studied and our own observations. The paper introduces a detailed model of assembly works planning and is an example of using a computer application developed on the basis of the described model. The example confirms the correctness of the algorithm and indicates its usefulness in the scope of analyzing many decision variants. Further research on labor productivity rates for assembly works, implementation of digital databases of assembly machines and prefabricated elements, as well as integration of the proposed application with the BIM environment should make it easier to commercialize the developed application.

Keywords: assembly works; computer planning; Monte Carlo method; selection; construction; application

1. Introduction

The construction industry is always changing and evolving. There are some trends and aims that are noticeable in this development. Sustainable construction is one of them. This is a set of activities subordinated to the requirements of sustainable development—a process aimed at satisfying the needs of the present generation in a way that allows the same generations to pursue the next generations. The construction industry is strongly associated with the concept of sustainable development as its impact on the environment is huge. According to the idea of sustainable construction, the life cycle of a building should comply with the requirements of sustainable development.

One of the important measures of sustainability in construction is the amount of energy used in construction. There are three types of energy use in the construction performance: embodied, operational and transformation (utilization) energy. All of them should be minimized through appropriate spatial, construction, material, technological, functional and organizational solutions.

A promising response to the needs of the imminent challenges of sustainability in construction is prefabrication. This technology fits very well in the economic, quality and social criteria of sustainability, and does not set limitations on the architectural form. From the point of view of the common interests of present and future generations, prefabricated constructions usually enable the improvement to a high-energy efficiency of the building, and compared to the on-site constructions, emit fewer pollutants at the production stage. Prefabrication also gives specific restrictions on energy consumption for building and demolishing facilities. It increases in importance because about 50% of all the materials utilized are materials used in construction.

Prefabrication technology offers many development opportunities arising from the automation of production and construction processes. It can be strongly influenced by 5G and IoT (Internet of Things)

technology development. Intelligent construction facilities require standardization and standardized solutions to be implemented in a highly specialized construction production.

There is a good chance that prefabrication will only be a key technology for the sustainable construction if the innovative techniques of assembly works planning are developed. Such techniques should involve assembly automation, digital models of construction and planning data.

Digitized information on both the construction site and machines used for the assembly, as well as construction elements to be mounted, simplify effective construction planning. The work of many assembly machines should be planned so that their effective utilization on site is taken into account, which is a partial goal of sustainable construction.

There is therefore a need for a computer-assisted planning method that enables the planner (scheduler) to search for energy-optimal technological and organizational solutions, especially if the work of many assembly machines is required.

In the literature, a variety of solutions to problems associated with the operation of cranes in the assembly planning can be found. A fast development in the methods applied to solve assembly planning problems has its beginning in the 1980s and it was associated with the dynamic growth in construction (including prefabrication) and construction equipment. However, work on these methods was triggered by the development of information technology, which is supposed to be a key tool to improve all technical and organizational systems.

One of the originators of improving the organization of assembly works were, among others, Gray and Little, Furusaka, Warszawski, Peled, Van Dijk, Van Gassel and Schaefer [1–5]. Their models of supporting the selection of crane's size were related to the cost of crane's work, and among the others, involved expert systems.

A number of applications that support assembly works planning are based on graphic methods and enable the selection of the size of the crane [6–8] or even a group of tower cranes [9] required for the particular location. The more advanced tools worked as crane operation simulators [10], which selected the size of the crane and the type of slings [11] required, or estimated the costs and time of work. The presented methods, however, do not offer the possibility of optimizing the cranes' operation. It applies to both commercial applications that can be used in order to select the type and location of the lifting device [12,13] and advanced non-commercial applications that are still valid today [14–16]. This type of application works well when planning the organization of the construction site and assembly of heavy-weight units. Other applications, usually available as plug-ins for the CAD software, enable assembly works planning, as well as their subsequent simulation that helps to eliminate potential collisions. In summary, contemporary computer aided assembly planning is usually limited to checking the possibility of mounting the load by the selected crane. It does not provide scheduling, assembly cost estimation, hardly ever compares individual assembly situations with each other, and does not provide tools for optimization of assembly works.

The problems of crane selection were discussed in Reference [17] where the fuzzy logic theory was used, in Reference [18] where neural networks were involved, and in Reference [19] where genetic algorithms were used to select the type of crane. The important role of soft factors in the crane selection problem and a procedure for crane selection in relation to these factors was presented in Reference [20]. In turn, Reference [21] proposes the AHP method (analytic hierarchy process) to analyze crane type selection and takes into account both hard and soft factors. In each case, the problem boiled down to the problem of the lifting device selection, which does not meet the current needs of work planning where a variety of assembly machines are used.

An original crane planning system was developed in References [22–24], where, based on the weight of the mounted elements, reach and lifting height, and using the defined databases, the required crane sizes and locations are determined. The final result is an assembly plan that excludes collisions between cranes. However, this system is dedicated to monolithic works as it analyses tower cranes only. Similarly, in Reference [25], the crane optimization problem was mathematically formulated as an NP-complete problem and was solved as the TSP problem (traveling salesman problem). However, it is also dedicated to monolithic works, where one machine supports several assembly brigades.

A significant contribution to the development of methods that support assembly planning was made by researchers from the Faculty of Engineering of the University of Alberta in Edmonton, Canada [26–29]. They proposed, in Reference [30], a method that incorporates object information modeling (BIM) and external databases to plan mobile crane positions and react to changes in the project site layout during modular structure assembly in the extraction of crude oil from the oil sands.

With the above in mind, worldwide literature presents advanced knowledge and proposes various ways to solve decision problems related to construction work planning. In most cases, the proposed solutions relate to selected planning issues, without a comprehensive approach to assembly planning as a whole. This is why the advanced support systems cannot be applied to assembly works directly. The modern method of planning assembly works should then present comprehensive solutions that combine the problem of crane selection, their size, number or location, take into account the assembly of each element together with the assembly schedule, time and cost of works and the optimization of these variables. All these issues are incorporated into the proposed method of assembly works planning.

2. The Proposed Method of Assembly Works Planning

2.1. Prefeace

The main problems in assembly works planning focus on the selection of cranes, their number and locations. The complexity of organizational dilemmas occurring at the initial stage of planning and during assembly works indicate that simulation methods are advisable to solve the assembly planning problems. In the proposed model, the Monte Carlo method and logical algorithms based on observational studies related to assembly works are used.

The main purpose of the article is to present a method of assembly works planning, which focuses on the simulation of assembly works over time. The planning process is run with the interaction of the planner, who controls the planning attempts, defines constraints (i.e., on resources) and decides on the final solution. A similar simulation approach to works planning was presented in Reference [31], where an interactive simulation was applied to plan concrete works and formwork utilization.

The assembly planning method has been algorithmized and pre-programmed in order to check its correctness. The application's effectiveness has been confirmed by simulations run for a real assembly works of the office building.

2.2. Model of Planning Assembly Works

Lets consider the set of prefabricated elements $Q = \{q_1, q_2, \ldots, q_a, \ldots, q_A\}$, and their characteristics defined by the matrix $\mathbf{L} = [l_{ab}]_{A \times 13}$, where:

$l_{a,1}$—stands for the weight of the a-th element, in tons,

$l_{a,2}$—stands for the maximum width of the a-th element, in meters,

$l_{a,3}$—stands for the maximum length of the a-th element, in meters,

$l_{a,4}$—stands for the maximum height of the a-th element, in meters,

$l_{a,5}$—stands for the duration of loading the a-th element, in minutes,

$l_{a,6}$—stands for the duration of assembly of the a-th element, in minutes,

$l_{a,7}$—stands for the shape coefficient for the a-th element,

$l_{a,8}$—stands for the required date and time of delivery of the a-th element,

$l_{a,9}$—stands for the weight of the sling used in the assembly of the a-th element, in tons,

$l_{a,10}$—stands for the height of the sling used in the assembly of the a-th element, in meters,

$l_{a,11}$—is the a-th element location-coordinate x,

$l_{a,12}$—is the a-th element location-coordinate y,

$l_{a,13}$—is the a-th element location-coordinate z.

For the assembly of prefabricated elements, various types of assembly machines can be used. They are divided into two sets: a set of stationary machines (which do not change their location throughout the assembly works) $M_{st} = \{m_{st,1}, m_{st,2}, \ldots, m_{st,c}, \ldots, m_{st,C}\}$ and mobile machines (which change their location if necessary) $M_{nst} = \{m_{nst,1}, m_{nst,2}, \ldots, m_{nst,d}, \ldots, m_{nst,D}\}$. Possible locations of assembly machines (coordinates in a flat coordinate system) create the set $P_{st}(x,y) = \{p_{st,1}(x,y), p_{st,2}(x,y), \ldots, p_{st,e}(x,y), \ldots, p_{st,E}(x,y)\}$ for stationary machines and the set $P_{nst}(x,y) = \{p_{nst,1}(x,y), p_{nst,2}(x,y), \ldots, p_{nst,f}(x,y), \ldots, p_{nst,F}(x,y)\}$ for mobile machines.

To start the assembly cycle, it is necessary to define: the number of planning attempts (which are also the number of assembly planning solutions, I_L, start and end times of work of the assembly teams $<T_{START,W}, T_{END,W}>$ and to set the priority for stationary machines. When the priority of stationary machines is applied, the construction components will be assigned to stationary machines first, and only if there is no technical possibility of assembly, mobile machines will be considered. Such a solution allows for a better utilization of stationary cranes, with a unit cost of work that is usually lower, and has breakdowns that cause less loss due to their underutilization. In addition, such an approach reduces the work of more expensive mobile cranes, which, if the breakdown is long enough, can be used in other project (or task) performances. If the priority of stationary cranes is not applied, the algorithm assigns the subsequent a-th construction element to the crane, without taking into account the mobility aspect.

The start of the simulation procedure initiates the determination of the next k-th solution number $R = \{r_1, r_2, \ldots, r_k \ldots, r_{I_L}\}$. For each rk solution, the number of stationary cranes (n) is drawn from the set M_{st} and the number of mobile ones (o) is drawn from the set M_{nst}. The variables have a value of at least 0, and at most, C for stationary machines, $n \in \langle 0; C \rangle \wedge n \in \mathbb{N} \wedge n \leq E$, and at least 0, and at most, D for mobile machines $o \in \langle 0; D \rangle \wedge o \in \mathbb{N}$. The number of randomly selected stationary cranes is further limited by their number of possible locations. This condition does not apply to mobile machines, which have locations that may change. The draw begins with a value of 0, so the cranes can be selected from both sets of stationary and mobile machines. For each solution, r_k is the random order of stationary and/or the mobile machine is drawn. In each k-th planning simulation, they form a set $M_{st}^k = \{m_{st,1}, m_{st,2}, \ldots, m_{st,c}, \ldots, m_{st,n}\}$ and $M_{nst}^k = \{m_{nst,1}, m_{nst,2}, \ldots, m_{nst,d}, \ldots, m_{nst,o}\}$. Each c-th stationary crane is assigned to the e-th location, and for every d-th mobile crane, the f-th location is assigned. If the number of stationary cranes n and mobile cranes o is greater than the number of their potential locations, E and F, respectively, crane assignment to their location ends with the last free place $p_{st,E}$ or $p_{nst,F}$. Other stationary cranes will not be used in the k-th simulation run, while mobile cranes may be used if there is no technical possibility of assembly of the a-th element by any cranes in their assumed locations. It brings about the need to search for the f-th location of the d-th crane. In each of the k simulation runs, sets of assembly machines can be described with the function as below. The argument of this function is the crane's location:

$$f : P_{st,e} \Rightarrow M_{st,c}^k \text{ for } c \in \langle 0; \min\{E, n\}\rangle$$
$$f : P_{nst,f} \Rightarrow M_{nst,d}^k \text{ for } d \in \langle 0; \min\{F, o\}\rangle, \tag{1}$$

The planning goal is to specify, in each simulation, the work plan for assembly machines $M_{st}^k(P_{st,e})$ and $M_{nst}^k(P_{nst,f})$ that minimizes the cost of assembly operations for the elements defined by the set Q and matrix $\mathbf{L} = [l_{ab}]_{A \times 12}$ as well to minimize the duration of works and the number of unmounted elements. If there is no technical possibility of assembly of the a-th element, either by crane $M_{st}^k(P_{st})$ or crane $M_{nst}^k(P_{nst})$, from their locations available in the k-th simulation run, the element is labelled unmounted. The inability to mount a few or a dozen of elements does not discredit the solution. If the cost and duration of the task are acceptable, the planner should accept it and assume that the unmounted elements will be assembled individually. The number of unmounted elements indicate the quality of solutions and can be a guide to better assembly planning. The described above decision situation is presented in Figure 1.

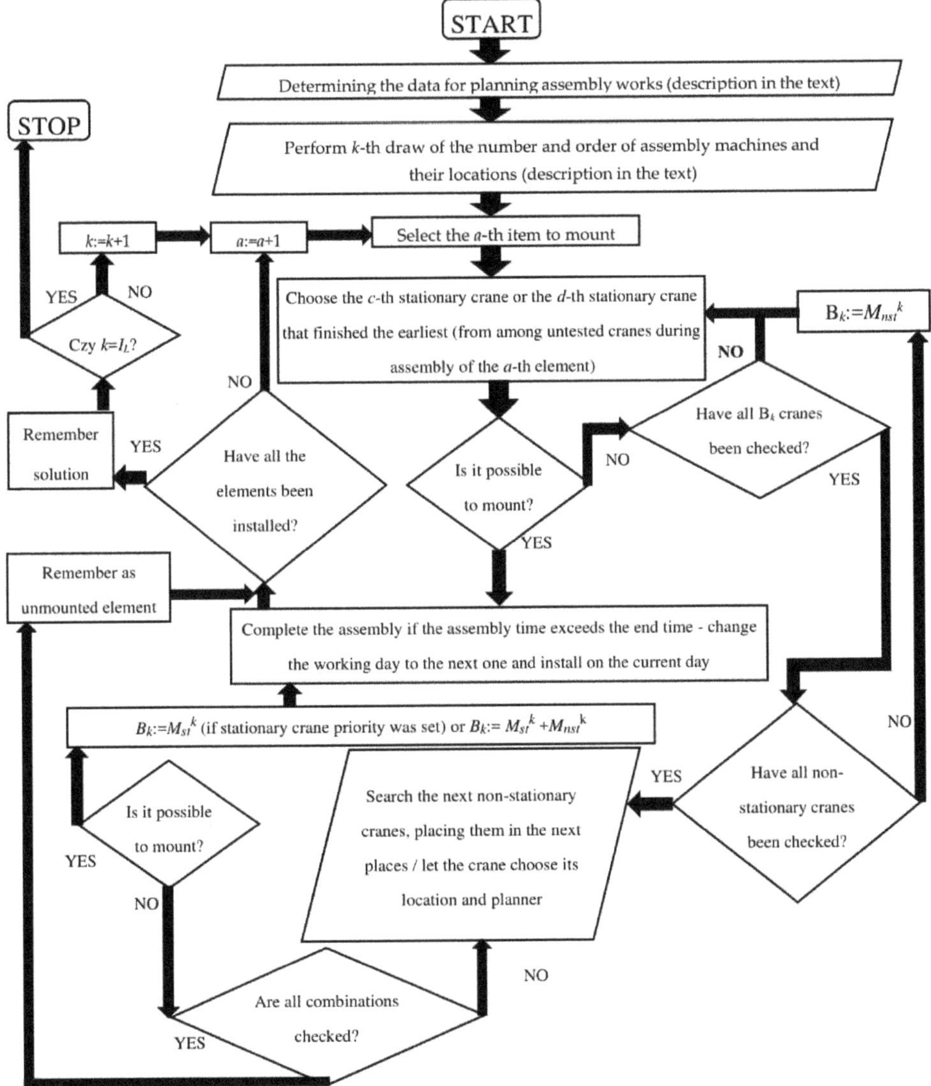

Figure 1. A block diagram of the assembly planning system. Sourced from our own study.

The following assumptions are made:

- the sets of available cranes M_{st} and M_{nst} as well as their locations $P_{st}(x, y)$ and $P_{nst}(x, y)$ are finite and can be modified before assembly planning (before simulation);
- In each simulation k, the deadline for the a-th element assembly is one that results from the crane selection. The crane that is being considered is the crane that finished work the earliest and is sought at the time when the assembly of a preceding $(a-1)$ element or element labelled as unmounted ends;
- If the assembly date (as described above) is ahead of an actual delivery date, it should be set on the date of delivery. Therefore, if delivery dates are defined as the prefabricated elements'

characteristics, the assembly is being planned in correlation with their delivery. If the delivery dates correspond at most to the assembly commencement date, this issue is not taken into account.

In order to differentiate solutions within the priority of stationary cranes, the B set, which may include stationary cranes M_{st} or both types of cranes, M_{st} and M_{nst}, is created.

If there is no technical possibility of assembling the a-th element by the c-th stationary crane from the e-th position (and if the priority of stationary machines is not applied-also by the d-th crane from the f-th position), the algorithm attempts to find the d-th mobile crane in the f-th location (if the priority of stationary machines was set).

If a crane capable of assembling the a-th element still has not been found, the simulator checks (in the order that results from the initial ordering) every d-th mobile crane in each possible location f, starting from the first crane and the first defined location f. If a proper crane is found, the duration of the a-th element assembly is increased by the duration of the d-th crane transfer to the f-th location. At the same time, the assignment of mobile cranes M_{nst}^{k} to their locations $P_{nst}^{k}(x, y)$ is updated.

The cranes' working times are calculated periodically and take into account normal daily breaks. Every j-th day of the assembly brigade's work starts at $T_{START,W}$. If the assembly completion time for each construction element exceeds the assembly completion time of the brigade ($T_{END,W}$), the assembly deadline is set on the next day for the assembly ($T_{START,W}$).

The planning task boils down to setting assembly schedules, which take into account a cranes cyclic work. The type, size, number and placement of cranes affect the costs, time and efficiency of the assembly operations. It is expressed in the number of unmounted construction elements.

Deriving an optimal solution requires a large number of planning tests in which the number and locations of cranes are changing. The problem can be solved through simulations. The cranes' numbers, ordering and locations may be drawn randomly. For each draw, the characteristic schedules and costs of cranes' work are forecasted. The number of schedules is, in turn, equal to the number of expected solutions (I_L), which are determined by the planner. The planner is the one who selects the number and size of the cranes and indicates their potential locations. The simulator is a planning tool that presents different time and cost solutions and the number of unmounted construction elements.

For each simulation k, the planning procedure begins with the selection of c-th stationary crane $M_{st,c}^{k*}(P_{st,e})$ from the B_k set (if the priority of stationary cranes is set) or the crane from the group of c-th stationary cranes $M_{st,c}^{k*}(P_{st,e})$ and d-th mobile cranes $M_{st,d}^{k*}\left(P_{nst,f}\right)$ (in the priority is not set). Only the cranes that finished their work at the given moment "*" shall be considered.

If there are cranes that have finished their work within the same deadline, the order of their work is determined based on ordering on the list of stationary cranes $M_{st}^{k}(P_{st})$ (if the priority is set), or list of all machines together $M_{st}^{k}(P_{st})$ and $M_{nst}^{k}(P_{nst})$) (if the priority is not applied). It requires a temporary crane to be defined. It is a crane that has completed its operation at the earliest $T_{END,minM^k}$ among the dynamically changing crane base:

$$T_{END,minM^k} \quad \overset{\min}{c} \quad \left\{ T_{END,\, M_{st,c}^{k}(P_{st,e})} \right\} \text{ for } c = 1, 2, \ldots, \min\{E; n\} \tag{2}$$

where $T_{END,M_{st,c}^{k}(P_{st,e})}$ is the date of assembly completion for the c-th stationary crane in e-th location in the k-th simulation.

Formula (2) does not allow for an unambiguous choice when at least two assembly machines have finished their work at the same time. It happens at the beginning of the assembly works. Therefore, it is necessary to search for the crane that has finished its work the earliest and has the lowest number in the set, according to the relationship:

$$\text{If } T_{END,M_{st,c}^{k}(P_{st,e})} = T_{END,minM^k},$$
$$\Rightarrow T_{END,minM^k} T_{END,\, M_{st,c}^{k}(P_{st,e})}, \text{ for } c = 1, 2, \ldots, \min\{E; n\}, \tag{3}$$

Thus, as the first at the moment, the c-th stationary crane $M^k_{st,c}(P_{st,e})$ will finish its work. If in the B_k set also contains mobile cranes $M^k_{nst}(P_{nst})$, then Formulas (2) and (3) take the form:

$$T_{END,minM^k} \min\left\{T_{END,\,M^k_{st,c}(P_{st,e})};\, T_{END,\,M^k_{st,c}(P_{st,e})}\right\}$$
$$\text{for } c = 1, 2, \ldots, \min\{E; n\};\ d = 1, 2, \ldots, \min\{F; o\}, \tag{4}$$

$$\text{If } T_{END,M^k_{st,c}(P_{st,e})} = T_{END,minM^k},$$
$$\Rightarrow T_{END,minM^k} T_{END,\,M^k_{st,c}(P_{st,e})},\ \text{for } c = 1, 2, \ldots, \min\{E; n\},$$
$$\text{or, if } T_{END,M^k_{nst,d}(P_{nst,f})} = T_{END,minM^k},$$
$$\Rightarrow T_{END,minM^k} T_{END,\,M^k_{nst,d}(P_{nst,f})},\ \text{for } d = 1, 2, \ldots, \min\{F; o\}, \tag{5}$$

where $T_{END,M^k_{nst,d}(P_{nst,f})}$ is the date of assembly completion for the d-th mobile crane in the f-th location in the k-th simulation.

In this case, the crane that will finish its work first at the given moment will be the c-th stationary crane or the d-th mobile crane, respectively, $M^k_{st,c}(P_{st,e}) \vee M^k_{nst,d}(P_{nst,f})$.

However, if the selection is only made from a group of mobile cranes, Formulas (2) and (3) take the following form:

$$T_{END,minM^k} \underset{d}{\min}\, T_{END,\,M^k_{nst,d}(P_{nst,f})} \ \text{for } d = 1, 2, \ldots, \min\{F; o\} \tag{6}$$

$$\text{If } T_{END,M^k_{nst,d}(P_{nst,f})} = T_{END,minM^k},$$
$$\Rightarrow T_{END,minM^k} T_{END,M^k_{nst,d}(P_{nst,f})},\ \text{for } d = 1, 2, \ldots, \min\{F; o\}, \tag{7}$$

It happens if the priority of stationary cranes is set and if there is no possibility of assembling the a-th element by stationary cranes.

It is verified in the next step, whether it is possible to assemble the a-th element by the selected crane $M^k_{st,c}$ or $M^k_{nst,d}$ from the location $P_{st,e}(x, y)$ or $P_{nst,f}(x, y)$. It is done by checking the minimum required load capacity, as well as the minimum reach and lift height in this location $\left\{Q^{a,e\vee f}_{min}, L^{a,e\vee f}_{min}, H^{a,e\vee f}_{min}\right\}$. The assembly is possible if the lifting capacity of the c-th or d-th crane is at least equal to the minimum load capacity required at a reach equal to $L^{a,e\vee f}_{min}$ and a lifting height equal to $H^{a,e\vee f}_{min}$ according to the following relationships:

$$Q^e_{M^k_{st,c}(P_{st,e})} \geq Q^{a,e}_{min} \ \text{for } \left\{L^{a,e}_{min}, H^{a,e}_{min}\right\},\ \text{for stationary cranes} \tag{8}$$

$$Q^f_{M^k_{nst,d}(P_{nst,f})} \geq Q^{a,f}_{min} \ \text{dla } \left\{L^{a,f}_{min}, H^{a,f}_{min}\right\},\ \text{for mobile cranes} \tag{9}$$

where $Q^e_{M^k_{st,c}(P_{st,e})}$ is the load capacity of c-th stationary crane in e-th location at a reach equal to $L^{a,e}_{min}$ and a lifting height equal to $H^{a,e}_{min}$.

$Q^f_{M^k_{nst,d}(P_{nst,f})}$ is the load capacity of the d-th mobile crane in f-th location at a reach equal to $L^{a,f}_{min}$ and a lifting height equal to $H^{a,f}_{min}$.

There are different types of cranes that may require a different approach to the collision check. Moreover, the checking procedure is frequently repeated. This is why the calculation block has been distinguished as a separate "collision checking procedure".

Verifying the assembly possibilities for a tower crane is not a problem. In the case of a mobile crane, it is necessary to check additional conditions that result from the possible collisions between the boom and the construction or the assembled element. The checking procedure in such case is described in References [32,33].

If the assembly is technically impossible, the temporary crane is removed from the set B^k, $M_{st,c}^k(P_{st,e}) \notin B^k$ lub $M_{nst,d}^k(P_{nst,f}) \notin B^k$. This operation is repeated until an appropriate crane is found or until the set is empty, $B^k = \varnothing$.

If the set B_k is empty and it had contained only stationary cranes $M_{st}^k(P_{st})$, (the priority of stationary cranes was set) and there are unmounted construction elements, the supplementation of set B^k with mobile cranes, $B^k M_{nst,d}^k(P_{nst,f})$ begins. The first available machine is then selected again according to Formulas (6) and (7).

As before, if there is a possibility of assembly according to Formula (9), the selected d-th crane $M_{nst,d}^k(P_{nst,f}))$ is temporarily removed from the set B^k. The operation is repeated until an appropriate crane is found or until the set is empty, $B^k = \emptyset$.

If none of the stationary cranes $M_{st,c}^k(P_{st,e})$, as well as mobile cranes $M_{nst,d}^k(P_{nst,f})$, from the assigned locations are able to assemble the a-th element, the algorithm, for a d-th mobile crane ($d \in \langle 1;o \rangle$), searches for a location $P_{nst,f}^k(x,y)$, $f \in \langle 1;F \rangle$, in which the assembly conditions are met. If such a crane exists and it is the d-th crane $M_{nst,d}^k{}'$ at the f-th location $P_{nst,f}^k{}'(x,y)$, its location changes into $P_{nst,f}^k{}'(x,y)$. The assembly duration for the a-th element is then increased by the duration of a possible crane disassembly and transfer from the current to the new location:

$$l_{a,6}l'_{a,6} = \begin{cases} l_{a,6} + m_{5,1(M_{nst,d})}, & \text{when } M_{nst,d}^k{}' \notin f : M_{nst}^k(P_{nst}) \\ l_{a,6} + m_{5,1(M_{nst,d})} + m_{6,1(M_{nst,d})}, & \text{when } M_{nst,d}^k{}' \in f : M_{nst}^k(P_{nst}) \end{cases} \tag{10}$$

where $l'_{a,6}$ is the a-th element assembly duration increased by the duration of a mobile crane assembly at the f-th location.

$m_{5,1(M_{nst,d})}$ is the time required to assemble the d-th crane at its location.

$m_{6,1(M_{nst,d})}$ is the time required for disassembly and departure of the d-th crane from its location.

However, if at the f-th location, the d-th crane $M_{nst,d}^k$ was located, it is disassembled now and removed from this location. This crane is held ready to assembly subsequent elements from the other locations. The time of work of this crane is increased by the duration of disassembly from the f-th location:

$$T_{d,k,j} := T_{d,k,j} + m_{6,1(M_{nst,d})} \tag{11}$$

At the same time, the f-th location is left by the d-th crane $M_{nst,d}^k$ and replaced by the d-th crane $M_{nst,d}^k{}'$, which capable of assembling the a-th element:

$$\begin{aligned} M_{nst,d}^k(P_{nst,f}) &\notin f : M_{nst,}^k(P_{nst}) \\ M_{nst,d}^k{}'(P_{nst,f}) &\in f : M_{nst,}^k(P_{nst}) \end{aligned} \tag{12}$$

In each of the above cases, if it is possible to assemble the a-th element q_a by the c-th stationary crane or the d-th mobile crane (in each k-th simulation), then the assembly duration of the a-th element should be added to the crane's working time. If the assembly time exceeds the assembly brigade's end time, the assembly should be scheduled for the next j-th day of work. The date of assembly commencement results from the completion of work by the c-th or d-th crane or it is the date of delivery of the a-th element to the construction site. For stationary cranes, the deadline for assembly completion by the c-th crane, on the j-th day of work, in the k-th simulation and at a given moment "*" should be calculated from the formula:

$$T_{c,k*,j} = \begin{cases} T_{c,k,j} + l_{a,6} \text{ gdy } T_{c,k,j} + l_{a,6} \leq T_{END,W} \wedge l_{a,8} \leq T_{c,k,j} \\ l_{a,8} + l_{a,6} \text{ gdy } l_{a,8} + l_{a,6} \leq T_{END,W} \wedge l_{a,8} > T_{c,k,j} \\ T_{START,W,j+1} + l_{a,6} \text{ gdy } T_{c,k,j} + l_{a,6} > T_{END,W} \vee l_{a,8} + l_{a,6} > T_{END,W}, \end{cases} \tag{13}$$

where

$T_{c,k,j}$ is deadline for the assembly of the element that precedes the a-th element.

$T_{START,W,j+1}$ is the time (in hours) when the assembly teams start their work on day $j + 1$.

In the case of mobile cranes, the deadline assembly completion by the d-th crane, on the j-th day of work, in the k-th simulation and at a given moment "*" should be calculated from the formula:

$$T_{d,k*,j} = \begin{cases} T_{d,k,j} + l_{a,6} \text{ when } T_{d,k*,j} + l_{a,6} \le T_{END,Z} \wedge l_{a,8} \le T_{d,k,j} \\ l_{a,8} + l_{a,6} \text{ when } l_{a,8} + l_{a,6} \le T_{END,Z} \wedge l_{a,8} > T_{d,k,j} \\ T_{START,W,j+1} + l_{a,6} \text{ when } T_{d,k,j} + l_{a,6} > T_{END,W} \vee l_{a,8} + l_{a,6} > T_{END,W} \end{cases} \tag{14}$$

If the a-th element cannot be mounted by either the c-th stationary crane from the e-th location or the d-th mobile crane from the f-th location, the element remains unmounted and the assembly procedure for the next element begins:

$$\underset{M^k_{st,c} \in M^k_{st}}{\underbrace{\nexists}} Q^e_{M^{k*}_{st,c}} \ge Q^{a,e}_{min} \ dla \left\{ L^{a,e}_{min}, H^{a,e}_{min} \right\} \wedge \underset{M^k_{nst,d} \in M^k_{nst}}{\underbrace{\nexists}} Q^f_{M^{k*}_{nst,d}} \ge Q^{a,f}_{min} \ for \ \{L^{a,f}_{min}, H^{a,f}_{min}\},$$
$$where \ f \in \langle 0, o \rangle \Rightarrow a \in Q^k_n \wedge aa + 1 \tag{15}$$

where

Q^k_n is a set of elements unmounted during the k-th simulation.

The procedure described above is repeated until all the prefabricated elements are mounted or until it is proven that some elements technically cannot be mounted. For each k-th simulation and its solution, the following data are remembered:

- cost of the assembly operations,
- duration of the assembly tasks,
- the assembly schedule,
- the assembly machines work schedule,
- the number of assembly machines used,
- the number of unmounted elements.

The procedure is repeated k times. After that, the planner can analyze the aggregate as well as detailed results for each simulation, sort them by cost, duration or the number of unmounted elements. He (or she) also has the ability to make changes in the input data, such as the number, type or size of stationary/mobile cranes and the coordinates of their possible locations. The simulation is, therefore, interactive and allows the planner to find the most suitable solution for him (her).

For the model of the assembly performance, due to its complexity and numerous simulations required, a computer application was developed. Its graphical interface was created with the WPF graphical system. The application was developed in Visual Studio 2015 in C# and is presented in Reference [33].

2.3. Example

In order to present the possibilities and benefits of using the method of assembly planning described above, let us follow an example of a real office building assembly. The results of the simulations will be compared to the data obtained from a real schedule.

The subject of assembly planning is the "Green2Day" office building located at 11 Szczytnicka Street in Wrocław and built between 2015 and 2017 by SKANSKA JSC. The building has a two-level underground garage for 231 vehicles and consists of seven floors above the ground with a usable area of 17,000 sq. m. The "Green2Day" building was designed in accordance with the LEED certification system at the Gold level, which matches the name of the building. Figure 2 presents a front view of the building.

Figure 2. "Green2Day" office building front view. Investor: Tenali Investments Ltd. and Trikala SCA. Designers: Maćków Pracownia Projektowa LP & Grupa Projektowa Konstruktor Ltd. Own study.

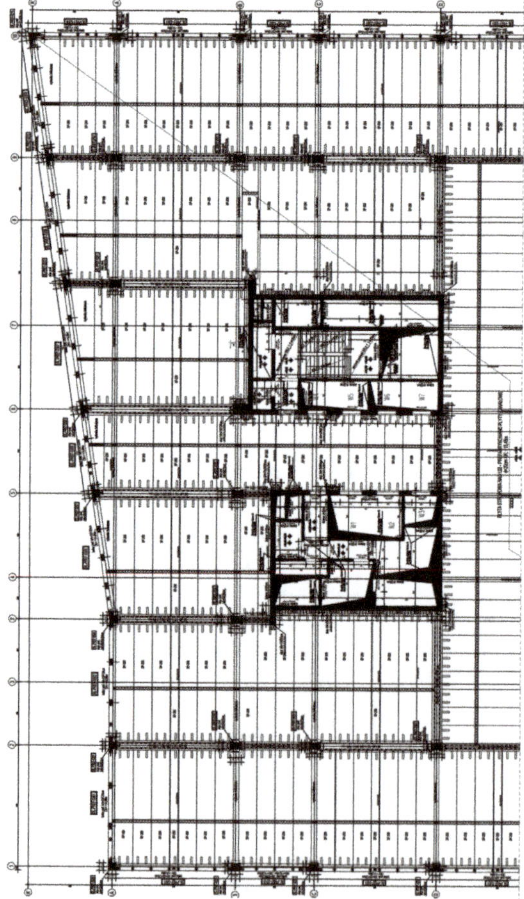

Figure 3. The Green2Day building slabs' layout—Part 1. The design drawing was obtained from Grupa Projektowa Konstruktor Ltd.

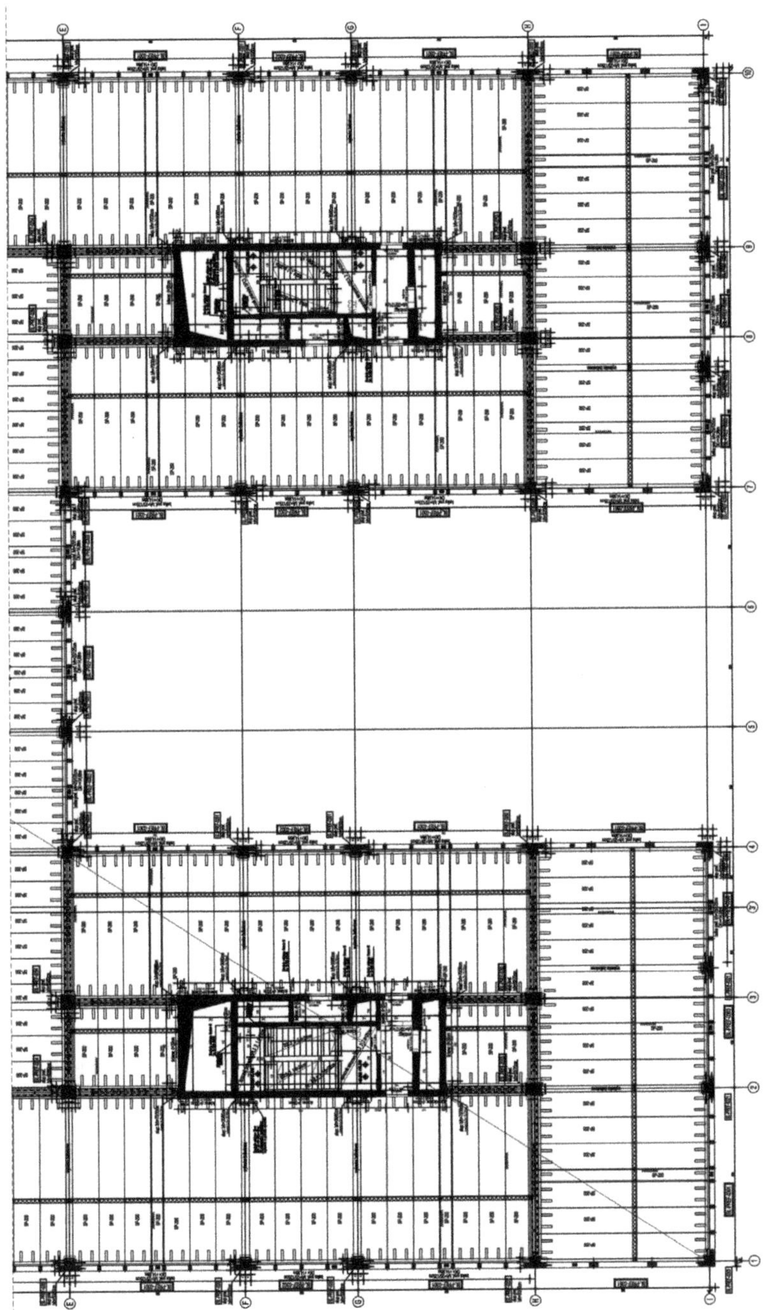

Figure 4. The Green2Day building slabs' layout—Part 2. The design drawing was obtained from Grupa Projektowa Konstruktor Ltd.

The underground part of the building was designed as cast-in-place concrete construction with diaphragm walls. The structure is stiffened with four monolithic concrete shafts and staircases, which provide the support for the prefabricated ceilings. The above-ground construction is predominantly prefabricated. There are concrete columns founded on external diaphragm walls and slab on grade foundation, DELTABEAM beams supported by the columns inside the building, prefabricated reinforced concrete beams on the perimeter of the building and hollow core slabs. The layout of the above-ground floor structure is shown in Figures 3 and 4.

There were 59 prefabricated columns, 37 DELTABEAMs, 42 perimeter beams and 275 hollow core slabs on each repetitive floor above the ground. The total number of large-size prefabricated elements to be assembled was close to 3000 (see Table 1).

Table 1. List of reinforced concrete prefabricated elements to be mounted in the "Green2Day" building. Our own study based on the design by Grupa Projektowa Konstruktor Ltd.

Type of Item	Total Number of Items
Column 50 × 70 cm	98
Column 50 × 50 cm	287
Column 60 × 60 cm	28
Peripheal beam (length 5.0–6.0 m)	182
Peripheal beam (length 8.0–8.5 m)	112
Peripheal beam (length 1.8–4.0 m)	84
DELTABEAM (length 4.0–6.0 m)	49
DELTABEAM (length 6.0–8.0 m)	91
DELTABEAM (length 8.0–10.0 m)	35
Floor slab HC width 50 cm, length 3.5–6.0 m	14
Floor slab HC width 120 cm, length 3.5–6.0 m	217
Floor slab HC width 50 cm, length 6.0–9.0 m	182
Floor slab HC width 120 cm, length 6.0–9.0 m	1512
Element of stairs	42
Total:	2933

The construction was assembled using two Potain high-speed tower cranes with a load capacity of 10 tons. The third crane (crane No. 3 in Figure 5) was reserved for unloading materials other than prefabricated elements. The scheme of the construction site layout is shown in Figure 5. Due to the small area available, the assembly was planned as "just in time". The fast pace of work required work in two shifts. On the second shift, prefabricated elements were mounted. The available number of cranes and large dimensions of the building allowed for a simultaneous work of two assembly brigades. The assembly works lasted over 3.5 months.

The assembly planning simulations, with the developed application, were carried out in six organizational variants.

The first variant reflects the actual schedule and it was done in order to check the correctness of the application. The origin of the coordinate system, the coordinates of the building's corners and the location of cranes No. 1 and No. 2 were defined. Prefabricated elements, along with their characteristics, were also introduced (elements' dimensions, weights and coordinates of their location in construction were obtained from the detailed design of the building). Slings' weights and heights were determined by the planner. Durations of their assembly were calculated based on previous research on work processes. The chronological order of assembly was developed based on the actual schedule of works. The organizational layout of the assembly is shown in Figure 5.

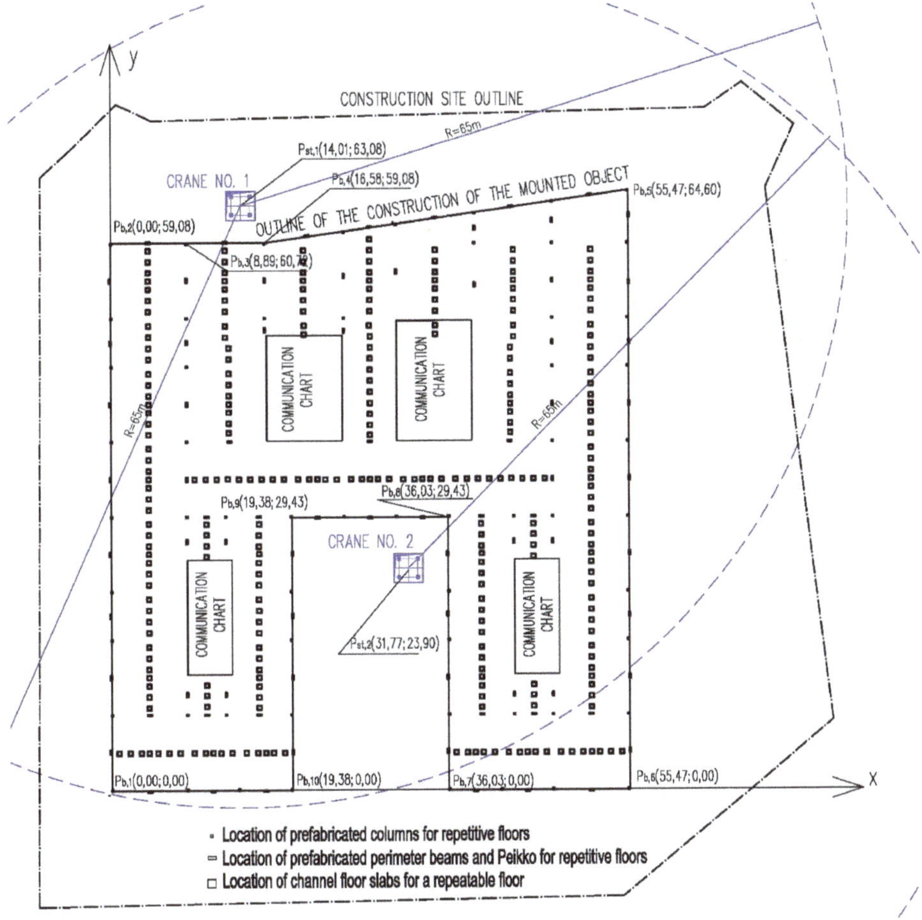

Figure 5. The organizational layout of assembly works the actual way of carrying out assembly works at the "Green2Day" facility.

Two high-speed Liebherr 202 EC-B 16 Litronic tower cranes with a maximum load capacity of 10 tons, similar to the Potain cranes, were used in the simulation. Logistic costs related to the site preparation, cranes' transportation on site, assembly and dismantling were estimated at 10,000 PLN. The unit labor costs of one brigade include crane operation, mobile scaffolding and a 6-person assembly brigade. The unit costs of the brigade's work were estimated at PLN 360.00 per hour. Separate costs of crane delivery and transportation characteristics were omitted. These costs were included, along with the assembly and disassembly of the cranes.

The works commencement date was set on 18 June 2016. Work began at 3.00 p.m. and ended at 11.00 p.m. The lifting height was increased by safety margin of 1.0 m. The number of simulations was set at 10.

For each of 10 simulations, its results are the same (excluding the variants in which only one crane is used). This is due to the lack of random factors. The simulation algorithm analyses two locations of cranes and two cranes of equal parameters. For such conditions, the completion date was calculated for 13 October 2016, which is similar to the actual schedule of works (10 October 2016) and indicates that the algorithm works correctly. The calculated assembly time is longer by three days in relation to the actual time spent on works due to overtime work. The cost of the assembly operations, according to calculations, is 552,576.80 PLN.

In order to obtain other assembly schedule variants, the initial assumptions were modified: additional possible locations for stationary cranes (in elevator shafts) and mobile cranes (on the site) were defined in accordance with Figure 6. The size of cranes' pool was increased by two Liebherr 150 EC-B 8 litronic high-speed cranes, two Liebherr 172 EC-B 8 litronic cranes and three mobile cranes with a maximum load capacity of 50, 70, and 100 tons. In addition, the priority of stationary machines was set and the number of simulation increased to 100. The costs due to cranes utilization are collected in Table 2. The results of simulations are gathered in Table 3.

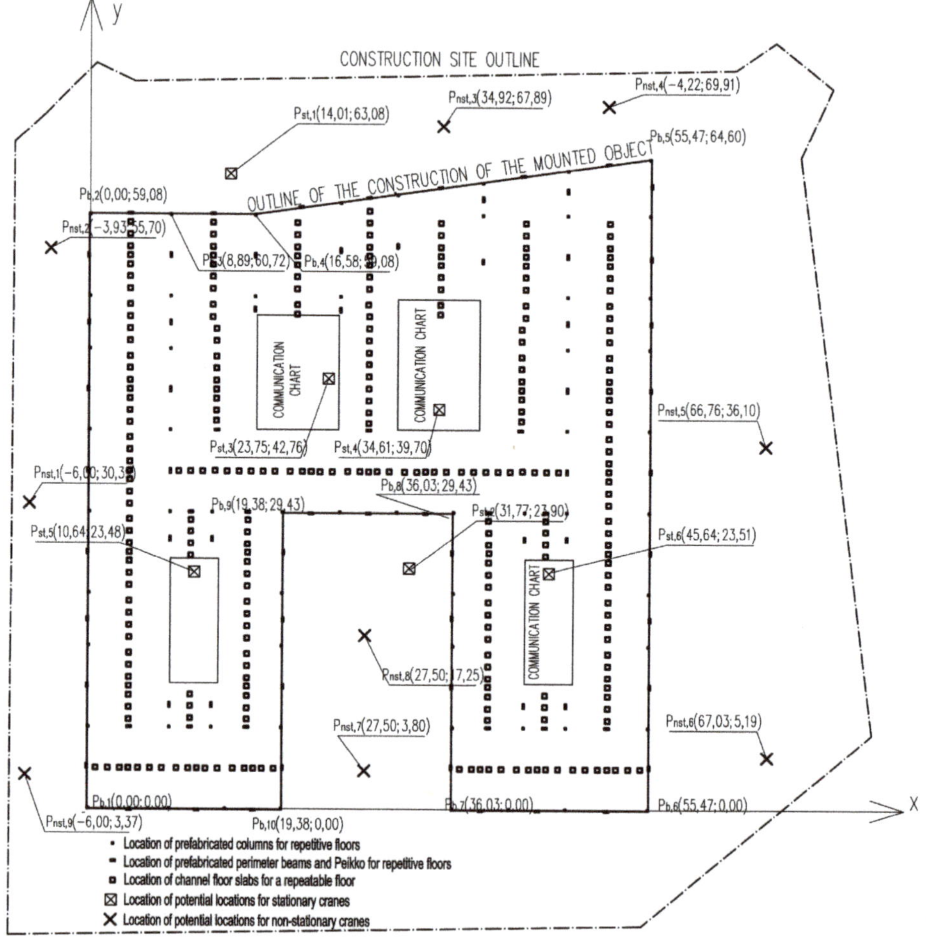

Figure 6. The organizational layout of assembly works-the potential locations of cranes for the assembly of the "Green2Day" facility.

Table 2. Cost characteristics of the assembly machines shown in the example. Sourced from our own study.

	Crane Delivery Cost, PLN/km *	Distance between Crane's Base and the Construction Site, km	Crane Assembly and Disassembly Costs, PLN **	Unit Operating Costs; Includes Crane's Work and Assembly Brigade Work PLN/h	The Time Limit of Crane Underutilization, h ***
Top-slewing tower cranes					
Liebherr 150 EC-B 8 litronic	0.00	0	9,000.00	320.00	0
Liebherr 170 EC-B 8 litronic	0.00	0	9,500.00	340.00	0
Liebherr 202 EC-B 16 litronic	0.00	0	10,000.00	360.00	0
Mobile cranes					
Liebherr LTM 1050-3.1 (50T)	4.00	20	0.00	370.00	8
Liebherr LTM 1070-4.2 (70T)	6.00	20	0.00	400.00	8
Liebherr LTM 1100-4.2 (100T)	12.00	20	0.00	470.00	8

* crane delivery cost is included in the costs of assembly and disassembly. ** crane assembly and disassembly costs for mobile cranes are included in cost of assembly works; they are designated as the product of machine assembly and disassembly duration, the number of necessary crane relocations within the site and unit operating costs that include the work of the assembly brigade. *** the time when the assembly team cannot work; it happens if the crane has no lifting capacity for several subsequent elements, or the crane awaits for the assembly of the preceding elements mounted by other assembly team, or if the priority of stationary crane is set. This time influences on the loss cost due to the underutilization of assembly. This cost may be negligible if the time of underutilization allows the crane (and the assembly brigade) to carry out another construction project. In the example, it was assumed that, in the case of mobile cranes, the minimum negligible time of crane underutilization is 0 or 1 working day (8 h).

Table 3. The results of t computer simulations-description in the text. Sourced from our own study.

No.	Assembly Machines	Assembly Cost, PLN	Assembly Duration, in Days	Number of Unmounted Elements
1	2 x Liebherr 202 EC-B 16 litronic	552,576.80	98	0
2	2 x Liebherr 170 EC-B 8 litronic Liebherr LTM 1070-4.2 (70T)	531,660.20	81	0
3	2 x Liebherr 170 EC-B 8 litronic Liebherr LTM 1070-4.2 (70T)	533,851.10	79	0
4	Liebherr 202 EC-B 16 litronic Liebherr 150 EC-B 16 litronic Liebherr LTM 1070-4.2 (70T)	533,974.11	78	0
5	Liebherr 202 EC-B 16 litronic Liebherr 150 EC-B 16 litronic	528,989.20	86	103
6	Liebherr 170 EC-B 8 litronic Liebherr LTM 1070-4.2 (70T) Liebherr LTM 1100-4.2 (100T)	607,169.52	69	0
7	Liebherr 202 EC-B 16 litronic 2 x Liebherr 150 EC-B 8 litronic	520,929.42	69	17
8	2 x Liebherr 150 EC-B 8 litronic 2 x Liebherr 170 EC-B 8 litronic	525,384.48	50	0
9	2 x Liebherr 150 EC-B 8 litronic 2 x Liebherr 170 EC-B 8 litronic Liebherr 202 EC-B 16 litronic	544,071.68	41	0

The perimeter beams played a key role in assembly works planning. It was necessary to mount them using large load capacity and large reach cranes. Stationary cranes' size reduction (solution No. 2), mounting them in elevator shafts and using a mobile crane with a maximum load capacity of 70 tons allowed for the saving of approx. 19,000 PLN and 17 days of work. As the priority for stationary machines was set, most of the construction elements were mounted by these cranes. Mobile cranes were only used in the case of heavy perimeter beams.

Solution No. 3 was equally beneficial. In this solution, stationary cranes were located just like on the construction site. For this reason, they were able to assemble a smaller number of elements and a couple of tasks more was left for mobile cranes. The solution was slightly less cost-profitable, while the savings were c.a. 17,000 PLN, but assembly time shortened by 19 days compared to solution No 1.

Another favorable option was solution No. 4, in which two stationary cranes with a different maximum load capacity and a mobile crane was used. Cost savings were 18,000 PLN, but time-savings reached 20 days, in comparison to solution No 1.

The described case studies above prove that the cooperation between stationary and mobile cranes bring savings and improves assembly productivity. Using only tower cranes, without mobile crane support, as was the case in solution No 5, made the heaviest perimeter beams unmounted. Moreover, it increased assembly costs significantly, like in solution No. 6.

Sometimes it is a good idea to increase the number of stationary cranes, like in solutions No 7, No. 8 and No. 9. The optimal number of assembly machines was 3 (solution No. 7). Thanks to the simultaneous operation of all machines, the savings reached c.a. 31,000 PLN and the assembly time shortened by 29 days, in comparison to solution No. 1. However, such a solution might be dangerous, as overlapping operation areas of these cranes would cause numerous collisions and downtimes. Using more stationary cranes obviously shortened assembly duration, but did not bring financial benefits (see solutions No. 8 and No. 9).

The example shows that stationary cranes should constitute the main equipment for the assembly works. It is strictly related to the construction itself—its shape, area, number of floors or the number of elements to be mounted. In practice, as each assembly case is different, computer simulations can efficiently aid assembly planning.

When can mobile cranes be more profitable that the stationery ones?

If the Green2Day building consisted of only three floors, then the high costs of tower cranes mounting would not be balanced by lower unit costs of their work. Thanks to the lower height of the building, the required load capacity of mobile cranes could be smaller, which would lower unit assembly costs. This condition is particularly important for small space construction sites where cranes must be located close to the external walls of the building. If we considered a three-story building, the assembly costs for two mobile cranes with a maximum load capacity of 50 tons would be reduced by c.a. 14,000 PLN, in comparison to solution No. 1 (solution applied in practice, see Table 4).

The conclusions that can be drawn from the above examples are only binding for the analyzed object. In practice, construction projects differ in the number of elements to be assembled, their dimensions, weights, locations, as well as the availability of the construction site, machines and assembly brigades. Therefore, each project requires individual planning, which can be supported by the proposed computer application. Selecting the type of cranes' construction, defining their potential locations, and setting the priority of stationary cranes allow for ongoing interaction between planner and computer, and hence, searching for a sub-optimal solution. The use of the presented simulative approach to assembly planning is not limited to planning prior to works commencement. If the pace of assembly works needs to be increased, it is possible to reorganize works at each stage of assembly, only taking into account yet unmounted elements.

Table 4. Selected results of the computer aided assembly planning for the first 3 floors of the building-description in the text. Sourced from our own study.

No.	Assembly Machines	Assembly Cost, PLN	Assembly Duration, in Days	Number of Unmounted Elements
1	2 x Liebherr 202 EC-B 16 litronic	248,247.20	42	0
2	2 x Liebherr 170 EC-B 8 litronic Liebherr LTM 1070-4.2 (50T)	237,496.30	35	0
3	2x Liebherr LTM 1050-4.2 (50T)	234,587.40	43	0
4	3x Liebherr LTM 1050-4.2 (50T)	235,241.80	29	0

3. Discussion and Conclusions

In the proposed method of assembly works planning, an active and key role is played by the planner who, based on their experience, and the results of historical simulations, has the ability to change the input assumptions. The size, first in regards to the load capacity of cranes, is effectively assessed by the number of unmounted elements in the solutions of subsequent simulations. If the number is large, it is a signal to the planner to change the size of the cranes. The potential locations of cranes can be effectively determined by the planner, assuming that he has elementary knowledge about the relationships between load capacity, reach and lifting height. The planner also does the final crane selection and assembly schedule, however, the developed method supports the decision-maker by presenting many acceptable solutions.

It should also be mentioned that the aim of planning is not always to choose the best solution for a selected construction project, but to achieve the smallest possible cost of works through the prism of the production potential. Therefore, it is not their goal to carry out works quickly on one particular construction site, but to use the production potential, construction equipment and workers effectively. The method of assembly planning cannot therefore focus on choosing one optimal solution that presents the most favorable ratio of the pace of works to their cost. The planner's support consists of indicating available solutions, and assessing their quality. The final decision should always be left to the decision maker.

The method does not explicitly include collisions between cranes. Algorithmically, this problem is difficult to solve, because construction situations are unstable-they depend on many random factors. Therefore, other solutions are advised, e.g., planning cranes' locations at a safe distance from each other or using anti crane collision systems and limit switches. This should solve the problem of work safety and collision-free assembly. As the spans of prefabricated elements are large, the work zones of cranes may overlap, but it does not affect the pace of work significantly if the planner or crane operator locates the machine skillfully.

Such an approach to the assembly planning problem was made in this paper—the planner indicates the potential locations of cranes in a way that reduces the risk of collision, while the possible risk of collision is eliminated using the already available techniques.

There are still several open issues in the problem of assembly planning that should be considered in the course of further research. These include, but are not limited to, the following:

1. In the current formula, the planning tool prioritizes the order of assembly in accordance with the order proposed by the planner (order according to the list of elements to be mounted). Each assembled element is the predecessor of the next element to be assembled. In fact, this relationship primarily stems from the technological conditions and lets the planer decide freely about the order of assembly. The order of the elements to be mounted should be enriched by a matrix of

dependency relations between the assembled elements and their predecessors, while decisions to move a mobile crane should be made after checking the possibility of assembling all elements for which the required predecessors have already been mounted. Such a solution would increase the effectiveness of the proposed method by reducing the number of necessary adjustments of mobile cranes.

2.   Introducing the characteristics of the mounted elements is tedious and time consuming. The proposed planning tool should be integrated with the BIM-based models in order to create the list of elements to be assembled and their characteristics automatically. The environment can also be used to import technological dependencies of assembly priority between individual elements, in accordance with point 1 above.

3.   The practical use of the proposed planning tool should be preceded by a comprehensive research on labor rates for prefabricated elements assembly. This should be the basis for the division of construction elements by the assembly duration (expected average value and standard deviation). The results obtained with the abovementioned research should be the basis for extending the proposed planning tool in order to include analyses in probabilistic conditions. The wide-range implementation of the research may be possible thanks to the constant reading of operating parameters and assigning mounting hook locations to the mounted elements. In the light of the emerging 5G technology and developing Internet of Things technology, the proposed method can constantly evolve.

4.   The proposed planning tool could be commonly used if the above mentioned problems are investigated. The graphical interface of the described application needs to be improved so it could be commercialized. The above presented possibilities of improving the computer application prove its potential and complexity of the problem as a whole.

The authors intend to use the proposed planning tool to plan assembly works in real planning situations. The practical application of the planning tool will be the subject of future publications

Author Contributions: Conceptualization, R.M. and M.B.; methodology, R.M. and M.B.; software, M.B.; validation, R.M. and M.B.; formal analysis, R.M.; resources, M.B.; writing—original draft preparation, M.B.; writing—review and editing, R.M.; supervision, R.M.; project administration, R.M.; funding acquisition, R.M. and M.B. All authors have read and agreed to the published version of the manuscript.

Funding: This research received no external funding.

Conflicts of Interest: The authors declare no conflict of interest.

References

1.   Gray, C.; James Little, J. A Systematic Approach to the selection of an appropriate crane for a construction site. *Constr. Manag. Econ.* **1987**, *3*, 121–144. [CrossRef]

2.   Gray, C. Crane location and selection by computer. In Proceedings of the 4th Int. Symp. Robotics and Artificial Intelligence in Building Construction, Haifa, Israel, 22–25 June 1987; Volume 1, pp. 163–167.

3.   Furusaka, S.; Gray, C. Model for the selection of the optimum crane for construction sites. *Constr. Manag. Econ.* **1984**, *2*, 157–176. [CrossRef]

4.   Warszawski, A.; Peled, N. An expert system for crane selection and location. In Proceedings of the 4th Int. Symp. Robotics and Artificial Intelligence in Building Construction, Haifa, Israel, 22–25 June 1987; Volume 1, pp. 64–75.

5.   Warszawski, A. Expert system for crane selection. *Constr. Manag. Econ.* **1990**, *82*, 179–190. [CrossRef]

6.   Cooper, C.N. Cranes—A rule-based assistant with graphics for construction planning engineers. In *The Application of Artificial Intelligence Techniques to Civil and Structural Engineering*; Civil Comp Press: Edinburgh, UK, 1987; pp. 47–54.

7.   Farrell, C.W.; Hover, K.C. Computerized crane selection and placement for the construction site. In *Proceedings of the 4th Int. Conf. on Civiland Structural Engineering Computing*; Topping, B.H.V., Ed.; Civil Comp Press: Edinburgh, UK, 1989; Volume 1, pp. 91–94.

8.   Choi, C.W.; Harris, F.C. A model for determining optimum crane position. *ICE Proc.* **1991**, *90*, 627–634.

9. Zhang, P.; Harris, F.C.; Olomolaiye, P.O.; Holt, G.D. Location optimization for a group of tower cranes. *J. Constr. Eng. Manage.* **1999**, *125*, 115–122. [CrossRef]

10. Williams, M.; Bennett, C. ALPS: The Automated Lift Planning System. In Proceedings of the ASCE Third Congress on Computing in Civil Engineering, Anaheim, CA, USA, 17–19 June 1996.

11. Ito, K.; Kano, Y. 3-D Graphical Simulation for Crane Planning using Object-Oriented Building Product Model. In Proceedings of the CIB proceedings Information Technology Support for Construction Process Re-engineering, Cairns, QLD, Australia, 9–11 July 1997.

12. Meehan, J. Computerize to organize. *Cranes Today* **2005**, *369*, 50.

13. North Cascade Industrial (NCI). Compu-Crane CSPS/LPS: Crane Selection and Lift Planning Software, Seattle. Available online: www.ncisoftware.com (accessed on 7 July 2018).

14. Cranimax. *Cranimation and TowerManagement: Software for Crane Job Site Planning*; Cranimax GmbH: Zweibrücken, Germany; Available online: www.cranimation.org (accessed on 7 July 2018).

15. LiftPlanner Software. LiftPlanner: 3D Crane and Rigging Lift Planning Software, Burnsville, MN, USA. Available online: www.liftplanner.net (accessed on 10 August 2018).

16. Progistik. MéthoCAD, Progistik, Bagnolet, France. Available online: www.methocad.com (accessed on 12 July 2018).

17. Hanna, A.S.; Lotfallah, W.B. A fuzzy logic approach to the selection of cranes. *Autom. Constr.* **1999**, *8*, 597–608. [CrossRef]

18. Sawhney, A.; Mund, A. Adaptive probabilistic neural network-based crane type selection system. *J. Constr. Eng. Manage.* **2002**, *128*, 265–273. [CrossRef]

19. Tam, C.M.; Tong, T.K.L.; Chan, W.K.W. Genetic algorithm for optimizing supply locations around tower crane. *J. Constr. Eng. Manage.* **2001**, *127*, 315–321. [CrossRef]

20. Shapira, A.; ASCE, F.; Goldenberg, M. "Soft" Considerations in Equipment Selection for Building Construction Projects. *J. Constr. Eng. Manag. Asce* **2007**, *133*, 749–760. [CrossRef]

21. Dalalah, D.; AL-Oqla, F.; Hayajneh, M. Application of the Analytic Hierarchy Process (AHP) in Multi-Criteria Analysis of the Selection of Cranes. *Jordan J. Mech. Ind. Eng.* **2010**, *4*, 567–578.

22. Günthner, A.; Kessler, S.; Tölle, S. *BKT Turmdrehkran-Einzatzplaner*; Leaflet Munich University of Technology: Munich, Germany, 1998.

23. Günthner, A.; Kessler, S.; Tölle, S. *Entwicklung eines Turmdrehkran-Einsatzplaners*; Research report; Technische Universität München: Munich, Germany, 2002.

24. Günthner, A.; Kessler, S.; Frenz, T.; Hefele, R.; Walter, M. Digital Tower Crane Deployment Planner, BauPortal 6-7. Available online: www.baumaschine.de/Krane (accessed on 10 August 2018).

25. Tork, A.Z. A real-time crane service scheduling decision suport system (CSS-DSS) for construction tower cranes. Ph.D. Thesis, Department of Civil, Environmental and Construction Engineering in the College of Engineering and Computer Science at the University of Central Florida, Orlando, FL, USA, 2013.

26. Lei, Z.; Behzadipour, S.; Al-Hussein, M.; Hermann, U. Application of robotic obstacle avoidance in crane lift path planning. In Proceedings of the 28th International Symposium on Automation and Robotics in Construction, ISARC, Seoul, Korea, 29 June–2 July 2011.

27. Han, S.H.; Al-Hussein, M.; Hasan, S.; Gökçe, K.U. Simulation of mobile crane operations in 3d space. In Proceedings of the 2012 Winter Simulation Conference, Berlin, Germany, 9–12 December 2012.

28. Han, S.H.; Hasan, S.; Lei, Z.; Sadiq Altaf, M.; Al-Hussain, M. A framework for crane selection in large-scale industrial construction projects. In Proceedings of the 30th International Symposium on Automation and Robotics in Construction and Mining (ISARC), Montreal, QC, Canada, 11–15 August 2013.

29. Lei, Z.; Taghaddos, H.; Hermann, U.; Al-Hussein, M. A methodology for mobile crane lift path checking in heavy industrial projects. *Autom. Constr.* **2013**, *31*, 41–53. [CrossRef]

30. Han, S.H. BIM-based Motion Planning of Mobile Crane Operation in Modular-based Heavy Construction Sites. Ph.D. Thesis, Department of Civil and Environmental Engineering University of Alberta, Edmonton, AB, Canada, 2014.

31. Krawczyńska-Piechna, A. Comprehensive Approach to Efficient Planning of Formwork Utilization on the Construction Site. *Procedia Eng.* **2017**, *182*, 366–372. [CrossRef]

32. Banach, M. Model wyboru wielkości urządzenia dźwigowego do wykonania robót montażowych. In *Aktualne Problemy Naukowo-Techniczne Budownictwa*, 1st ed.; Krawczyńska-Piechna, A., Ed.; Warsaw University of Technology: Warsaw, Poland, 2016; pp. 75–83.
33. Banach, M. Interaktywna Metoda Planowania Robót Montażowych w Budownictwie Prefabrykowanym. Ph.D. Thesis, Warsaw University of Technology, Warsaw, Poland, 2019.

 *buildings*

Article

Investigation of the Effective Use of Photovoltaic Modules in Architecture

Waclaw Celadyn [1],* and Pawel Filipek [2]

[1] Faculty of Architecture, Cracow University of Technology, 30-084 Krakow, Poland
[2] Architecture Studio Filipek, 31-423 Krakow, Poland; paulusfilipek@gmail.com
* Correspondence: wceladyn@pk.edu.pl

Received: 25 June 2020; Accepted: 3 August 2020; Published: 21 August 2020

Abstract: The application of photovoltaic systems is becoming a dominant feature in contemporary buildings. They allow for the achievement of zero-energy constructions. However, the principles of this strategy are not yet sufficiently known among architects. The purpose of this study is to enhance their expertise, which cannot be widened due to the shortage of targeted publications. The issue presentation was structured in a way that follows the typical design stages, beginning with large-scale urban problems up to the scale of building forms and components. Different types of photovoltaic (PV) systems are considered, based on their efficiency, relations with building fabrics, potential for thermally protecting buildings and their impact on esthetic values. The focus was mainly on the most popular PV modules. The application of these systems requires in-depth analyses which should be carried out by designers at the initial stage and through the next stages of the design. A method to analyze zoning plan regulations and site planning in view of PV modules' efficiency is novel. This paper also contains considerations with regard to some other untypical applications of these systems. There is need for changing attitudes in architects and investors regarding the issue of promoting the systems through further elucidations.

Keywords: architecture; architectural design; photovoltaic modules in architecture

---

1. Introduction

Among the few methods of gaining energy from renewable sources set in the paradigm of sustainable architecture, it is the photovoltaic (PV) modules (further termed PV panels) which have become the most promoted and used in contemporary buildings. Of the solar electric systems currently available, photovoltaic technology is the most advanced and mature [1] (p. 75). The relative simplicity of installation, skyrocketing electric rates and a comfortable accessibility of generated electrical power for household appliances make PV technology a reasonable option for homeowners as well as for an array of facilities to solve energy problems. Solar panels are becoming gradually more popular as their costs are falling, and this form of electricity generation is growing fast [2] (p. 260). Another reason for the increasing use of this technology in the building market is because solar thermal electricity costs more than PV electricity [3]. The green electrical generating systems, because of their sustainable and ecological nature, are classified as a nonmaterial tecnofact [1] (p. 68). They bring many ecological advantages not only for using renewable energy, but also for reducing carbon dioxide emissions and fostering a recommended decentralized electrical power production, which thereby achieves a higher security level, e.g., because of an increased resilience to power outages.

Architecture based on a solar energy concept is sometimes called regenerative. A regenerative system provides the continuous replacement, through its own functional process, of the energy and materials used in the operation [4] (p. 10). The energy is replaced primarily by incoming solar radiation; thus, photovoltaic systems make buildings ecological. The processes for converting solar energy to electrical power are the most efficient regenerative energy-conversion processes. Most regenerative technologies are modular in nature, just like photovoltaic cells [4] (p. 74).

Given the constant increase in thermal requirements, PV panels have become an indispensable element of nearly zero-energy buildings (NZEBs), which will shortly be a pervasive standard model. However, the application of photovoltaic systems to buildings requires in-depth analyses to make this energy option perform well in terms of its energy efficiency, economic issues, spatial effects, as well as the esthetic values of buildings, their components, and even the plot layout.

The basic and detailed knowledge of solar systems is ample and constantly increasing. New solar technologies concerning PV panels and other electricity generating systems are extensively covered in publications. However, the complexity of the related problems is rarely considered in publications in a way that is useful for architects. Therefore, architects' viewpoints, hardly present therein, were the main reason for this research. This is what makes this paper covering these issues significant. Many other professionals analyze the problem of photovoltaic systems in a sectional scope of view, adding to the knowledge but missing the overall image of this multidisciplinary issue. A relatively complex, however, incomplete, picture of these systems can be found in some larger publications, like *"Building-Integrated Solar Technology: Architectural Design with Photovoltaics and Solar Thermal Energy"* [5] or [6]. This deals with the historical aspects of PV systems and contains a series of study cases. There are also some other published papers related to the issue [7–12], to mention a few, but none are of full practical use for architects. The purpose of this work is to enhance the knowledge of the designers of architecture by issuing a research-based study, which would encourage them to approach their creative activity in a less intuitive way, in terms of solar electricity generation. This is important as the relations between the building components and installed PV panels must be analyzed at the first concept sketches for designed buildings. To do this in a competent way, an architect should possess an appropriate knowledge before he eventually analyzes the problem with specialists in the field, who, in turn, are less knowledgeable on spatial and esthetic issues. Contributing to improvements in architects' competence makes this study purposeful.

2. Materials and Methods

This research is by its nature, and the character of targeted professionals, a complex task. Its scope is closely linked to the consecutive stages of architectural design to be of practical use for designers. Therefore, the method and structure of the paper were determined, by the design procedures, as outcomes of every consecutive step; this philosophy has a very distinctive impact on the next stage and related analysis. The idea underlying the design procedure usually consists of the gradual passage from large-scale decisions to small-scale solutions. In order to deal properly and systematically with the matter of this research, which is PV panels, the first step was to carry out an analysis of a wide range of accessible PV components offered on the market. Among many solar electricity-generating systems, the technology using PV panels is crucial, as it is the most popular and frequently opted for among building investors. It also requires in-depth analyses of multiple aspects of their installation. This is the reason for which this study has been overwhelmingly focused on this technology. The study is mainly concentrated on PV panels installed on roofs, as this location is presently more exigent for them in some positions, i.e., mainly facades.

The purpose of this research is to define the conditions assuring a rational choice of devices, and a method of their use to achieve the optimum energy and esthetic efficiency, coupled with the best spatial results. Large-scale considerations, being the next stage of this procedure, are the multifaceted analyses of the location characteristics aiming at the formulation of proposals for the optimum energy generation-related layout of a given land plot. This initial task is not an easy problem because of different spatial and some other restrictions comprised within zoning plans for defined areas of cities or villages.

The determination of the mutual spatial configuration of PV panels mounted on buildings, and their position in relation to incident solar radiation assuring their efficiency, is the subsequent stage of this logical procedure. A method of deciding on the location of these devices on buildings and their parts is the next consideration. Esthetical issues are usually not considered a scientific domain. However, in the case of architectural discussions and analyses, it cannot be excluded from the relevant research. It also makes a significant consecutive part of this study. Finally, the research deals with some other functions that the PV panels used in architecture take on, and thus contribute to a better energy efficiency of contemporary buildings. A chart with the energy yield of an analyzed house was generated by sunnyportal.com. A simulation of the insolation and shading by PV panels was made with the use of Sketchup program and the application CuricSun.

The presented multifaceted and multistaged method permits the logically interrelate subsequent stages of research-based approaches to design buildings fitted with PV panels. The use of other PV technologies has been mentioned in these considerations, but they require a somewhat different approach.

3. Analyses of PV Technologies and Important Design Issues

3.1. PV Panel Systems and Their Development

PV panels, the most popular system for providing renewable energy to buildings on the market, have developed rapidly, bringing down their prices. The low performance of PV panels has increased gradually up to a 17% energy effectiveness, on average. Some of the newest systems boast a 40% energy effectiveness. The development of these systems enhanced not only the higher energy efficiency of panels, but also made some breakthroughs in terms of basic materials. The systems are classified into three generations [13] (p. 24). The first one is the most popular monocrystalline silicon cells (SCPV, m-Si) which have the highest efficiency (13–26.1% [14]), and are durable and expensive, as well as polycrystalline silicon cells (PCPV, p-Si) which have a lower energy efficiency of 10–23.3%, typically 11–14%, and are less expensive, as evidenced by [14,15] (p. 291), and [1] (pp. 76–77). The second generation are the transparent thin-film cells, which offer flexible systems, are less durable, less expensive and have lower efficiencies (5–7%), as evidenced by [1] (pp. 77–78), due to other sources having a maximum of 14% for amorphous Si:H stabilized solar cells [14]. They can be installed on curved surfaces. An attractive option is the possibility of their integration with membrane flexible constructions. The third-generation systems are of very different technologies. One novel solution is a polymer-based system (DSSC—dye-sensitized solar cells) which has a 12.3% efficiency (as can been seen in [14] and [16] (pp. 1–11)). Perovskite/Si tandem (monolithic) cells achieve a 29.1% energy yield [14]. The modules of mass production have somewhat lower efficiencies. The available maximum energy efficiency of multijunction solar cells is about 47.1% [14]. The solar cells at more than 40% efficiencies are mostly of very small dimensions and are used mostly in concentrated solar modules and in rigid frames (not flexible), which require heliostats for installations and operations. It should be mentioned that some current top solar cells are not ready or appropriate for applications in buildings now. The focus of the innovative third-generation PV systems is on thin-film technologies that combine the high electrical efficiency of monocrystalline cells with the flexibility and lower costs of thin-film manufacturing [8] (p. 106). Monocrystalline panels are black or deep-blue, which is a serious drawback from an architect's viewpoint (Figure 1a). Polycrystalline panels offer a wider palette of accessible colors, which was welcomed by architects (Figure 1b). However, an even better and more attractive solution is a new generation of PV cells that come in a much more diversified scale of colors (Figure 1c). In this case, there is an interrelation between a given color and the efficiency of a panel. Generally, it has been proved that the most energy efficient panels are black PV panels, and lighter colors reduce the efficiency [17].

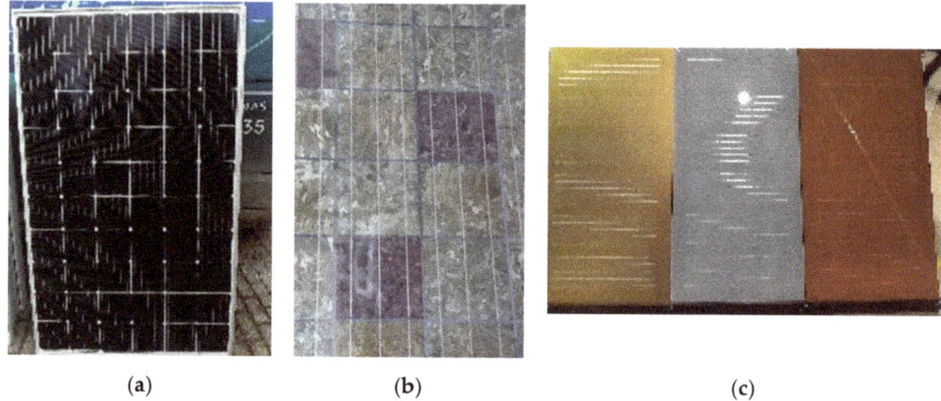

(a)   (b)   (c)

Figure 1. Photovoltaic panels: (**a**) monocrystalline, (**b**) polycrystalline, (**c**) dye-sensitized colored cell (photo by Celadyn, W.).

PV panels can be mounted on roofs, facades or be stand-alone frames on the ground. When they are in some way integrated with building components, they are termed building-integrated PV (BIPV) systems. Especially exciting are the building-integrated photovoltaic technologies integrating solar cells directly into building materials, such as semitransparent insulated glass windows, skylights, spandrel panels, flexible shingles, and raised-seam metal roofing [18] (p. 309), [19] (p. 3) and [20]. Well-integrated PV modules are suitable to contribute to the comfort of the building: they serve as weather protection, heat insulation, shading modulation, noise protection, thermal isolation and electromagnetic shielding, etc. [11] (p. 126). Holistically designed BIPV systems will reduce a building's energy demand from the electric utility grid while generating electricity on-site and performing as the weathering skin of the building [6] (p. 2). The first pioneering building with a BIPV installation was a multifamily residence designed by T. Herzog and B. Schilling and constructed in Munich in 1982 [21].

In future cities, solar cells and BIPV systems will evermore play an increasingly significant role in facade forming and electrical energy generation in the residential and other types of objects [8] (p. 104). The comparable prices between BIPV systems and conventional building materials confirm this assumption [20,22] (p. 9).

Retrofitting historical buildings, which aims at improving their energy-related parameters, can make use of photovoltaic systems. However, in this case the installation of these systems on facades or rooftops can be more difficult and controversial than in contemporary buildings as it can involve interventions in a building's valuable historic appearance. Therefore, instead of considering BIPV as a technical constraint for designers, a new approach based on the integration of BIPV solutions as a new "raw material" for architectural renewal projects is a good option avoiding conspicuous disfigurations of the building envelope [10] (pp. 1–2).

PV solar systems can store the produced energy, locally converted from a DC to AC current, in home batteries, or send it into the utility grid—the community's electrical wires—to be distributed to others [23] (p. 223). Panel installations can be fixed or track the sun, usually on one axis only.

Not everyone likes the appearance of typical solar electric systems, which are usually mounted on roofs. Therefore, solar manufacturers have begun to produce less conspicuous systems like thin-film solar electric materials using a noncrystalline sun-absorbing layer, which use a fraction of the semiconductor material of their predecessors. This amorphous silicon is, however, only 5% efficient. Despite its evident positive characteristics permitting its use in windows and skylights to produce electricity, it has some significant disadvantages [23] (p. 221).

Meeting the sustainability paradigm requires building resiliency, which can be achieved using a diversity of energy sources. This option enhances the system's ability to function under a wide variety of conditions and withstand many kinds of disturbances. Individual homeowners and businesses are encouraged to install small-scale wind turbines, photovoltaic panels, and other devices to produce renewable energy [24] (p. 158). PV panels are considered an indispensable, easy to install, and relatively inexpensive solution to such systems. Therefore, they play a significant role in increasing the sustainability of buildings of all kinds.

3.2. Impact of the Local Zoning Plans and Building Location on the Energy Efficiency of PV Panels

Decisions concerning the implementation of solar systems at the urban scale should be based on the local conditions of solar irradiation and these values defined in solar maps. Solar maps provide data about the position and system size of PV systems on roofs, the produced amount of electricity, the installation size, and the financial payback time. In this case, the output of the PV system is, besides the efficiency and additional losses, calculated by considering the air temperatures near urban rooftops [25] (p. 44). The appropriate configuration of PV panels with a building and its components is a basic requirement for the energy efficiency of PV systems. In many cases, it is the built or natural elements surrounding a building that may be crucial for solar harvesting on building facades and roofs. They can impair the solar radiation incident on the building and PV panels. Therefore, the building designers should carefully analyze the proposed site plan in terms of the potential obstacles to the undisturbed flow of radiation toward a building. There are two determining factors in this regard: (1) regulations in development (zoning) plans and (2) site planning solutions unrestricted by building regulations.

Zoning plans contain various regulations that must be respected by architects when designing buildings and implementing site plans. Especially important are stipulations regarding the orientation of buildings, their position on the lot, and their relation to the urban grid. They all impact the efficiency of PV panels installed on buildings. Even if respecting the local ordinance does not result in some impairments to solar systems, other elements of site plans, of which their location is unrestricted or undefined therein, may conflict with the efficiency of panels. The built structures located on the same or adjacent tract of land, as well as vegetation, can aggravate their yield or even thoroughly render them inefficient or useless if they are an obstacle intercepting the path of solar rays. There are some typical situations in zoning plans that could be indicated as conflicting with the rules for PV panel positioning to ensure an acceptable electricity yield in these installations. In the case of the pitched roofs of houses, the optimum location for the installation of panels is the south-inclined roof surface. A problem appears when the zoning plan determines the north–south axis of a building as compulsory for the building's orientation. The panels, for obvious reasons, must be exposed to the east or west, which are not optimal situations (Figure 2). Similar problems can occur in the case of a north–south street orientation (Figure 3). NW–SE orientations can also be disadvantageous for a similar reason (Figure 4).

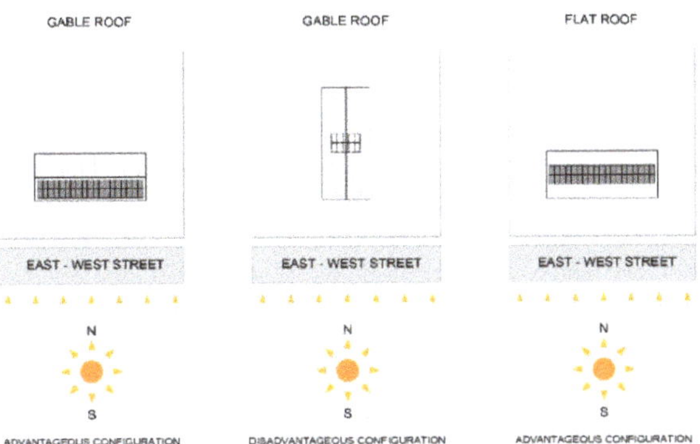

Figure 2. Possible positions of photovoltaic (PV) panels as a function of a W–E street orientation (diagram by W. Celadyn, P. Filipek).

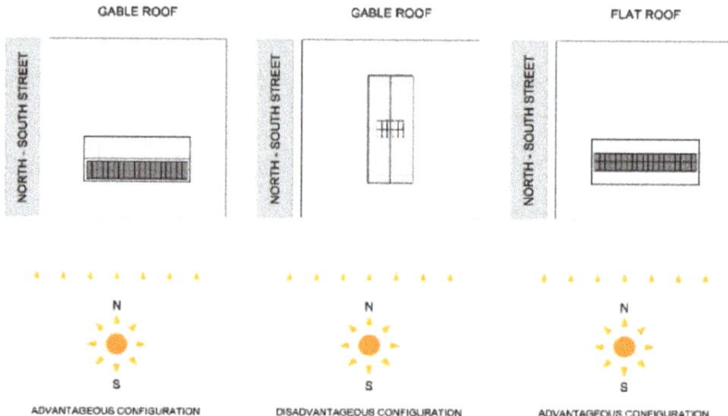

Figure 3. Possible positions of PV panels as function of an N–S street orientation (diagram by W. Celadyn, P. Filipek).

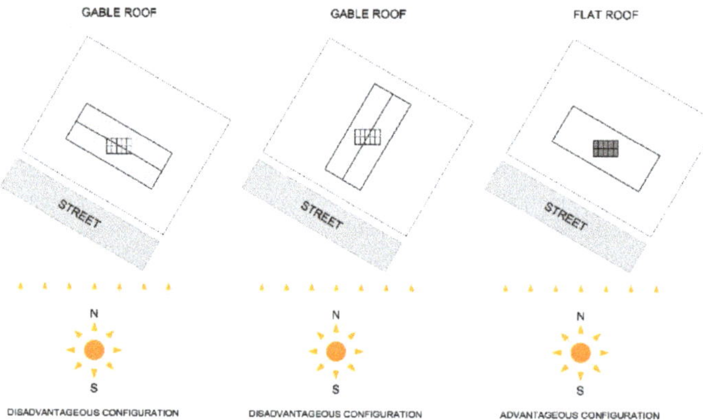

Figure 4. Possible positions of PV panels as function of an NW–SE street orientation (diagram by W. Celadyn, P. Filipek).

Another aspect of zoning plan regulations and their consequence on PV panels' efficiency is the obligation to respect the building lines comprised therein. There are usually two basic types of building lines that determine the location of constructions on building lots: the build-to line and unsurpassable building line. In the first case, a building should be located on a building lot so that its main facade is contiguous to the line. The compulsory character of a building's location permits the design of surrounding vegetation, if present, to ensure that the intensity of solar radiation incident on PV panels is not impaired (Figure 5). In the second variant, a building can be located to ensure the defined line is not surpassed. The first option ensures that the spatial situation is controllable, whereas, in the second, it is impossible to predict the final location, which is dependent on the architect's decision. This occurrence makes the issue of the reasonable configuration of adjacent buildings and the vegetation existing on-site prior to their construction unpredictable, as it does for the efficiency of the potential solar systems installed on buildings (Figure 6).

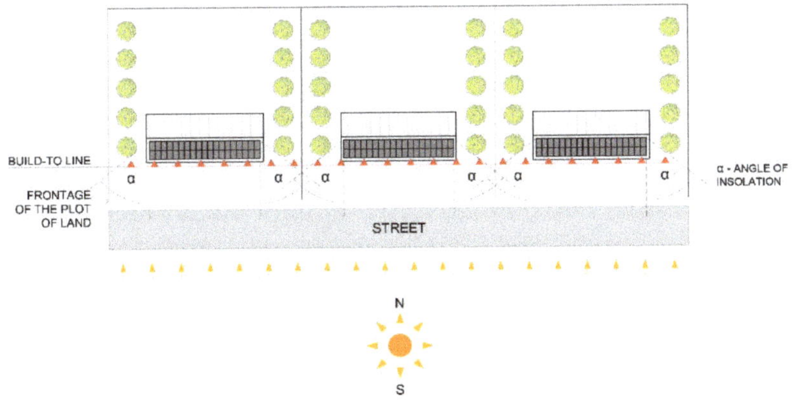

Figure 5. Predictable positions of buildings and PV panels because of a compulsory built-to line defined in the zoning plan (diagram by W. Celadyn, P. Filipek).

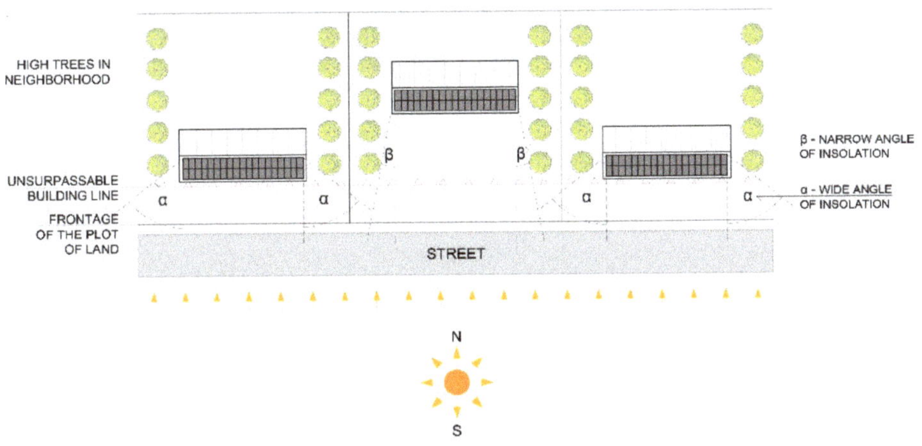

Figure 6. Probable positions of buildings and PV panels because of noncompulsory unsurpassable building line defined in the zoning plan (diagram by W. Celadyn, P. Filipek).

This analysis indicates that the zoning plan regulations have meaningful relationships with building solar systems. Their mutual dependence should be seriously considered and carefully analyzed by both urban planners and architects to ensure the systems work in terms of the potential electrical energy generation. Zoning plans generally do not envisage such analyses. This creates challenges for the installation of PV panels, rendering them frequently useless. This occurs with existing buildings and their surroundings. The property relations and adjacent parcels being built and arranged with high vegetation make the situation difficult to resolve. So far, the awareness of planning officers and urban planners in this regard appears insufficient, if not absent.

3.3. Spatial Position of PV Panels and Their Energy Efficiency

Whereas solar thermal systems have always been closely tied to the planning of buildings, developments in photovoltaic technology allowed photovoltaic elements to be integrated in the building envelope since the early 1980s [26] (p. 106). However, the yield from a vertical facade panel is much lower. Unlike with thermal collectors, even an incident energy <200 W/m^2 can still contribute to generating electricity [15] (p. 291). This is why photovoltaic systems are less dependent on the orientation of building components. However, to be energy efficient, the mutual configuration of the panel surfaces and the angles of incidence of solar rays must be optimized. This is a factor that, in practice, determines the possibility of installing effective solar systems on buildings. Therefore, it is important for the configuration to avoid any disturbances in the accessibility of solar rays. In contrast to solar thermal applications, in photovoltaics, even relatively little shading of the solar cells can lead to a considerable reduction in the energy yield [26] (p. 106).

The highest transmission of solar radiation through glass occurs when the angle of incidence of the solar rays on the glass surface is perpendicular. Research indicated that within the range of 0–60°, the deviation from perpendicular gives a transmission loss of energy between 8 and 10% [27], or even more. This loss (technically an incident angle modifier) is due to the glass internal transmission (due to a longer light path length) and glass surface reflection, not attributed to the total irradiance on the panel and PV electricity generation. The solar radiation (the beam component) is reduced to 50% for an incident angle of 60°, and abruptly drops down to 0% by the direct radiation angle of incidence approaching 90°. Behind this range, the intensity of the transmitted solar energy abruptly drops down to 0% when the angle of incidence approaches 90°. This does not mean that below this range that the electricity is not generated. About 50% of the radiation occurs in the form of diffuse radiation [15] (p. 291), so it still can generate some amount of energy; this even occurs under an overcast sky. PV cells can be mounted on movable panels programmed to track the sun so that the cells are always perpendicular to the sun's rays for the maximum interception of solar radiation [4] (p. 64).

The issue of the relationship between the angle of incidence of solar rays and the plane of photovoltaic panels is less important in the case of photovoltaic cells mounted on the membrane that absorb all incident sun rays from any direction at any time of the year without the need for any manual or automatic override [28] (p. 34) (Figures 7 and 8).

Solar radiation varies widely over the course of a day and a year and is strongly influenced by the prevailing weather conditions. Radiated energy can differ up to a factor of 10 on two consecutive days, being, at times, up to 50 times higher values on a clear summer day than on an overcast winter day [29] (p. 49). Some sources suggest that the highest annual radiation volume in Central Europe is available to south-facing fixed systems installed at an angle of 30 degrees or less to the horizontal [15] (p. 291). A south–west orientation by the same inclination reduces the yield to only 96% [29] (p. 54). The case of a house in Figures 7 and 8 proves that PV panels with an energy efficiency of 5.5 kWp installed on its roof and deviating by 37° from the north–south towards the south–east can still produce electricity in significant quantities (Figure 9). On the winter day of 15 March, the electricity generation reached its peak of 4.2 kW, and was registered as late at 6 pm.

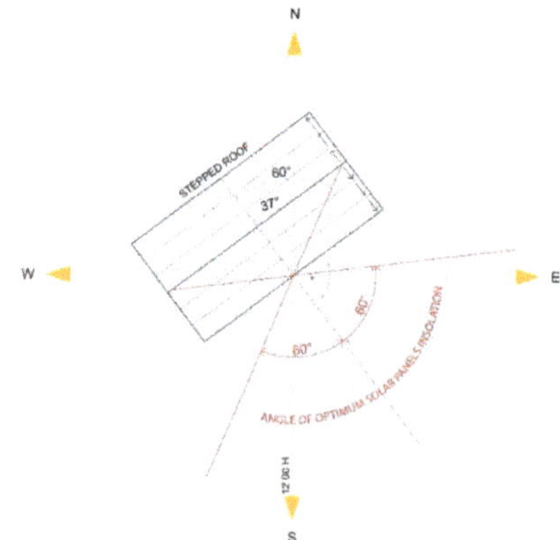

Figure 7. A house oriented at an angle of 37° from the N–S direction with PV panels installed on a stepped roof (diagram by W. Celadyn, P. Filipek).

Figure 8. Stepped roof with PV monocrystalline photovoltaic panels installed at the recommended angle of 15°. Location: 49°57′ N, 19°55′ E (photo by W. Celadyn).

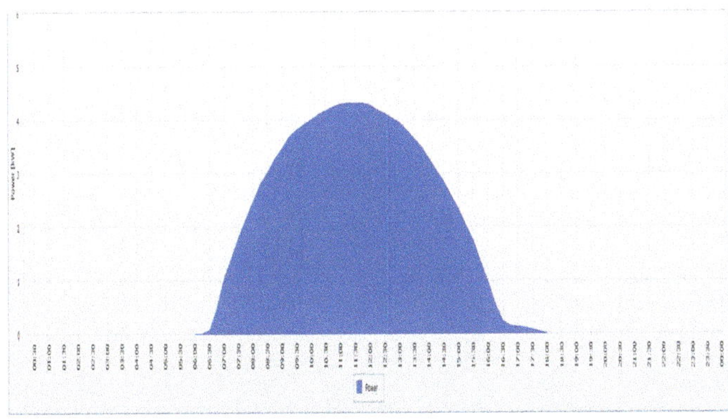

Figure 9. A chart indicating the solar efficiency of the above house on a sunny day (15 March 2020). Location: 49°57′ N, 19°55′ E (source: sunnyportal.com).

Computer tools are available for the calculation of the efficiency of photovoltaic systems; an example is PVSYST, used by, e.g., Fartaria [30] (pp. 93–101), to calculate the mutual shading of direct normal and diffuse radiation. Building envelopes can lend walls and roofs to photovoltaic installations, and substantial differences exist between their solar conditions. In the latter, pitched and flat roofs are also differentiated in this regard. Flat roofs are a particularly good place for the location of PV panels as their arrangement is independent of the roof pitch. However, the problem of self-shading occurs due to adjacent tilt panels on flat surfaces or due to them tilting away from low sloped roofs. As a rule of thumb, there must not be any shading on 21 December. The calculated minimal module spacing (in the Northern Hemisphere) is defined by the equation previously reported by [31] (p. 227). To accommodate as much PV power as possible, the optimum angle of attack, β, equal to 15° to the horizontal, is often changed to 20°, as the energy yield is then only reduced by 2% [31] (p. 227). This angle varies with the latitude of installation. The optimal tilt angle is within the latitude angle plus or minus 10–15°. The lower the angle of inclination of the module surface, the higher the usable incident radiation. When the modules are installed over the entire roof surface, almost horizontal, the overall energy yield is maximized [26] (p. 107). The energy output of PV panels can be compromised by their low tilt as the cleaning of their surfaces by rainwater is less efficient.

Most of the absorbed solar radiation becomes thermal energy that can heat up the PV panels. An increase in the temperature of the panels over 25 °C is disadvantageous as the PV panel will produce less than the rated generation efficiency. This efficiency loss due to an increased temperature depends on the types of solar cells involved. An effective ventilation of the back of panels or the use of generated thermal energy to heat the interiors or water can be helpful.

3.4. Spatial Relations of Buildings and PV Panels

Photovoltaic panels are installed on buildings in two basic configurations with respect to building components [29] (p. 59):

(1)   vertical, horizontal, or angled installations directly on top of water-bearing layers;
(2)   vertical, horizontal, or angled installations with a distance from water-bearing layers.

These relations between building components and photovoltaic array mounting systems can be also classified as BIPV and BAPV. BIPV is considered a functional part of the building structure as it is architecturally integrated into the building's design. This category includes designs that replace the conventional roofing materials, such as shingles, tiles, slate, and metal roofing. BAPV is considered an add-on to the building, not directly related to the structure's functional aspects (Figures 10–12). It relies on a superstructure that supports conventional framed modules. Standoff and rack-mounted arrays are the two subcategories of BAPV systems. Standoff arrays are mounted above the roof surface and are parallel to the slope of a pitched roof. Rack-mounted arrays are typically installed on flat or pitched roofs. In the second case, the tilt is either parallel to or different from the roof inclination to be more suitable to the angle of solar incidence. From the above definition, the main difference between BIPV and BAPV is the extent of tightness of the integration of photovoltaic systems and buildings [9] (p. 3593). The multitude of possible relations between these two components is illustrated in Figure 10.

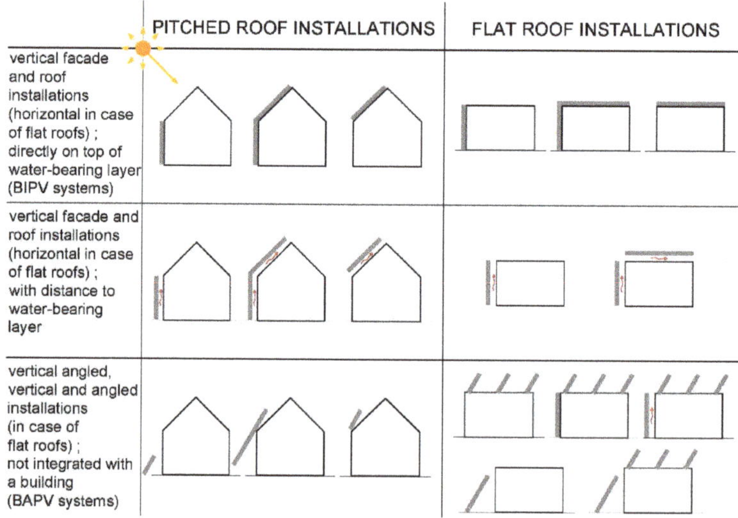

Figure 10. Types of relations between the photovoltaic solar systems and building forms and components (diagram by W. Celadyn, P. Filipek).

Figure 11. Solar electricity generating thin-film system independent of the building (photo by W. Celadyn).

Figure 12. PV modules installed on a structure added to an office building (photo by W. Celadyn).

An analysis of a building demonstrated the direct link between the functional requirements and external appearance. The drastic changes in the energy sector have had a lasting impact on this traditional link. The relationship between local conditions and their impact on the built environment is mostly nullified [29] (p. 38). Contrary to collectors that are mounted onto the building skin, these systems allow for a full integration both in terms of construction and design [32] (p. 262).

3.5. Photovoltaic Panels and Esthetical Issues in Buildings

The growing popularity of solar photovoltaic systems has created new problems regarding esthetic values. The unstoppable trend toward the installation of solar systems on building envelopes, especially the first-generation PV panels, evoked ambiguous opinions from architects. The esthetic effects of PV assemblies mounted on buildings were criticized. Two decades ago, the issue of esthetics concerning the "solar design" was raised. Then, some steps should have been undertaken to modify the appearance of PV panels so that they would not only generate electrical energy but also have visual appeal and blend in with their surroundings [33] (p. 1). At that time, voices stated that the esthetic qualities of the buildings continued to be a largely unsolved problem.

Typological studies of building skins are still lacking, which would be an important basis and evaluation tool for the visual integration of solar technical systems. They should be visually integrated into the overall architectural concept [29] (p. 60). Innovative PV systems of the second and third generations are much less controversial than conventional PV panels, as they can be easily integrated with building components, being available as roof tiles, window glass, or facade finish panels replacing other typical materials. This assortment has changed the attitude of designers toward solar systems because they can be hidden or visually blended in with the background. The esthetics issues in the case of solar buildings require a wider perspective to be discussed. A new kind of esthetic for the built environment has been suggested that explicitly teaches people about the potentially symbiotic relationship between culture, nature, and design. This is a powerful approach since new ideas are learned most rapidly when they can be expressed visually and experienced directly. This esthetic is called visual ecology and is opposed to the method of designing that hides natural processes and related technology out of the public view [34] (pp. 188–189).

Well-articulated ethics have not been developed for sustainable designs, neither has the environmental movement in general, with its various ethics on biodiversity, animal rights, stewardship, intergenerational ethics, and holism [35] (p. 83). A sustainable design gives people a beautiful experience of nature through highlighting the elegance of its processes. It may also forward the interaction of these processes with the patterns of space in the design, revealing the beauty of their connections [35] (p. 115). Something can be considered beautiful if it reveals how it changes over time, especially toward a greater integration, order, and complexity [35] (p. 119).

Solar systems fall into this philosophy. Many examples exist of buildings with conspicuous disharmoniously contrasting surfaces of installed PV modules and the covering materials of pitched roofs or facades. The rectangular or even irregular compositions of PV modules are negatively assessed against the backdrop of roof materials in colors significantly different from the black or deep-blue tones of typical monocrystalline panel materials (Figure 13). Such disharmonized color compositions usually appear on existing houses covered with ceramic tiles. Novel photovoltaic panel systems, as innovative organic cells and nanocrystals, offer interesting color effects for the panels [36]. They are available in a much richer palette of colors and tones and can offer a remedy to this problem as an appropriate color choice can reduce unwanted contrasts, but unfortunately, they are of a much lower energy efficiency (Figure 14). Darker colors of panels absorb solar radiation better and therefore are better in this regard. Multicolored glass–glass (MCGG) and crystalline silicon cell (c-Si) PV laminates are an approach to overcome some of these issues and achieve aesthetically pleasing, yet technically and economically viable, building-integrated PV systems [37] (p. 2).

Figure 13. Disharmonious color composition on the building's roof covered with red ceramic tiles and monocrystalline PV panels (photo by W. Celadyn).

Figure 14. Harmonized color composition on the building's roof with the application of a new generation of color PV panels (computer simulation by P. Filipek).

Some new methods are available for computationally matching the color of PV panels with roof covering materials. This procedure can achieve a perfect match [33] (p. 9). It is not only the issue of disharmonious colors; controversies exist over the excessive differences in the texture of shiny PV panels and matte roofing materials, which can substantially modify the originally well-matched colors.

Products are, by their very materiality, transient; their usefulness is unavoidably a function of time. They become obsolete for a variety of reasons, all of which help fix a product in a specific timeframe [38] (p. 139). PV panels are automatically associated with recent times and their installation on old-looking roof coverings or facades also creates a disharmonious and anachronic image. Photovoltaic panels are conspicuous technological components on building facades and roofs. This initiates an interaction between architecture and technology, which produces much controversy over the issue. Certain viewpoints state that they continually redefine each other. Depending on the type of applied solar systems and the way they are installed, they can define a high-tech or eco-tech aspect of a building, both considered opposite to each other [39] (p. 7). As the novel color PV modules offer many new unconventional esthetic opportunities, the possible integration of this technology and building envelope facilitates the approach of the built environment to the promoted eco-tech esthetics.

3.6. Other Functions of PV Panels for Buildings

In addition to the generation of electricity, PV modules are taking on more additional functions and are hence achieving numerous synergy effects—photovoltaic panels can provide protection from the weather, sun shading, and privacy functions, or, as insulating units, even constitute the thermal envelope. In addition, they can characterize the architecture [26] (p. 106). Some authors considered sun–shade systems, in addition to roof and facade systems, as one of the three basic photovoltaic systems [7] (p. 8). PV panels as sun protective devices, if they are strategically located, and due to adequate shading analyses, can effectively fulfill this role. There are many examples thereof. This strategy can lower the costs of construction through the dematerialization effect. Among such solutions, the shading role seems especially interesting due to the constantly increasing role of overheating in buildings. It relates mainly to office buildings, where this issue is important.

The southward-oriented building volume can be designed with a complementing facade module geometry. The overall irradiation of the building can be translated to a system of modules that allows external shading from solar radiation while permitting daylight entry and unimpeded views of the surrounding landscape from inside the building [40] (p. 3608). If the application of PV panels as shading systems on south-facing elevations is comprehensible, their use on other facades can be debatable, as is the case of conventional sun protective systems.

Sun shading on east or west facades is a difficult task. Solar radiation can be especially disturbing on office buildings with north–south orientations, with the longest facades exposed to low-angle of incident sunlight. The problem was illustrated with an example of an office building for which the incident solar beams and the shading pattern by PV modules were simulated (Figure 15). The set of diagrams presented below shows the path of solar rays and the resulting pattern of dark shadow patches on the elevation assigned to every hour between 7 a.m. and 12 p.m. (Figure 16). The simulation was carried out with the computer program Sketchup with a precisely defined geolocation and application CuricSun.

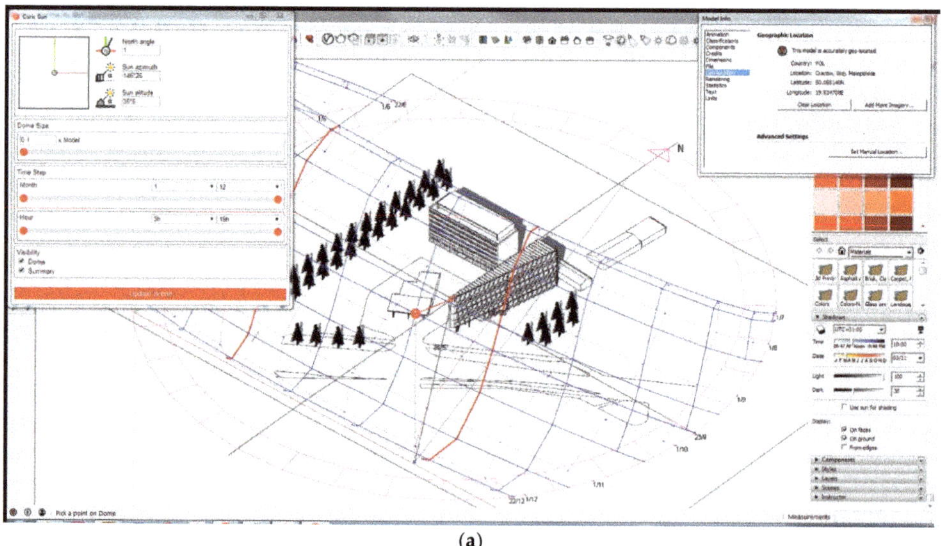

(a)

Figure 15. *Cont.*

(b)

Figure 15. Study of the sun shading of an office building fitted with PV panels located on geographical coordinates 50°5.118′ N and 19°56.096′ E (author: P. Filipek). (**a**) building's location, (**b**) building model and its shading pattern.

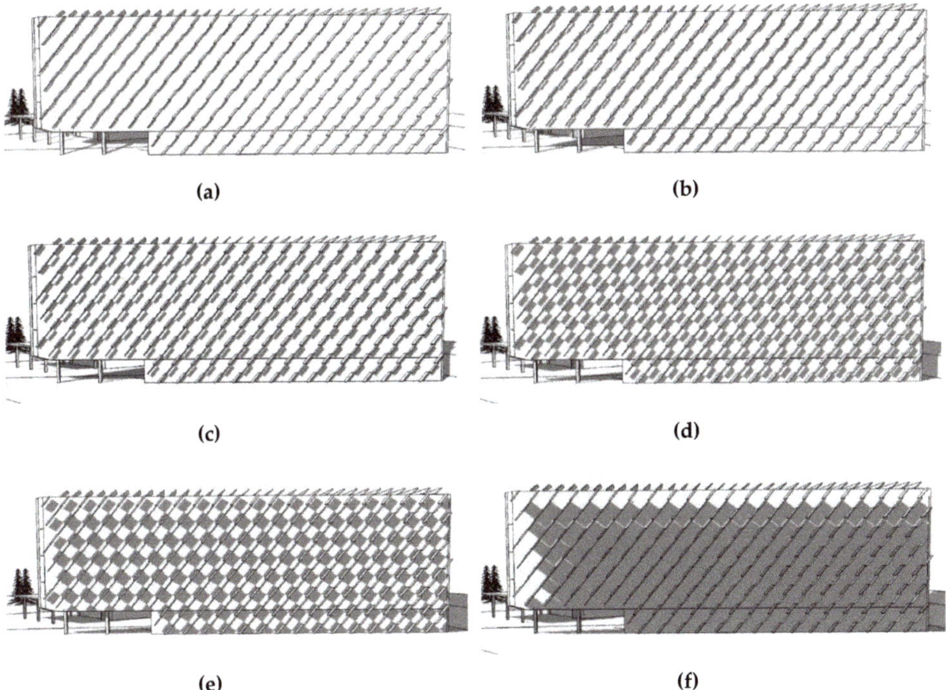

(a) **(b)**

(c) **(d)**

(e) **(f)**

Figure 16. Hourly sequential shading of an office building fitted with PV panels on the east façade (**a**) At 7.00 a.m.; (**b**) At 8.00 a.m.; (**c**) At 9.00 a.m.; (**d**) At 10.00 a.m.; (**e**) At 11.00 a.m.; (**f**) At 12.00 a.m. (by P. Filipek).

Photovoltaic solar systems are applied on buildings in the form of panels that come in glossy and shiny finishes and are either opaque or semi-transparent. Reflections on this surface may make the modules highly visible at a distance and occasionally cause undesirable glare. There are some reports of blinding people in their vicinity [41] (p. 70), [23] (p. 221) and [33] (p. 1). Blinding by the reflective surfaces of PV panels can be reduced by the application of an antireflexive layer on top, and this increases the panels efficiency. This problem can be neglected in the presented analyzed office building. The location, layout, and orientation of the two main facades facing the west and east could potentially be reflective enough to blind the drivers approaching the building from the west or east along the street. However, the regular and allover application of PV panels on these two facades and the rational arrangement at an angle resulting from the analysis of sun path diagram significantly reduce this problem. A possibility exists of blinding but only from the southern approach, which is impracticable for vehicles. There are methods of reducing the glare of glazed elevations. This effect can be achieved by a nonreflective film applied to panel surfaces [42] (p. 2). This method is practical for photovoltaic façade systems. Various designs were developed for prototypical applications to integrate PV systems into rooftop gardens, with a specific focus on retrofitting flat roofs. The concurrent integration of PVs and green roofs into the same surface area can be achieved with lightweight construction, which is particularly suitable for existing buildings. Such solutions for retrofitting existing roofs must be sought to transform the current building stock into energy generating green habitats [43] (pp. 1–2).

4. Discussion

This research covered issues that are now rapidly changing due to new developments in the photovoltaic industry. Given the large discrepancies concerning the energy performance and material efficiency of these systems, systematic updates of designers' knowledge is required for choosing an appropriate option. Despite the novel third-generation systems offered by producers, conventional monocrystalline panels are still the most popular choice, especially for residential buildings. However, the second- and third-generation systems are becoming increasingly popular on the market, as they are more versatile in terms of their location on buildings or their components. They also offer more opportunities in terms of their variety—i.e., flexibility and colors. Along with their improved energy efficiency, they will be implemented more frequently. Their potential for retrofitting historic buildings seems especially promising. Although all the considered PV systems are on the market in a wide variety, they are undergoing constant improvements. This applies to increasing the economy and the options for architectural integration [44] (p. 68). The applied technology covers a wide range of problems pertaining to the technical durability of buildings. PV systems may conflict with other building systems in this regard. The potential problems of their longevity may appear in the case of BIPV. The materials or components can perform satisfactorily for a long time if they are autonomous within the structure, but coupled with other materials, they might form a new and less-stable system [45] (p. 21).

Zoning plans, as a rule, do not consider the prospective use of solar systems in buildings despite such applications not being new. Their impact on planning procedures has not yet been noted or recognized. Given the analyzed spatial configurations of buildings and access streets, as well as other elements of various arrangements, this problem should be raised and broader discussions among specialists and architecture authorities responsible for planning should be inspired. Some design-established procedures practiced among professionals in the field exhibiting a traditional and meaning-limited scope of analyzed aspects during their work on zoning plans should be modified and supplemented with energy-related issues. This means that the building location on the lot and a multitude of parameters related to buildings and their parts should be an indispensable part of plans, as well as other regulations related to energy. So far, this is not the case and no serious discussions are being conducted related to this subject. The reasons for this may be the complexity of the issue, the diversity of urban grids, difficulty in properly fitting the buildings, and the building orientation.

Another meaningful cause is property colliding with the optimum vegetation patterns. Harmonizing all these factors is hardly a feasible task. Discussions must be undertaken if a sustainable building and land use are to find logical solutions.

Conventional black photovoltaic panels installed on red, brown, or similar roofing materials are a frequently seen picture in landscapes. The replacement of red tiles or steel panels on existing buildings to harmonize them with the color of modules is rarely performed. New buildings offer opportunities to achieve satisfactory outcomes in this regard, as the option can be selected during the design stage. Matching the roof and panel, both in color and texture, is considered the most desirable solution from the esthetical point of view. However, when comparing the energy-related parameters of monochromatic black monocrystalline panels with the next-generation systems, which offer an increased esthetic potential, unavoidably a trade-off must be considered. As indicated earlier, monocrystalline panels are still the best option in terms of the ratio of electrical energy yield to costs. When a typical investor is faced with such a dilemma, they would, in most cases, opt for a less expensive and more efficient solution. This situation mainly refers to residential buildings. Only this segment of construction is responsible for the aforementioned controversies. The buildings of other functions are, in most cases, covered with flat roofs, which make photovoltaic roof installation invisible from the ground level. Small houses pose yet some other related problems. The more articulated the layout and roof form of a building, the less appropriate it is for the installation of PV panels, as fewer plane roof surfaces are offered, making the investment less rational. The color integration of panels and roof coverings can be considerably improved once more efficient and more affordable photovoltaic systems appear on the market and gain popularity. This could substantially contribute to an increased harmonization of the built landscape and improvement of its esthetic values.

Photovoltaic systems applied as sun protection modules, as analyzed earlier, are steadily appearing on more office buildings. They generate the most energy when exposed and inclined to the south. This occurs both on roofs and facades. East and west elevations are less efficient as their insolation is reduced merely to the half of the incident on the south-exposed walls. However, it does not make such applications useless. A reduction in the heat load due to the use of solar protective PV panels could at least partly compensate for the extra expenditure on the panels. This disadvantageous orientation of a building from the analyzed point of view, as an alternative installation of solar systems on flat roofs characteristic of office buildings, is not an encouraging solution. The reason for this is the longitudinal roof layout extended along the north–south axis, which is disadvantageous due to the limited amount of PV modules that would fit in the space given the necessary long distancing of adjacent panels in the space-consuming row arrangement mentioned earlier. A remaining question is the installation of PV panels on facades of multistory buildings. They are not yet economically feasible, as was calculated for a commercial building in the 10- to 20-story range (USA) [7] (p. 104).

The problem of blinding in the case of large glazed facades is potentially important when they are perpendicular to the direction of pedestrian or driver movements. Therefore, installed exterior sun protective systems can be an effective solution to reduce glare. A reasonable proportion of PV panels play that role, provided they are mounted on facades in the configuration depicted in Figure 15.

Findings from surveys on public educational barriers showed various reasons for a poor public understanding of the cost perceptions of BIPV systems and their financial benefits, and a lack of enough knowledge by clients and the public in general. Additionally reported was a high negative perception of the system price and costs associated with aesthetic BIPV options. The lack of knowledge on how to ensure the most efficient choice of BIPV design was also noted [46] (p. 5). All of this highlights the need for further relevant written contributions. The application of BIPV systems will progress in the near future, but some practical barriers remain, such as investment costs and the payback times of the solar energy technology, which are highly important for real estate developers.

5. Conclusions

Photovoltaic systems are an indispensable part of contemporary low-energy buildings. Their increasing popularity is linked to them being the cheapest and easiest method of making new and existing buildings at least partly sustainable due to using a renewable source of energy (solar energy), hence the financial support offered for their installation in special state programs in many countries. Their application requires specific knowledge from designers and building specialists. The systemic approach to the issue of suitable design decisions, which was presented in this study, entails the need for analyses of factors like: the spatial and technical parameters of a building, an in-depth study of site features including the orientation of building and its relation to the access street and other built structures on the building lot, as well as the position and type of vegetation. This broad view facilitates the choosing of the optimal solution to obtain the desired energy efficiency of the applied PV system.

Another problem of increasing significance is the esthetics concerning features such as building materials and their color and texture in terms of their harmony with PV panels, which are frequently considered inconsistent with buildings' traditional esthetic values. Building designers and investors have offered various systems that feature different characteristics covered in this research. This study was designed for architects to enhance their knowledge on this subject and to systematize the knowledge, providing a step-by-step process with the proposed procedure of suitably coupling designed buildings with photovoltaic systems. Notably, obstacles remain on this path, including an insufficient and obsolete knowledge of these systems, potential problems with their implementation, and mistrust of investors wary of potential excessive costs. However, some positive experiences with photovoltaics and imaginative thinking would help further the application of photovoltaic solutions in building developments and related industries. This promising vision should encourage further studies on the subject.

Author Contributions: The co-author contributed actively to the illustrations, discussion of this research and in reviewing the article. All authors have read and agreed to the published version of the manuscript.

Funding: This research received no external funding.

References

1. Attman, O. *Green Architecture (Green Source Book): Advanced Technologies and Materials*; McGraw-Hill Education: New York, NY, USA, 2009; ISBN-13 978-0071625012, ISBN-10 0071625011.

2. McMullan, R. *Environmental Science in Building*, 7th ed.; Palgrave Macmillan: London, UK, 2012; ISBN 978-0-230-29080-8.

3. Solar Energy: Mapping the Road Ahead. Available online: https://www.iea.org/reports/solar-energy-mapping-the-roadahead (accessed on 16 June 2020).

4. Lyle, J.T. *Regenerative Design for Sustainable Development*; John Wiley and Sons Inc.: New York, NY, USA, 1994; ISBN 978-0-471-17843-9.

5. Becker, G.; Hauger, S.; Haselhuhn, R.; Hemmerle, C.; Kampfen, B.; Krippner, R.; Kuhn, T.E.; Maurer, C.; Reinberg, G.W.; Seltmann, T. *Building-Integrated Solar Technology: Architectural Design with Photovoltaics and Solar Thermal Energy*; Krippner, R., Ed.; Detail Business Information GmbH: Munich, Germany, 2017; ISBN 97883955533632.

6. Eiffert, P.; Kiss, G.J. A sourcebook for architects. In *Building-Integrated Photovoltaic Designs for Commercial and Institutional Structures*; National Renewable Energy Laboratory: Golden, CO, USA, 2000.

7. Kayal, S. Application of PV Panels in Large Multi-Story Buildings. Master's Thesis, California Polytechnic State University, San Luis Obispo, CA, USA, 2009.

8. Cekić, N.; Milosavljević, D.; Pavlović, T.; Mirjanić, D. Application of solar cells in contemporary architecture. *Contemp. Mater. (Renew. Energy Sources)* **2015**, *VI–2*, 104–114.

9. Peng, C.; Huanga, Y.; Wub, Z. Building-integrated photovoltaics (BIPV) in architectural design in China. *Energy Build.* **2011**, *43*, 3593. [CrossRef]

10. Aguacil, S.; Lufkin, S.; Rey, E. Architectural design scenarios with building-integrated photovoltaic solutions in renovation processes: Case study in Neuchâtel (Switzerland). In Proceedings of the PLEA 2016 Los Angeles—36th International Conference on Passive and Low Energy Architecture, Los Angeles, CA, USA, 11–13 July 2016.

11. Heinstein, P.; Ballif, C.; Perret-Aebi, L.E. Building Integrated Photovoltaics (BIPV): Review, Potentials, Barriers and Myths. *Green* **2013**, *3*, 125–156. [CrossRef]

12. Thomas, R.; Fordham, M. *Photovoltaics and Architecture: An Introduction for Architects and Engineers*; Spon Press: London, UK, 2001.

13. Podsiadlo, S. *New Materials for Photovoltaics*; Wydawnictwo Naukowe PWN: Warsaw, Poland, 2018; ISBN 978-83-01-19658-5.

14. NREL Transforming Energy. Available online: https://www.nrel.gov/pv/cell-efficiency.html (accessed on 24 July 2020).

15. Herzog, T.; Krippner, R.; Lang, W. *Façade Construction Manual*; Birkhauser-Publishers for Architecture: Basel, Switzerland, 2004; ISBN 3-7643-7109-9.

16. Green, M.A. Solar Cell Efficiency Tables (Version 41). *Prog. Photovolt. Res. Appl.* **2013**, *21*, 1–11. [CrossRef]

17. Røyset, A.; Kolås, T.; Jelle, B.P. Coloured building integrated photovoltaics: Influence on energy efficiency. *Energy Build.* **2020**, *208*, 109623. [CrossRef]

18. Kibert, C.J. *Sustainable Construction: Green Building Design and Delivery*, 4th ed.; John Wiley and Sons Inc.: Hoboken, NJ, USA, 2016; ISBN 978-1-119-05517-4.

19. Ogbeba, J.E.; Hoskara, E. The Evaluation of Single-Family Detached Housing Units in Terms of Integrated Photovoltaic Shading Devices: The Case of Northern Cyprus. *Sustainability* **2019**, *11*, 593. [CrossRef]

20. IEA PVPS Task 15. Development of BIPV Business Cases Guide for Stakeholders 2020; Report IEA-PVPS T15-10: June 2020. Available online: https://iea-pvps.org/wp-content/uploads/2020/06/Task-15-STB-B3-Report_Final.pdf (accessed on 24 July 2020).

21. Baum, R. Studies on Light-Transmissive Photovoltaics (LTPV): Patterns of Integration into Architectural Design. Ph.D. Thesis, The University of Tokyo, Tokyo, Japan, 2012.

22. Krawietz, S.; Poortmans, J. Building Integrated Photovoltaics (BIPV) as a Core Element for Smart Cities—Integrated Research, Innovation and Competitiveness Strategies within the Energy Union, BIPV Position Paper; European Technology and Innovation Platform, December 2016. Available online: www.etip-pv.eu (accessed on 1 June 2020).

23. Chiras, D.D. *The New Ecological Home*; Chelsea Green Publishing Company: Hartford, VT, USA, 2004; ISBN 978-1-931498-16-6.

24. *New Directions in Sustainable Design*; Parr, A.; Zaretsky, M. (Eds.) Routledge: Abingdon, UK, 2011; ISBN 978-0-415-78037-7.

25. Kanters, J.; Wall, M.; Dubois, M.C. Development of a Façade Assessment and Design Tool for Solar Energy (FASSADES). *Buildings* **2014**, *4*, 43–59. [CrossRef]

26. Hegger, M.; Fuchs, M.; Stark, T.; Zeumer, M. *Energy Manual. Sustainable Architecture*; Birkhauser Edition Detail: Munich, Germany, 2008; ISBN 978-3-7643-8830-0.

27. Froelich, H. Warmeschutz mit Verglasungen und Fenstern. *Bauphysik* **1997**, *19*, 79–89.

28. Clements-Croome, D.J. (Ed.) *Intelligent Buildings. Design, Management and Operation*; ICE Publishing: London, UK, 2013.

29. Schittich, C. (Ed.) *Buidling Skins. Concepts, Layers, Materials*; Birkhauser Edition Detail: Munich, Germany, 2001; ISBN 3-7643-6465-3.

30. Fartaria, T.O.; Pereira, M.C. Simulation and computation of shadow losses of direct normal, diffuse solar radiation and albedo in a photovoltaic field with multiple 2-axis trackers using ray tracing methods. *Sol. Energy* **2013**, *91*, 93–101. [CrossRef]

31. Mertens, K. *Photovoltaics. Fundamentals, Technology and Practice*; John Wiley and Sons Inc.: Hoboken, NJ, USA, 2014.

32. Daniels, K. *Gebaudetechnik. Ein Leitfaden fur Architekten und Ingenieure*; R. Oldenbourg Verlag: Munich, Germany, 1996.

33. Schregle, R.; Krehel, M.; Wittkopf, S. Computational Colour Matching of Laminated Photovoltaic Modules for Building Envelopes. *Buildings* **2017**, *7*, 72. [CrossRef]

34. Van der Ryn, S.; Cowan, S. *Ecological Design, Ten Anniversary Edition*; Island Press: Washington, DC, USA, 2007; ISBN-13 978-1-59726-141-8, ISBN-10 1-59726-141-6.

35. DeKay, M. *Integral Sustainable Design*; Earthscan: London, UK, 2011; ISBN 978-1-84971-202-6.

36. Des Panneaux Solaires de Toutes les Couleurs. Available online: https://www.lesoir.be/305941/article/2020-06-09/des-panneaux-solaires-de-toutes-les-couleurs (accessed on 16 June 2020).

37. Park, J.; Hengevoss, D.; Wittkopf, S. Industrial Data-Based Life Cycle Assessment of Architecturally Integrated Glass-Glass Photovoltaics. *Buildings* **2019**, *9*, 8. [CrossRef]

38. Walker, S. *Sustainable by Design: Explorations in Theory and Practice*; Earthscan: London, UK, 2006; ISBN 1-84407-353-X.

39. Slessor, C.; Linden, J. *Eco-Tech. Sustainable Architecture and High Technology*; Thames and Hudson: London, UK, 2001; ISBN 978-0500341575.

40. Van Berkel, B.; Minderhoud, T.; Piber, A.; Gijzen, G. Design Innovation from PV-module to Building Envelope: Architectural Layering and Non Apparent Repetition (conference materials). In Proceedings of the 29th European Photovoltaic Solar Energy Conference and Exhibition, Amsterdam, The Netherlands, 22–26 September 2014.

41. Basnet, A. Architectural Integration of Photovoltaic and Solar Thermal Collector Systems into Buildings. Master's Thesis, Norwegian University of Science and Technology, Trondheim, Norway, June 2012.

42. Schregle, R.; Renken, C.; Wittkopf, S. Spatio-Temporal Visualisation of Reflections from Building Integrated Photovoltaics. *Buildings* **2018**, *8*, 101. [CrossRef]

43. Sattler, S.; Zluwa, I.; Österreicher, D. The "PV Rooftop Garden": Providing Recreational Green Roofs and Renewable Energy as a Multifunctional System within One Surface Area. *Appl. Sci.* **2020**, *10*, 1791. [CrossRef]

44. Cremers, J. *The Potential of Building Envelopes to Actively Provide Renewable Energy—A Review and Outlook, in Building Technologies and Energy*; Celadyn, W., Kuc, S., Eds.; Cracow University of Technology: Krakow, Poland, 2017; pp. 59–71.

45. Celadyn, W. Durability of Buildings and Sustainable Architecture. *Tech. Trans. Archit.* **2014**, *111*, 17–26.

46. Attoye, D.E.; Adekunle, T.O.; Aoul, K.A.T.; Hassan, A.; Attoye, S.O. A Conceptual Framework for a Building Integrated Photovoltaics (BIPV) Educative-Communication Approach. *Sustainability* **2018**, *10*, 3781. [CrossRef]

 *buildings*

Article

Technological Advances and Trends in Modern High-Rise Buildings

Jerzy Szolomicki [1],* and Hanna Golasz-Szolomicka [2]

[1] Faculty of Civil Engineering, Wroclaw University of Science and Technology, 50-370 Wroclaw, Poland
[2] Faculty of Architecture, Wroclaw University of Science and Technology, 50-370 Wroclaw, Poland
* Correspondence: Jerzy.Szolomicki@pwr.edu.pl; Tel.: +48-505-995-008

Received: 29 July 2019; Accepted: 22 August 2019; Published: 26 August 2019

Abstract: The purpose of this paper is to provide structural and architectural technological solutions applied in the construction of high-rise buildings, and present the possibilities of technological evolution in this field. Tall buildings always have relied on technological innovations in engineering and scientific progress. New technological developments have been continuously taking place in the world. It is closely linked to the search for efficient construction materials that enable buildings to be constructed higher, faster and safer. This paper presents a survey of the main technological advancements on the example of selected tall buildings erected in the last decade, with an emphasis on geometrical form, the structural system, sophisticated damping systems, sustainability, etc. The famous architectural studios (e.g., for Skidmore, Owings and Merill, Nikhen Sekkei, RMJM, Atkins and WOHA) that specialize, among others, in the designing of skyscrapers have played a major role in the development of technological ideas and architectural forms for such extraordinary engineering structures. Among their completed projects, there are examples of high-rise buildings that set a precedent for future development.

Keywords: high-rise buildings; development; geometrical forms; structural system; advanced materials; damping systems; sustainability

---

1. Introduction

High-rise buildings play an increasingly important role in contemporary architecture. Their raising is a necessity for the process of population growth and its concentration in cities, as well as for the high demand for areas in city centers [1]. It can be observed the dynamic development of their construction in terms of both quantity and quality [2]. There are plans to build 219 high-rise buildings worldwide in 2019. According to the Global Tall Buildings Database of the CTBUH (Council on Tall Buildings and Urban Habitat) until now were erected 1647 buildings taller than 200 m. The high-rise building construction is characterized by high demand of construction technology and complex engineering works [3].

In contemporary architecture, designers go beyond the framework of standard codified construction assumptions in order to provide additional and unusual aesthetic experiences [4]. Geometric shapes, impressive in terms of body and scale, are used for this purpose, as well as the newest material technologies, thanks to which skyscrapers can be classified as eco-buildings.

The change in the approach in building design in the last two decades is reflected in the models for shaping a sustainable, energy-saving environment, which are specified in the context of comparable methods for assessing buildings with various criteria (quality assessment tools, including Leed). These changes are evidenced by many documents, including the Aalborg Charter [5], the European Charter for Solar Energy in Architecture and Urban Planning [6], and the White Book of the Architects' Council of Europe [7]. Energy-efficient architecture is promoted by such architects as Norman Foster [8], Renzo Piano [9], Thomas Herzog [10] and Gilles Perraudin [11].

The main trend among new high rise buildings is the striving to achieve zero energy, which is associated with Leed certification [12]. Obtainment of Leed v4 certification at the Platinum level means the highest green building standard in the world. Bryant Park (New York, NY, USA) became the first high-rise building in the world to attain this certificate. Other buildings to achieve the Leed v4 certificate include, among others, Shanghai Tower (Shanghai, China), Taipei 101 (Taipei, Taiwan) and Hearst Tower (New York, NY, USA).

One of the pro-ecological ideas is the design of bioclimatic skyscrapers, in which users' comfort is increased by greenery inside the buildings through the use of public terraces or multi-level atrias (Oasia Hotel, Singapore).

The problem of high building design particularly concerns problems related to the limitation of horizontal displacements of the building and ensuring its spatial rigidity, proper foundation and resistance to dynamic wind action and seismic effects. The key design challenge associated with acting loads is the appropriate selection of the structural system, while at the same time optimizing its geometrical dimensions. The existing construction solutions mainly differ in their way of transmitting horizontal forces from the wind and seismic impacts on the foundations. A sophisticated construction system allows building in seismic areas with strong wind (Tokyo Sky-tree, Tokyo) and artificially created land (United Tower, Manama; Marina Bay Sands complex, Singapore).

The paper presents the architectural and constructional characteristics of selected modern high-rise buildings, which follow the trend constitutes an architectural paradigm that focuses on sustainable design.

2. Methods

The high-rise buildings implemented today are astonishing in terms of the multitude of their architectural and constructional solutions, as well as their technology. Conducting a comprehensive analysis of technological innovations used in these buildings, due to the size of the issue, requires a special approach. Therefore, a methodology was developed that includes the following elements:

-   Gathering information on innovative technologies of modern high-rise buildings and completing photos and videos documenting their erection,
-   Interpretation of collected information on the basis of literature, generally addressing the problem of advanced technologies used in completed buildings,
-   Conducting, according to a structured diagram, architectural and construction analysis of selected high-rise buildings, in which the applied technologies were significantly more advanced than those of previous projects.

Such a system facilitated the conducting of a structured analysis, with particular emphasis being put on the building's body, construction system, vibration damping system, ultra-strong concrete and steel, low-emission glass, double or triple skin facades and elements decisive for energy saving of the building.

3. Technological Innovations in High-Rise Buildings

3.1. New Design Trends in Geometrical Forms

High-rise buildings were often designed in the form of rectangular blocks with glass façades. Such buildings, although practical and aesthetic, are somewhat monotonous. Contemporary architecture is trying to face this monotony. Apart from the mass execution of rational high-rise buildings, the appearance of another trend has been noted. This is the phenomenon of erecting "iconic" buildings, which are distinguished by their shape and scale. Based on the information gathered in the CTBUH database and also looking at new created high-rise buildings, it is reasonable to believe that the next generation of tall buildings will be more towards aerodynamic and curvilinear shapes and forms. Analysis of wind action on tall buildings shows the importance of the effect of form and geometry of high-rise buildings. For instance, in Taipei 101, corner modifications provide 25% reduction in the base moment when compared to the original square section.

Geometric solids (such as polyhedra, cones, cylinders, spheres, ellipsoids and toruses) and curved surfaces appear as the components of each modern skyscraper [13].

Analyzing the form of a building can identify the specific types of basic solids or surfaces used in all or part of the building. In addition to creating the composition of a building, different kinds of distortion of the solids or surfaces are used. Generally, spatial forms can be geometrically divided into polyhedra, solids of revolution and surfaces [14]. Polyhedra are solids limited by a closed surface and constructed with a finite number of flat polygons. They are divided into: Prismatoids (pyramids, prisms, anti-prisms and octahedrons), polyhedra (platonic solids), semiregular polyhedra (the Archimedean solids) and other polyhedra compounds (Catalan, Johnson and toroidal).

The second group of geometric forms is the solids of revolution, which are limited by a closed surface of revolution or toroidal based on a circle, ellipse or other closed figure. In them are a sphere, ellipsoid or torusoid of revolution with a normal section in the shape of a circle or ellipse.

The last group of spatial forms is surfaces, which include: Ruled surfaces (Catalana, conical and cylindrical), curved surfaces of a constant generatix (rotary, torusoidal and translational) and curved surfaces of a varying generatix (wedge, parabolic-elliptic and minimum). From an architectural point of view, modern skyscrapers can be categorized into the following groups: Extruder, rotor, twister, tordos and also free form.

3.1.1. Extruder

Buildings of this type have the same cross section for their full height. An example can be a rectangular or cylindrical solid (Figure 1a). Within this group the following modifications may exist:

- Individual stories are arranged one on the other in a constant slope and the floor plans may have a straight or curved contour ("anglers", Figure 1b);
- Stories are arranged one on the other, sometimes with a different angle of inclination and often in the form of straight segments inclined in different directions that are smoothly connected to the curved segments ("sliders", Figure 1c), which may be narrowing with an increasing height ("tapering sliders", Figure 1d);
- Single buildings of this type can be connected together in groups in order to provide increased rigidity or fire protection exit ("slider assemblies", Figure 1e).

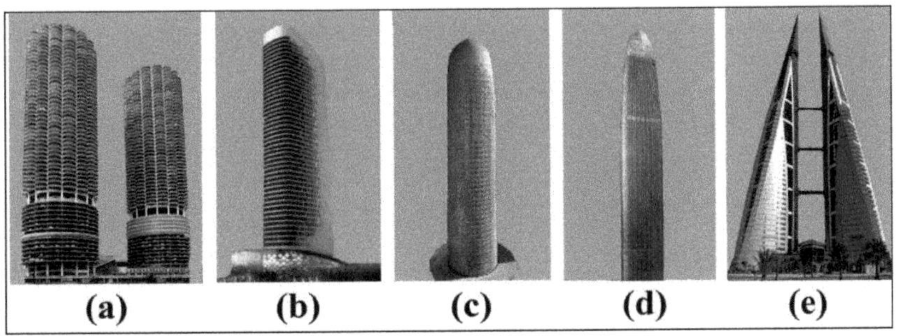

Figure 1. Extruder type of high-rise buildings: (**a**) Marina City towers (Chicago, cylindrical central core system), (**b**) Bay Gate (Dubai, wall frame system), (**c**) Doha tower (Doha, tube frame system), (**d**) Greenland Puli Center (Jinan, core and outrigger system) and (**e**) the World Trade Center (Bahrain, shear wall frame system), figure by authors.

3.1.2. Rotor

Buildings of this type have the same cross section for their full height. An example can be a rectangular or cylindrical solid (Figure 2a). Within this group the following modifications may exist:

- Individual stories are arranged one on the other in a constant slope and the floor plans may have a straight or curved contour ("anglers", Figure 2b);
- Stories are arranged one on the other, sometimes with a different angle of inclination and often in the form of straight segments inclined in different directions that are smoothly connected to the curved segments ("sliders", Figure 2c), which may be narrowing with an increasing height ("tapering sliders", Figure 2d);
- Single buildings of this type can be connected together in groups in order to provide increased rigidity or fire protection exit ("slider assemblies", Figure 2e).

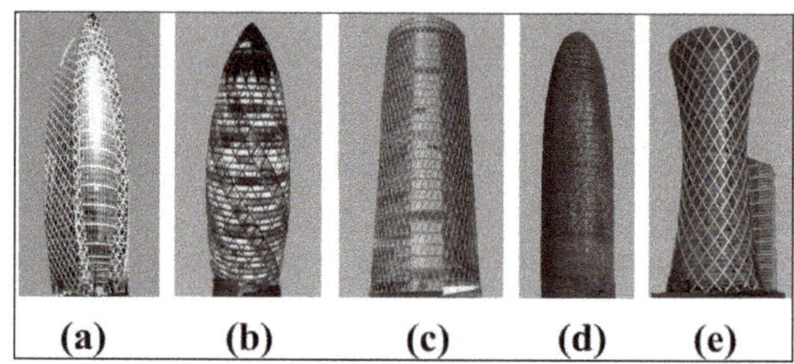

Figure 2. Rotor type of high-rise buildings: (**a**) Mode Gakuen Cocoon Tower (Tokyo, tube system with concrete columns), (**b**) Swiss Re (London, diagrid frame tube system), (**c**) Westhafen Tower (Frankfurt, diagrid frame tube system), (**d**) Torre Agbar (Barcelona, diagrid frame tube system) and (**e**) Tornado Tower (Doha, concrete core system with an external tubular steel diagrid), figure by authors.

3.1.3. Twister and Tordos

Buildings of this type are in the form of a twisted solid with the "twister" facade repeated on all floors (Figure 3a). Buildings with an orthogonal core and one or two twisted towers belong to the category of "toros" (Figure 3b). The conversion tower axis of the helical form the Revolution Tower (Figure 3c) may be derived from orientation of asymmetric floors not through the center of the circular segment stories, but the center of gravity of the floor. The body of the building belongs to the category of "sliding twister" (Figure 3c), where the floors are moved upward along the 2D or 3D curve and rotation is added to the outer structure. When the 3D curve rotation has the shape of a spiral it belongs to the category of "helical twister" (Figure 3d,e). The intersecting body of the building, in the shape of a twisted spiral, has an internal vertical zone dedicated to the lift shaft.

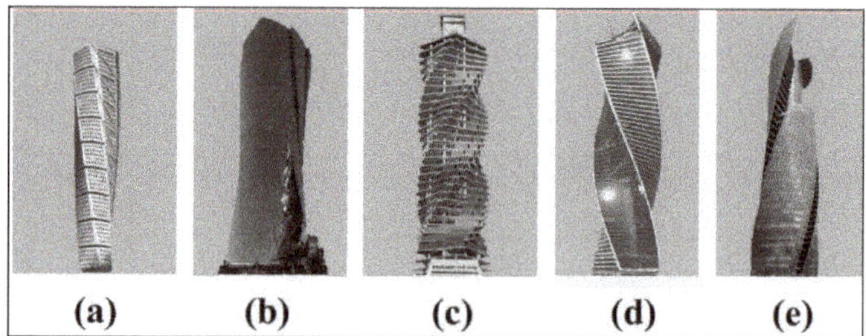

Figure 3. Twister and Tordos types of high-rise building: (**a**) Turning Torso (Malmö, mega core system), (**b**) Al Bidda Tower (Doha, wall frame system), (**c**) Revolution Tower (Panama City, core system), (**d**) Evolution Tower (Moscow, core system)) and (**e**) Mode Gakuen Spiral Towers (Nagoya, tube system), figure by authors.

3.1.4. Free Form

The free geometry building form is constructed using a combination of geometrically simple objects (lines, surfaces and solids), when the sequence of the architect's actions is not obvious and the form does not fit into any other category. In this category we can distinguish the subcategory "slicer". It includes buildings that have a curved facade with balconies and other extended elements. Figure 4a shows the curved outer surface obtained by the contours of winding balconies around a rectangular solid. Alternatively, the curved segments of the balconies can be repeated on the upper floors with their rotation (Figure 4b). This vertical twisting of the outer surface is formed in a cross-section that is not a straight line but a curve. The receding facade of the building in Figure 4b is decorated with flat elements. The smooth surface of the building in Figure 4c is obtained by a large number of blinds. The verticality of railings is less obvious in high-rise buildings than in low buildings (Figure 4d), where the facade is rather stepped and does not create a smoothed curve. The building is classified in the "sliced twister" category (Figure 4b) when it has repeated vertical floors with horizontal rotation.

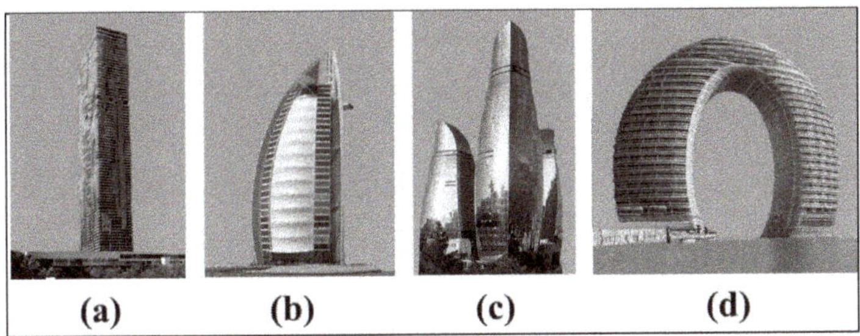

Figure 4. Free form type of high-rise buildings: (**a**) Aqua Tower (Chicago, core and outrigger system), (**b**) Burj Al Arab (Dubai, composite frame system with diagonal steel trusses), (**c**) Flame Towers (Baku, frame tube system) and (**d**) Sheraton Huzhou Hot Spring Resort (Huzhou, core system), figure by authors.

High-rise shaping is largely related to the numerical modeling tools that architects have available. Simple modeling procedures enable intuitive shaping of complex geometry, however mathematical analysis is required because the consequences for the structure are considerable.

3.2. Innovations in Structural Systems

3.2.1. Structural Systems

The relationship between structure and architectural form has reached its peak in present times. Form and structure have become inseparable and complementary [15]. The primary structural skeleton of a high-rise building can be visualized as a vertical cantilever beam with its base fixed in the ground. The structure has to carry vertical gravity loads, the lateral wind and also earthquake loads. The building must therefore have adequate shear and bending resistance and must not lose its vertical load-carrying capability.

Structural systems of tall buildings can be divided into various types due to different criteria (e.g., internal and external). The choice of system and application of constructional material is affected by many factors, in particular:

- The height of the building,
- The ratio of height to width (slenderness),
- The required spatial rigidity for the transfer of lateral forces (wind, seismic),
- The formation of the building's body,
- The conditions of the layout of the lower floor and foundation.

The structural system of high-rise buildings can be divided on the following types [16]: Rigid frame, shear frame (shear trussed frame, shear walled frame), flat plate, mega column (frame, truss), core, mega core, outriggered frame and tube (framed tube, truss tube).

3.2.2. Innovative Diagrid System

Currently the diagrid system is one of the most innovative and adaptable approaches to structuring high-rise building (Capital Gate Tower (Abu Dhabi, UAE), Swiss Re (London, UK), Hearst Tower (New York, NY, USA) and CCTV headquarters (Beijing, China)). This kind of structure has evolved from a diagonalized tube. A diagrid is a special form of spatial truss. The difference between a conventional braced-tube structure and the current diagrid structure is that the diagrid system has almost completely eliminated the use of columns [17,18]. This is possible because diagonal elements in the diagrid system can carry gravity loads as well as horizontal loads due to their triangular configuration. The constructional function is realized by transfer lateral loads through the axial action of structural components. The bending stiffness is obtained by a diagonal grid, which also gives the shear stiffness. Adoption of such forms is very beneficial for reasons of dynamic impacts. As the height of a building increases, the lateral strength becomes more important than the load-bearing system that carries gravity loads. Therefore, any modifications to the geometric form of tall buildings generally reduce the adverse effects of the wind, which is an additional reason for the greater creativity of architects.

The diagonal grid module has a trapezoid shape and its height is several floors. Depending on the number of stories, the modules are divided into small (2–4 stories), medium (6–8 stories) and large (over 8 stories). Modules and diagonal angles play a key role in the structural, architectural and aesthetic concept of the design of the building. Due to the form, they may be flat, crystalline or multi-curved. The steel construction expresses regular diagonals in the facade of the building, is easier and quicker to assemble and is highly compatible with the concept of a sustainable building. In the design of the diagrid construction, an important factor is to choose the right diagonal angle. If the diagonal angle deviates from the optimum value, the required amount of steel is substantially increased. Since the optimum angle of placement of the columns for maximum bending stiffness is 90 degrees, and diagonals achieve maximum shear stiffness at an angle of 350, the optimum angle for diagrid construction elements is therefore taken between the values of these angles. The arrangement of diagonal elements with larger angles in the corners of the building increases its bending stiffness. High-rise buildings with a high ratio (height/width) behave like bent beams. Therefore, when a building's height rises, the optimum diagonal angle also increases. Buildings in this construction system are designed on a circle, ellipse, or other curved geometric form. The diagrid system is perfectly matched in the modification of the classical geometrical form. In this system, the following forms are known: Hyperboloidal, cylindrical, twisted, tilted and free [15].

3.3. Advanced Vibration Damping Systems

The development of the advanced damping system has been characterized on the basis of Japan, which has the most active seismic zone in the world and which paradoxically occupies third place in terms of the number of skyscrapers.

An essential aspect of designing tall buildings is their dynamic reaction to earthquakes and counteracting wind vortices. Moreover, high buildings are sensitive to wind-induced vibrations, and the impact of such vibrations becomes dominant for buildings higher than 200 m. Under the action of the wind, a building not only deflects statically from its vertical position in the direction of wind pressure, but can also fall into the vibrations, which are transverse to the wind direction. These vibrations become dangerous, and even resonant when taking place with the frequency of air vortexing from the sides of the building. A simple measure of the quality of a skyscraper, i.e., its resistance to oscillatory swaying and resonance with the wind, and also its dynamic stiffness under bending is the fundamental (minimum) natural frequency of the building. Possible oscillations form a spectrum of waves with different vibration frequencies—higher or shorter. However, more diverse waves occur when the building bends. The longest wave with the smallest frequency, called the fundamental frequency, represents the shaking of the entire building. Such oscillations are most easily created at the lowest wind speed—hence their fundamental importance for the comfort of use and safety of the structure (risk of resonance with the wind). A better building has a higher fundamental frequency of vibrations, is stiffer, vibrates faster and introducing it into dangerous resonance vibrations requires a wind with a higher speed, which occurs less often and is less likely.

While the strength of building materials, such as steel, has doubled in the last few decades, its stiffness has not increased significantly. This has led to an elastic-based approach to design in which lateral deflections and accelerations are the dominant structural constraints for tall buildings. Vibrations can be partially damped by the structure itself. Increasing the stability of the structure causes an increase in the natural frequency. According to the numerical simulation of a construction's response to wind, if the natural frequency is greater, the maximum acceleration decreases in proportion to half of the natural frequency.

The light steel structure used in high-rise buildings has little natural damping or natural dissipation of energy and is sensitive to dangerous accelerations in conditions close to resonance. The dynamic reinforcement of load conditions can be reduced by redistributing stiffness in order to avoid resonance, or by the implementation of a damping system in the building (Figure 5). The need for motion control has led to the development of various methods and devices for dissipating energy. Damping devices can be passive, which do not require an additional energy supply, or active (AMD), which suppress the reaction with input energy, usually through the use of actuators [19,20]. Although there are many effective applications for active dampers, the increased complexity, maintenance and cost and lower reliability of passive dampers means that they are more often used. In addition to passive and active systems, there are mixed hybrid systems.

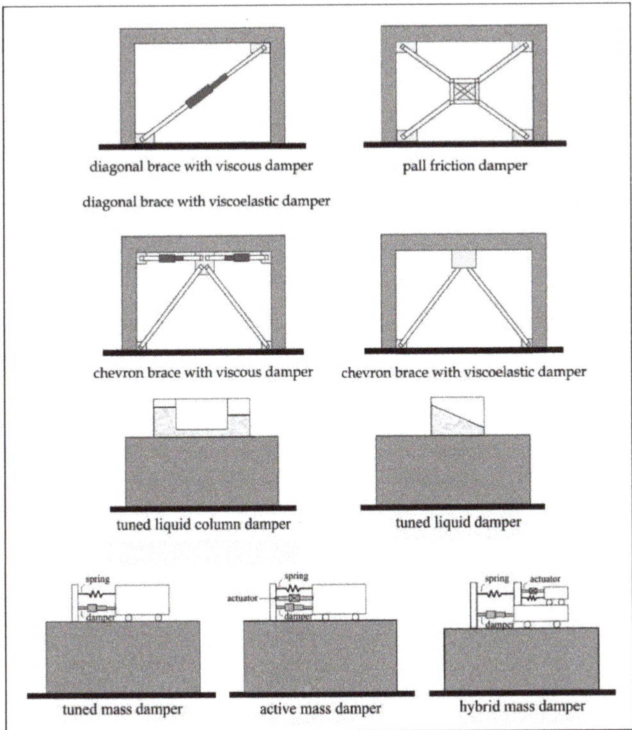

Figure 5. Vibration damping systems (figure by authors).

3.3.1. Passive Damping—Material Based Dissipation System

The material based dissipation dampers are an integral part of primary structural systems, and they are positioned in optimal locations (e.g., in bracing systems). There are different types of devices that belong to this category, and among all, the most important are: Hysteretic dampers and viscous dampers.

Hysteretic damping uses steel vibration absorbers SD (steel damper) and SJD (steel joint damper), as well as viscoelastic dampers (VED), lead dampers (LD) and friction dampers (FD), which are used to reinforce material interactions at the FD connections. Steel vibration absorbers dissipate energy through the cyclic inelastic deformation of materials. These damping systems are often designed in the form of a triangular plate, or are X-shaped. Due to this shape, plastic deformations appear in a much larger area, which leads to a more efficient dissipation of energy. This system was used in the Ohjiseishi Building (Tokyo, Japan), Art Hotels Sapporo (Sapporo, Japan) and Kobe Fashion Plaza (Kobe, Japan).

In friction dampers, energy dissipation occurs as a result of friction between two solids moving in relation to each other. There are two types of friction dampers used in steel framed buildings: Rigid frame friction dampers and braced frame friction dampers. For example, friction dampers were used in the Sonic City Office Tower (Ohmiya, Japan) and Asahi Beer Tower (Tokyo, Japan).

Viscous dampers (VD) and oleo-dynamic dampers (OD) use viscous materials in which the resistance force acting on the body moving in the material is proportional to the speed of the body [21]. In this case, high viscosity chemicals such as silicone oil are used. The thermal effect is also significant. VDs are particularly effective in the high frequency range and low vibration levels against moderate earthquakes and strong winds. This type of damper, consisting of steel plates, is installed as a part of a diagonal brace, where it can dissipate vibrational energy by the shearing action of the VE material.

Viscoelastic dampers were used in the TV-Shizuoka Media City buildings (Tokyo, Japan) and in the Torishima Riverside Hill Tower (Osaka, Japan) to counteract the vibrations caused by extremely large earthquakes.

3.3.2. Passive Damping–Additional Mass System

This passive system is based on the counteracting inertial force created by an additional mass allocated at the top of a building. There are two main categories of devices belonging to this group: Tuned mass dampers (TMD) and tuned liquid dampers (TLD). A TMD is an additional mass [22], usually in the order of two percent of the total weight of the building, which is attached to the structure by means of springs and dashpots. The inertia force of the mass damps the reaction of the building. However, TMDs are mostly effective only when they are excited by the resonant frequency for which they have been designed. Sometimes, spacing limitations do not permit a traditional TMD system, which requires the installation of alternative configurations such as pendulums, hydrostatic bearings or laminated rubber bearings. A TMD damper was used in Fukuoka Tower (Fukuoka, Japan), Higashimyama Sky Tower (Nagoya, Japan) and Huis Ten Bosch Domtoren (Nagasaki, Japan).

Another type of mass damping system is tuned liquid dampers (TLCD). This damping system uses the movement of liquids in special containers to absorb the energy of building vibrations. The vibration frequency of TLCDs can be controlled by the water depth and the size of the container. TLCDs are preferred because of their simplicity, low maintenance price and the possibility of including water for emergency fire protection. The TLCD system was used in the Rokko-Island P and G Building (Kobe, Japan), Crystal Tower (Osaka, Japan) and Sea Hawk Hotel and Resort (Fukuoka, Japan).

Different from passive systems that are tuned to work on some range of loading conditions, active system perform more efficiently over a wider range. The most prominent active devices are active mass dampers (AMD) and active variable stiffness devices (AVSD). Both devices rely on the same principles of mass and material based dissipation but their properties are adjusted from a computer control system. The AMD system was used in the Applause Tower (Osaka, Japan) and AVSD in ORC 200 Bay Tower (Osaka, Japan).

3.3.3. Hybrid Damping

In recent years, hybrid dampers have appeared, which are a combination of a mass damper with an additional active element, which aims to improve the efficiency of passive damping [23]. The forces from the active actuator increase the effectiveness of the mass damper and are very effective in the event of changes in the dynamic characteristics of a structure. The active portion of the system is only used under excitation of a high-rise building, otherwise, it behaves passively. The hybrid system was used in the Landmark Tower building (Yokohama, Japan), the Ando Nishikicho building (Tokyo, Japan) and in Osaka World Trade Center (Osaka, Japan).

3.4. Use of Advanced Materials

The development of high-rise buildings is inextricably linked to the search for efficient construction materials, Figure 6. Technological achievements in material engineering have gradually shaped the form, height and construction, as well as energy efficiency of buildings. Initially, steel was the leader in building constructions, as the technology of concrete was not sufficiently developed, and because the produced concrete had a much lower strength than steel. At present, there is a growing interest in concrete as the main structural material in this type of buildings [24]. In the construction of high-rise buildings are also developing mixed steel-concrete technologies, such as the Petronas Twin Tower (Kuala Lumpur, Malaysia), Burji Khalifa (Dubai, UAE), Princess Tower (Dubai, UAE), One57 (New York, NY, USA) and Kingdom Center (Riyadh, Saudi Arabia). Currently, among the 100 highest buildings in the world, nine are built as steel structures, 30 as reinforced concrete, 5 as steel and reinforced concrete and 56 as composite structures. Advances in physical science have led to a new generation of intelligent materials, especially those that improve the acoustic, light, electrical and thermal environment of buildings [25].

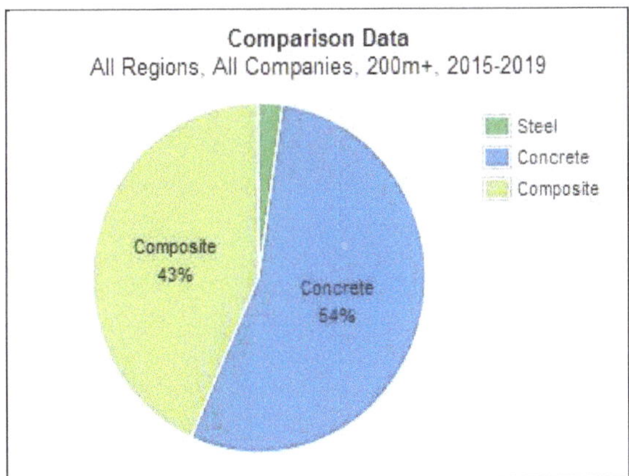

Figure 6. Comparison of material system applied in high-rise buildings (above 200 m) in the years 2015–2029 (on the basis of the global tall buildings database of CTBUH (Council on Tall Buildings and Urban Habitat)).

3.4.1. Concrete

Over the last years, there has also been significant progress in the field of modeling the physical and rheological properties of concrete. Added admixtures allow for a significant increase in strength, accelerate the curing of concrete and enable construction works at both very low and very high temperatures. Very high strength concrete (VHSC), self-consolidating (SCC) led the concrete to be the most appropriate structural material for super tall buildings, such as Burj Khalifa (828 m, Dubai, UAE) and Kingdom Tower (1000 m, Jeddah, Saudi Arabia). VHSC has a compressive strength of

240 MPa with steel fiber reinforcement incorporated in the mixture and achieves a flexular tensile strength of 40 MPa [26]. Moreover, the development of construction technology (moving formworks with high accuracy and speed of assembly and disassembly, vertical transport systems—pumps, etc.), high susceptibility to shaping, the faster growth in strength than prices, and high fire resistance are further advantages for the use of concrete. The development of concrete technology and methods of construction organization has not only allowed for the construction of higher and higher skyscrapers, but also for the diversification of their forms and shapes.

3.4.2. Steel

Despite the increase in the number of high-rise buildings made of ultrahigh-strength concrete, steel is still an irreplaceable material in seismic areas. Currently, Japan is one of the most advanced countries in terms of the development of steel structures. Despite the unfavorable geographical location associated with the occurrence of frequent earthquakes, Japan can boast of having such building structures as: Akashi Kaiyko bridge (one of the longest suspension bridges in the world, longest span 1991 m) and Tokyo Sky-tree (the highest free-standing tower, 634 m). Undoubtedly, the factor that influenced the creation of these structures was the improvement of the efficiency of steel materials, which favored their development. TMCP (Thermo-Mechanical Control Process) technology was used to obtain high-strength steel. This technology is a combination of "controlled rolling", which favors the refining of the microstructure by introducing dislocation in a high temperature range, and also "accelerated cooling", which realizes the quenching effect while suppressing grain growth. With such limited carbon technology, high-performance steel materials with excellent weldability and efficiency can be produced. The use of high-strength steel in high-rise buildings was a consequence of their earlier use in bridge constructions. At present, steel with a tensile strength of over 1200 N/mm^2 is available [27]. This high strength was achieved by the development of a dual-phase steel (DP steel), which has a structure composed of hard and soft material and TRIP steel in which the plasticity effect of unstable austenite is caused by martensitic transformation. In addition to the tendency to increase the strength of steel, there was also a demand for steels with a low yield stress (yield strength 100 N/mm^2 class and yield strength 225 N/mm^2 class), which were first used for the construction of vibration dampers. Since the plastic deformability of high strength steel is lower than for conventional steels, the performance of an entire building is achieved with a combined use of dampers.

3.4.3. Smart Material and Nanotechnology

Smart materials can be divided into the following groups: Piezoelectric, electroactive, photostrictive, thermostrictive, magnetostrictive, chemostrictive materials and fiber optic sensors [28]. These smart materials can constitute the components of a smart structure, which is an electronically enhanced physical framework. For example piezoelectric materials convert mechanical energy into electrical energy after strained. Piezoelectric dampers have been developed as an example of controllable materials. There are other forms of smart materials, such as shape memory alloys, which can be used as temperature sensors for ventilation systems or as actuators for sensing and monitoring devices. With nanotechnology can be improved properties of glass by self-cleaning, antimicrobial and reducing pollution properties. Titanium dioxide nanoparticles with a smooth surface create an anti-adhesive coating.

3.4.4. Glass

Technologically advanced high-strength glazing is equally important as steel and concrete for the building of high-rise buildings. In this case, the main challenges are related to wind load, temperature and altitude differences, and also the condensation of water vapor. Other important factors are light and heat. In the case of high-rise buildings, there is always the possibility of condensed steam appearing on the outside glass, which results from the temperature difference between its internal and external part. The use of low-emission glass as an internal pane prevents the passage of heat from inside the building to the outside. Low-E glass helps to reflect long-wave radiation and minimizes its transmission. Heat treatment of the glass through hardening or heat strengthening causes the glass to be many times stronger and able to withstand extreme wind load and temperature difference. In high-rise buildings, a wide variety of glass types are used depending on the climate zone, Table 1. To fully characterize glass system, it is necessary to specify the following characteristics: U-value, solar heat gain coefficient (SHGC) and glass visible transmittance [29,30].

Table 1. Indicative characteristics of different glass types [30].

Glass Type	Glass Thickness (cm)	Visible Transmittance (% daylight)	U-Factor (Winter)	Solar Heat Gain Coefficient
Single Pane	0.63	89	1.09	0.81
Single White Laminated	0.63	73	1.06	0.46
Double Pane Insulated	0.63	79	0.48	0.70
Double Bronze Reflective	0.63	21	0.48	0.35
Triple Pane Insulated	0.32	74	0.36	0.67
Pyrolitic Low-e Double	0.32	75	0.33	0.71
Soft-coat Low-e Double	0.63	73	0.26	0.57
High Efficiency Low-e	0.63	70	0.29	0.37
Suspended Coated Film	0.32	55	0.25	0.35
Suspended Coated Film Argon gas fill	0.32	53	0.19	0.27
Double Suspended Coated Film	0.32	55	0.10	0.34

U-value indicates the rate of heat flow due to conduction, convection and radiation through a glass as a result of the temperature difference between the inside and outside. The higher the U-factor the more heat is transferred through the window in winter.

SHGC indicates how much of the sun's energy striking the glass is transmitted through the glass as heat. As the SHGC increases, the solar gain potential through the window increases.

Visible transmittance indicates the percentage of the visible portion of solar spectrum that is transmitted through a glass.

3.5. Innovative Energy Systems in High-Rise Buildings

The achievement of high energy efficiency in modern high-rise buildings requires many environmental conditions to be taken into account at the stages of design and construction. Satisfying these requirements allows the maximum use of available ambient energy, the reduction of heat loss from the building, and also a smaller demand for heat and electricity. One of the most finance-intensive requirements is the ventilation and heating of buildings, accounting for about 30% of the energy demand in high-rise buildings. The use of natural ventilation is an increasingly popular solution that reduces these costs. The inner atrium allows light to be supplied to the interior of the building, Figure 7. The full height of the windows causes the amount of light reaching inside to be sufficient for work, and there is therefore no need to use artificial lighting for most of the day. The ventilation of the rooms is also ensured by specially designed windows, constructed of a three-layer facade system with an air gap, allowing air to circulate.

The use of free energy from renewable sources, such as sun, wind, biomass and low-temperature geothermal energy, is also becoming more and more popular. This is especially the domain of passive buildings, and also sometimes of energy-saving buildings. Among the activities preceding the implementation of a project, the selection of the right location is of particular importance and results in the efficient use of available renewable energy sources. The next elements are: Adaptation of the architectural design to local microclimatic conditions, proper location of the building, accurate orientation towards the sun and correct shaping of the surroundings of the nearest building. The location of buildings should provide good insolation conditions and the maximum number of hours of sunshine per year.

Figure 7. Atrium in the Marina Bay Sands Hotel (Singapore, photograph by authors).

Such a situation is beneficial for bioclimatic reasons, as well as for the possibility of using solar energy in active and passive photothermal and photoelectric conversion systems. Direct conversion includes:

- Photothermal methods, implemented in low-temperature active solar systems (solar collectors, Figure 8) and in passive systems (solar architecture of buildings) [31],
- Photoelectric methods, implemented in photovoltaic systems (cells) [32], Figure 9.

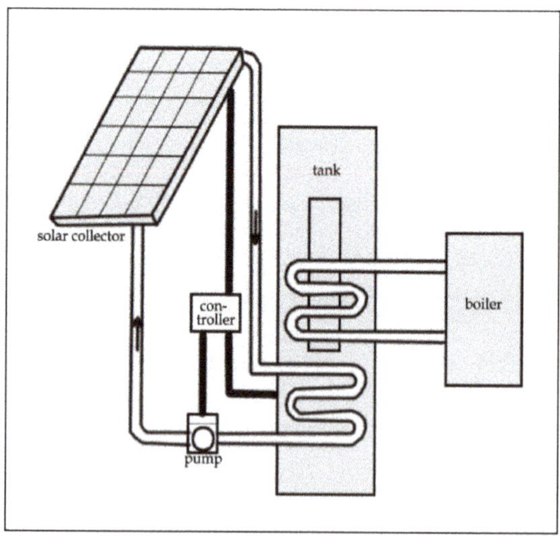

Figure 8. Photothermal solar technology (figure by authors).

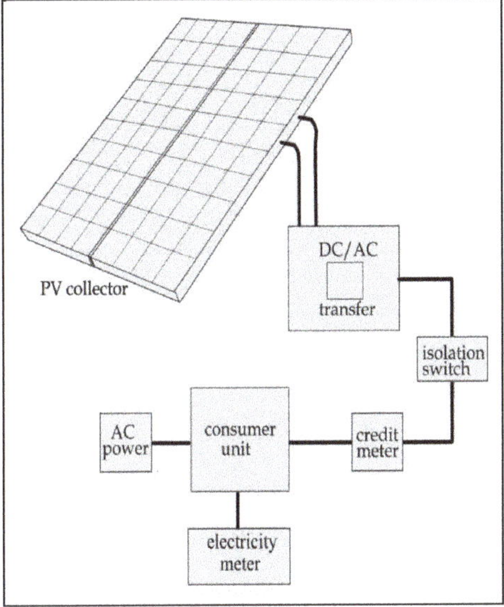

Figure 9. Gallium arsenide photovoltaic cells technology (figure by authors).

In the active solar system, energy acquisition and also its separation and storage takes place through the use of such elements as: Solar collectors, storage tanks, safety devices and elements of control and measurement automation. In addition to the proper selection of system components, it is important to properly arrange the collectors by setting the right angle of deviation from the southern direction (declination), and also the angle of inclination with the ground plane (inclination). Passive solar heating systems for generating heat in a building use solar radiation energy directly or indirectly. To achieve a more sustainable design gallium arsenide photovoltaic cells combined with a rain screen in the southeast facade are often used [33].

3.6. New Technologies of Facades

One of the most significant changes in technical solutions and the aesthetics of high-rise buildings was caused by the role of the contemporary glass facade of the building. Architects Norman Foster and Thomas Herzog [34] in the European Charter for Solar Energy in Architecture and Urban Planning stated that the building's exterior walls in terms of light, heat, air and transparency must be susceptible to change and ultimately be controllable to respond to changing local climate conditions. It is noticeable in the last few years of development of new advanced facade solutions integrated with plants to combine architectural features and trends to reduce carbon emissions. The concept of vertical gardens in the form of green facades of buildings is now a trend of sustainable design, in which the ecological facade material offers an unlimited number of patterns and colors that change both during the day and in different seasons (Oasia Hotel downtown (Singapore). The presented structure, in the form of a wall covered with plants, aroused the interest of architects and finally resulted in cooperation. Numerous ecological green designs were implemented by Jean Nouvel (One Central Park (Sydney, Australia)), Herzog and Pierre de Meuron (Beirut terraces (Beirut, Libanon)) and Stefano Boeri (Bosco Verticale (Milan, Italy), Nanjing Tower (Nanjing, China)).

Currently ventilated double skin facades represent a most valid technology (Figure 10) [35,36]. The principle of ventilated double skin facade is to position the shading devices between two layers of glazing, capturing the energy trapped in the cavity (Figure 11). Among the technologically advanced facades, it can be distinguished by two types: Active wall facade (Manulife Financial, Boston; Jiu Shi Headquarters, Shanghai) and interactive wall facade (Al Bahr Towers, Abu Dhabi).

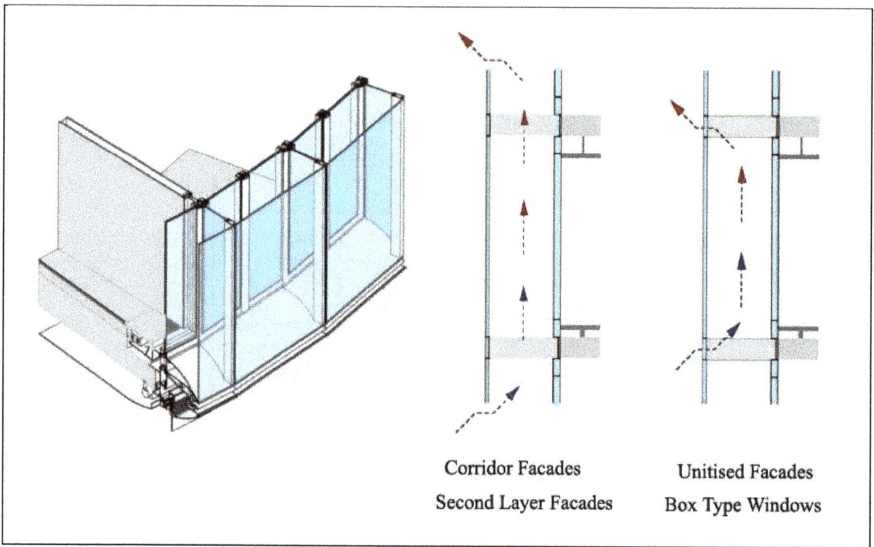

Figure 10. Natural ventilation of double skin facades (figure by authors).

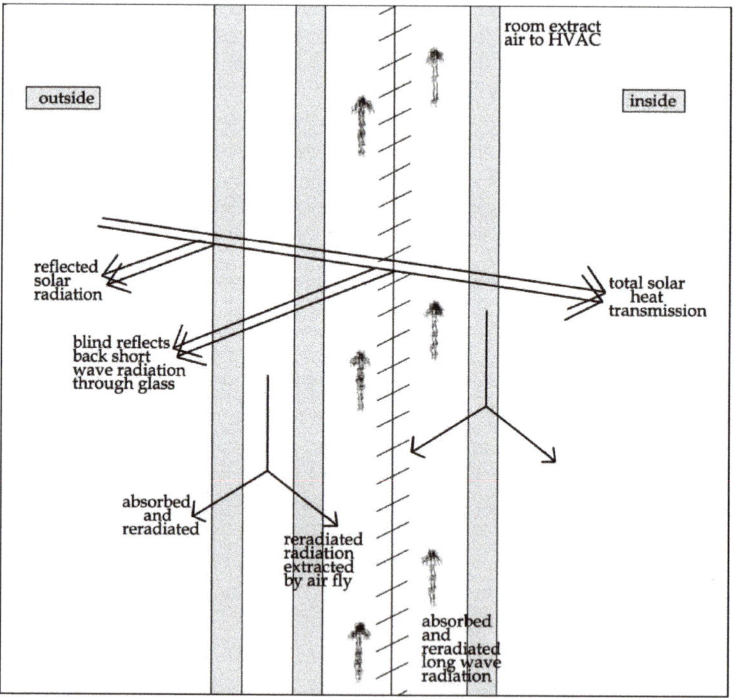

Figure 11. The active facade system (figure by authors).

The system performance depends on numerous design parameters (light transmission, solar factor, thermal transmittance and acoustic insulation), which values for different facade typology are shown in Table 2. The integration between double facade and environmental systems generally results in a reduction of installed power for heating and cooling due to a reduced U-value, a lower solar factor for hot climate (SHGC) and a potential heat recovery. The solar factor of the facade is a key variable to control both the reduction of overall energy and separate the cooling demand from the orientation of the building.

Table 2. Indicative characteristics of different facade typology [37].

Façade Typology	Light Transmission τ_{vis} (-)	Solar Factor SHGC^2 (-)	Thermal Transmittance U–Value (W/m^{2}K)	Acoustic Insulation R_w (dB)
Externally shaded	0.50–0.70	0.05–0.25	2.0	32–36
Naturally ventilated	0.60–0.70	0.10–0.20	1.4	38–44
Active Wall	0.60–0.70	0.15–0.25	1.0	38–44
Interactive Wall	0.60–0.70	0.10–0.20	1.3	38–44

High-rise buildings with traditional glazed curtain walls that allow sunlight to penetrate indoors cause unwanted heat gains and losses, thereby increasing cooling or heating. In a naturally ventilated facade, the cavity between the two skins is ventilated with outdoor air. An active wall facade is composed of an external insulating glazing unit and an internal single layer of glass. The cavity between the two skins is ventilated with return room air, which is extracted from the room at the base of the glazing and returned to the air-handling unit at the top.

Interactive wall facade has a digital, mechanical adaptive solution system, which can react. The concept is that the interior side of the facade interacts with its inhabitants, perceiving their body movement and adjusts its form accordingly. The exterior side interacts with the movement of the sun, working as a shade, reacting to environmental changes.

In some type of interactive wall facade, the cavity between two skins is ventilated with outdoor air at the base of glazing and returned to the outside at the top by means of temperature-regulated radial fans located in the upper part of the facade.

To manage sunlight appropriately and provide occupant comfort, innovative window systems and glazing are developed in order to regulate the sunlight. Different types of glass and film coatings, such as low-E, are used to enhance the performance of the façade [38].

At present, systems where the air exchange is limited to the height of one story are considered optimal, and this serves to prevent the possibility of the release of used air into rooms located on higher floors. For this purpose, glass ribs are used to separate the individual sections of the facade vertically. An integral part of the contemporary double skin facade is a sunblind placed in the space between the glazing layers. Its task is to reduce the penetration of direct sunlight into the rooms so as to reduce the amount of heat accumulating in them. This new generation of high-performance envelopes have contributed to the emergence of sophisticated assemblies that combine a real-time environmental response, advanced materials, dynamic automation with embedded microprocessors, wireless sensors and actuators and design-for-manufacture techniques.

4. Examples of Selected High-Tech High-Rise Buildings

4.1. Burj Khalifa (Dubai, UAE)—The Highest Mega-Structure System

Burj Khalifa is a mixed use skyscraper with a steel and reinforced concrete structure designed by the architectural studio Skidmore, Owings and Merrill. The building is 829 m high and contains 163 floors at the above-ground level and one underground floor, Figure 12. The initial concept for geometry of the Burj Khalifa originated from the hymenocallis flower, characteristic element in Islamic architecture. The form was designed based on how to counteract the effect of the wind on the structure. Burj Khalifa is designed on Y-shaped plan (Figure 13), which is ideal for residential usage, allowing maximum outward views and inward natural light.

Figure 12. Burj Khalifa (photograph by authors).

To achieve a very strong and stiff building the latest technology in materials, analysis and construction methods were used. The skyscraper has a raft foundation that is 3.7 m thick, and cooperates with 192 piles that are 47 m in length [39]. A cathodic protection system creates a barrier to protect structure from polluted ground water affecting the concrete [40].

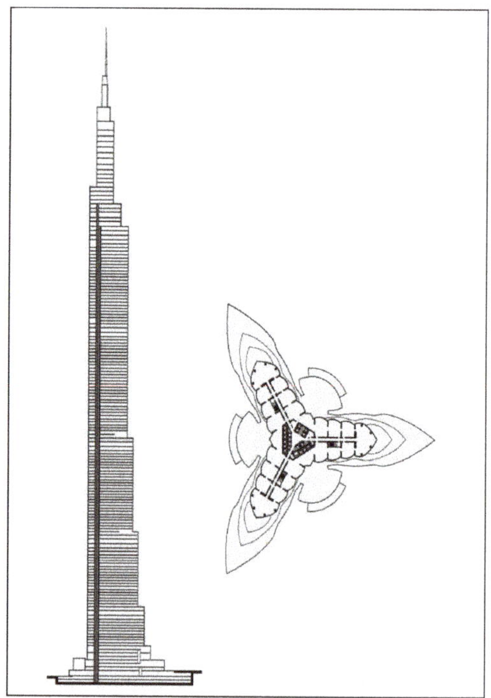

Figure 13. Burj Khalifa: Floor plan and section (developed by authors on the basis of [41]).

The main structural system for this mega-structure is the buttressed core, which allows a very big increase in height. The structural system consists of three-winged structure anchored to a hexagonal central core. The central core provides the torsional resistance for the building, and wings provide the shear resistance and increased moment of inertia. With the addition of floor RC (Reinforced Concrete) slabs and perimeter columns, the entire structural system acts like a single unit creating the tower.

The skyscraper is set on the podium, which provides a base anchoring the structure to the ground, allowing a grade access from three different sides to three different levels. The steel spire crowns the building has a diagonally braced lateral system.

The exterior cladding is comprised of reflective glazing with aluminum and textured stainless steel spandrel panels (Figure 14) and stainless steel vertical tubular fins. The cladding system is designed to withstand Dubai's extreme temperature, and to further ensure its integrity.

Figure 14. Burj Khalifa: Facade-reflective glazing with aluminum and textured stainless steel spandrel panels (photograph by authors).

Fire safety and speed of evacuation were prime factors in the design of Burj Khalifa. The skyscraper was the first mega-high rise in which certain elevators are programmed to permit controlled evacuation for fire or security events.

Burj Khalifa is a pro-ecological building with many innovative technological solutions. The solar heating system located on the roof of the offices is installed, which serve as solar collectors. Among other key sustainable energy and water use, the condensate from all the air-conditioning equipment is reclaimed to cool the potable water. The condensate is then collected in an on-site irrigation tank and used for the skyscraper's landscaping.

4.2. Sky Tree Tower (Tokyo, Japan)—Advanced Damping System

Tokyo Sky tree is a radio-television and observation tower with a steel and reinforced concrete structure as shown in Figure 15. The tower is the highest in the world with a height of 634 m and was designed by the architectural studio Nikhen Sekkei. The tower has two observation areas, the Tembo Deck and the Tembo Gallery. Tokyo Sky tree employs advanced technologies and never before adopted design approaches. Configuration of towers seems simple, but in actuality it contains extremely complex curves. The building at the base was designed on a triangular plan, which progressively changes from a triangular shape to a circular form higher up, Figure 16. This unique configuration does not occur anywhere else in the world.

Figure 15. Tokyo Sky tree (photograph by authors).

Tokyo Sky tree is located on the banks of the river, where the surface layer is soft silt. The foundation of the tower consists of steel piles filled with concrete, and also reinforced concrete walls with a thickness of 1.2 m that are located at a depth of 35 m on the load-bearing layer under the surface of soft silt. A set of cylindrical steel and thin-walled piles reaches up to a depth of 50 m [42]. This system of rigid foundation construction and vulnerable ground uses a relative displacement that is used to damp vibrations. For the tower to be able to withstand the uplift and compressive forces of earthquakes and strong winds, piles have nodule-protrusions that work to hold the piles firmly in the ground, greatly increasing their strength in supporting the tower. The foundation must not only ensure horizontal stiffness, but also vertical stiffness, as well as counteract the overturning moment.

The tower's structure consists of two separate parts, one of which is a steel truss, the other an internal reinforced concrete core. Both parts can move independently. To minimize seismic energy, a central core or so-called shin-bashira [43], utilized for centuries in traditional Japanese architecture in pagodas, was used. The core has a diameter of 8 m, a thickness of 6 m, a height of 375 m and operates on a stationary pendulum that balances seismic waves by reducing vibrations. Additionally, the elements supporting the reduction of vibrations are viscous oil dampers attached to the upper part of the core [44]. An independent steel truss structure employing wide-bore high strength steel pipes was used, which can be rarely found in building construction, Figure 17a. The largest section applied at the foot of the tower is 2.3 m in diameter, made from 10 cm thick steel plates. The truss is not only light and strong, which is necessary from the design point of view in seismic areas but is also effective in a wind-resistant construction, reducing the frontal area and not causing an unstable aerodynamic reaction due to the absence of an external wall. There are two types of steel structure in the tower. One structure is a truss, and the other one is a mega truss with a lattice core and girder, known as the Kanae truss [18]. These trusses are composed of four main members and are located in each corner of the equilateral triangle of the tower base. The three-level Tembo Deck features an observation floor that is situated at height 350 m. The Tembo Gallery has two floors, connected to each other by a circumferential ramp, Figure 17b. The higher of these floors is 451 m above the ground.

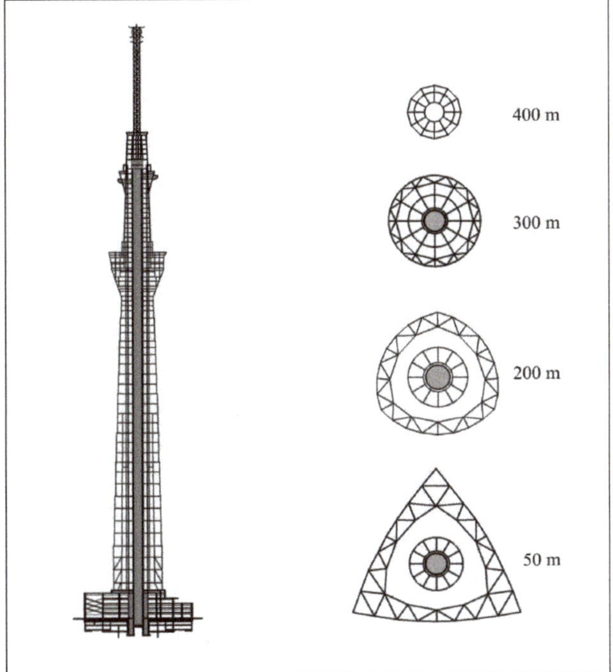

Figure 16. Tokyo Sky Tree: Structural outline (developed by authors on the basis of [44]).

Figure 17. Tokyo Sky Tree: (**a**) An independent steel truss structure employing wide-bore high strength steel pipes; and (**b**) the spiral ramp of the Tembo Gallery in the second observation deck (photograph by authors).

To minimize the impact of wind in the upper part of the tower, a system of tuned mass dampers was installed. In practice, this is characterized by two massive ballast weights, weighing 25 and 40 tons, which were supplied by Mitsubishi Heavy Industries and hung close to the top with large springs and vibration absorbers. As in the case of structures with a reinforced concrete core and an external truss, these two counterweights work on shifting any lateral movement.

4.3. Capital Gate (Abu Dhabi, UAE)—Complex Geometrical Form and Advanced Structural System

Capital Gate Tower is a mixed use high-rise building (hotel and office) with steel and reinforced concrete structure an original very complex geometric form, designed by architectural studio RMJM

(Robert Matthew, Johnson Marshal). The building is 164.7 m high and contains 36 floors at the above-ground level and one underground floor, Figure 18.

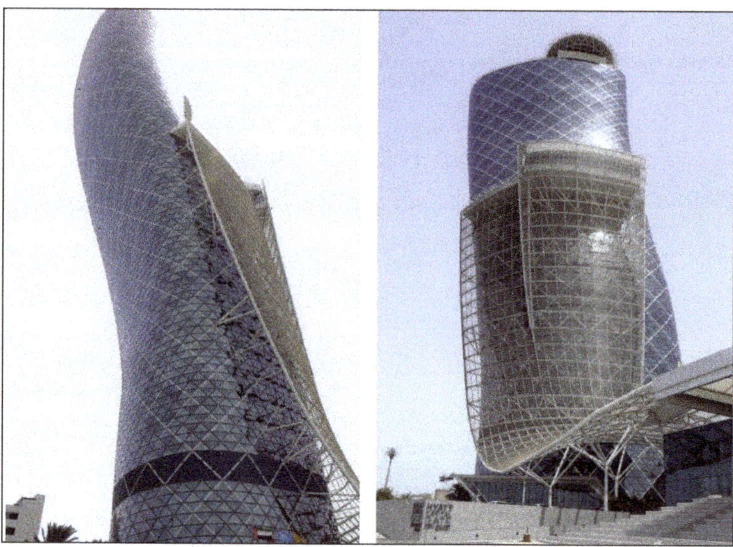

Figure 18. Capital Gate Tower (photograph by authors).

The geometrical form (Figure 19) is meant to represent a swirling spiral of sand, while the curved canopy that runs over the adjoining grandstand creates a wave-like effect reflecting the building's proximity to water and the city's sea-faring heritage.

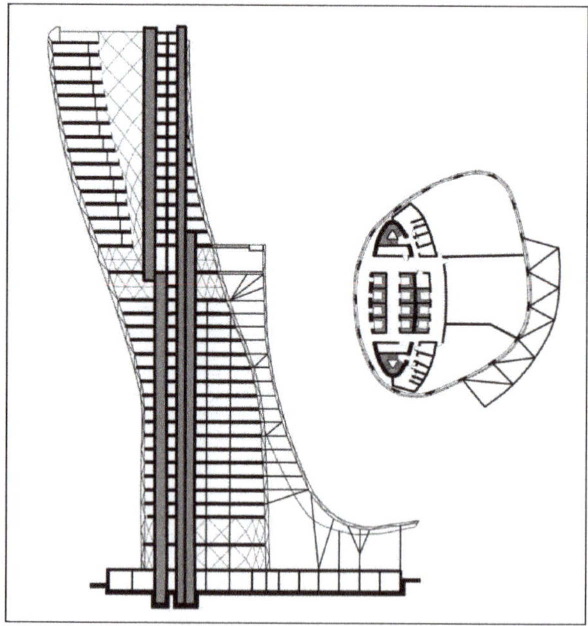

Figure 19. Capital Gate Tower: Floor plan and section (developed by authors on the basis of [45]).

Capital Gate has a raft foundation that is 2 m thick, and cooperates with 490 piles and drilled 30 m underground to withstand wind, seismic and gravitational forces caused because of its inclination [45].

The tower is a two-layer design. Capital Gate's base structure is a vertical concrete core surrounded by an advanced steel diagrid that determines the external form of the tower, Figure 20.

Figure 20. Capital Gate Tower: Diagrid structure—view from the interior (photograph by authors).

On the ground floor, a massive concrete ring beam transfers the thrust of the diagrid into the foundations. The central pre-cambered core leans in the opposite direction to the inclination of the building and straightens with the height. The building inclines by 18 degrees and holds the record for the furthest leaning tower in the world. A triangular structure with a diagonal support beam is applied for the formation of the diagrid.

Steel beams span between the two, and support metal deck and concrete composite floor slabs. An 80 m long cone atrium is formed above the base with an internal steel diagrid attached to the core. The post-tensioned core was designed with vertical cables on one side that are tensioned to counteract the lean on the other side. Steel girders span directly between the external and internal diagrids. The Capital Gate's wind bracing is designed as a separate system.

Capital Gate's most visible sustainable feature is geometrical twist around the building towards the south, to shield the building from direct sunlight. The double skin facade is made of glass panels and steel elements, which are matched to the curvature and form a diamond shape. The special glass minimizes the intense summer glare through the use of ant-glare elements in two silver coatings.

4.4. One World Trade Center (New York, NY, USA)—The Most Safe Building Structure

One World Trade Center is an office tower of reinforced concrete and steel structure designed by David Childs from the architectural studio Skidmore, Owings and Merrill. The building is 541 m high and contains 94 floors at the above-ground level and five under-ground floors, Figure 21. The first floor is occupied by a spacious atrium with a height of 15 m. The skyscraper is topped with a 124 m high spire [46]. There are antennas serving as a broadcasting tool for radio and television transmission in the ring of the spire. The One World Trade Center has a rectangular body with cut corners based on a rectangular base. As the height of the building increases from the base level, its edges form a geometrical form, which consisted of eight elongated isosceles triangles (four up and four down, alternately) [47]. In the middle part, the plan has the form of an ideal octagon. Then it is topped with a glass attic, which in the plan has a square shape with a size of 45 m and is turned 45 degrees to the base. The form of the building refers to the shape of the crystal and similarly to it, breaks the sun's rays.

Figure 21. One World Trade Center (photograph by authors).

The base of the skyscraper is designed on a square plan (61 m), whereas the structure is a square with cut corners. In the central part there is a reinforced concrete core on a square plan (33.5 m) in which staircases and elevators are placed, Figure 22. The reinforced concrete base has a height of 19 floors, Figure 23. From 20 floors, the plan of individual floors and cross-sections of the core change with the shape of the body.

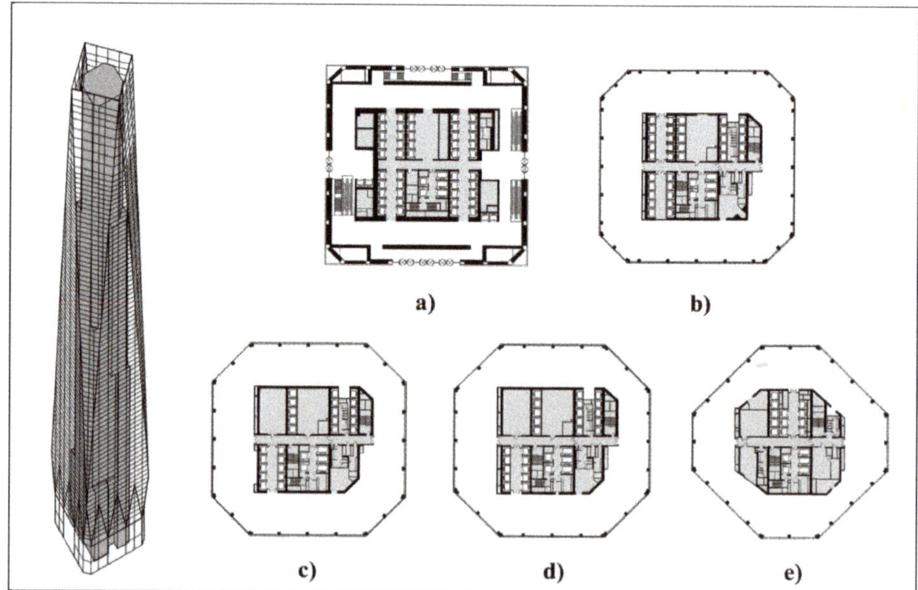

Figure 22. One World Trade Center: Main structure and floor plans: (**a**) Ground plan, (**b**) 45–49 floors, (**c**) 56–59 floors, (**d**) 60–63 floors and (**e**) 80–89 floors (developed by authors on the basis of [46]).

Figure 23. One World Trade Center: Reinforced concrete base with aluminum claddings and glass panels mounted in special profiles (photograph by authors).

The One World Trade Center tower was founded in granite rock with the use of long strip footings and spot footings with a capacity of six tons per square meter. Due to spatial constraints caused by the proximity of existing railway lines, a deeper foundation was necessary to obtain a higher load capacity. The anchorage of the foundation in the rock reached a depth of 24 m to counter the effect of the overturning moment as a result of extreme wind action.

The geometric shape of the One World Trade Center, with the body narrowing together with the height, in combination with cut corners effectively reduces the impact of the wind [47]. The structure of the tower consists of a hybrid system combining a massive reinforced concrete core with a peripheral steel frame [48]. An important element of the construction is a nineteen-story reinforced concrete base, whose massive reinforced concrete walls serve as a hidden safety barrier.

The reinforced concrete wall core in the middle of the building is the main supporting element that carries gravitational loads and counteracts horizontal loads from the wind and seismic effects. Due to its very high stiffness, there was no need for a special vibration damper. The core has a square plan with a length of 33.5 m, which is enough to constitute an independent building. The slab floor system without a column extends between the core and the peripheral steel frame.

Ultra-high-strength concrete used for thick concrete walls of the core, referred to as mass concrete, required a concrete mix to fulfill the most stringent requirements. All mixtures, depending on the needs, contained additional cement materials, fly ash, granulated lump slag cement and silica fume [49]. Due to the high height and slenderness of the building structure, the proportions of the concrete mix for the core were designed taking into account creep, shrinkage and modulus of elasticity.

To increase the safety of the building, designers used much less steel in the construction, and more composite materials. The foundation was made of concrete, but with an admixture of a substance that increases resistance to shocks (green concrete [50]). The 56 m high base is a windowless concrete wall designed to absorb a shock wave from a possible explosion. In building were applied triple laminated Viracon glass, whose high-performance coating reduces the amount of penetrating heat, UV radiation and infrared, while maintaining maximum visible light transmission.

Each lift in 1WTC (One World Trade Center) is protected in the central structure of the core, which is basically a vertical concrete bunker. Partition walls are reinforced with concrete and non-combustible materials. The concrete, which was used, is practically fireproof and the stairwell is subjected to pressure to prevent smoke from entering escape stairs.

The ventilation system was secured with special filters in the event of a terrorist attack using chemical or biological substances. There is also a special staircase for rapid response, used by rescue teams when public stairs are not available. Separate water and fireproof elevators are provided for firemen and security personnel [46]. All escape routes, e.g., stairs, have independent systems: Ventilation, radio and lighting. Air quality controls three thousand sensors. If one detects an excessive amount of carbon dioxide, it immediately transmits a signal to the computer, which automatically pushes more oxygen into the room. In the case of a fire, the building has double-capacity water tanks in relation to the standard requirements for New York buildings. In addition the sprinklers and emergency call buttons are protected by concrete shields.

The One World Trade Center is one of the greenest office buildings in the world, belonging to the fifth generation in terms of energy consumption [51]. Over 30% of the used materials come from regional sources and 25% from post-industrial recycled materials. The building is partially secured by the supply of energy through 12 hydrogen-powered batteries. They generate 4.8 MW of power for themselves and other buildings in the complex. One World Trade Center incorporates not only new architectural and safety standards but also new environmental standards setting a new level of social responsibility in urban design.

4.5. Bahrain World Trade Center (Manama, Bahrain)—Advanced Sustainable Building with Large-Scale Integration of Wind Turbines

Bahrain World Trade Center is a complex of two twin office towers (Figure 24) with a reinforced concrete and steel structure designed by Architectural Studio Atkins. The buildings have an intelligent and environmentally responsive design. The two towers are 240 m high and contain 45 floors at the above-ground level and one underground floor, as shown in Figure 25. The towers have the shape of a sail and support three wind turbines (Figure 26) with a diameter of 29 m, which are supported on three different levels by bridges stretching between them [52]. The Bahrain World Trade Center is the world's first large-scale integration of wind turbines into a building. The towers are integrated on top of a three-story podium and basement. Each tower has a separate continuous piled raft foundation. The raft slabs have a different thickness according to loading and also incorporate lift pits. The raft thickness is 3 m beneath the main cores and the piles are 1.2 m in diameter. Away from the main core, the raft thickness reduces progressively to 2 m, and the piles to 1.05 m. The primary structure comprises two reinforced concrete cores. The main core houses lifts, escape stairs, plant rooms and toilets, and the secondary core houses escape stairs for the MEC (Mechanical) rooms. The floor plates typically have a story height of 3.6 m and are framed with reinforced vertical concrete columns on an 8 m grid and raking columns, which follow the sloping face of the building as it tapers in elevation.

Figure 24. Bahrain World Trade Center (photographs by authors).

The elliptical form of the plan of towers and their profile cause the wind to act on them like the wings of an aircraft, creating a negative pressure that results in an increase in wind speed by up to 30%. This phenomenon has been effectively used in three wind turbines installed in buildings, which are oriented toward the extremely dominant prevailing wind. In conjunction with the shape of the towers and the velocity profile of the wind, the upper and lower turbines produce 109% and 93% of energy when compared to 100% for the middle turbine [53].

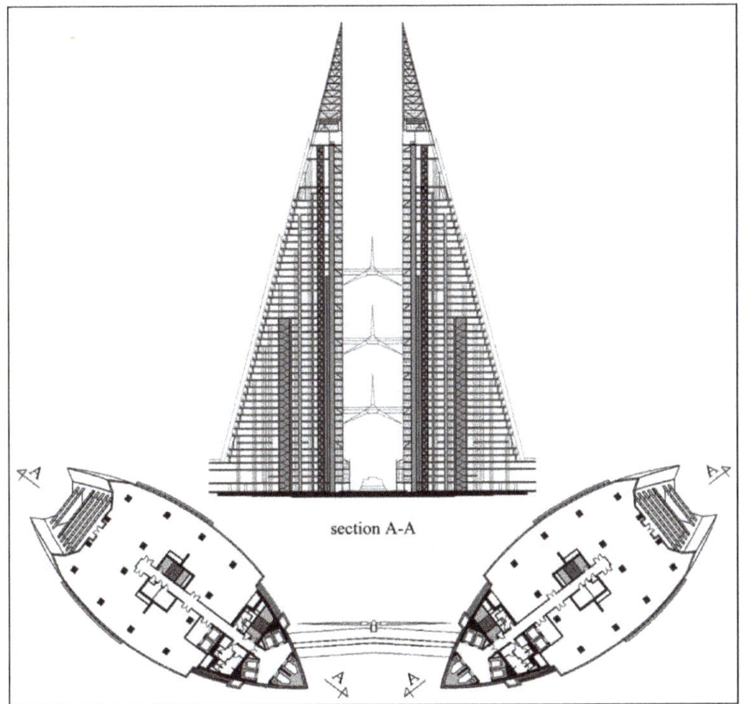

Figure 25. Bahrain World Trade Center: Floor plan and section (developed by authors on the basis of [54]).

Figure 26. Bahrain World Trade Center: Three wind turbines supported on three different levels by bridges stretched between the towers (photographs by authors).

Besides wind energy, the Bahrain World Trade Centre building has other sustainable architecture elements [55]. The glass covering the building is high-quality solar glass with low shading to reduce the building's air temperature. The building is also connected to a district cooling system. The Bahrain World Trade Center takes seawater from the Persian Gulf, which is pumped through a pipeline to chilling units. The units then pass the chilled water through air conditioning units, which cool the air. There are also reflection pools at the entry points of the building, providing local evaporative cooling. Additionally, low-leakage, openable windows installed in the building support the mixed-mode operation in winter months. All these designs are cost-effective and reduce carbon emissions, as opposed to traditional heating and cooling systems.

4.6. Oasia Hotel Downtown (Singapore)—Advanced Green Facade System

Oasia Hotel Downtown is a mixed use high-rise building with reinforced concrete structure designed by studio WOHA Architects. The building is 193.3 m high and contains 27 floors at the above-ground level, Figure 27. Oasia Hotel Downtown has a cylindrical form, and is characterized by a new typology of a tropical high-rise with many terraces with gardens and vertical vegetation. Sky terraces on levels 6, 12, 21 and 27 offer ample public space for recreation social interactions throughout the tower [56].

Figure 27. Oasia Hotel Downtown (photograph by authors).

The building has raft foundation that cooperates with piles. The load-bearing structure is a reinforced concrete slab-column, with four cores located in the corners of the truncated square plan, Figure 28. With the structural cores, which are located in the corners of the building, "sky terraces" enable a unique 360-degree view through the gardens to the city. This would not be possible with a typical centrally located core. By dividing the skyscraper into vertical segments, the sky terraces, together with the green facade, provide an ecological surface area of over 1000% in relation to the surrounding buildings. Sky terraces also serve as huge overhangs, directly shading the terrace below. The openness allows breezes to pass through the building for effective cross-ventilation.

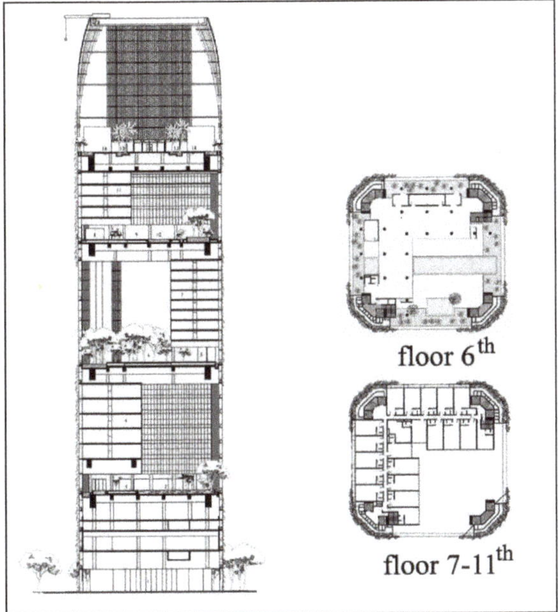

Figure 28. Oasia Hotel Downtown Floor plan and section (developed by authors on the basis of [56]).

The characteristic elements of the building are the L-shaped atriums appearing on every sky terrace that have a height of 21 to 35 m. Each atrium achieves an approximate ratio of height to depth of 1:1, providing a bright and airy environment during daylight. The external aluminum grid in five shades of red, which is attached to the facade of the building, allows the integration of various biological forms and the creation of a green cover (Figure 29). In addition, it also creates a contrast with the lush greenery and blue sky and enables the building to stand out among the numerous skyscrapers in the city center.

Figure 29. Oasia Hotel Downtown: Green facade (photograph by authors).

5. Discussion

Several dozen years ago, architects expressed their visions of an urban future based on new skyscraper typology. Le Corbusier was among them with his 1923 proposal of a series of 60-story office buildings and next for the Radiant City, which extended the concept more by specifying zones for working, living and rest. Le Corbusier's ideas for reshaping city centers have been for many years the foundation for high-rise public housing complexes in United States and Europe [57]. Another tall building visionary was Frank Lloyd Wright, who proposed Illinois Tower with a high of 1 mile. While the idea was physically impossible then, Wright's vision became a touchstone in the ongoing race of skyscrapers height. The twenty first century has brought its own visions of the future [58]. The "Mile-Height" Kingdom Tower built near Jeddah (Saudi Arabia), reflects Wright's own scheme. The nearly mile-height Nakheel Tower proposed for Dubai replicates the features of the emirate's the highest building in the world Burj Khalifa. The Burj Mubarak al-Kabir in Kuwait is proposed to reach 1001 m, symbolizing "Arabian Nights" collection of stories.

The evolution of skyscrapers has become a trend that defines the nature of twenty first century cities. Most notable is the fact that skyscrapers are no longer an American phenomenon. Currently, high-rise buildings have been registered in 72 countries. Each city tries to build a skyscraper as an element of prestige and wealth. At the same time, it becomes an important landmark in the city and plays a major role in the technological development of modern architecture. After a tragic terrorist attack on the twin towers of the World Trade Center in New York, engineers have intensified their efforts to develop a super-safe construction. Technological progress has been clearly targeted, as exemplified by the safest building in the World.

The skyscraper is today in the most common form an Asian phenomenon. Tall buildings have spread well beyond Asia. Mirroring recent changes in the global economy, the Middle East has also adopted the new urban form. Hundreds of residential and mixed use high-rise buildings have been erected in Middle East over the last decade, many incorporating advanced designs and technology.

How high the next generation of skyscrapers will go is difficult to determine. Dubai's Burj Khalifa has 828 m of height, it is 60% taller than Taipei 101, the previous tallest building in the world. A number of supertall buildings will be completed in the next few years. Supertall buildings are extremely complicated to design, require a very robust leasing and sales market, and take more time to construct than most lenders can accept.

An important topic of discussion is how sustainable a high-rise tower can be. Although the trend to achieve net zero energy buildings, for which a balance between energy flow and renewable supply is established, the way to reach the goal is long. New skyscrapers in dense urban areas are generally greener than other types of commercial and residential buildings. They are typically located near mass transit, minimizing negative environmental impacts associated with road traffic. Vertical living also requires less energy for heat. Designers of skyscrapers continue to go to great lengths to minimize the environmental footprint of new buildings. These efforts take many forms: Orienting the building better to the sun and the wind, expanding the use of natural light and ventilation, providing thermal barriers in curtain wall design, maximizing the use of renewable energy (solar and wind), ensuring better collection and utilization of rainwater, and conserving energy through intelligent building managements systems.

The shape of today's skyscraper is particularly notable. The advances in technology and materials have allowed erection not only of very high buildings but also allowed them to take on new and exciting shapes. Today high-rise buildings can twist, lean and turn back on themselves. These shapes are chosen for visual effect, but occasionally they contribute to minimizing wind loads by improving a building's aerodynamic properties.

The skyscraper of the future will have a mixed-use function. The increasing popularity of mixed-use complexes, and in particular the growth in residential towers, has left its mark on every aspect of skyscraper design and construction. In terms of structure, concrete has now overtaken steel as the most prevalent skyscraper material. In terms of construction, mixed-use buildings are more difficult and costly to erect than single-purpose one. In terms of design, mixed-use buildings present the added complexity of segregating users and uses, taken into account pedestrian flows, vertical transportation, loading and other services. In designing these buildings, architects must often deal with multiple building code provisions, as standards for commercial and residential occupancy often differ.

Finally, it must be stated that knowledge about the advantages of high-rise buildings should remind us that although they are structures with advanced technology, they also cause shade that prevents light from penetrating surrounding areas. Among other things, this problem is widely discussed by urban planners in New York and concerns the shading of Central Park by newly built skyscrapers. Therefore, a question arises regarding the future and direction of the development of high-rise buildings. However, one thing is certain, regardless of their future, they will always be a catalyst for technological development.

6. Conclusions

Technological innovations used in high-rise buildings can be manifested in many areas: Geometric form, construction, materials, vibration damping systems and energy efficiency. The development of computer technology has facilitated the design of high-rise buildings with complicated structural and functional solution forms. Increased computing power has allowed the creation of more advanced engineering programs, which for building models better simulate the actual behavior of a structure. This can especially be seen in high-rise buildings erected in the last years. Modern designs have broken the stereotypes of high-rise buildings in terms of history and tradition. An important aspect in the design of various architectural forms is the determination of the relationship between the shape of a building and the quality of its construction. Very often curvaceous shapes are inspired by various forms, which can occur in nature Capital Gate (Abu Dhabi, UAE) and Burj Khalifa (Dubai, UAE). A tall building, due to its shape, can be a very distinctive landmark in its environment and thus an easily recognizable building.

A simple measure of building quality is resistance to oscillating sway, resonance with the wind and also dynamic flexural stiffness. Complex shapes and requirements resulting from the height of buildings cause an increase in the load of constructional elements. Enhanced static and dynamic effects must be reflected in a properly selected construction system. At the end of the 19th century, the efficiency of diagonally braced elements that counteract lateral forces was taken into account when designing the first high-rise buildings. The use of the diagrid construction system is not new, but there is now a noticeable increase in the interest and application of this system in the design of tall buildings with large spans, especially concerning complex geometry. Diagrid structures do not require a core with high shear stiffness, because shear forces can be carried by the diagonal elements located on the perimeter of the structure. Perimeter diagrids carry horizontal and gravity loads and are used to support the edges of slab floors. This system is part of the trend of spectacular aesthetics, which can be exemplified by very iconic buildings (Hearst Tower, Capital Gate Tower, Doha Tower, The Bow, Swiss Re, etc.).

Undoubtedly, the least-resistant construction for an earthquake is a skyscraper, which is a certain paradox in comparison with their number in the world. The most modern skyscrapers in Tokyo are able to withstand earthquakes of over seven degrees on the Richter scale. Of course, more forces affect a building with a larger earthquake, and its construction therefore experiences larger displacements. A building's response to earthquakes is vibrations in the form of sinusoidal motion. In order to counteract both these forces and the impact of wind, apart from a rigid construction, very advanced technologies of damping devices are used. For example, the foundations of these buildings (Maison Hermes Tokyo) are mounted with a system of spring or elastomer vibration dampers, due to which tectonic movements affect the upper part of the building to a lesser extent. In addition, as presented by the characteristics of high-rise buildings, viscous oil dampers (Mode Gakuen Cocoon), anti-buckling steel stabilizers (Midtown Tower, Roppongi Hills, Kabukiza Tower) and tuned mass dampers (Tokyo Tree Tower) are used in various levels of these buildings. When using all these supporting elements, it is most important that the location of the center of gravity of the building does not change during earthquakes.

A very important aspect associated with the technological development of high-rise buildings is the safety of their users. One World Trade Center in New York is the most advanced building in the world when it comes to security technology, setting new standards for the design of high-rise buildings.

Sustainability is also a major issue concerning high-rise buildings [59]. It is strongly required to use sustainable concepts and applicable technology for reducing energy consumption and CO_2 emissions.

Covering the walls of a building with greenery affects the changing of microclimate, produces oxygen, absorbs CO_2 and captures particles of pollution [60,61]. Currently, plants are becoming an appropriate facade material in the creation of architecture. Their use is planned and dedicated to achieving both a specific aesthetic and ecological effect.

By the nature of high-rise buildings, it is very difficult to achieve a low energy building. High energy consumption in high-rise buildings has influenced the search for innovative solutions aimed at improving energy efficiency in this area. The research was focused on solutions based on renewable energy sources. Currently, photovoltaic panels and wind turbines are primarily used to produce electricity for a building's own needs. Designing the building together with an integrated wind turbine constituted a major design challenge for the Bahrain World Trade Center building. The project had to take into account the wind speed and direction, which occur in a given area, and as a result change parameters depending on the geometry of the building. There are many factors that affect the flow of wind in these installations. Among them are not only the location and occurring terrain, but also the shape of the building and its dimensions. Skyscrapers not only favor the development of innovative solutions, but also aim to improve human comfort when visiting a building or the safety of people residing in it. For example, the HMS (home management system) system is used, which integrates the majority of installations in an apartment. The organization of operation and modern equipment of high-rise buildings means that they belong to the category of "smart buildings".

Author Contributions: Both authors contributed the same to the analysis of the problem, discussion and writing the paper.

Funding: This research received no external funding.

Conflicts of Interest: The authors declare no conflict of interest.

References

1. Ali, M.M.; Moon, K.S. Advances in Structural Systems for Tall Buildings: Emerging Developments for Contemporary Urban Giants. *Buildings* **2018**, *8*, 104. [CrossRef]
2. Rychter, Z. Influence of shape of the skyscrapers on the quality of construction. *Archit. Artibus* **2013**, *2*, 33–38. (In Polish)
3. Dai, L.; Liao, B. Innovative High Efficient Construction Technologies in Super High Rise Steel Structure Buildings. *Int. J. High-Rise Build.* **2014**, *3*, 205–214.
4. Al-Kodmany, K. The Sustainability of Tall Building Developments: A Conceptual Framework. *Buildings* **2018**, *8*, 7. [CrossRef]
5. Charter of European Cities & Towns towards Sustainability. Available online: Portal.uur.cz/pdf/aalborg-charter-1994.pdf/ (accessed on 10 May 2018).
6. *Solar Energy in Architecture and Urban Planning; Architecture & Design*; Herzog, T. (Ed.) Prestel Verlag: Munchen, German; New York, NY, USA, 1996.
7. Survey on Architectural Policies in Europe. Available online: www.efap-fepa.org/docs/EFAP_Survey_Book_2012.pdf (accessed on 10 May 2018).
8. Foster, N. *Architecture and Sustainability*; Foster + Partners: London, UK, 2003; pp. 1–12.
9. Renzo Piano Building Workshop—Piece by Piece. Available online: https://www.snf.org/media/.../Exhibition-Guide.pdf/ (accessed on 10 June 2019).
10. Eteghad, A.N.; Raviz, S.R.H.; Guardiola, E.U.; Aira, A.A. Energy Efficiency in Thomas Herzog's Architecture: From Interdisciplinary Research to Performance Form. In Proceedings of the International Conference on Architecture and Civil Engineering, Singapore, 13–14 April 2015; pp. 8–14.
11. Moe, K. *An Architectural Agenda for Energy*; Routledge Publisher: New York, NY, USA, 2013; pp. 1–301.
12. Amiri, A.; Ottelin, J.; Sorvari, J. Are Leed-Certified Buildings Energy—Efficient in Practice? *Sustainability* **2019**, *11*, 1672. [CrossRef]
13. Vahedi, A. Nature as a Source of Inspiration of Architectural Conceptual Design. Master's Thesis, Eastern Mediterranean University, Gazimağusa, Cyprus, 2009.
14. Nassery, F. Geometric forms of contemporary architecture. *Czas. Tech.* **2010**, *107*, 284–289. (In Polish)

15. Moon, K.S. Structural Design and Construction of Complex Shaped Tall Buildings. *Int. J. Eng. Technol.* **2015**, *7*, 30–35. [CrossRef]
16. Gűnel, M.H.; Ilgin, H.E. *Tall Buildings: Structural Systems and Aerodynamic Form*; Routledge Publisher: London, UK, 2014; pp. 1–214.
17. Yadav, S.; Garg, V. Advantage of Steel Diagrid Building over Conventional Building. *Int. J. Civ. Struct. Eng. Res.* **2015**, *3*, 394–406.
18. Varsani, H.; Pokar, N.; Gandhi, D. Comparative Analysis of Diagrid Structural System and Conventional Structural System for High Rise Building. *Int. J. Adv. Res. Eng. Sci. Technol.* **2015**, *2*, 1–4.
19. Kawecki, J.; Maslowski, R. Application of quasi-active and hybrid passive dampers to reduce seismic and paraseismic vibrations of buildings—Overview of solutions. *Czasopismo Techniczne* **2010**, *1*, 59–67. (In Polish)
20. Lago, A.; Wood, A.; Trabucco, D. *Damping Technologies for Tall Buildings: New Trends in Comfort and Safety*; Elsevier: Amsterdam, The Netherlands, 2015.
21. Infanti, S.; Robinson, J.; Smith, R. Viscous Dampers for High-Rise Buildings. In Proceedings of the World Conference on Earthquake Engineering, Beijing, China, 12–17 October 2008; pp. 1–8.
22. Kareem, A.; Kijewski, T.; Tamura, Y. Mitigation of Motions of Tall Buildings with Specific Examples of Recent Applications. *Wind Struct.* **1999**, *2*, 201–251. [CrossRef]
23. Asai, T.; Chang, C.M.; Phillips, B.M.; Spencer, B.F. Real-time hybrid simulation of a smart outrigger damping system for high-rise buildings. *Eng. Struct.* **2013**, *57*, 177–188. [CrossRef]
24. Bester, N. Concrete for high-rise buildings: Performance requirements, mix design and construction considerations. *Struct. Concr. Prop. Pract.* 2013, pp. 1–4. Available online: https://www.researchgate.net/publication/263889511_Concrete_for_high-rise_buildings_Performance_requirements_mix_design_and_construction_considerations (accessed on 10 April 2019).
25. Sev, A.; Çirpi, M.E. Innovative Technologies and Future Trends in Tall Building Design and Construction. *ISITES*. 2014, pp. 1114–1123. Available online: akademikpersonel.kocaeli.edu.tr/ (accessed on 10 May 2019).
26. Marshall, R.; Knapp, G. Recent Advances in Technologies, Techniques and Materials. In *The Tall Building Reference Book*; Routledge: New York, NY, USA, 2013; pp. 167–180.
27. Tsujii, M.; Kanno, R. Advances in Steel Structures and Steel Materials in Japan. In *Nippon Steel & Sumitomo Metal Technical Report*; Nippon Steel & Sumitomo Metal Corporation: Chiyoda Tokyo, Japan, 2016; pp. 3–12.
28. Ritter, A. *Smart Materials in Architecture. Interior Architecture and Design*; Birkhaűser Architecture: Basel, Switzerland, 2007; pp. 1–191.
29. Lee, C.; Hong, T.; Lee, G.; Jeong, J. Life-Cycle Cost Analysis on Glass Type of High-Rise Buildings for Increasing Energy Efficiency and reducing CO_2 Emissions in Korea. *J. Constr. Eng. Manag.* **2012**, *138*, 897–904. [CrossRef]
30. Windows and Glazing. Whole Building Design Guide. Available online: https://www.wbdg.org/resources/windows-and-glazing (accessed on 10 May 2018).
31. Kuhn, T.E. State of the art of advanced solar control devices for buildings. *Sol. Energy* **2017**, *154*, 112–133. [CrossRef]
32. Wehle-Strzelecka, S. Solar Architecture—Contemporary Technologies and Aesthetics. *Czas. Tech.* **2007**, *104*, 313–320. (In Polish)
33. Lotfabadi, P. Solar considerations in high-rise buildings. *Energy Build.* **2015**, *89*, 183–195. [CrossRef]
34. Raczynski, M. The Types of Modern Glass Façade. In Search of The Idea of Continuity of The Cuboidal Form. *Przestrz. Forma* **2013**, *19*, 121–132.
35. Barnas, J. Double-Skin Façades—The Shaping of Modern Elevations—Technology and Materials. *Tech. Trans. Archit.* **2014**, *111*, 5–15.
36. Knaack, U.; Klein, T.; Bilow, M. *Façades*; 010 Publishers: Rotterdam, The Netherlands, 2008; pp. 6–128.
37. Zobec, M.; Colombari, M.; Kragh, M. Introduction of advanced façade technology. In Proceedings of the World Renewable Energy Congress, Cologne Germany, 29 June–5 July 2002; pp. 1–10.
38. Yadav, R.; Sarkar, J.; Jadhav, K. Innovative Façade Design Strategies. *Archit. Time Space People* **2014**, *9*, 18–27.
39. Burj Khalifa—Structura Elements. Available online: https://www.burjkhalifa.ae/en/the-tower/structures/ (accessed on 10 May 2017).
40. Laurens, S.; François, R. Cathodic protection in reinforced concrete structures affected by macrocell corrosion: A discussion about the significance of the protection criteria. *RILEM Tech. Lett.* **2017**, *2*, 27–32. [CrossRef]

41. Berheimer, C.; Clariday, M.; Lawley, S.; Mengers, C.; Robalino, C. Burj Khalifa. Available online: https://www.scribd.com/document/398939620/Burj-Khalifa (accessed on 10 May 2017).

42. Konishi, A. Structural Design of Tokyo Sky Tree. In Proceedings of the CTBUH World Conference, Seoul, Korea, 10–12 October 2011; pp. 513–520.

43. Abe, M.; Kawaguchi, M. Structural Mechanism and Morphology of Timber Towers in Japan. *J. Asian Archit. Build. Eng.* **2002**, *32*, 25–32. [CrossRef]

44. Keii, M.; Konishi, A.; Kagami, Y.; Watanabe, K.; Nakanishi, N.; Esaka, Y. Tokyo Sky Tree—Structural Outline of Terrestrial Digital Broadcasting Tower. *Steel Constr. Today Tomorrow* **2010**, *31*, 5–9.

45. Schofield, J. Case Sudy: Capital Gate, Abu Dhabi. *CTBUH J.* **2012**, *II*, 13–17.

46. Lewis, K.; Holt, N. Case Study: One World Trade Center, New York. *CTBUH J.* **2011**, *III*, 14–19.

47. Golasz-Szolomicka, H.; Szolomicki, J. One World Trade Center in New York—A modern ecological office building with a hybrid structure. *Bud. Archit.* **2018**, *17*, 141–158. (In Polish)

48. Rahimian, A.; Eilon, Y. The Rise of One World Trade Center. Global Interchanges: Resurgence of the Skyscraper City. In Proceedings of the CTBUH International Conference, New York, NY, USA, 26–30 October 2015; Wood, A., Malott, D., Eds.; pp. 66–71.

49. Li, G.; Zhao, X. Properties of concrete incorporating fly ash and ground granulated blast-furnace slag. *Cem. Concr. Compos.* **2003**, *25*, 293–299. [CrossRef]

50. Pietrzak, A. Proecological technologies in building structure on the example of "green concrete". *Bud. Zoptymalizowanym Potencjale Energetycznym* **2014**, *1*, 86–93. (In Polish)

51. Oldfield, P.; Trabucco, D.; Wood, A. Five Energy Generations of Tall Buildings: A History Analysis of Energy Consumption in High Rise Buildings. *J. Archit.* **2009**, *14*, 591–613. [CrossRef]

52. Smith, R.; Killa, S. Bahrain World Trade Center (BWTC): The First Large-Scale Integration of Wind Turbines in a Building. Struct. *Des. Tall Spec. Build.* **2007**, *16*, 429–439. [CrossRef]

53. Chaudhry, H.N.; Calutit, J.K.; Hughes, B.R. Numerical Analysis of the integration of Wind Turbines into the Design of the Built Envirinment. *Am. J. Eng. Appl. Sci.* **2014**, *7*, 355–365. [CrossRef]

54. Bahrain World Trade Center. Available online: htpps://bahrainwtc.wordpress.com/drawings-diagrams/ (accessed on 15 July 2017).

55. Alnaser, N.W. Towards Sustainable Buildings in Bahrain, Kuwait and United Arab Emirates. *Open Constr. Build. Technol. J.* **2008**, *2*, 30–45. [CrossRef]

56. Wong, M.S.; Hassel, R.; Phua, H.W. Oasia Hotel Downtown: A Tall Prototype for the Tropics. *CTBUH J.* **2018**, *III*, 12–19.

57. Ascher, K. *The Heights: Anatomy of a Skyscraper*; Penguin Book: New York, NY, USA, 2013; pp. 1–207.

58. Dupre, J. *Skyscrapers. A History of the World's Most Extraordinary Buildings*; Black Dog and Leventhal Publishers: New York, NY, USA, 2013; pp. 1–176.

59. Navaei, F. An Overview of Sustainable Design Factors in High-Rise Buildings. *Int. J. Sci. Technol. Soc.* **2015**, *3*, 18–23. [CrossRef]

60. Widiastuti, R.; Prianto, E.; Budi, W. Performance Evaluation of Vertical Gardens. *Int. J. Archit. Eng. Constr.* **2016**, *5*, 13–20.

61. Kmieć, M. Green Wall Technology. *Tech. Trans. Archit.* **2014**, *111*, 47–60.

 *buildings*

Article

Welds Assessment in K-Type Joints of Hollow Section Trusses with I or H Section Chords

Miroslaw Broniewicz * and Filip Broniewicz

Faculty of Civil and Environmental Engineering, Bialystok University of Technology, Wiejska 45 E, 15-351 Bialystok, Poland; f.broniewicz@doktoranci.pb.edu.pl
* Correspondence: m.broniewicz@pb.edu.pl

Received: 15 December 2019; Accepted: 27 February 2020; Published: 3 March 2020

Abstract: The use of hollow section structures has received considerable attention in recent years. Since the first publication of CIDECT (International Committee for the Development and Study of Tubular Structures), additional research results became available, especially concerning the design of welds between members of trusses joints. To assess the capacity of welded joints of trusses between braces made of hollow sections and I-beam chords, the effective lengths of the welds should be estimated and their location around the braces and the forces acting on individual weld's sections. The objective of this paper is to present the most up-to-date information to designers, teachers, and researchers according to the design of welds for certain K and N overlapped joints between rectangular hollow section (RHS) braces and I- or H-section chord.

Keywords: steel trusses; semi-rigid joints; RHS braces; H-section chords; overlapped joints; resistance of welds

1. Introduction

The structural elements made of circular or rectangular (square) hollow sections are usually used for lattice structures (roof trusses and lattice frames) and less often to Vierendeel beams. Welds between such elements (e.g., between brace members and chords) are designed as butt welds or fillet welds. Tubular steel structures are characterized by numerous advantages, among which the most important are low weight, favorable aerodynamic shape, aesthetic appearance, and very good strength properties [1,2].

The current European standards concerning the design of steel construction contain many of the principles and recommendations referring to the design of welded connections in nominally pinned or rigid joints. However, in the range of semi-rigid joints, made of hollow sections, principles are general and recommendations too simplistic [3,4].

In the case of truss structures, the general principle is to design welds of such thickness that their resistance is not less than the resistance of joining member walls [5]. This principle is satisfied by full butt welds, which cannot be performed in all cases, or the thick fillet weld, without specifying what their thickness must be taken. As a specific recommendation it is indicated that we can take welds of thickness less than mentioned in general rule, but without any information on how to determine their value.

Nowadays, there are basic recommendations for assessing the effective lengths of fillet welds in K-type gap joints made of rectangular hollow sections [6,7]. These recommendations were also extended to the T-type of joints [8].

According to the general rule, we should always use thick welds in all design situations, even when it is totally unnecessary. The use of thick fillet welds is often a reason for the introduction of large welding stresses, preventing proper execution of construction and increasing labor costs. However, the use of butt welds is often not advisable, because it requires chamfering the edges of joining members.

The shaping of overlapped joints of trusses has been widely discussed in [9,10]. The basis for calculating their capacity provides European codes and other standardization documents [11,12].

In the European standards, rules to determine the fillet weld strength in welded joints made of hollow sections have been presented in a general way without giving detailed design recommendations.

The Canadian publication written by Packer and Henderson [9] presents information on determining effective lengths only for K joints with a spacing between the braces, whereas in the case of K-type overlapped joints no design recommendations have been presented so far. An uncomplicated procedure of assessing effective lengths has been presented in IIW recommendations [13] and in publications [2,14], as well as repeated in the ISO standard [12].

In this paper, the authors based on the cited references, suggest an estimated assessment of the resistance of the weld of K-type overlapped joints with rectangular hollow section (RHS) braces and the chords made of I- or H-section, thereby extending the use of the calculation method shown in [15–17]. The method of evaluating both design cases is the same, but there are some differences in determining effective widths of the welds, resulting from different flexibility of the chord walls. These differences will be presented further and their impact on the strength of effective lengths of welds will be discussed.

2. Method of Determining the Resistance of Fillet Welds

Welded joints made of hollow sections should be made using fillet or butt welds, or combination of the two, laid on the perimeter of the profile. In overlapped joints, the covered part of the member need not be welded, when the components of axial forces in the brace members perpendicular to the chord axis do not differ by more than 20% [18].

According to Dexter and Lee [19], the resistance of the overlapped CHS joints with the hidden part welded was about 10% higher.

To simplify the calculation of the welds in hollow section structures European standard recommends the design of the welds in such a way, that the weld resistance per unit length of the member perimeter should not be less than the resistance of that member also calculated per unit length of perimeter. This condition is met when butt or fillet welds which are used have such thickness, that their resistance is equal to the resistance of connected members. Methods of estimating the thickness of fillet welds that meet this requirement are given in [16].

The European standard [20] suggests, in cases where the design of a full butt weld or the adequate fillet weld is not required, that the thickness of the weld may be reduced, on condition that the resistance of such weld and its rotation capacity are checked, considering only the weld effective lengths.

In Figure 1, using the IIW guidance [13] the layout of fillet welds in the K-type overlapped joint between the RHS braces and the H chord is shown. An assumption was made that the value of the component force perpendicular to the chord transferred directly through the welds connecting the brace members is equal to (Figure 1a):

$$\Delta K_i = \alpha K_i \sin \theta_i \text{ when } 25\% \leq \lambda_{ov} \leq 80\%,$$
$$\Delta K_i = K_i \sin \theta_i \text{ when } \lambda_{ov} > 100\%, \tag{1}$$

where $\alpha = q/p$ and $0.25 \leq \alpha \leq 0.80$, $\lambda_{ov} = (q/p) \cdot 100\%$ in %, q—the overlapping surface of braces projected on the face of the chord, p—length of contact area between the overlapping brace and the chord (the procedure of assessment of welds resistance in K-type joints made of hollow section is presented in [21]).

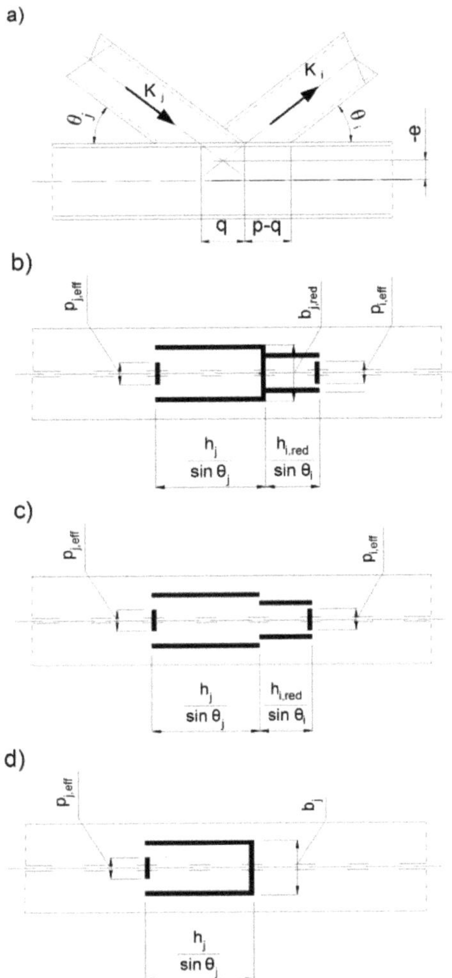

Figure 1. The connection of hollow section brace members to I- or H-section chords: (**a**) The standard joint: (**b**) The layout of fillet welds joining the rectangular hollow section (RHS) braces to the H chords when $25\% \leq \lambda_{ov} \leq 80\%$ and the hidden part of the connection is welded; (**c**) the layout of fillet welds joining the RHS braces to the H chords when $25\% \leq \lambda_{ov} \leq 80\%$ and the hidden part of the connection does not have to be welded; (**d**) layout of fillet welds joining the RHS braces to the H chords when $\lambda_{ov} > 100\%$.

The remaining part of the load component perpendicular to the chord is

$$red\Delta K_j = K_j \sin\theta_j - \alpha K_i \sin\theta_i \text{ when } 25\% \leq \lambda_{ov} \leq 80\%,$$
$$red\Delta K_j = K_j \sin\theta_j - K_i \sin\theta_i \text{ when } \lambda_{ov} > 100\%, \tag{2}$$
$$red\Delta K_j = 0 \text{ in the case of no external load applied to the joint.}$$

The values of effective lengths of welds are determined as follows (Figure 2):

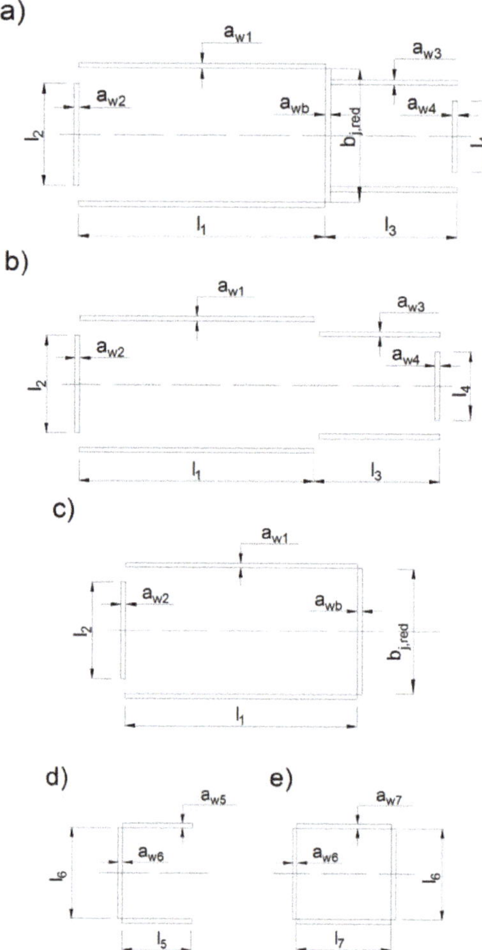

Figure 2. Effective lengths of welds: (**a**) the layout of fillet welds joining the RHS braces to the H chords when $25\% \leq \lambda_{ov} \leq 80\%$ and the hidden part of the connection is welded; (**b**) the layout of fillet welds joining the RHS braces to the H chords when $25\% \leq \lambda_{ov} \leq 80\%$ and the hidden part of the connection is not welded; (**c**) the layout of fillet welds joining the RHS braces to the H chords when $\lambda_{ov} > 100\%$; (**d**) welds between the braces in the case of the partial overlap; (**e**) welds between the braces in the case of the full overlap.

1. In connections of braces with the flange of the chord (Figure 2a–c):

$$l_1 = h_j / \sin\theta_j, \quad b_{j,red} = b_j - 2a_w,$$
$$l_2 = p_{j,eff} = t_w + 2r + 7t_f f_{y0} / f_{yj}, \text{ but } p_{j,eff} \leq b_j,$$

$$l_3 = h_{i,red} / \sin\theta_i = (1 - \alpha)h_i / \sin\theta_i,$$
$$l_4 = p_{i,eff} = t_w + 2r + 7t_f f_{y0} / f_{yi}, \text{ but } p_{i,eff} \leq b_i.$$

$$(3)$$

2. In the direct connection between braces (Figure 2d,e):

- In the case of the partial overlap:

$$l_5 = \frac{q}{\left(1 + tg\theta_j/tg\theta_i\right)\cos\theta_j}, \quad l_6 = b_i \tag{4}$$

- In the case of the complete overlap:

$$l_7 = h_i/\sin\left(\theta_i + \theta_j\right), l_6 = b_i \tag{5}$$

where θ_i, θ_j—inclination angles of the overlapping and overlapped braces in relation to the chord, f_{y0}—the yield strength of the chord, b_i, h_i—respectively, the width and the height of the section of the overlapping brace, b_j, h_j—respectively, the width and the height of the section of the overlapped brace, t_i—the wall thickness of the overlapping brace, t_j—the wall thickness of the overlapped brace, t_f—the flange thickness of the I-section, t_w—the web thickness of I-section, r—the radius of the I-section.

Areas of cross-sections of effective lengths of welds are (Figure 1): $A_1 = l_1 a_{w1}$, $A_2 = l_2 a_{w2}$, $A_3 = l_3 a_{w3}$, $A_4 = l_4 a_{w4}$, $A_{j,red} = b_{j,red} a_{wb}$, $A_5 = l_5 a_{w5}$, $A_6 = l_6 a_{w6}$, $A_7 = l_7 a_{w7}$, where a_{w1}, a_{w2}, a_{w3}, a_{w4}, a_{wb}, a_{w5}, a_{w6}—thicknesses of welds.

In the case of $25\% \leq \lambda_{ov} \leq 80\%$ design situation, all welds are made (also in hidden part) in the place of the splice of braces with the chord. The sections of fillet welds are loaded by the component forces in braces parallel to the chord (Figure 3a).

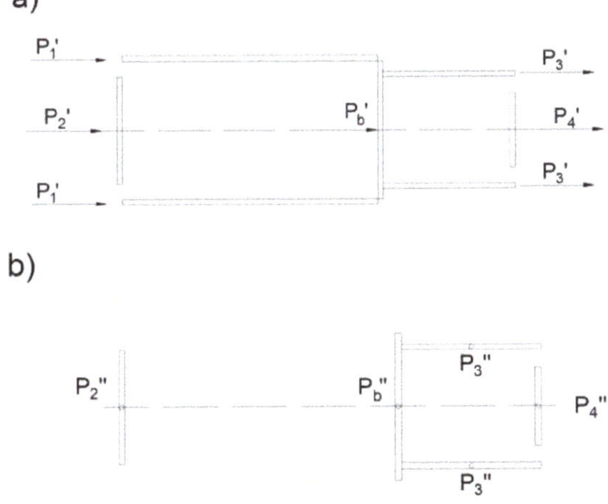

Figure 3. Component loads in the welds of braces: (**a**) parallel to the chord; (**b**) perpendicular to the chord.

Loads for individual effective lengths are

$$P_1' = \left(K_j \cos\theta_j + K_i \cos\theta_i\right)A_1/\sum A, \tag{6a}$$

$$P_2' = \left(K_j \cos\theta_j + K_i \cos\theta_i\right)A_2/\sum A, \tag{6b}$$

$$P_3' = \left(K_j \cos\theta_j + K_i \cos\theta_i\right)A_3/\sum A, \tag{6c}$$

$$P_4' = \left(K_j \cos\theta_j + K_i \cos\theta_i\right)A_4/\sum A, \tag{6d}$$

$$P_b' = \left(K_j \cos\theta_j + K_i \cos\theta_i\right) A_{j,red} / \sum A, \tag{6e}$$

where K_j and K_i are the designed axial loads acting respectively in overlapped and overlapping braces and

$$\sum A = 2A_1 + A_2 + 2A_3 + A_4 + A_{j,red}. \tag{7}$$

In the same design situation, the fillet weld sections are loaded by load components in braces perpendicular to the chord, as shown in Figure 3b. Forces loading effective lengths can be calculated based on equations:

$$P_1'' = 0, \tag{8a}$$

$$P_2'' = red\Delta K_j \cdot A_2 / \left(A_2 + A_{j,red}\right), \tag{8b}$$

$$P_3'' = (1 - \alpha)\Delta K_i \cdot A_3 / (2A_3 + A_4), \tag{8c}$$

$$P_4'' = (1 - \alpha)\Delta K_i \cdot A_4 / (2A_3 + A_4), \tag{8d}$$

$$P_b'' = red\Delta K_j \cdot A_{j,red} / \left(A_2 + A_{j,red}\right), \tag{8e}$$

Components of loads in welds directly between the braces, in the case of the partial overlap are determined from equations (Figure 2d):

- Loads parallel to the overlapped brace axis:

$$P_5' = \Delta K_i \cdot A_5 \sin\theta_j / (2A_5 + A_6), \tag{9}$$

$$P_6' = \Delta K_i \cdot A_6 \sin\theta_j / (2A_5 + A_6). \tag{10}$$

- Loads perpendicular to the overlapped brace axis:

$$P_5'' = \Delta K_i \cdot A_5 \cos\theta_j / (2A_5 + A_6), \tag{11}$$

$$P_6'' = \Delta K_i \cdot A_6 \cos\theta_j / (2A_5 + A_6). \tag{12}$$

Components of loads in welds directly between the braces, in the case of the full overlap are determined from equations (Figure 2e):

- Loads parallel to the overlapped brace axis:

$$P_6' = \Delta K_i \cdot A_6 \sin\theta_j / (2A_6 + 2A_7), \tag{13}$$

$$P_7' = \Delta K_i \cdot A_7 \sin\theta_j / (2A_6 + 2A_7). \tag{14}$$

- Loads perpendicular to the overlapped brace axis:

$$P_6'' = \Delta K_i \cdot A_6 \cos\theta_j / (2A_6 + 2A_7) \tag{15}$$

$$P_7'' = \Delta K_j \cdot A_7 \cos\theta_j / (2A_6 + 2A_7). \tag{16}$$

Stresses in welds caused by the force parallel to the chord at the partial overlap of $25\% \leq \lambda_{ov} \leq 80\%$:
1. In longitudinal welds (Figure 4a):

- In the case of welds placed by the overlapped brace's walls:

$$\sigma' = 0, \ \sigma_\perp' = \tau_\perp' = 0, \ \tau_{II}' = P_1' / (a_{w1} \cdot l_1). \tag{17}$$

- In the case of welds placed by the overlapping brace's walls:

$$\sigma' = 0, \; \sigma'_\perp = \tau'_\perp = 0, \; \tau'_{II} = P'_3 / (a_{w3} \cdot l_3) \tag{18}$$

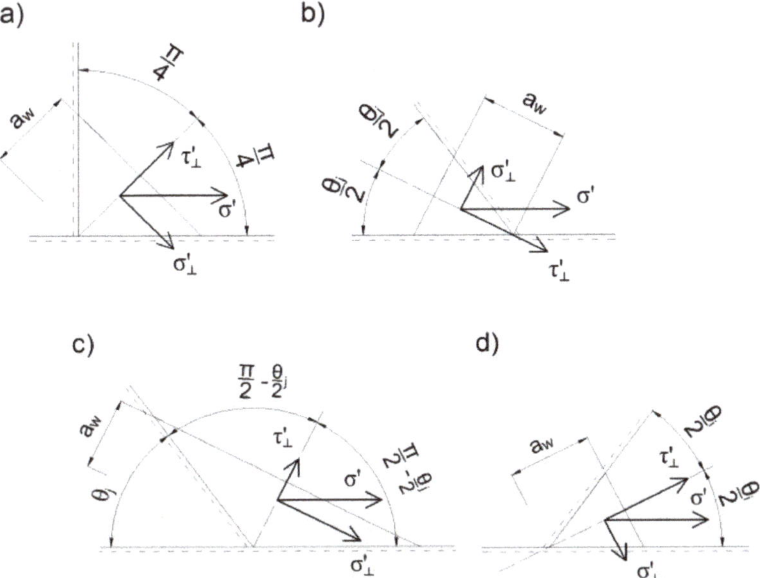

Figure 4. Stresses in welds caused by the load parallel to the chord: (**a**) in longitudinal welds; (**b**) on the not fully cooperating overlapped brace's transverse length; (**c**) on the fully cooperating transverse length of the overlapped brace; (**d**) on the not fully cooperating transverse length of the overlapping brace.

2. On the not fully cooperating overlapped brace's transverse length (Figure 4b):

$$\sigma' = P'_2 / (a_{w2} l_2), \; \sigma'_\perp = \sigma' \sin(\theta_j/2), \; \tau'_\perp = \sigma' \cos(\theta_j/2), \; \tau'_{II} = 0. \tag{19}$$

3. On the fully cooperating transverse length of the overlapped brace (Figure 4c):

$$\sigma' = P'_b / (a_{wb} b_{j,red}), \; \sigma'_\perp = \sigma' \cos(\theta_j/2), \; \tau'_\perp = \sigma' \sin(\theta_j/2), \; \tau'_{II} = 0. \tag{20}$$

4. On the not fully cooperating overlapping brace 's transverse length (Figure 4d):

$$\sigma' = P'_4 / (a_{w4} l_4), \; \sigma'_\perp = \sigma' \sin(\theta_i/2), \; \tau'_\perp = \sigma' \cos(\theta_i/2), \; \tau'_{II} = 0. \tag{21}$$

Stresses in welds of load perpendicular to the chord at the partial overlap of $25\% \le \lambda_{ov} \le 80\%$:

1. In longitudinal welds (Figure 5a):

- In the case of welds placed by the walls of the overlapped brace:

$$\sigma'' = \frac{P_1''}{a_{w1} l_1}, \; \sigma_\perp'' = \frac{\sigma''}{\sqrt{2}}, \; \tau_\perp'' = -\frac{\sigma''}{\sqrt{2}}, \; \tau_{II}'' = 0. \tag{22}$$

- I the case of welds placed by the walls of the overlapping brace:

$$\sigma'' = \frac{P_3''}{a_{w3}l_3}, \; \sigma_\perp'' = -\frac{\sigma''}{\sqrt{2}}, \tau_\perp'' = \frac{\sigma''}{\sqrt{2}}, \; \tau_{II}'' = 0. \tag{23}$$

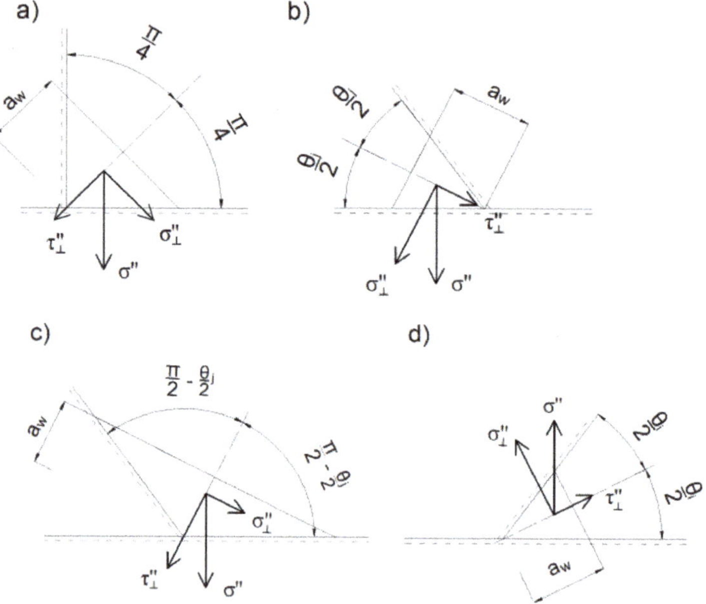

Figure 5. Stresses in welds of the load perpendicular to the chord: (**a**) in longitudinal welds; (**b**) on the not fully cooperating the overlapped brace's transverse weld; (**c**) on the fully cooperating the overlapped brace's transverse weld; (**d**) On the not fully cooperating the overlapping brace's transverse weld.

2. On the not fully cooperating the overlapped brace's transverse weld (Figure 5b):

$$\sigma'' = \frac{P_2''}{a_{w2}l_2}, \; \sigma_\perp'' = -\sigma'' \cos\frac{\theta_j}{2}, \tau_\perp'' = \sigma'' \sin\frac{\theta_j}{2}, \; \tau_{II}'' = 0. \tag{24}$$

3. On the fully cooperating the overlapped brace's transverse weld (Figure 5c):

$$\sigma'' = \frac{P_b''}{a_{wb}b_{j,red}}, \; \sigma_\perp'' = \sigma'' \cos\frac{\theta_j}{2}, \tau_\perp'' = -\sigma'' \sin\frac{\theta_j}{2}, \; \tau_{II}'' = 0. \tag{25}$$

4. On the not fully cooperating the overlapping brace's transverse weld (Figure 5d):

$$\sigma'' = \frac{P_4''}{a_{w4}l_4}, \; \sigma_\perp'' = -\sigma'' \cos\frac{\theta_i}{2}, \tau_\perp'' = \sigma'' \sin\frac{\theta_i}{2}, \; \tau_{II}'' = 0. \tag{26}$$

Stresses in the welds made directly between the brace members at the partial overlap of $25\% \leq \lambda_{ov} \leq 80\%$ from the force parallel to the overlapped brace axis:

1. In longitudinal welds (Figure 6a):

$$\sigma' = 0, \; \sigma_\perp' = \tau_\perp' = 0, \; \tau_{II}' = \frac{P_5'}{a_{w5}l_5}. \tag{27}$$

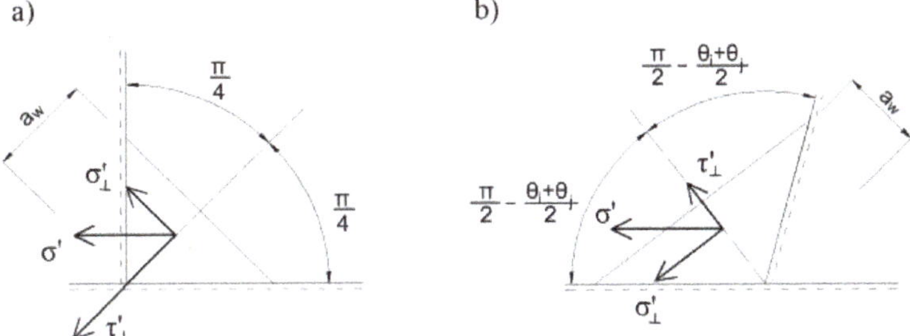

Figure 6. Stresses in welds between the brace members from the force parallel to the overlapped brace:
(**a**) in longitudinal welds; (**b**) in the fully cooperating transverse weld.

2. In the fully cooperating transverse weld (Figure 6b):

$$\sigma' = \frac{P_6'}{a_{w6}l_6}, \ \sigma_\perp' = -\sigma' \cos\frac{\theta_i + \theta_j}{2}, \ \tau_\perp' = \sigma' \sin\frac{\theta_i + \theta_j}{2}, \ \tau_{II}' = 0. \tag{28}$$

Stresses in welds placed between the brace members at the partial overlap of $25\% \leq \lambda_{ov} \leq 80\%$ from the force perpendicular to the overlapped brace:
1. In longitudinal welds (Figure 7a):

$$\sigma'' = \frac{P_5''}{a_{w5}l_5}, \ \sigma_\perp'' = -\frac{\sigma''}{\sqrt{2}}, \ \tau_\perp'' = \frac{\sigma''}{\sqrt{2}}, \ \tau_{II}'' = 0. \tag{29}$$

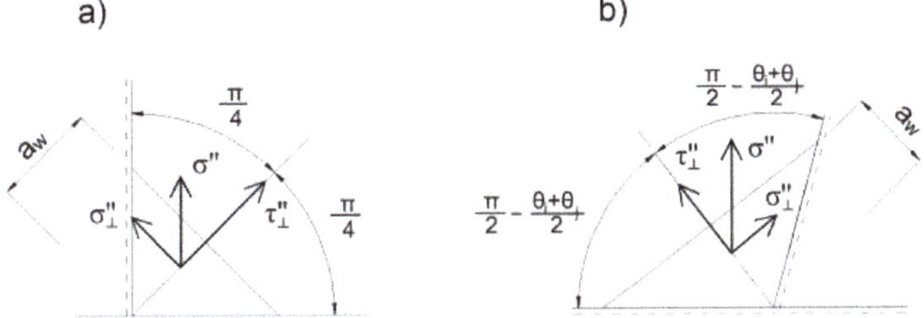

Figure 7. Stresses in welds between the brace members from the force perpendicular to the overlapped
brace: (**a**) in longitudinal welds; (**b**) in the fully cooperating transverse weld.

2. In the fully cooperating transverse weld (Figure 7b):

$$\sigma'' = \frac{P_6''}{a_{w6}l_6}, \ \sigma_\perp'' = \sigma'' \cos\frac{\theta_j + \theta_i}{2}, \ \tau_\perp'' = -\sigma'' \sin\frac{\theta_j + \theta_i}{2}, \ \tau_{II}'' = 0. \tag{30}$$

The procedure of checking of a design situation with the full overlap of braces $\lambda_{ov} = 100\%$ is analogous. In that case, the stresses in the transverse weld located near the connection of the overlapped brace with the chord should be examined (Figure 8), using the loads expressed by the Equations (24) and (25). The components of stress are:

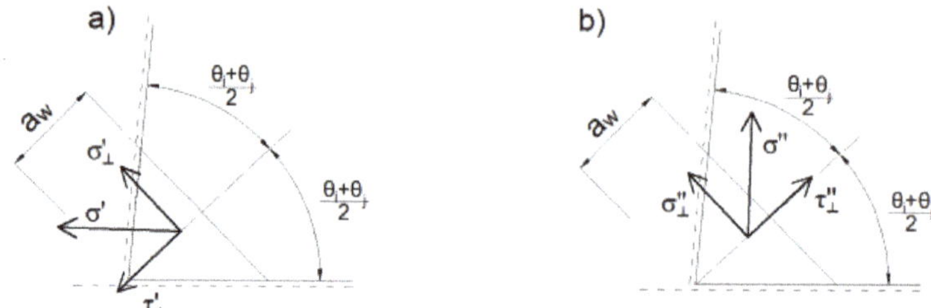

Figure 8. Stresses in the transverse weld with full overlap: (**a**) in the transverse weld from the force parallel to the overlapped brace; (**b**) in the transverse weld from the force perpendicular to the overlapped brace.

1. In the transverse weld from the force parallel to the overlapped brace (Figure 8a):

$$\sigma' = \frac{P_6'}{a_{w6}l_6}, \ \sigma_{\perp}' = -\sigma' \sin\frac{\theta_j + \theta_i}{2}, \ \tau_{\perp}' = -\sigma'' \cos\frac{\theta_j + \theta_i}{2}, \ \tau_{II}' = 0. \tag{31}$$

2. In the transverse weld from the force perpendicular to the overlapped brace (Figure 8b):

$$\sigma'' = \frac{P_6''}{a_{w6}l_6}, \ \sigma_{\perp}' = -\sigma' \cos\frac{\theta_j + \theta_i}{2}, \ \tau_{\perp}'' = \sigma'' \sin\frac{\theta_j + \theta_i}{2}, \ \tau_{II}'' = 0. \tag{32}$$

The component stresses occurring in cross-section of the welds should be added using the formulas:

$$\tau_{II} = \tau_{II}' + \tau_{II}''; \sigma_{\perp} = \sigma_{\perp}' + \sigma_{\perp}''; \tau_{\perp} = \tau_{\perp}' + \tau_{\perp}''. \tag{33}$$

Standardized formulas for checking the safety of fully or partially cooperating transverse and longitudinal welds are:

$$\left[\sigma_{\perp}^2 + 3\left(\tau_{\perp}^2 + \tau_{II}^2\right)\right] \le f_u / (\beta_w \gamma_{M2}), \tag{34}$$

$$\sigma_{\perp} \le 0.9 f_u / \gamma_{M2}, \tag{35}$$

where: β_w—the correlation coefficient, f_u—the tensile strength of steel, $\gamma_{M2} = 1.25$—the safety factor.

3. Conclusions

RHS joints are generally semi-rigid, mainly because of the preferred technologies for their production, i.e., the direct welding of members. It implements a significant load from braces to the relatively slender front walls of the chord. The basic information for calculating joint resistance is given in many standards and references, but European standards contain a general recommendation on the calculation of capacity of welded joints and do not provide detailed design guidelines. The information contained is random and concerns only Y, X, K, and N joints with the gap. Additionally, in the case of the K and N types of joints with partially overlapped brace members, there is no indication of how to calculate the capacity of welds between the members.

This paper presents the method of assessment of the welded K and N type overlapped joints between RHS brace members and I or H section chords. This method comprises determining the stress components in welds in different load cases based on their effective lengths.

Design Example:

Check the resistance of welds of an overlap K joint with a chord made of HEB 120 and SHS braces (Figure 9). Steel grade of S355H, $f_y = 355$ N/mm^2, $f_u = 490$ N/mm^2. Chord: $h_0 = 120$ mm, $b_f = 120$ mm, $t_w = 6.5$ mm, $t_f = 11.0$ mm, $r = 12.0$ mm, $A_0 = 34.0$ cm^2, $W_{pl.y,0} = 165.2$ cm^3, $N_0 = -159.9$ kN. Brace loaded with compressive force: $h_2 = 80$ mm, $b_2 = 60$ mm, $t_2 = 4$ mm; $N_{2.Ed} = -136.1$ kN. Brace loaded with tension force: $h_1 = 60$ mm, $b_1 = 50$ mm, $t_1 = 3$ mm, $N_{1.Ed} = +103.2$ kN. The angles: $\theta_1 = 50.34° > 30°$, $\theta_2 = 40.02° > 30°$ (sin $\theta_1 = 0.7698$, cos $\theta_1 = 0.6382$, sin $\theta_2 = 0.6431$, cos $\theta_2 = 0.7658$, sin $(\theta_1 + \theta_2) \cong 1.0$). The overlapped transverse weld is done $-c_s = 2$.

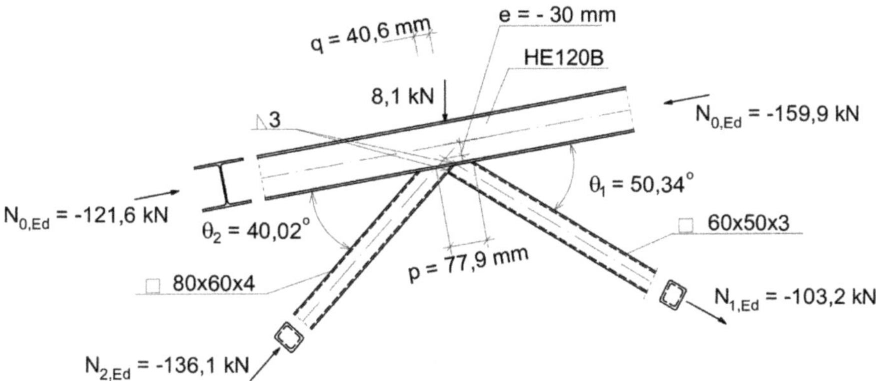

Figure 9. The overlap joint made of H-section chord and SHS braces.

A. An Overlap Value.

$$e = -30 \text{ mm},$$

$$q = \left(e + \frac{h_0}{2}\right)\frac{\sin(\theta_1+\theta_2)}{\sin\theta_1\sin\theta_2} - \frac{h_1}{2\sin\theta_1} - \frac{h_2}{2\sin\theta_2} =$$
$$\left(-30 + \frac{120}{2}\right)\frac{1.0}{0.7698\cdot0.6431} - \frac{60}{2\cdot0.7698} - \frac{80}{2\cdot0.6431} = -40.6 \text{ mm},$$

$$p = h_1 / \sin\theta_1 = 60/0.7698 = 77.9 \text{ mm},$$

$$\lambda_{0v} = (q/p) \cdot 100\% = (40.6/77.9) \cdot 100\% = 52.1\% < \lambda_{0v,\text{lim}} = 80\%.$$

B. The Design Conditions.

$$\frac{h_1}{t_1} = \frac{60}{3} = 20.0 < 35, \quad \frac{b_1}{t_1} = \frac{50}{3} = 16.7 < 35, \quad \frac{h_2}{t_2} = \frac{80}{4} = 20.0 < 35,$$

$$\frac{b_2}{t_2} = \frac{60}{4} = 15.0 < 35,$$

$$\frac{h_1}{b_1} = \frac{60}{50} = 1.2 > 1.0, \quad \frac{h_2}{b_2} = \frac{80}{60} = 1.33 > 1.0.$$

C. The Design Resistance of the Joint.

In the case of $50\% < \lambda_{0v} = 52.1\% < 80\%$:

$$p_{1,eff} = t_w + 2r + 7t_f f_{y0}/f_{y_1} = 6.5 + 2\cdot12 + 7\cdot11\cdot355/355 = 107.5 \text{ mm} > b_1 = 50 \text{ mm}$$

Adopted $p_{1,eff} = 50$ mm.

$$b_{e,ov} = \frac{10}{b_j/t_j}\frac{f_{yj}t_j}{f_{yi}t_i}b_i = \frac{10}{60/4}\frac{355\cdot4}{355\cdot3}\cdot50 = 44.4 \text{ mm} < b_1 = 50 \text{ mm}.$$

The brace failure.

$$N_{1,Rd} = f_{y1}t_1\left(p_{1,eff} + b_{e,ov} + h_1 - 2t_1\right)/\gamma_{M5} =$$
$$355 \cdot 3(50 + 44.4 + 60 - 2 \cdot 3)/1.0 = 158000 \text{ N} = 158.0 \text{ kN},$$

$$N_{2,Rd} = N_{1,Rd}\frac{\sin\theta_1}{\sin\theta_2} = 158.0 \cdot \frac{0.7698}{0.6431} = 189.1 \text{ kN}.$$

Checking the braces resistance.

$$\frac{103.2}{158} = 0.65 < 1.0, \quad \frac{136.1}{189.1} = 0.72 < 1.0.$$

The brace bending and the axial force resistance.

$$M_0 = \pm 0.5(N_{02.Ed} - N_{01.Ed})e = \pm 0.5\,(-159.9 + 121.6) \cdot 30 = \pm 574.5 \text{ kNmm},$$

$$M_{0.pl} = W_{pl.0} \cdot f_y/\gamma_{M1} = 165.2 \cdot 10^3 \cdot 355/1.0 = 586.5 \cdot 10^2 \text{ kNmm},$$

$$N_{0.pl} = A_0 \cdot f_y/\gamma_{M1} = 34.0 \cdot 10^2 \cdot 355/1.0 = 1207 \cdot 10^3 \text{ N} = 1207 \text{ kN},$$
$$\frac{N_0}{N_{0.pl}} + \frac{M_0}{M_{0.pl}} = \frac{159.9}{1207} + \frac{574.5}{58650} = 0.14 < 1.0.$$

D. The Shear Resistance of Thin Fillet Welds.

Effective parts of welds, $\left(\alpha = \frac{p}{q} = 0.521\right)$:

$$l_1 = h_2/\sin\theta_2 = 80/0.6431 = 124.4 \text{ mm},$$
$$l_2 = p_{2,eff} = t_w + 2r + 7t_f f_{y0}/f_{y2} = 6.5 + 2 \cdot 12 + 7 \cdot 11 \cdot 355/355 = 107.5 \text{ mm} > b_2 = 60 \text{ mm}$$

Assumed $l_2 = 60$ mm.

$$b_{j.red} = b_2 - 2a_w = 60 - 2 \cdot 3 = 54 \text{ mm},$$
$$l_3 = h_{i.red}/\sin\theta_i = (1 - \alpha)h_i/\sin\theta_i = (1 - 0.521)60/0.7698 = 37.3 \text{ mm},$$
$$l_4 = p_{1.eff} = t_w + 2r + 7t_f f_{y0}/f_{y1} = = 107.5 \text{ mm} > b_1 = 50 \text{ mm}.$$

Assumed $l_4 = 50$ mm.

$$l_5 = \frac{q}{(1+\tan\theta_2/\tan\theta_1)\cos\theta_2} = \frac{40.6}{(1+0.8397/1.2062)0.7658} = 31.3 \text{ mm},$$

$$l_6 = b_1 = 50 \text{ mm}.$$

Cross-section areas of effective lengths of welds.
Assumed the thickness of welds $a_{w1} = a_{w2} = a_{w3} = a_{wb} = a_{w4} = a_{w5} = a_{w6} = 3.0$ mm,

$$A_1 = l_1 a_{w1} = 124.4 \cdot 3 = 373.2 \text{ mm}^2, \quad A_2 = l_2 a_{w2} = 60 \cdot 3 = 180.0 \text{ mm}^2,$$

$$A_3 = l_3 a_{w3} = 37.3 \cdot 3 = 111.9 \text{ mm}^2, \quad A_{2,red} = b_{2,red}a_{wb} = 54 \cdot 3 = 162.0 \text{ mm}^2,$$

$$A_4 = l_4 a_{w4} = 50 \cdot 3 = 150.0 \text{ mm}^2, \quad A_5 = l_5 a_{w5} = 31.3 \cdot 3 = 93.9 \text{ mm}^2,$$

$$A_6 = l_6 a_{w6} = 50 \cdot 3 = 150.0 \text{ mm}^2,$$

$$\sum A = 2A_1 + A_2 + 2A_3 + A_{j,red} + A_4 = 2 \cdot 373.2 + 180 + 2 \cdot 111.9 + 162 + 150 = 1462.2 \text{ mm}^2.$$

Forces acting on individual weld's sections ($K_1 = N_{1.Ed}$; $K_2 = N_{2.Ed}$):

$$\Delta K_1 = \alpha K_1 \sin\theta_1 = 0.521 \cdot 103.2 \cdot 0.7698 = 41.4 \text{ kN},$$

$$red\Delta K_2 = K_2 \sin \theta_2 - \alpha K_1 \sin \theta_1 = 136.1 \cdot 0.6431 - 0.521 \cdot 103.2 \cdot 0.7698 = 46.1 \text{ kN.}$$

Loads in effective lengths parallel to the chord.

$$P_1' = (K_2 \cos \theta_2 + K_1 \cos \theta_1)A_1 / \sum A = (136.1 \cdot 0.7658 + 103.2 \cdot 0.6382)373.2/1462.2 = 43.4 \text{ kN,}$$

$$P_2' = (K_2 \cos \theta_2 + K_1 \cos \theta_1)A_2 / \sum A = (136.1 \cdot 0.7658 + 103.2 \cdot 0.6382)180.0/1462.2 = 20.9 \text{ kN,}$$

$$P_3' = (K_2 \cos \theta_2 + K_1 \cos \theta_1)A_3 / \sum A = (136.1 \cdot 0.7658 + 103.2 \cdot 0.6382)111.9/1462.2 = 13.0 \text{ kN,}$$

$$P_4' = (K_2 \cos \theta_2 + K_1 \cos \theta_1)A_4 / \sum A = (136.1 \cdot 0.7658 + 103.2 \cdot 0.6382)150.0/1462.2 = 17.4 \text{ kN,}$$

$$P_b' = (K_2 \cos \theta_2 + K_1 \cos \theta_1)A_{j,red} / \sum A = (136.1 \cdot 0.7658 + 103.2 \cdot 0.6382)162/1462.2 = 18.8 \text{ kN,}$$

$$P_5' = \Delta K_1 \cdot A_5 \cdot \sin \theta_2 / (2A_5 + A_6) = 41.4 \cdot 93.9 \cdot 0.6431/(2 \cdot 93.9 + 150) = 7.4 \text{ kN,}$$

$$P_6' = \Delta K_1 \cdot A_6 \cdot \sin \theta_2 / (2A_5 + A_6) = 41.4 \cdot 150 \cdot 0.6431/(2 \cdot 93.9 + 150) = 11.8 \text{ kN.}$$

Loads in effective lengths perpendicular to the chord.

$$P_1'' = 0,$$

$$P_2'' = red\Delta K_2 \cdot A_2 / (A_2 + A_{2,red}) = 46.1 \cdot 180/(180 + 162) = 24.3 \text{ kN,}$$

$$P_3'' = (1 - \alpha)\Delta K_1 \cdot A_3 / (2A_3 + A_4) = (1 - 0.521) \cdot 41.4 \cdot 111.9/(2 \cdot 111.9 + 150) = 5.9 \text{ kN,}$$

$$P_4'' = (1 - \alpha)\Delta K_1 \cdot A_4 / (2A_3 + A_4) = (1 - 0.521) \cdot 41.4 \cdot 150/(2 \cdot 111.9 + 150) = 8.0 \text{ kN,}$$

$$P_b'' = red\Delta K_2 \cdot A_{2,red} / (A_2 + A_{2,red}) = 46.1 \cdot 162/(180 + 162) = 21.8 \text{ kN,}$$

$$P_5'' = \Delta K_1 \cdot A_5 \cos \theta_2 / (2A_5 + A_6) = 41.4 \cdot 93.9 \cdot 0.7658/(2 \cdot 93.9 + 150) = 8.8 \text{ kN,}$$

$$P_6'' = \Delta K_1 \cdot A_6 \cos \theta_2 / (2A_5 + A_6) = 41.4 \cdot 150 \cdot 0.7658/(2 \cdot 93.9 + 150) = 14.1 \text{ kN.}$$

Stresses on the throat section of a fillet welds from the force component parallel to the chord:

- longitudinal welds between the lower (overlapped) brace and the chord:

$$\sigma' = 0,$$

$$\sigma'_\perp = \tau'_\perp = 0,$$

$$\tau'_{II} = \frac{P_1'}{a_{w1}l_1} = \frac{43.4 \cdot 10^3}{3.0 \cdot 124.4} = 116.3 \text{ MPa,}$$

- the not fully effective transverse weld between the lower brace and the chord:

$$\sigma' = \frac{P_2'}{a_{w2}l_2} = \frac{20.9 \cdot 10^3}{3.0 \cdot 60} = 116.1 \text{ MPa,}$$

$$\sigma'_\perp = \sigma' \sin \frac{\theta_2}{2} = 116.1 \cdot 0.3422 = 39.7 \text{ MPa,}$$

$$\tau'_\perp = \sigma' \cos \frac{\theta_2}{2} = 116.1 \cdot 0.9396 = 109.1 \text{ MPa,}$$

$$\tau'_{II} = 0,$$

- longitudinal welds between the upper (overlapping) brace and the chord:

$$\sigma' = 0,$$

$$\sigma'_\perp = \tau'_\perp = 0,$$

$$\tau'_{II} = \frac{P'_3}{a_{w3}l_3} = \frac{13.0 \cdot 10^3}{3.0 \cdot 37.3} = 116.2 \text{ MPa},$$

- the fully effective transverse weld between the lower brace and the chord:

$$\sigma' = \frac{P'_b}{a_{wb}b_{j.red}} = \frac{18.8 \cdot 10^3}{3.0 \cdot 54} = 116.0 \text{ MPa},$$

$$\sigma'_\perp = \sigma' \cos \frac{\theta_2}{2} = 116.0 \cdot 0.9396 = 109.0 \text{ MPa},$$

$$\tau'_\perp = \sigma' \sin \frac{\theta_2}{2} = 116.0 \cdot 0.3422 = 39.7 \text{ MPa},$$

$$\tau'_{II} = 0,$$

- the not fully effective transverse weld between the upper brace and the chord:

$$\sigma' = \frac{P'_4}{a_{w4}l_4} = \frac{17.4 \cdot 10^3}{3.0 \cdot 50} = 116.0 \text{ MPa},$$

$$\sigma'_\perp = \sigma' \sin \frac{\theta_1}{2} = 116.0 \cdot 0.3849 = 44.6 \text{ MPa},$$

$$\tau'_\perp = \sigma' \cos \frac{\theta_1}{2} = 116.0 \cdot 0.9050 = 105.0 \text{ MPa},$$

$$\tau'_{II} = 0,$$

- longitudinal welds between the upper (overlapping) brace and the lower (overlapped) brace:

$$\sigma' = 0,$$

$$\sigma'_\perp = \tau'_\perp = 0,$$

$$\tau'_{II} = \frac{P'_5}{a_{w5}l_5} = \frac{7.4 \cdot 10^3}{3.0 \cdot 31.3} = 78.8 \text{ MPa},$$

- the transverse weld between the upper brace and the lower brace:

$$\sigma' = \frac{P'_6}{a_{w6}l_6} = \frac{11.8 \cdot 10^3}{3.0 \cdot 50} = 78.7 \text{ MPa},$$

$$\sigma'_\perp = -\sigma' \cos \frac{\theta_1 + \theta_2}{2} = -78.7 \cdot 0.7049 = -55.5 \text{ MPa},$$

$$78.8 \cdot 0.7093 = 55.9 \text{ MPa},$$

$$\tau'_{II} = 0$$

Stresses on the throat section of a fillet welds from the force component perpendicular to the chord:

- longitudinal welds between the lower (overlapped) brace and the chord:

$$P_1'' = 0,$$

$$\sigma_\perp'' = \tau_\perp'' = 0,$$

$$\tau_{II}'' = 0,$$

- the not fully effective transverse weld between the lower brace and the chord:

$$\sigma'' = \frac{P_2''}{a_{w2}l_2} = \frac{24.3 \cdot 10^3}{3.0 \cdot 60.0} = 135.0 \text{ MPa},$$

$$\sigma_\perp'' = -\sigma'' \cos \frac{\theta_2}{2} = -135.0 \cdot 0.9396 = -126.8 \text{ MPa},$$

$$\tau_\perp'' = \sigma'' \sin \frac{\theta_2}{2} = 135.0 \cdot 0.3422 = 46.2 \text{ MPa},$$

$$\tau_{II}'' = 0,$$

- longitudinal welds between the upper (overlapping) brace and the chord:

$$\sigma'' = \frac{P_3''}{a_{w3}l_3} = \frac{5.9 \cdot 10^3}{3 \cdot 37.3} = 52.7 \text{ MPa},$$

$$\sigma_\perp'' = -\frac{\sigma''}{\sqrt{2}} = -\frac{52.7}{\sqrt{2}} = -37.2 \text{ MPa},$$

$$\tau_\perp'' = \frac{\sigma''}{\sqrt{2}} = \frac{52.7}{\sqrt{2}} = 37.2 \text{ MPa},$$

$$\tau_{II}'' = 0,$$

- the fully effective transverse weld between the lower brace and the chord:

$$\sigma'' = \frac{P_b''}{a_{wb}b_{j.red}} = \frac{21.8 \cdot 10^3}{3.0 \cdot 54} = 134.6 \text{ MPa},$$

$$\sigma_\perp'' = \sigma'' \cos \frac{\theta_2}{2} = 134.6 \cdot 0.9396 = 126.4 \text{ MPa},$$

$$\tau_\perp'' = \sigma'' \sin \frac{\theta_2}{2} = 134.6 \cdot 0.3422 = 46.1 \text{ MPa},$$

$$\tau_{II}'' = 0,$$

- the not fully effective transverse weld between the upper brace and the chord:

$$\sigma'' = \frac{P_4''}{a_{w4}l_4} = \frac{8.0 \cdot 10^3}{3.0 \cdot 50} = 53.3 \text{ MPa},$$

$$\sigma_\perp'' = -\sigma'' \cos \frac{\theta_1}{2} = -53.3 \cdot 0.9050 = -48.3 \text{ MPa},$$

$$\tau_\perp'' = \sigma'' \sin \frac{\theta_1}{2} = 53.3 \cdot 0.3849 = 20.5 \text{ MPa},$$

$$\tau_{II}'' = 0,$$

- longitudinal welds between the upper brace and the lower brace:

$$\sigma'' = \frac{P_5''}{a_{w5}l_5} = \frac{8.8 \cdot 10^3}{3.0 \cdot 31.3} = 93.7 \text{ MPa},$$

$$\sigma_\perp'' = -\frac{\sigma''}{\sqrt{2}} = -\frac{93.7}{\sqrt{2}} = -66.3 \text{ MPa},$$

$$\tau_\perp'' = \frac{\sigma''}{\sqrt{2}} = \frac{93.7}{\sqrt{2}} = 66.3 \text{ MPa},$$

$$\tau_{II}'' = 0,$$

- the transverse weld between the upper brace and the lower brace:

$$\sigma'' = \frac{P_6''}{a_{w6}l_6} = \frac{14.1 \cdot 10^3}{3.0 \cdot 50} = 94.0 \text{ MPa},$$

$$\sigma_\perp'' = \sigma'' \cos \frac{\theta_1 + \theta_2}{2} = 94 \cdot 0.7049 = 66.3 \text{ MPa},$$

$$\tau_\perp'' = -\sigma'' \sin \frac{\theta_1 + \theta_2}{2} = -94 \cdot 0.7093 = -66.7 \text{ MPa},$$

$$\tau_{II}' = 0$$

.

The normal and shear stresses in welds.

- longitudinal welds between the lower brace and the chord:

$$\tau_{II} = \tau_{II}' + \tau_{II}'' = 116.3 + 0 = 116.3 \text{ MPa},$$

$$\sigma_\perp = \sigma_\perp' + \sigma_\perp'' = 0 + 0 = 0 \text{ MPa},$$

$$\tau_\perp = \tau_\perp' + \tau_\perp'' = 0 - 0 = 0 \text{ MPa},$$

- the not fully effective transverse weld between the lower brace and the chord:

$$\tau_{II} = \tau_{II}' + \tau_{II}'' = 0 + 0 = 0,$$

$$\sigma_\perp = \sigma_\perp' + \sigma_\perp'' = 39.7 - 126.8 = -87.1 \text{ MPa},$$

$$\tau_\perp = \tau_\perp' + \tau_\perp'' = 109.1 + 46.2 = 155.3 \text{ MPa},$$

- longitudinal welds between the upper brace and the chord:

$$\tau_{II} = \tau_{II}' + \tau_{II}'' = 116.2 + 0 = 116.2 \text{ MPa},$$

$$\sigma_\perp = \sigma_\perp' + \sigma_\perp'' = 0 - 37.2 = -37.2 \text{ MPa},$$

$$\tau_\perp = \tau_\perp' + \tau_\perp'' = 0 + 37.2 = 37.2 \text{ MPa},$$

- the fully effective transverse weld between the lower brace and the chord:

$$\tau_{II} = \tau_{II}' + \tau_{II}'' = 0 + 0 = 0,$$

$$\sigma_\perp = \sigma_\perp' + \sigma_\perp'' = 109.0 + 126.4 = 235.4 \text{ MPa},$$

$$\tau_\perp = \tau_\perp' + \tau_\perp'' = 39.7 + 46.1 = 85.8 \text{ MPa},$$

- the not fully effective transverse weld between the upper brace and the chord:

$$\tau_{\parallel} = \tau'_{\parallel} + \tau''_{\parallel} = 0 + 0 = 0 \text{ MPa,}$$

$$\sigma_{\perp} = \sigma'_{\perp} + \sigma''_{\perp} = 44.6\text{--}48.3 = -3.7 \text{ MPa,}$$

$$\tau_{\perp} = \tau'_{\perp} + \tau''_{\perp} = 105.0 + 20.5 = 125.5 \text{ MPa,}$$

- longitudinal welds between the upper brace and the lower brace:

$$\tau_{\parallel} = \tau'_{\parallel} + \tau''_{\parallel} = 78.8 + 0 = 78.8 \text{ MPa,}$$

$$\sigma_{\perp} = \sigma'_{\perp} + \sigma''_{\perp} = 0\text{--}66.3 = -66.3 \text{ MPa,}$$

$$\tau_{\perp} = \tau'_{\perp} + \tau''_{\perp} = 0 + 66.3 = 66.3 \text{ MPa,}$$

- the transverse weld between the upper brace and the lower brace:

$$\tau_{\parallel} = \tau'_{\parallel} + \tau''_{\parallel} = 0 + 0 = 0 \text{ MPa,}$$

$$\sigma_{\perp} = \sigma'_{\perp} + \sigma''_{\perp} = -55.5 + 66.3 = 121.8 \text{ MPa,}$$

$$\tau_{\perp} = \tau'_{\perp} + \tau''_{\perp} = 55.9\text{--}66.7 = -10.8 \text{ MPa.}$$

E. The design resistance of the fillet welds.

- longitudinal welds between the lower brace and the chord:

$$\left[\sigma_{\perp}^2 + 3\left(\tau_{\perp}^2 + \tau_{\parallel}^2\right)\right]^{0.5} = \left[0^2 + 3\left(0^2 + 116.3^2\right)\right]^{0.5} = 201.4 \text{ MPa} < \frac{f_u}{\beta_w \gamma_{Mw}} = \frac{490}{0.9 \cdot 1.25} = 435.6 \text{ MPa,}$$

$$\sigma_{\perp} = 0 \text{ MPa} \ < \ \frac{0.9 \cdot 490}{1.25} = 352.8 \text{ MPa,}$$

- the not fully effective transverse weld between the lower brace and the chord:

$$\left[\sigma_{\perp}^2 + 3\left(\tau_{\perp}^2 + \tau_{\parallel}^2\right)\right]^{0.5} = \left[(-87.1)^2 + 3\left(155.3^2 + 0^2\right)\right]^{0.5} = 282.7 \text{ MPa} < \frac{f_u}{\beta_w \gamma_{Mw}} = \frac{490}{0.9 \cdot 1.25} = 435.6 \text{ MPa,}$$

$$\sigma_{\perp} = 87.1 \text{ MPa} \ < \ \frac{0.9 \cdot 490}{1.25} = 352.8 \text{ MPa,}$$

- longitudinal welds between the upper brace and the chord:

$$\left[\sigma_{\perp}^2 + 3\left(\tau_{\perp}^2 + \tau_{\parallel}^2\right)\right]^{0.5} = \left[(-37.2)^2 + 3\left(37.2^2 + 116.2^2\right)\right]^{0.5} = 247.6 \text{ MPa} < \frac{f_u}{\beta_w \gamma_{Mw}} = \frac{490}{0.9 \cdot 1.25} = 435.6 \text{ MPa,}$$

$$\sigma_{\perp} = -72.1 \text{ MPa} < \frac{0.9 \cdot 490}{1.25} = 352.8 \text{ MPa,}$$

- the fully effective transverse weld between the lower brace and the chord:

$$\left[\sigma_{\perp}^2 + 3\left(\tau_{\perp}^2 + \tau_{\parallel}^2\right)\right]^{0.5} = \left[235.4^2 + 3\left(85.8^2 + 0\right)\right]^{0.5} = 278.4 \text{ MPa} \ < \ \frac{f_u}{\beta_w \gamma_{Mw}} = \frac{490}{0.9 \cdot 1.25} = 435.6 \text{ MPa,}$$

$$\sigma_{\perp} = 235.4 \text{ MPa} \ < \ \frac{0.9 \cdot 490}{1.25} = 352.8 \text{ MPa,}$$

- the not fully effective transverse weld between the upper brace and the chord:

$$\left[\sigma_{\perp}^2 + 3\left(\tau_{\perp}^2 + \tau_{\parallel}^2\right)\right]^{0.5} = \left[(-3.7)^2 + 3\left(125.5^2 + 0^2\right)\right]^{0.5} = 217.4 \text{ MPa} < \frac{f_u}{\beta_w \gamma_{Mw}} = \frac{490}{0.9 \cdot 1.25} = 435.6 \text{ MPa,}$$

$$\sigma_\perp = -3.7 \text{ MPa} < \frac{0.9 \cdot 490}{1.25} = 352.8 \text{ MPa},$$

- longitudinal welds between the upper brace and the lower brace:

$$\left[\sigma_\perp^2 + 3\left(\tau_\perp^2 + \tau_\parallel^2\right)\right]^{0.5} = \left[(-66.3)^2 + 3\left(66.3^2 + 78.8^2\right)\right]^{0.5} = 190.3 \text{ MPa} < \frac{f_u}{\beta_w \gamma_{Mw}} = \frac{490}{0.9 \cdot 1.25} = 435.6 \text{ MPa},$$

$$\sigma_\perp = -66.3 \text{ MPa} < \frac{0.9 \cdot 490}{1.25} = 352.8 \text{ MPa},$$

- the transverse weld between the upper brace and the lower brace:

$$\left[\sigma_\perp^2 + 3\left(\tau_\perp^2 + \tau_\parallel^2\right)\right]^{0.5} = \left[121.8^2 + 3\left((-10.8)^2 + 0^2\right)\right]^{0.5} = 123.2 \text{ MPa} < \frac{f_u}{\beta_w \gamma_{Mw}} = \frac{490}{0.9 \cdot 1.25} = 435.6 \text{ MPa},$$

$$\sigma_\perp = 121.8 \text{ MPa} < \frac{0.9 \cdot 490}{1.25} = 352.8 \text{ MPa}.$$

The safety coefficient is equal: $\frac{435.6 - 282.7}{435.6} \cdot 100\% = 35.1\%$.

Author Contributions: Conceptualization, M.B.; methodology, M.B. and F.B.; analysis, M.B. and F.B; writing, M.B. and F.B. All authors have read and agreed to the published version of the manuscript.

Funding: This research was funded by grants numbers: S/WBiIŚ/2/2017; W/WBiIŚ/12/2019.

Conflicts of Interest: The authors declare no conflict of interest.

References

1. Eekhout, M. *Tubular Structures in Architecture*, 2nd ed.; Delft University of Technology: Delft, The Netherlands, 2011; 127p.
2. Wardenier, J.; Packer, J.A.; Zhao, X.L.; Van der Vegte, G.J. *Hollow Sections in Structural Applications*; John Wiley & Sons: Hoboken, NJ, USA, 2010; 232p.
3. Brodka, J.; Broniewicz, M. *Steel Structures from Hollow Sections*; Arkady: Warsaw, Poland, 2001; 383p.
4. Dutta, D. *Structures with Hollow Sections*, 1st ed.; Ernst & Sohn: Berlin, Germany, 2002; 600p.
5. Packer, J.A.; Wardenier, J. Design rules for welds in RHS K, T, Y and X connections. In Proceedings of the IIW International Conference on Engineering Design in Welded Constructions, Madrid, Spain, 7–8 September 1992; pp. 113–120.
6. Frater, G.S.; Packer, J.A. Weldment design for RHS truss connections. I: Applications. *J. Struct. Eng.* **1992**, *118*, 2784–2803. [CrossRef]
7. Frater, G.S.; Packer, J.A. Weldment design for RHS truss connections. II: Experimentation. *J. Struct. Eng.* **1992**, *118*, 2804–2820. [CrossRef]
8. Teh, L.H.; Rasmussen, K.J.R. *Strength of Butt Welded Connections between Equal Width Rectangular Hollow Sections*; Research Report No. R817; Department of Civil Engineering, University of Sydney: Sydney, Australia, 2002; 27p.
9. Packer, J.A.; Henderson, J.E. *Hollow Structural Section Connections and Truss—A Design Guide*, 2nd ed.; Canadian Institute of Steel Construction: Toronto, ON, Canada, 1997; 447p.
10. Packer, J.A.; Sherman, D.R.; Lecce, M. *HSS Connections. Steel Design Guide No. 24*; American Institute of Steel Construction: Chicago, IL, USA, 2009; 146p.
11. IIW. *Recommendations for Fatigue Design of Welded Joints and Components*; Commission XV. IIW Docs, IIW-1823-07; International Institute of Welding: Paris, France, 1823.
12. ISO 14346. *Static Design Procedure for Welded Hollow Section Joints—Recommendations*; International Institute of Welding: Paris, France, 2013.
13. IIW. *Static Design Procedure for Welded Hollow Section Joints—Recommendations*, 3rd ed.; Commission XV. IIW Doc. XV-1329-09; International Institute of Welding: Paris, France, 2009.
14. Packer, J.A.; Wardenier, J.; Zhao, X.L.; Van der Vegte, G.J.; Kurobane, Y. *Design Guide for Rectangular Hollow Section (RHS) Joints under Predominantly Static Loading*; LSS Verlag, CIDECT: Koln, Germany, 2009; pp. 149p.

15. Brodka, J.; Broniewicz, M. Design of welded joints of trusses made of hollow sections. In *Steel Construction*; No 1; Polish Chamber of Steelwork: Warsaw, Poland, 2002; p. 29.
16. Brodka, J.; Broniewicz, M. Assessing of resistance of connections according to EN 1993-1-8:2006. In *Steel Construction*; No 1; Polish Chamber of Steelwork: Warsaw, Poland, 2007; pp. 32–38.
17. Brodka, J.; Broniewicz, M. Calculation of welding trusses overlap joints. In *Archives of Civil Engineering*; Committee for Civil Engineering, Polish Academy of Sciences: Warsaw, Poland, 2013; 24p.
18. Davies, G.; Crockett, P. Effect of the hidden weld on RHS partial overlap K joint capacity. In Proceedings of the 6th International Symposium on Tubular Structures, Melbourne, Australia, 14–16 December 1994; Balkema: Rotterdam, The Netherlands, 1994; pp. 573–579.
19. Dexter, E.M.; Lee, M.M.K. Effect of overlap on the behaviour of axially loaded CHS K-joints. In Proceedings of the 8th International Symposium on Tubular Structures, Singapore, 26–28 August 1998; Balkema: Rotterdam, The Netherlands, 1998; pp. 249–258.
20. EN 1993-1-8. *Eurocode 3: Design of Steel Structures. Part 1–8: Design of Joints*; European Committee For Standardization: Brussels, Belgium, 2006.
21. Brodka, J.; Broniewicz, M. Calculation of welding trusses overlap joints made of rectangular hollow sections. *Arch. Civ. Eng.* **2013**, *59*, 441–468. [CrossRef]

MDPI
St. Alban-Anlage 66
4052 Basel
Switzerland
Tel. +41 61 683 77 34
Fax +41 61 302 89 18
www.mdpi.com

Buildings Editorial Office
E-mail: buildings@mdpi.com
www.mdpi.com/journal/buildings